Rigorous Atomic and
Molecular Physics

NATO ADVANCED STUDY INSTITUTES SERIES

A series of edited volumes comprising multifaceted studies of contemporary scientific issues by some of the best scientific minds in the world, assembled in cooperation with NATO Scientific Affairs Division.

Series B. Physics

Recent Volumes in this Series

Volume 65 – Nonequilibrium Superconductivity, Phonons, and Kapitza Boundaries
edited by Kenneth E. Gray

Volume 66 – Techniques and Concepts of High-Energy Physics
edited by Thomas Ferbel

Volume 67 – Nuclear Structure
edited by K. Abrahams, K. Allaart, and A. E. L. Dieperink

Volume 68 – Superconductor Materials Science: Metallurgy, Fabrication, and Applications
edited by Simon Foner and Brian B. Schwartz

Volume 69 – Photovoltaic and Photoelectrochemical Solar Energy Conversion
edited by F. Cardon, W. P. Gomes, and W. Dekeyser

Volume 70 – Current Topics in Elementary Particle Physics
edited by K. H. Mütter and K. Schilling

Volume 71 – Atomic and Molecular Collision Theory
edited by Franco A. Gianturco

Volume 72 – Phase Transitions: *Cargèse 1980*
edited by Maurice Lévy, Jean-Claude Le Guillou, and Jean Zinn-Justin

Volume 73 – Scattering Techniques Applied to Supramolecular and Nonequilibrium Systems
edited by Sow-Hsin Chen, Benjamin Chu, and Ralph Nossal

Volume 74 – Rigorous Atomic and Molecular Physics
edited by G. Velo and A. S. Wightman

This series is published by an international board of publishers in conjunction with NATO Scientific Affairs Division

A Life Sciences B Physics	Plenum Publishing Corporation London and New York
C Mathematical and Physical Sciences	D. Reidel Publishing Company Dordrecht, Boston, and London
D Behavioral and Social Sciences E Applied Sciences	Sijthoff & Noordhoff International Publishers Alphen aan den Rijn, The Netherlands, and Germantown, U.S.A.

Rigorous Atomic and Molecular Physics

Edited by

G. Velo

Institute of Physics
University of Bologna
Bologna, Italy

and

A. S. Wightman

Princeton University
Princeton, New Jersey

PLENUM PRESS • NEW YORK AND LONDON

Published in cooperation with NATO Scientific Affairs Division

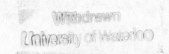

Library of Congress Cataloging in Publication Data

International School of Mathematical Physics (4th: 1980: Erice, Italy)
Rigorous atomic and molecular physics.

(NATO advanced study institutes series (B-Physics); v. 74)
"Proceedings of the Fourth International School of Mathematical Physics, held June 1-15, 1980 in Erice, Sicily, Italy."
Published in cooperation with NATO Scientific Affairs Division."
Bibliography: p.
Includes index.
1. Nuclear physics—Congresses. 2. Molecules—Congresses. I. Velo, G, (Giorgio), 1940- . II. Wightman, A. S. III. North Atlantic Treaty Organization. Division of Scientific Affairs. IV. Title. V. Series: NATO advanced study institutes series. Series B. Physics; v. 74.
QC170.I59 1980 539.7 81-12059
ISBN 0-306-40829-5 AACR2

Proceedings of the Fourth International School of Mathematical Physics,
held June 1-15, 1980, In Erice, Sicily, Italy

A Division of Plenum Publishing Corporation
233 Spring Street, New York, N.Y. 10013

Printed in the United States of America

PREFACE

One of the goals of mathematical physics is to provide a rigorous derivation of the properties of macroscopic matter starting from Schrödinger's equation. Although at the present time this objective is far from being realized, there has been striking recent progress, and the fourth "Ettore Majorana" International School of Mathematical Physics held at Erice, 1-15 June 1980 with the title Rigorous Atomic and Molecular Physics focussed on some of the recent advances.

The first of these is the geometric method in the theory of scattering. Quantum mechanical scattering theory is an old and highly cultivated subject, but, until recently, many of its fundamental developments were technically very complicated and conceptually rather obscure. For example, one of the basic properties of a system of N particles moving under the influence of appropriately restricted short-range plus Coulomb forces is asymptotic completeness: the space of states is spanned by the bound states and scattering states. However, the proof of asymptotic completeness for N bodies was achieved only with physically unsatisfactory restrictions on the nature of the interaction and even for $N = 2$ required an involved argument rather more subtle than the physical circumstances seemed to warrant. The reader will find in the present volume a very simple and physical proof of asymptotic completeness for $N = 2$ as well as an outline of the geometrical ideas which are currently being used to attack the problem for $N > 2$. (See the lectures of Enss.) A second striking insight appears in the lectures of Pearson which, to be brief and picturesque, we may describe as Pearson's wave trap and Pearson's singular spectrum generator. The wave trap is a potential with the property that a wave sent toward it from infinity never reappears. It shows one of the ways in which asymptotic completeness can fail. The singular spectrum generator illustrates a second possible failure in asymptotic completeness. Here the wave is not trapped in any fixed region but engages in a recurrent motion in which it spends an appreciable fraction of its time arbitrarily far from the scattering center: it cannot

make up its mind whether to be a really localized bound state or a scattering state. At last, we know the kind of thing that has to be shown not to happen in real atoms.

A third new development centers around the Thomas-Fermi atom. For half a century, this model has been useful in providing a simple approximation to the Schrödinger description of the ground state. Lieb and Simon recently showed that the theory when put in rigorous mathematical form becomes exact in the limit $Z \to \infty$. (See the lectures of Lieb for an introduction.) Furthermore, Lieb and Thirring have shown that the model also provides exact lower bounds on the ground state energy for all Z. (See the lectures of W. Thirring.) Thus, the Thomas Fermi model has a new role in providing rigorous bounds on the exact ground energies of atoms.

A fourth development is in the theory of Coulomb systems. It was realized half a century ago that the phenomenon of screening of charge plays an important role there. (Since Debye based a theory of electrolytes on the phenomenon, it is customarily called Debye screening.) Nevertheless, until the recent work of Brydges and Federbush it had not been shown from first principles that the phenomenon actually is a consequence of Schrödinger's equation. It is hoped that the lectures of Brydges will provide convenient access to this difficult but important derivation.

We will refrain from a further recitation of the themes of the lectures, hoping that this foretaste will encourage the reader to explore for himself.

This course, a NATO Advanced Study Institute, was sponsored by the Italian Ministry of Public Education, the Italian Ministry of Scientific and Technological Research, the Italian National Research Council, and the Sicilian Regional Government.

G. Velo
A.S. Wightman
Directors of the School

CONTENTS

Geometric Methods in Spectral and Scattering Theory of Schrödinger Operators 1
V. Enss

Stationary Scattering Theory 71
J.M. Combes

Spectral Properties and Asymptotic Evolution in Potential Scattering 99
D.B. Pearson

Schrödinger Operators with External Homogeneous Electric and Magnetic Fields 131
I.W. Herbst

The Born-Oppenheimer Approximation 185
J.M. Combes, P. Duclos, and R. Seiler

Thomas-Fermi and Related Theories of Atoms and Molecules 213
E. Lieb

The Stability of Matter 309
W. Thirring

Phase Diagrams and Critical Properties of (Classical) Coulomb Systems 327
J. Fröhlich and T. Spencer

Debye Screening in Classical Coulomb Systems 371
D.C. Brydges and P. Federbush

Internal Structure of Coulomb Systems in One Dimension 441
M. Aizenman

Free Energy and Correlation Functions of Coulomb Systems 467
J.L. Lebowitz

List of Seminars 489

Index 491

GEOMETRIC METHODS IN SPECTRAL AND SCATTERING THEORY OF SCHRÖDINGER OPERATORS

Volker Enss

Institut für Mathematik
Ruhr-Universität
D-4630 Bochum 1, Fed. Rep. Germany

ABSTRACT

An extensive introductory survey is given of *geometric, time-dependent methods* in spectral and scattering theory of Schrödinger operators. We prove asymptotic completeness in potential scattering for short-range and some long-range interactions (including Coulomb forces), and we derive weaker statements about asymptotically free motion for wider classes of potentials. Some properties of bound states and their energies are given as well. The emphasis is on the concepts and methods of this approach and we present various alternative routes to obtain the results.

1. INTRODUCTION

In this report we describe a variety of results in spectral and scattering theory of Schrödinger operators which have in common that they can be derived with geometric time-dependent methods. We will treat systems of two particles which interact with forces depending only on the relative position of the two particles. Then one can seperate off the (trivial) free center of mass motion and one is left with the (interesting) relative motion of the particles. Thus one has reduced the problem to that of a system in which one particle is influenced by a potential V. Typically the Hamiltonian is a self-adjoint operator given as

$$H = H_o + V(\vec{x}), \ H_o = \vec{P}^2/2m = -(1/2m)\ \Delta.$$

(In terms of the two body system $\vec{x}$ is the relative position and m the reduced mass.) We have chosen our units such that Planck's constant $\hbar = 1$.

We will restrict ourselves to potentials which tend to zero as $|\vec{x}| \to \infty$ (in a sense to be specified below) and which are not too singular locally. Among the results which we discuss are

1. The physical bound states (characterized by their localization uniform in time) are the states in the point-spectral subspace of the Hamiltonian (the closed linear span of the eigenvectors of H).

2. Bound state energies (eigenvalues of H) can accumulate only at 0.

3. Nonzero eigenvalues cannot be infinitely degenerate.

Scattering states are the vectors which belong to the orthogonal complement of the bound states, the continuous spectral subspace of H. Depending on the decay assumptions for the potential we prove various degrees of asymptotically free motion for scattering states.

4. A scattering state leaves every bounded region in space in the time mean.

5. It leaves every bounded region in space in the limit $|t| \to \infty$.

6. For the (selfadjoint) generator of the dilation group $D = (1/2)[\vec{x} \cdot \vec{p} + \vec{p} \cdot \vec{x}]$ the positive spectral subspace is absorbing for scattering states as $t \to +\infty$ (i.e. the component of the scattering state in the negative spectral subspace vanishes as $t \to +\infty$). Similarly the negative

spectral subspace is absorbing as $t \to -\infty$. Since the negative (resp. positive) spectral subspace of D corresponds to incoming (resp. outgoing) states, scattering states have been incoming in the remote past and will be outgoing in the far future.

7. Asymptotic completeness holds (in its strong form). For every scattering state the true (= interacting) time evolution asymptotically coincides with the free time evolution. (Modifications are necessary for Coulomb- and other long range forces.) Equivalently the ranges of both wave operators are equal to the continuous spectral subspace of the Hamiltonian. In particular H does not have a singular continuous spectrum.

Until recently various time-independent methods have been the dominant tool for mathematically rigorous work in spectral and scattering theory. A lot of highly sophisticated technology and impressive hard analysis has been developed to solve the main problems. The geometric, time-dependent approach uses very simple mathematical methods to answer the main questions which are physically motivated, often under weaker assumptions. Besides its simplicity another main advantage of our approach is that even single steps of the proofs are motivated by physical considerations and thus can be easily understood. One translates the physical intuition and folk lore into mathematically precise statements which have a simple proof. There is a price to pay: Some intermediate results of the traditional treatment which are of independent mathematical interest are lost. Moreover time independent methods seem to be better suited for actual computations of relevant data. (One may view [24-26] as a first step to fill this gap a bit.)

To preserve the simplicity in our exposition we will sometimes prove a theorem under more restrictive conditions than necessary and delete the refinements and extensions to the remarks. Moreover we will rephrase many mathematical statements in physical terms to stress the connections. We collect some notation for easy reference in the next section. The assumptions on the decay rate of the potential vary from section to section, see there for details, but everything is true for Kato-bounded (see (2.3)) short-range potentials (see (2.15) or (2.18), (2.19)) which are multiplication operators in $\vec{x}$-space.

Local compactness is introduced as a local condition on the interaction in Section 3 and its connection with mutual subordination of H and H_o is pointed out. In the following section it is applied to the geometric characterization of bound states and scattering states which is equivalent to the spectral one. Sections 5 and 6 are pure kinematics. We give several characterizations of states which fly away from the origin, essentially this is a

localization property in phase-space. In particular we use the generator of the dilation group $D = (1/2)[\vec{x} \cdot \vec{p} + \vec{p} \cdot \vec{x}]$. The corresponding subspaces of states are then used to derive uniform estimates for the propagation under the free time evolution and it is pointed out that quantum wave packets behave asymptotically like classical ones.

Then we return to dynamical questions and study the asymptotic behavior of the time evolution on the states which are orthogonal to the bound states, the scattering states. For potentials with sufficient decay one expects that the time evolution is asymptotically free. In Section 7 this is tested with the asymptotic evolution of certain observables which is the same as in the free case. Thus one knows for a wide class of interactions that the scattering states asymptotically have been incoming and will be outgoing. Moreover they will be localized far from the scatterer. A stronger notion of asymptotically free motion uses the wave operators and compares the free and interacting unitary evolution groups themselves. We show existence and completeness of the wave operators for short-range potentials in Sections 8 and 9. The modifications for long-range forces are given in Section 10. Finally in Section 11 we discuss results on bound state energies and their degeneracy. Other applications of the method, in particular to two-cluster scattering of multiparticle systems, are mentioned in the last section.

In this report we have almost completely ignored other methods in the study of Schrödinger operators. We refer the reader to the monographs [2, 36, 54-56] where also references to the research papers can be found. For stationary geometric methods see the lectures of Combes [7]. Since the ideas in geometric, time dependent scattering theory are so natural, many of them developed independently at different places and it was often accidental who published them first. Therefore I have to apologize for omitted or improperly assigned credits. The first geometric, time dependent proof of asymptotic completeness was given in [18]. Variations, simplifications, and extensions appeared soon thereafter, e.g. [61, 46, 12, 51, 66] and other references given later.

The mathematical prerequisits are rather modest, we have collected many of them in the next section. The degree of mathematical sophistication in our arguments varies considerably but it is generally rather low.

I am indebted to G. Velo and A.S.Wightman for organizing this School with its very stimulating atmosphere. The preparation of this report profited from numerous discussions and remarks which I greatfully acknowledge.

2. SETUP AND NOTATION

Our state space is the Hilbert space $\mathcal{H} = L^2(\mathbb{R}^\nu, d^\nu x)$ for a scalar particle moving in ν-dimensional space. For multicomponent wave functions to describe particles with spin or solutions of classical wave equations see Section 12 on refinements and extensions. The free dynamics are generated by the *free Hamiltonian*

$$H_o = -(1/2\,m)\,\Delta \tag{2.1}$$

which is self-adjoint on its domain $\mathcal{D}(H_o)$. The full, interacting time evolution is generated by the self-adjoint *Hamiltonian*

$$H = H_o + V \tag{2.2}$$

with domain $\mathcal{D}(H)$ which is a perturbation of the free Hamiltonian. (see e.g.[36, 54] for details about paragraphs 1. - 3.)

1. Kato-bounded perturbations: If V is H_o-bounded with H_o-bound smaller than 1, i.e. $\mathcal{D}(H_o) \subset \mathcal{D}(V)$ and for all $\Psi \in \mathcal{D}(H_o)$

$$\|V\Psi\| \leq a\|H_o\Psi\| + b\|\Psi\| \tag{2.3}$$

for some $a < 1$, then by the Kato-Rellich theorem H is bounded below and $\mathcal{D}(H) = \mathcal{D}(H_o)$. In particular H and H_o are bounded relative to each other:

$$(H-z)(H_o-z)^{-1}\,,\quad (H_o-z)(H-z)^{-1}, \tag{2.4}$$

$$V(H_o-z)^{-1}\,,\ \text{and}\quad V(H-z)^{-1} \tag{2.5}$$

are all bounded operators for z in the resolvent sets. For simplicity we will use this assumption in some of our proofs, although most proofs have an immediate extension to the more general case, see the various remarks.

2. The form bounded case: Assume for the form domains $\mathcal{Q}(V) \supset \mathcal{Q}(H_o)$ and for all $\Psi \in \mathcal{Q}(H_o)$

$$|(\Psi, V\Psi)| \leq a(\Psi, H_o\Psi) + b\,\|\Psi\|^2 \tag{2.6}$$

for some $a < 1$. Then by the KLMN-Theorem H is self-adjoint with form domain $\mathcal{Q}(H) = \mathcal{Q}(H_o)$,

$$(H-z)^{1/2}\,(H_o-z)^{-1/2}\,,\quad (H_o-z)^{1/2}\,(H-z)^{-1/2}, \tag{2.7}$$

$$\text{and } (H-z)^{-1/2} V (H_o-z)^{-1/2} \tag{2.8}$$

are all bounded operators, and H is semibounded. Since (2.3) implies (2.6) (with different b's) case 1 is included here.

3. Strong positive singularities: Split the potential into its positive and negative parts $V = V_+ - V_-$ and let the negative part be relatively form bounded (2.6). Then $\mathcal{Q}(H_o - V_-) = \mathcal{Q}(H_o)$. If V_+ is a closed positive quadratic form on $\mathcal{Q}(V_+)$ and

$$\mathcal{Q}(H) = \mathcal{Q}(H_o) \cap \mathcal{Q}(V_+) \tag{2.9}$$

is dense then there is a unique self-adjoint semibounded Hamiltonian H with this form domain (2.9). In this case the following operators are bounded for z in the resolvent sets:

$$(H_o-z)^{1/2} (H-z)^{-1/2} , \tag{2.10}$$

$$(H-z)^{-1} V (H_o-z)^{-1} = (H_o-z)^{-1} - (H-z)^{-1} \tag{2.11}$$

and in particular for functions $g, \overline{g} \in C_o^\infty (\mathbb{R})$ (and larger classes of functions)

$$\overline{g}(H)\ V\ g(H_o). \tag{2.12}$$

So far both the free and the interacting dynamics act on the same Hilbert space $\mathcal{H} = L^2 (\mathbb{R}^\nu, d^\nu x)$. In $\nu = 3$ dimensions let V have a local singularity behaving as

$$V(\vec{x}) \sim |\vec{x}|^{\alpha} \text{ for } |\vec{x}| \to 0, \tag{2.13}$$

then for $\alpha < 3/2$ $V \in L^2_{loc}(\mathbb{R}^3)$ and case 1 applies. For $\alpha < 2$ $V \in L^{3/2}_{loc}(\mathbb{R}^3)$ and it is in the Rollnik class, both cases 2 and 3 apply. For the positive part V_+ $\alpha < 3$, $V_+ \in L^1_{loc}(\mathbb{R}^3)$ is permitted in case 3.

For scattering off hard cores and acoustical obstacle scattering the Hilbert space for the interacting dynamics is different from that for the free dynamics, only wave functions with support outside the obstacles are admitted. Then one has to use the two Hilbert space formulation with a proper identification operator. If the obstacles lie within a bounded region then the methods and results have an obvious extension to that case which we won't discuss in detail.

With regard to local singularities all of the above conditions are stronger than necessary. Except for Section 7 the main results

are true if one assumes only the weaker local compactness condition given in the next section.

4. Spectral Projections: By $F(\cdot)$ we denote spectral projections of various self-adjoint operators corresponding to the regions of spectral values as specified in the parentheses. Thus, e.g., $F(|\vec{x}| \leq R)$ is in $\vec{x}$ - space multiplication with the characteristic function of the given region and $F(H > E_{min})$ projects to states with energy support above E_{min}.

We use freely functions of self-adjoint operators defined by the functional calculus [54].

5. Decay at Infinity: Physically realistic forces become weak for far separated particles (except possibly quark confining potentials if there are such). Therefore we require that in some sense V tends to zero at infinity, although we admit the possibility that it is unbounded arbitrarily far out. A natural candidate for an extremely weak decay assumption is that the difference of the free and interacting resolvents (2.11) which is bounded, decays at infinity in norm:

$$\|(H-z)^{-1} V (H_o-z)^{-1} F(|\vec{x}| \geq R)\| \to 0 \qquad (2.14)$$

as $R \to \infty$. The resolvents on both sides of V are there to smooth out local peaks of V, the convolution kernel in $\vec{x}$-space of the free resolvent decays rapidly and thus respects localization properties. We will discuss consequences and equivalent assumptions in the next section.

6. Short-Range Potentials: This class of potentials has a particularly simple scattering theory. The decay requirements are strong enough to exclude the Coulomb force of charged particles with its well known difficulties. Therefore we single out this class although in most sections long-range forces are included in our treatment.

V is called a *short-range potential* [22] if for any function $g \in C_o^\infty(\mathbb{R})$

$$\|g(H)\ V\ g(H_o)\ F(|\vec{x}| \geq R)\| =: h(R) \in L^1(\mathbb{R}_+, dR). \qquad (2.15)$$

The function h depends on g; it is finite for $R = 0$ by (2.12). The integrability and monotonicity imply $R \cdot h(R) \to 0$ as $R \to \infty$.

For simplicity we will sometimes use the stronger condition

$$\|V(H_o + 1)^{-1} F(|\vec{x}| \geq R)\| \in L^1(\mathbb{R}_+, dR) \qquad (2.16)$$

for Kato-bounded potentials which implies both (2.14) and (2.15). Other special cases of (2.15) are potentials which fulfill

$$\|(H - z)^{-\beta} V(H_o - z)^{-\alpha} F(|\vec{x}| \geq R)\| \in L^1(\mathbb{R}_+, dR) \tag{2.17}$$

for suitable α and β as proposed independently by Davies [12] and Perry [51], or potentials with a factorization property proposed by Simon [61]: $V = W^* U$ where $W(H + i)^{-1/2}$ is bounded and U fulfills (2.16).

We have not assumed that V is a multiplication operator in $\vec{x}$-space, it can be a pseudodifferential operator describing velocity dependent forces. However, most of the physically relevant potentials are multiplication with a function $V(\vec{x})$ which is continuous for $|\vec{x}| \geq R_o$. Then the regularization with $g(H_o)$ is not necessary and

$$V(\vec{x}) = O(|\vec{x}|^{-1-\varepsilon}), \ \varepsilon > 0 , \tag{2.18}$$

or a potential with

$$\int_{R_o}^{\infty} dr \sup_{|\vec{x}| \geq r} |V(\vec{x})| < \infty \tag{2.19}$$

are typical examples of short-range potentials. See also Problem 151 on page 403 in [55] for related properties.

7. Compact Operators are a very useful analytical tool because they are approximately finite dimensional in the uniform topology. For the convenience of the reader we collect some well known properties the proof of which can be found in any book on Hilbert space theory. Let C denote a compact operator.

(a) C has a representation

$$C = \sum_{k=1}^{\infty} \alpha_k |\Psi_k\rangle\langle\phi_k| \tag{2.20}$$

where $|\alpha_k| \to 0$, $\{\Psi_k\}$ and $\{\phi_k\}$ form orthonormal systems.

(b) For any $\varepsilon > 0$ there is a finite dimensional operator C_ε such that

$$\|C - C_\varepsilon\| < \varepsilon. \tag{2.21}$$

(c) Let F_n be a sequence of operators converging strongly to 0. Then

$$\|F_n C\| \to 0, \ \|C F_n^*\| \to 0. \tag{2.22}$$

(d) If B is a bounded operator then

BC and CB are compact. (2.23)

(e) Let C_n be a sequence of compact operators and let

$$\|K - C_n\| \to 0, \tag{2.24}$$

then K is compact.

(f) If K^*K is compact then K is compact. (2.25)

(g) If ϕ_n converges weakly to zero then $\|C\,\phi_n\| \to 0$. (2.26)

8. Spectral Subspaces of the Hamiltonian. We denote by $\mathcal{H}^p$ the point-spectral subspace of the Hamiltonian, i.e. $\mathcal{H}^p$ is spanned by the square integrable eigenfunctions of H. Its orthogonal complement is $\mathcal{H}^{cont}$, the continuous spectral subspace of H. In general it will split into a direct sum $\mathcal{H}^{cont} = \mathcal{H}^{sing} \oplus \mathcal{H}^{ac}$ where $\mathcal{H}^{ac}$ denotes the (Lebesgue-) absolutely continuous spectral subspace and $\mathcal{H}^{sing}$ the singular continuous one. As a byproduct of our analysis we will show absence of $\mathcal{H}^{sing}$ for potentials with suitable decay. By P^p, P^{cont}, etc. we denote the projections onto the corresponding spaces.

9. Strong Resolvent Convergence. [36,53] A sequence of (unbounded) self-adjoint operators A_n converges to A in the strong resolvent sense if suitable bounded functions thereof converge strongly, e.g. for Im $z \neq 0$

$$\text{s-lim}\,(A_n - z)^{-1} = (A - z)^{-1}, \tag{2.27}$$

or, equivalently, if for all $s \in \mathbb{R}$

$$\text{s-lim}\, e^{isA_n} = e^{isA}. \tag{2.28}$$

It follows that $f(A_n)$ converges strongly to f(A) for bounded continuous functions. If A_n and A have purely continuous spectrum then f may be discontinuous, e.g. a spectral projection. A useful sufficient condition is: If

$$A_n\,\Psi \to A\,\Psi \tag{2.29}$$

for all Ψ in a core for A then A_n converges to A in the strong resolvent sense.

3. LOCAL COMPACTNESS

In classical phase space (the 2ν-dimensional configuration-

and momentum-space) a system is confined to a finite volume if one restricts it in configuration space to a bounded region B and moreover restricts the kinetic energy to $p^2/2m \le E < \infty$. If the singularities of the potential are not too bad the same is true for the full energy:

$$\int_B d^\nu x \int_{(p^2/2m)+V(x)\le E} d^\nu p < \infty. \tag{3.1}$$

In $\nu = 3$ dimensions an elementary calculation shows that negative singularities

$$V(\vec{x}) \sim |x|^{-\alpha}, \ \alpha < 2$$

and arbitrary positive singularities are permitted. $\alpha = 2$ is a known boarderline where troubles arise, compare the discussion of local singularities (2.13). According to the correspondence principle a quantum state occupies a volume $(2\pi\hbar)^\nu$ of classical phase space, $\hbar$ Planck's constant. Thus there should be only finitely many different quantum states localized in a bounded region B and having finite energy. Since H and $\vec{x}$ do not commute one has to admit tails and a proper statement is *local compactness:*

$$F(|\vec{x}| \le R) \quad F(H \le E) \text{ is compact} \tag{3.2}$$

for any $R, E < \infty$, because this expresses an approximately finite dimension of the subset.

For any pair of bounded functions f and g with $|f(\vec{x})| \to 0$ as $|\vec{x}| \to \infty$ and $|g(\omega)| \to 0$ as $\omega \to \infty$ it follows from (2.23) and (2.24) that also

$$f(\vec{x}) \ g(H) \text{ is compact.} \tag{3.3}$$

If on the other hand for a continuous g as above which does not vanish (like e.g. $g(H) = (H - z)^{-1/2}$) one has that

$$F(|\vec{x}| \le R) \ g(H) \tag{3.4}$$

is compact for any $R < \infty$, then (3.2) holds.

A compactness criterion like (3.2) had been proposed by Haag and Swieca [30] to distinguish quantum field theories with a complete particle interpretation. Also the classical phase space argument to support it is due to them. Except for the free field it could not yet be verified in any field theoretic model and it has hardly been applied [16].

In non-relativistic quantum mechanics, however, the situation is much better. The verification is simple and these lectures are

full of its applications. First observe that

$$F(|\vec{x}| \leq R)\ (H_o + 1)^{-k} \tag{3.5}$$

is compact, it is even Hilbert-Schmidt for k large enough (depending on the dimension ν) because the kernel corresponding to (3.5)

$$\chi_R(\vec{x})\ (H_o + 1)^{-k}\ (\vec{x} - \vec{y}) \tag{3.6}$$

is square integrable in $\mathbb{R}^{2\nu}$. Consequently (3.5) is compact for $k = 1/2$. The boundedness of $(H_o + 1)^{1/2}\ (H - z)^{-1/2}$, (2.10) valid for all three cases mentioned explicitly in Section 2 implies compactness of

$$F(|\vec{x}| \leq R)\ (H - z)^{-1/2} \tag{3.7}$$

for any R, thus the compactness criterion (3.2) holds. Actually one could include e.g. hard cores, exterior electromagnetic fields etc. and (3.2) remains true, see Section III of [3] for a detailed discussion. In fact almost any method to construct a semibounded self-adjoint Hamiltonian as a perturbation of a reasonable free one automatically implies local compactness. In non-relativistic quantum mechanics it is an extremely weak condition. Except for Section 7 this is the only condition restricting local singularities which we actually need for the main results. It is sufficient to save us from falling into one of the holes D.B.Pearson is digging for us with his beautiful and instructive counterexamples [47 - 50] . In his model of local adsorption he solves the hard task to construct a mathematically decent Hamiltonian via its eigenfunction expansion which is physically misbehaved, in particular it violates local compactness, and Theorem 4.1 does not hold.

Local compactness became a standard tool in the scattering theory of classical wave equations since its introduction by Lax and Phillips [44]. Earlier Rellich used it in his study of the perturbation of point spectra.

The notion of a pair of operators H and H_o which are *mutually subordinate* has been introduced by Birman [5]. Roughly it says that the full energy will increase if the kinetic energy does so, and vice versa. Precisely,

$$\|G_1(H)\ G_2^{-1}\ (H_o)\| + \|G_1(H_o)\ G_2^{-1}\ (H)\| < \infty \tag{3.8}$$

for suitably chosen continuous functions $G_i \geq 1$ with $G_i(u) \to \infty$ as $u \to \infty$. We will show that this mathematically useful concept (see also Theorem 13 of [12] and [28]) is related to two physically motivated conditions: local compactness and decay at infinity

(2.14) together imply mutual subordination; on the other hand mutual subordination implies local compactness. To show the first statement we prove the stronger (3.9) which implies mutual subordination (see [12], also for other equivalent conditions).

Lemma 3.1 . For the pair H_o, $H = H_o + V$ local compactness (3.2) and decay at infinity (2.14) together are equivalent to

$$(H - z)^{-1} - (H_o - z)^{-1} \text{ is compact} \tag{3.9}$$

for z in the resolvent sets.

Proof. $(H - z)^{-1} - (H_o - z)^{-1}$

$$= (H - z)^{-1} F(|\vec{x}| \leq R) - (H_o - z)^{-1} F(|\vec{x}| \leq R)$$

$$- (H - z)^{-1} V(H_o - z)^{-1} F(|\vec{x}| \geq R).$$

The first two summands are compact for any $R < \infty$ by (3.2) and the third has arbitrary small norm for big enough R by (2.14). By (2.24) this implies (3.9). On the other hand $F(|\vec{x}| \geq R)$ is self-adjoint and converges strongly to zero. The compactness of (3.9) implies norm convergence to zero in (2.14) by (2.22). Moreover (3.9) implies compactness of $(H - z)^{-1} F(|\vec{x}| \leq R) - (H_o - z)^{-1} F(|\vec{x}| \leq R)$ for all R. The second summand is always compact and thus also the first, i.e. (3.2) holds.

□

The connection between compactness and decay at infinity had also been pointed out by Jörgens and Weidmann in Section 3 of [35]. Next observe that $F(|\vec{x}| \leq R)\, G_1^{-1}(H_o)$ is compact for any R and G_1 as used in (3.8). With (3.8) we obtain compactness of $F(|\vec{x}| \leq R)\, G_2^{-1}(H)$, thus mutual subordination implies local compactness.

The first spectral theoretic result now follows from (3.9) by Weyl's theorem

$$\sigma_{ess}(H) = \sigma_{ess}(H_o) = [0, \infty) . \tag{3.10}$$

The physical reason for this is very simple. The decay at infinity of the potential requires that the states with strictly negative energy must essentially be localized in a bounded region. By local compactness there are only finitely many of them, thus the negative spectrum of H is discrete. (For proofs along these lines see [17, 60, 64] where more general cases are treated.)

The compactness of the difference of the resolvents (3.9) implies compactness of the difference of other functions.

If $g \in C_o^\infty$ or $1-g \in C_o^\infty$ then

$$g(H) - g(H_o) \text{ is compact.} \tag{3.11}$$

One easily proves this by a Stone Weierstrass approximation or with several other methods. Moreover for $g \in C_o^\infty$

$$g(H)\ V\ g(H_o) \text{ is compact.} \tag{3.12}$$

4. GEOMETRIC CHARACTERIZATION OF BOUND STATES AND SCATTERING STATES

In the quantum mechanics textbooks bound states are usually defined by a spectral property, namely that they are eigenvectors of the Hamiltonian H, they belong to $\mathcal{H}^p$ (see paragraph 8 of Section 2 for notation). The orthogonal complement $\mathcal{H}^{cont}$ then should consist of scattering states.

Physically it is more natural to characterize bound states and scattering states by localization properties during their time evolution. Following Ruelle [57] the set of *geometric bound states* M_{bd} consists of the states which are localized uniformly in time: $\Psi \in M_{bd}$ if for any $\varepsilon > 0$ there is an R_ε such that

$$\sup_{t \in \mathbb{R}} \| F(|\vec{x}| \geq R_\varepsilon)\, e^{-iHt}\, \Psi \| < \varepsilon. \tag{4.1}$$

$\Psi \in M_{lv}$, the set of states which *leave any bounded region in the time mean* if for any R

$$\lim_{T \to \infty} \frac{1}{2T} \int_{-T}^{T} dt\, \| F(|\vec{x}| \leq R)\, e^{-iHt}\, \Psi \| = 0. \tag{4.2}$$

The following theorem is due to Ruelle [57] who proved it for a special class of interactions, Amrein and Georgescu [3] extended it to a wider class of potentials. Closely related is the proof of local energy decay of Lax and Phillips [44], see also Wilcox [65] for the absolutely continuous case. We add the estimate (4.3) which is uniform in $\Psi \in \mathcal{D}(H) \cap \mathcal{H}^{cont}$. See Theorem XI.115 in [55] for another proof.

<u>Theorem 4.1 .</u> Let $H = H_o + V$ obey local compactness. Then $M_{bd} = \mathcal{H}^p$, $M_{lv} = \mathcal{H}^{cont}$. Moreover for $\Psi \in \mathcal{H}^{cont} \cap \mathcal{D}(H)$

$$\frac{1}{T} \int_{o}^{T} dt\, \| F(|\vec{x}| \leq R)\, e^{-iHt}\, \Psi \| \leq f_R(|T|)\, \| (H+i)\Psi \| \tag{4.3}$$

with $f_R(|T|) \to 0$ as $T \to \pm\infty$ for any R.

<u>Proof.</u> It is easy to see that M_{bd} and M_{lv} are closed linear subspaces. They are orthogonal because for any pair of states $\Psi \in M_{bd}$, $\phi \in M_{lv}$ we have for $\varepsilon > 0$

$$|(\Psi,\phi)| = \frac{1}{2T}\int_{-T}^{T} dt\, |(e^{-iHt}\Psi, e^{-iHt}\phi)|$$

$$\leq \sup_t \| F(|\vec{x}| \geq R_\varepsilon)\, e^{-iHt}\Psi\|$$

$$+ \frac{1}{2T}\int_{-T}^{T} dt\, \|F(|\vec{x}| \leq R_\varepsilon)\, e^{-iHt}\phi\|$$

$$\leq 2\varepsilon \text{ for } T \geq T_\varepsilon .$$

The trajectory $e^{-iHt}\Psi$ for a state $\Psi \in \mathcal{H}^p$ is approximately contained in a subspace spanned by finitely many eigenstates uniform in time. Thus the whole trajectory is contained in a compact subset of $\mathcal{H}$ and (4.1) follows.

We have proved $\mathcal{H}^p \subset M_{bd} \perp M_{lv}$. If we show that (a dense subset of) $\mathcal{H}^{cont}$ is contained in M_{lv} by verifying (4.3), then the proof of the theorem is complete.

For a function $0 \leq h(t) \leq 1$ the time mean of h(t) goes to zero if and only if that of $h^2(t)$ tends to zero:

$$\frac{1}{2T}\int_{-T}^{T} dt\, h^2(t) \leq \frac{1}{2T}\int_{-T}^{T} dt\, h(t) \tag{4.4}$$

$$\leq \{\frac{1}{2T}\int_{-T}^{T} dt\, h^2(t)\}^{1/2} . \tag{4.5}$$

In the last step we have used the Schwarz inequality. For $\Psi \in \mathcal{D}(H) \cap \mathcal{H}^{cont}$

$$\frac{1}{T}\int_{0}^{T} dt\, \|F(|\vec{x}| \leq R)\, e^{-iHt}\Psi\|^2$$

$$= ((H+i)\Psi, \frac{1}{T}\int_{0}^{T} C_R(t)\, (H+i)\Psi) \tag{4.6}$$

$$\leq f_R^2(T)\, \|(H+i)\Psi\|^2$$

where $C_R(t)$ is the compact operator

$$C_R(t) = e^{iHt}(H-i)^{-1} F(|\vec{x}| \leq R)(H+i)^{-1} e^{-iHt} \tag{4.7}$$

and

$$f_R^2(T) = \| P^{cont} \frac{1}{T} \int_0^T dt\, C_R(t)\, P^{cont} \| . \tag{4.8}$$

The decay of $f_R(T)$ will be shown in the next Lemma.

□

For later use we have singled out

Lemma 4.2. Let C be a compact operator, then for

$$C(t) = e^{iHt}\, C\, e^{-iHt} \tag{4.9}$$

$$\| \frac{1}{T} \int_0^T dt\, C(t)\, P^{cont} \| \to 0 \text{ as } |T| \to \infty . \tag{4.10}$$

Remark. This Lemma follows from Wiener's theorem (Theorem XI.114 in [55]. We give the direct proof. By taking the adjoint one sees that P^{cont} could be on the left hand side.

Proof. Up to an arbitrarily small error C can be replaced by a finite sum of rank one operators (see (2.20)), and it is sufficient to estimate each contribution $|\Psi\rangle\langle\phi|$. With P^{cont} on the LHS only $\Psi \in \mathcal{H}^{cont}$ contribute.

$$\| \frac{1}{T} \int_0^T dt\, |\Psi\rangle\langle\phi|\,(t) \|^2$$

$$\leq T^{-2} \int_0^T dt \int_0^T d\tau\, \| |\phi\rangle\langle\Psi| e^{-iH(t-\tau)} |\Psi\rangle\langle\phi| \| \tag{4.11}$$

$$\leq \{T^{-2} \int_0^T dt \int_0^T d\tau |\langle\Psi| e^{-iH(t-\tau)} |\Psi\rangle|^2\}^{1/2} . \tag{4.12}$$

In the last step we have used (4.5). If $d\mu$ denotes the continuous normalized spectral measure of H in the state Ψ the curly bracket in (4.12) is

$$T^{-2} \int_0^T dt \int_0^T d\tau \iint \exp\{i(\lambda-\sigma)(t-\tau)\}\, d\mu(\lambda)\, d\mu(\sigma)$$

$$= \iint \{[2/T(\lambda-\sigma)]\, \sin(\lambda-\sigma)T/2\}^2\, d\mu(\lambda)\, d\mu(\sigma) . \tag{4.13}$$

The contribution to this integral from the region $|\lambda-\sigma| > 2\delta$ is bounded by $(T\,\delta)^{-2}$ and thus vanishes as $|T| \to \infty$ for any δ. The continuous function $\mu(\lambda)$ is monotone and bounded, therefore it is uniformly continuous. The diagonal contribution

$$\int_{|\lambda-\sigma| \leq 2\delta} d\mu(\lambda)\, d\mu(\sigma)$$

$$= \int d\mu(\lambda)\, (\mu(\lambda + 2\delta) - \mu(\lambda - 2\delta))$$

can be made arbitrarily small by choosing δ small, and the integrand is bounded by 1.

□

Remark. As is clear from the proof the RHS of (4.3) could be replaced by $f_R(T)\ \{\|\Psi\|\ \|G(H)\Psi\|\}^{1/2}$ where G is any strictly positive function which goes to infinity.

In this section the only assumption was local compactness, we did not require any decay at infinity. The crucial point was that eigenstates of the Hamiltonian and superpositions thereof have trajectories which stay inside a compact subset. On the other hand the trajectories of continuum states never stay inside a compact set, they even leave any compact set in the time mean. This is true for any unitary evolution group, it is a basic ingredient of ergodic theory. Local compactness allows to translate this mathematical fact into physical statements about localization. If local compactness is violated then Pearson's example mentioned above [47,50] shows that there may be continuum states which are geometrically bounded in the future or past and thus violate (4.2).

Corollary 4.3. Fix a sequence $R_n \to \infty$ and $\varepsilon_n \downarrow 0$. For any $\Psi \in \mathcal{H}^{cont}$ there is a sequence $\tau_n \to \infty$ (and one with $\tau_n \to -\infty$) such that

a) $$\|F(|\vec{x}| \leq R_n)\ e^{-iH\tau_n}\ \Psi\| < \varepsilon_n \tag{4.14}$$

b) $$w - \lim_{n\to\infty}\ e^{-iH\tau_n}\ \Psi = 0. \tag{4.15}$$

Proof. b) follows from a) because $\|F(|\vec{x}| \geq R_n)\ \phi\| \to 0$ for any ϕ. To schow a) pick a sequence E_n such that $\|F(H \geq E_n)\Psi\| < \varepsilon_n/2$. By (4.3)

$$\frac{1}{T}\int_0^T \|F(|\vec{x}| \leq R_n)\ e^{-iHt}\ F(H \leq E_n)\ \Psi\|$$

$$\leq f_{R_n}(|T|)\ (E_n + 1)\|\Psi\|.$$

For T_n such that $f_{R_n}(T_n)\ (E_n + 1) < \varepsilon_n/2$ there are a positive and a negative τ_n with $|\tau_n| \leq T_n$ such that

$$\| F(|\vec{x}| \leq R_n)\, e^{-iH\tau_n} F(H \leq E_n)\, \Psi \| < \varepsilon_n/2$$

which implies (4.14).

□

Note that T_n can be chosen independent of Ψ if e.g. $\| (H + i)\, \Psi \| \leq M$. Then one has a uniform bound on the size of $|\tau_n|$ which is sufficient.

Certainly one expects that physically reasonable scattering states have much better asymptotic behavior than is expressed in (4.2), (4.3), or (4.14). If one considers only states in the absolutely continuous spectral subspace then the time average is not necessary and a simple application of the Riemann-Lebesgue lemma shows

$$\lim_{|t| \to \infty} \| F(|\vec{x}| \leq R)\, e^{-iHt} P^{ac}\, \Psi \| = 0. \tag{4.16}$$

However, there is no stronger counterpart to the uniform estimate (4.3).

Even if one adds to local compactness our weak condition (2.14) of decay at infinity one should not expect good asymptotic behavior of continuum states. This is best illustrated by another of Pearson's examples, see [49,50]. For a one-dimensional system the potential consists of a sequence of far separated smooth bumps with slowly decreasing size. The particle has a non-zero probability to pass any finite number of bumps, thus there are no bound states. But no matter how far out the particle has travelled it will certainly be reflected by one of the infinitely many remaining bumps, it will never be free. In the time independent setup this means that the improper eigenfunctions of the Hamiltonian won't be approximated by plane waves asymptotically. In this case H has a purely singular continuous spectrum!!

We learn from this that in addition to local compactness we have to impose stronger decay requirements on the potential to guarantee that the time evolution becomes asymptotically free. We will return to the dynamical questions in Section 7 and discuss first some kinematics.

5. INCOMING AND OUTGOING STATES

Consider a classical point particle with position $\vec{x}$ at a given time and velocity $\vec{v} = \vec{p}/m$, $\vec{p}$ its momentum. We call the particle outgoing if the angle enclosed by $\vec{x}$ and $\vec{v}$ is acute, i.e. if $\vec{x} \cdot \vec{p} > 0$, because then the distance of the particle from the origin will increase (at least for some time). For $\vec{x} \cdot \vec{p} < 0$ we call it incoming because then the distance has been bigger in the past. Certainly

this distinction depends on the choice of the origin of the coordinate system. Nevertheless any particle with asymptotically straight trajectory (like the hyperbolas in the Coulomb potential) will be outgoing in the far future and incoming in the remote past for any choice of the origin. If the potential is strong in two separated regions then a particle between them may be incoming w.r.t. one part of the potential and outgoing w.r.t. the other. But if one is mainly interested in particles far out and the potential decays at infinity then the ambiguity is irrelevant.

When we transfer this notion to quantum mechanics we are faced with a difficulty. To determine whether a state is incoming or outgoing we have to know both (the directions of) $\vec{x}$ and $\vec{p}$ simultaneously, we have to know something about the localization of the quantum state in classical phase-space. Since our first encounters with quantum mechanics we have heard innoumerous times that this is in conflict with the uncertainty principle. We believe that this is the main psychological reason why this concept has not been introduced before. Because of the uncertainty principle and because of the ambiguity in the choice of the origin it won't be helpful to decompose all states of $\mathcal{H}$, but it will be very useful for the description of continuum states at large times. From the previous section we know already that they are localized far out most of the time. Typical wave packets spread over a region which increases as $|t| \to \infty$. For a given momentum distribution the spread in configuration space due to the uncertainty principle thus becomes irrelevant.

In our applications in later sections we will make precise the following intuitive argument. A far out localized particle which is outgoing won't come close to the region where the potential is strong, only the weak far reaching tail of the potential will act on it. Thus outgoing particles starting far from the origin will be influenced very little by the forces. The same argument applies if also obtuse angles between $\vec{x}$ and $\vec{p}$ are permitted, only a neighborhood of π has to be excluded. Therefore there are many essentially equivalent decompositions of the state space into its incoming and outgoing part, we will give three of them below . Here "essentially equivalent" means that the propagation properties hold which were mentioned above. We will prove them in the next section. Following [19], or [61,55], we first describe the

<u>1. Phase-space decomposition:</u> Let $\chi(\vec{v})$ be a "smoothed projection" obtained e.g. as a convolution of the characteristic function of the cube of side a centered at the origin with a positive smooth function of integral 1 with support in a ball of diameter a/4. Let $j \in a\,\mathbb{Z}^{\nu}$ be a point on the lattice of side a. Denote by χ_j the multiplication operator in momentum space with the function $\chi[(\vec{p}/m)-j]$, then we get a decomposition of the identity

$$\sum_{j \in a\mathbb{Z}^\nu} \chi_j = 1. \tag{5.1}$$

States in the range of χ_j have velocity support in the cube with side of length 3a/2 centered at j. In configuration space χ_j acts as a convolution with a bounded function of rapid decay uniformly in j.

Let the real function ψ with integral 1 have a smooth Fourier transform with support in a ball of diameter a/4m. Let f be the convolution of ψ with the characteristic function of a unit cube centered at the origin. For $i \in \mathbb{Z}^\nu$, the unit lattice, denote by f_i the multiplication operator in $\vec{x}$-space with the function $f(\vec{x} - i)$. This defines a decompositon of the identity

$$\sum_{i \in \mathbb{Z}^\nu} f_i = 1. \tag{5.2}$$

There is a bounded function $\zeta(r)$ of rapid decay such that uniformly in i, j, Ψ

$$|(f_i\ \chi_j\ \Psi)\ (\vec{x})| < \zeta(r)\ \|\Psi\| \tag{5.3}$$

and

$$|(\chi_j\ f_i\ \Psi)\ (\vec{x})| < \zeta(r)\ \|\Psi\| \tag{5.4}$$

if $|\vec{x} - i| > r$. The velocity support of states in the range of $f_i\ \chi_j$ is contained in the cube with side of length 2a centered at j. Thus the states in the ranges of $f_i\ \chi_j$ and their adjoints are well localized in configuration space near i and have compact velocity support near j. In particular this implies that any partial sum of such terms is bounded. Now we are ready to define the decomposition of the identity into two bounded operators

$$P_{out} = \sum_{i \cdot j \geq 0} \chi_j\ f_i, \tag{5.5}$$

$$P_{in} = \sum_{i \cdot j < 0} \chi_j\ f_i\ . \tag{5.6}$$

The states $P_{out}\ \Psi$ and $P_{out}^*\ \Psi$ can both be used as "outgoing", similarly for P_{in}. Up to rapidly decaying tails it coincides with the classical definition for non-vanishing momenta, e.g. for $|j| \geq 2a$. The parameter a can still be chosen suitably for applications, it describes how fine the cell decomposition in velocity space is made. For smaller a due to the uncertainty principle the tails in configuration space will extend farther but still decay faster than any inverse power of the distance.

This simple minded decomposition has the advantage of giving explicit control of the localization properties in configuration- and velocity-(momentum-) space. Moreover a wide class of free Hamiltonians is easy to treat, see the next section.

A partial summation

$$f_{j+} = \sum_{\substack{i \\ i\cdot j \geq 0}} f_i, \qquad f_{j-} = \sum_{\substack{i \\ i\cdot j < 0}} f_i$$

gives a decomposition with half spaces instead of small cells in $\vec{x}$-space which is almost identical to one proposed by Kato [39]

$$P_{out/in} = \sum_j \chi_j f_{j\pm} .$$

Instead of the approximate phase space cell decomposition labelled by $(i,j) \in \mathbb{Z}^\nu \times a\,\mathbb{Z}^\nu$ Davies [12] uses a continuous analogon, the generalized coherent states [11] labelled by $(\vec{x},\vec{p}) \in \mathbb{R}^\nu \times \mathbb{R}^\nu$.

An elegant alternative is to use

2. The Generator of the Dilation Group. The self-adjoint operator

$$D = \frac{1}{2}(\vec{x}\cdot\vec{p} + \vec{p}\cdot\vec{x}) \tag{5.7}$$

$$= \vec{x}\cdot\vec{p} - i\nu/2 = \vec{p}\cdot\vec{x} + i\nu/2$$

generates the unitary dilation group, e.g.

$$\widehat{(e^{-i\lambda D}\Phi)}(\vec{p}) = (e^\lambda)^{\nu/2}\,\hat{\phi}(e^\lambda\vec{p}). \tag{5.8}$$

With $p = |\vec{p}|$ we easily conclude from (5.8) that

$$e^{i\lambda D}\ln p\, e^{-i\lambda D} = \ln(e^\lambda p) \tag{5.9}$$

$$= \ln p + \lambda .$$

Thus D and ln p are a pair of canonically conjugate operators and one can diagonalize D by a Fourier transform in ln p. Observe that $\exp(i\lambda \ln p) = p^{i\lambda}$, thus one is led to the Mellin transform.[(5.10) differs by the sign of λ from the usual definition which has to be used in $\vec{x}$-space; see [51] for further details.] Write the momentum space wave function $\hat{\phi}$ as a function of p and the direction ω then

$$\mathring{\phi}(\lambda,\omega) = (2\pi)^{-1/2}\int\frac{dp}{p}\,p^{\nu/2}\,p^{i\lambda}\,\hat{\phi}(p,\omega) \tag{5.10}$$

fulfills

$$(f(D)\phi)^{\circ}(\lambda,\omega) = f(\lambda)\ \overset{\circ}{\phi}(\lambda,\omega). \tag{5.11}$$

The operator D is the direct quantum analog of the classical $\vec{p}\cdot\vec{x}$. Following Mourre [46] one then defines

$$P_{out} = F(D \geq 0), \tag{5.12}$$

$$P_{in} = F(D \leq 0). \tag{5.13}$$

All applications given below remain true if one chooses another finite cutoff point instead of zero. Also a smoothed cutoff could be used if the tails decay rapidly.

It is easy to verify that

$$\|\{(\vec{x} - i)(\vec{p} - j)+(\vec{p} - j)(\vec{x} - i)\}^N f_i\ \chi_j\| \leq M_N \tag{5.14}$$

uniform in i and j. The curly bracket in (5.14) can be written as

$$D - i \cdot j - i \cdot(\vec{p} - j) - j \cdot(\vec{x} - i). \tag{5.15}$$

This suggests that $f_i\ \chi_j\ \Psi$ is localized in the spectral representation of D mainly near $i \cdot j$ (as one expects) but the tails increase with growing i and j. This is clear intuitively because the cutoff using D gets "sharper" in $\vec{x}$-space for growing $\vec{p}$ and vice versa. But these fine points are not important for our applications below, both definitions are equally well suited. A third possibility which we discuss uses

<u>3. The Formal Time Operator.</u> On the classical level we were interested only in the sign of $\vec{p}\cdot\vec{x}$. For $p \neq 0$ this is the same as the sign of $(2\,H_o)^{-1}\,\vec{p}\cdot\vec{x}$. Its symmetric counterpart for quantum mechanics would be the operator

$$T = (4\,H_o)^{-1}\,\vec{p}\cdot\vec{x} + \vec{x}\cdot\vec{p}\,(4\,H_o)^{-1}. \tag{5.16}$$

It obeys formally the canonical commutation relations

$$[T, H_o] = i. \tag{5.17}$$

T cannot be selfadjoint because H_o is semibounded, the deficiency indices of the differential operator on the half line are (0,1). Equivalently shifts to the left in energy space are not isometric.

Let us restrict ourselves to the subspace $F(H_o > E_o)\mathcal{H}$

where $E_o > 0$ will be kept fixed. (Later we will have to exclude zero velocities anyway for physical reasons: particles at rest are notorious troublemakers because they do not run away from the potential.) Then the shifts

$$e^{i\varepsilon T} \tag{5.18}$$

are well defined for $|\varepsilon| < E_o$ and they are isometric, moreover sums and integrals thereof are admissible functions of T. To make this precise one could double the Hilbert space by adding a copy with negative energies. Then T extends naturally to a self-adjoint operator and (5.18) is unitary [62]. Or one could consider only those functions of T which map the subspace $F(H_o > E_o)\mathcal{H}$ into the physical Hilbert space of positive energy states.

One can easily give an explicit construction using the inverse Fourier transform w.r.t. the energy. For the state Φ with momentum space wave function $\hat{\phi}(p\,\omega)$ denote by

$$\phi(E,\omega) = \sqrt{m}\ (2m\,E)^{(\nu-2)/4}\ \hat{\phi}(\sqrt{2m\,E}\ \omega) \tag{5.19}$$

its (kinetic) energy representation which is normalized in $L^2(R_+ \times S^{\nu-1}, dE\, d\,\Omega)$. If $\Phi \in F(H_o > E_o)\mathcal{H}$ the wave function vanishes for $E < E_o$. Apply to $\phi(E,\omega)$ the inverse Fourier transformation in the variable E to obtain the "T-space representation"

$$\check{\phi}(\tau,\omega) = (2\pi)^{-1/2} \int dE\ \phi(E,\omega)\ e^{iE\tau} . \tag{5.20}$$

The physical state space is isometrically imbedded into $L^2(\mathbb{R} \times S^{\nu-1}, d\tau\, d\Omega)$ which we call the extended T-space. The natural extension of H_o is the self-adjoint operator $H_o = (-i)d/d\tau$. Then $F(H_o \geq 0)$ is the identity on the physical state space and it is the orthogonal projection of the extended T-space onto the physical Hilbert space which consists of those functions $\check{\phi}(\tau,\omega)$ whose Fourier transform with respect to τ has support on the positive half line.

Define f(T) as multiplication operator in T-space by $f(\tau)$ which is equivalent to a convolution of the energy representation with $(2\pi)^{-\nu/2}\hat{f}$:

$$(f(T)\Phi)(E,\omega) = (2\pi)^{-\nu/2} \int dE'\ \hat{f}(E-E')\phi(E',\omega) . \tag{5.21}$$

On the subspace $\Psi \in F(H_o > E_o)\mathcal{H}$ both f(T) and its adjoint are "admitted" if the support of $\hat{f}$ is contained in the interval $(-E_o/2, E_o/2)$, then $f(T)\,\Psi$ and $f^*(T)\,\Psi \in \mathcal{H}$. In general $f(T)\mathcal{H}$ lies in the extended T-space but $F(H_o \geq 0)\ f(T)\ F(H_o \geq 0)$ maps $\mathcal{H}$ into $\mathcal{H}$ and it is self-adjoint for real f.

For any $E_o > 0$ let the real function ψ have integral one

and $\hat{\psi}$ have support in $(-E_o/4, E_o/4)$. Denote by $f_{\pm}$ the convolution of ψ with the characteristic function of the positive/negative half axis, then the tails of $f_{\pm}$ decay rapidly and $f_{\pm}(T)$ are admitted functions on $F(H_o > E_o)\mathcal{H}$ with range in $F(H_o > E_o/2)\mathcal{H}$. They add up to the identity. Also $F(H_o > E_o)\, f_{\pm}(T)$ is defined. It is possible to set

$$P_{out} = f_+(T) \tag{5.22}$$

$$P_{in} = f_-(T) \tag{5.23}$$

as the third characterziation of incoming and outgoing states. It uses only quantities in the given Hilbert space. Or one may use the extended T-space explicitly and define as proposed by Yafaev [66]

$$P_{out} = F(H_o \geq 0)\, F(T \geq 0)\, F(H_o \geq 0), \tag{5.24}$$

$$P_{in} = F(H_o \geq 0)\, F(T \leq 0)\, F(H_o \geq 0). \tag{5.25}$$

Or one can replace the sharp cutoff $F(T \geq 0)$ by a smooth one.

Again the splitting is basically equivalent to the previous ones. There are numerous other possibilities. Recall that D and T were roughly the canonically conjugate operators to ln p and p^2 resp. . Ginibre [28] observed that any other sufficiently smooth function h(p) with positive derivative which tends to infinity as $p \to \infty$ is equally well suited. Replacing T in (5.24), (5.25) by A, the canonically conjugate of h(p) yields the same results in the next section. Formally

$$2A = (p\, h'(p))^{-1}\, \vec{p} \cdot \vec{x} + \vec{x} \cdot \vec{p}\, (p\, h'(p))^{-1},$$

and the sign of A determines whether a state is incoming or outgoing. If $h(0) > -\infty$ one will need an auxiliary space like the extended T-space above.

So far we have used that the momentum operator $\vec{p} = m\,\vec{v}$ is proportional to the velocity operator. This won't be the case for general free Hamiltonians, e.g. $H_o(\vec{p}) = (p^2 + m^2)^{1/2}$. The velocity operator is generally $\vec{v}(\vec{p}) = \vec{\nabla}_p H_o(\vec{p})$ which in the example of an underlying crystalline structure even need not be parallel to $\vec{p}$. In the "phase-space decomposition" one has to replace $\vec{p}/m$ by $\vec{v}(\vec{p})$ in the definition of χ_j. The formal time operator (replacing (5.16)) is

$$T = (2v^2)^{-1}\, \vec{v} \cdot \vec{x} + \vec{x} \cdot \vec{v}\, (2v^2)^{-1},$$

the energy representation has to be adjusted but (5.20) - (5.25) remain unchanged. The generator D of the dilation group is not so useful in this case. See the end of the next section for further details.

6. PROPAGATION OF STATES UNDER THE FREE TIME EVOLUTION

A classical state is given by a probability distribution on phase-space, e.g.

$$f(\vec{x},\vec{p})\; d^{\nu}x\; d^{\nu}p \tag{6.1}$$

where f could be a positive integrable function or a δ-function. For a point particle at $\vec{x}_o$ with momentum $\vec{p}_o$ the state is $\delta(\vec{x}-\vec{x}_o)\,\delta(\vec{p}-\vec{p}_o)\,d^{\nu}x\,d^{\nu}p$. The configuration space distribution is

$$f(\vec{x})\; d^{\nu}x = [\int f(\vec{x},\vec{p})\; d^{\nu}p]\, d^{\nu}x \tag{6.2}$$

and similarly for the momentum distribution. The free time evolution is

$$f(t;\,\vec{x},\vec{p}) = f(0;\,\vec{x} - t\,\vec{p}/m,\,\vec{p}). \tag{6.3}$$

Choose at time 0 a state at the origin with momentum distribution $g(\vec{p})\,d^{\nu}p$:

$$f(0;\,\vec{x},\vec{p}) = \delta(\vec{x})\; g(\vec{p}). \tag{6.4}$$

Its configuration space distribution at time t is

$$[\int d^{\nu}p\; f(t;\,\vec{x},\,\vec{p})\,]\; d^{\nu}x$$

$$= [\int d^{\nu}p\; \delta(\vec{x} - t\,\vec{p}/m)\;\; g(\vec{p})\,]\, d^{\nu}x$$

$$= (m/t)^{\nu}\;\; g(m\vec{x}/t)\;\; d^{\nu}x\,. \tag{6.5}$$

The $\vec{x}$-space distribution is determined by the momentum distribution, in particular it vanishes if $m\vec{x}/t \notin \operatorname{supp} g$. In a suggestive oversimplification we will call the region where $m\vec{x}/t$ is not contained in the momentum support of a state the *classically forbidden region* of that state. For any initial state (6.5) holds in the limit $|t| \to \infty$ where g is the conserved momentum distribution. If the initial configuration space distribution decays rapidly then the probability to find the particle in the classically forbidden region decreases faster than any inverse power of $|t|$. Let us call states with $f \in L^1(\mathbb{R}^{2\nu})$ *classical wave packets*.

The folk lore that freely evolving quantum states asymptoti-

cally look like classical wave packets has been particularly emphasized by Dollard [13, 14]. He used the explicitly known $\vec{x}$-space kernel of the free propagator $\exp(-i H_o t)$ to show a stronger result which implies

$$|(e^{-iH_o t}\Phi)(\vec{x})|^2 \to (m/t)^{\nu} |\hat{\phi}(m\vec{x}/t)|^2 \tag{6.6}$$

in the sense that the L^1-norm of the difference vanishes as $|t| \to \infty$. But already Brenig and Haag [6] identified (6.6) as the asymptotically leading contribution. Moreover they applied the stationary phase method and thus introduced it as a tool to estimate the asymptotics of freely moving wave packets. This method has been further developed and refined, notably by Hörmander [32], or see Appendix 1 of Section XI.3 in [55]. For example one can easily show that for suitable states the tails of the plot into the classically forbidden region decay rapidly:

<u>Lemma 6.1.</u> Let Φ have a momentum wave function $\hat{\phi}(\vec{p}) \in C_o^{\infty}(\mathbb{R}^{\nu})$. For some $\delta > 0$ consider those $\vec{x}$ and t with $|m\vec{x}/t - \vec{p}| > \delta$ for all $\vec{p} \in \text{supp}\,\hat{\phi}$ (i.e. slightly less than the set of all $\vec{x}$ and t in the classically forbidden region). Then

$$|(e^{-iH_o t}\Phi)(\vec{x})| \le C_N(\Phi)\,(1 + |t| + |\vec{x}|)^{-N}. \tag{6.7}$$

The Lemma is a special case of the theorems we prove below, or see [55] for a direct proof. The constants $C_N(\Phi)$ generally depend on Φ and it is easy to find examples where for a given N $C_N(\Phi)$ is arbitrarily large.

The new aspect is the uniformity of this constant on suitably chosen subsets of states. Again we are led by the classical intuition. The states which give rise to large $C_N(\Phi)$ are e.g. those which are at a late time well localized near $\vec{x} \approx \vec{0}$. At time zero such a state could have been far out to the left and travelling to the right, i.e. incoming; or it could have a very small velocity and almost be at rest. We have to exclude these two possibilities to obtain uniform bounds. Consider the states with kinetic energy support above $(m/2) v_o^2$ which in addition are outgoing (resp. incoming) then the region $|\vec{x}| \le v_o |t|$ is classically forbidden in the future (resp. past). In the quantum case we prove rapid decay in time of the part of the state in the classically forbidden region which is uniform for all states as characterized above. The precise statement will be given in several theorems below.

For later use we quote without proof a special case of the stationary phase estimate, see [32] or Theorem XI.14 in [55].

Lemma 6.2. Let $u(\vec{k}) \in C_o^\infty(\mathbb{R}^\nu)$ have support in the compact set K, $\mathcal{O}$ is an open neighborhood of K. If the phase function $\phi(\vec{k}) \in C^\infty(\mathcal{O})$ has non-vanishing gradient on K then for any $n \in \mathbb{N}$

$$|\int e^{i\omega\phi(\vec{k})} u(\vec{k})\, d^\nu k| \le c(n)\,(1+|\omega|)^{-n}\, \|u\|_{n,\infty} \tag{6.8}$$

where

$$\|u\|_{n,\infty} = \sum_{|\alpha|\le n} \|D^\alpha u\|_\infty .$$

If a family M of phase functions ϕ satisfies on $\mathcal{O}$:
$\|\phi\|_{n+1,\infty} \le \text{const.}(n)$ for all $\phi \in M$ and on K:
$|\text{grad}\,\phi(\vec{k})| \ge \text{const.} > 0$, then $c(n)$ can be chosen uniform for $\phi \in M$. The $c(n)$ are proportional to the volume of the set K.

Recall the definitions of the operators f_i and χ_j used for the phase-space decomposition in the previous section, see the paragraphs preceding (5.1) and (5.2). The states $\chi_j f_i \Psi$ are approximately localized at $i \in \mathbb{Z}^\nu \subset \mathbb{R}^\nu$ and have velocities near $j \in a\,\mathbb{Z}^\nu \subset \mathbb{R}^\nu$. A classical free point particle with these initial conditions will be at time t at i + jt. Denote by $\|\cdot\|$ the following norm in $\mathbb{R}^\nu$

$$\|\vec{a}\| = \max|a^{(l)}| \ge |\vec{a}|\,(\nu)^{-1/2} \tag{6.9}$$

where $a^{(l)}$ is the l-th component of the vector $\vec{a}$. Then the "classically forbidden region" for the state $\chi_j f_i \Psi$ is $\|\vec{x} - i - jt\| > (3a/2)|t|$.

Lemma 6.3. For any $N \in \mathbb{N}$ there is a C'_N with

$$|(e^{-iH_o t}\chi_j f_i \Psi)(\vec{x})| \le \|\Psi\|\; C'_N (1 + \|\vec{x} - i - jt\|)^{-N} \tag{6.10}$$

if $\|\vec{x} - i - jt\| \ge 2a|t|$. The constants C'_N depend on f and χ but are independent of Ψ, i,j; they can be chosen uniformly bounded for $a \ge \alpha_o > 0$.

Proof. The L.H.S. of (6.10) equals

$$|(2\pi)^{-\nu/2} \int d^\nu p\, \exp(i\vec{p}\vec{x} - itp^2/2m)\, \chi[(\vec{p}/m) - j]\,(f_i\Psi)^\wedge(\vec{p})|$$

$$= (2\pi)^{-\nu/2} |\int d^\nu q\, \exp[i\vec{q}(\vec{x} - jt) - itq^2/2m]\, \chi[(\vec{q}/m)]\,(f_i\Psi)^\wedge(\vec{q} + mj)|$$

$$= |(e^{-iH_o t}\chi f \Psi_{ij})(\vec{x} - i - jt)|$$

with the unitary mapping $\Psi \to \Psi_{ij}$, where

$$\hat{\psi}_{ij}(\vec{q}) = e^{i(\vec{q}\cdot i)}\, \hat{\psi}(\vec{q} + mj). \tag{6.11}$$

Thus it is sufficient to prove (6.10) for $i = j = 0$. We apply Lemma 6.2 with $u(\vec{p}) = (\chi\ f\ \psi)\hat{}(\vec{p})$, $\omega = \|\vec{x}\|$, $\phi(\vec{p}) = \phi(\vec{p};\vec{x},t) = \|\vec{x}\|^{-1}\{\vec{x}\cdot\vec{p} - t\ p^2/2m\}$. Observe that

$$\sup_{\vec{p}} |D^{\alpha}\ (f\ \Psi)\hat{}(\vec{p})|$$

$$\leq \text{const.}\ \| |\vec{x}|^{\alpha}\ f(\vec{x})\ \psi(\vec{x})\|_1$$

$$\leq \text{const.}\ \| |\vec{x}|^{\alpha}\ f(\vec{x})\|_2\ \|\Psi\|. \tag{6.12}$$

Thus

$$\|\chi\ f\ \Psi\hat{}\|_{n,\infty} \leq c'(n)\|\Psi\| \tag{6.13}$$

with $c'(n)$ depending on f and χ only. The phase function ϕ and all its derivatives are uniformly bounded for all parameter values $\vec{x},t$ which satisfy $|t|/\|x\| \leq 2a$ on any compact set in $\vec{p}$-space. Moreover for $\vec{p} \in \text{supp}\ \chi$, and $\vec{x},t$ as above

$$|\vec{\nabla}_p\ \phi(\vec{p};\ \vec{x},t)| \geq \|(\vec{x}/\|x\|)-(t\ \vec{p}/m\|x\|)\|$$

$$\geq 1 - (3a/2)(1/2a) = 1/4.$$

Thus Lemma 6.2 applies and (6.10) follows. The uniformity in $a \geq a_o > 0$ is evident from (6.12) and (6.13) because a bigger a allows to make the decrease of $f(\vec{x})$ faster and the derivative of $\chi(\vec{v})$ smaller.

□

Now we are prepared for our first uniform estimate. Recall the definitions (5.5), (5.6) $P_{in/out} = \Sigma\ \chi_j f_i$ where the double sum extends over $i \cdot j < 0$ (resp. ≥ 0), $i \in \mathbb{Z}^{\nu}$, $j \in a\,\mathbb{Z}^{\nu}$. The parameter $a > 0$ is a measure of the region in which the smooth momentum cutoff is made, it can be chosen conveniently such that for a given state Ψ with strictly positive kinetic energy $F(\|\vec{v}\| \leq 3a)\ \Psi = 0$.

<u>Theorem 6.4.</u> Let $\Psi = F(\|\vec{v}\| \geq 3a)\Psi$. With P_{in} and P_{out} as defined in (5.5), (5.6)

$$\sup_{\|\vec{x}\| \leq a|t|} \left|(e^{-iH_ot}\ P_{out}\ \Psi)\ (\vec{x})\right| \leq \|\Psi\|\ D_N\ a^{-N}\ |t|^{-N},\ t > 0; \tag{6.14}$$

and the same with P_{in} for $t < 0$. D_N may be chosen independent of a for $a \geq a_o > 0$.

Proof. For the specified states

$$\chi_j f_i \Psi = 0 \quad \text{if} \quad \|j\| \le 2a;$$

therefore the summation for $P_{in/out}$ extends only over $\|j\| \ge 3a$. For $t > 0$ (resp. < 0) the condition $i \cdot j \ge 0$ (resp. < 0) implies

$$\|i + j\,t\| \ge \max(\|i\|, \|jt\|).$$

For $\|\vec{x}\| \le at$ we have

$$\|\vec{x} - i - jt\| \ge \begin{cases} \|i\| \\ |t|\,(\|j\| - a). \end{cases} \tag{6.15}$$

The estimate (6.10) of Lemma 6.3 implies with (6.15)

$$\sup_{\|\vec{x}\| \le a|t|} \left| (e^{-iH_0 t} P_{out} \Psi)(\vec{x}) \right|$$

$$\le \|\Psi\| \, C_{N+\nu+1} \sum_i (1 + \|i\|)^{-\nu-1} \sum_{\|j\| \ge 3a} \{|t|\,(\|j\| - a)\}^{-N}$$

$$\le \|\Psi\| \, D_N \, a^{-N} \, |t|^{-N} \tag{6.16}$$

Similarly for $t < 0$.

□

Remark. If one wants to consider P_{out}^* and P_{in}^* as well one easily sees that the statement of Lemma 6.3 remains true for $\|x-i-jt\| \ge 3a|t|$ if $\chi_j f_i$ is replaced by $f_i \chi_j$. Theorem 6.4 and the Corollaries 6.5 and 6.6 below remain true for P_{out}^* and P_{in}^* if the small velocity cutoff $F(\|\vec{v}\| \ge 4a)$ is used.

The theorem holds because the tails into the classically forbidden region decay rapidly in $\vec{x}$ and t for each component from a phase-space cell. This permits to sum up infinitely many such contributions if the cone $\|\vec{x}\| \le a|t|$ belongs to regions where the tails are smaller and smaller. For a given j the "outgoing" region in $\vec{x}$-space was defined as the approximate half space where $i \cdot j \ge 0$. However, it is sufficient to exclude a suitable cone around the "backward direction" $-j$. For $\|j\| \ge 4a$, e.g., it would be sufficient to exclude in the definition of P_{out} those i where $\tan \measuredangle(i,-j) \le 1/2$. Theorem 6.4 then remains valid. Or one could damp the incoming part with rapidly decreasing coefficients. This illustrates why the splitting into incoming and outgoing part is rather arbitrary. A large part of the

state is both incoming and outgoing in the sense that the distance of its localization region from the origin grows linearly with time in the future and past.

The divergence of the bound (6.14) at $t = 0$ is necessary because $\psi(\vec{0})$ may be infinite. (In passing we have shown that singularities cannot propagate into the classically forbidden region). However an upper energy-cutoff excludes this possibility.

Corollary 6.5. Let $\Psi = F(3a \leq \|\vec{v}\| \leq b)\Psi$, then

$$\sup_{\|\vec{x}\| \leq a|t|} \left|(e^{-iH_0 t} P_{out} \Psi)(\vec{x})\right| \leq \|\Psi\| D'_N (1 + |t|)^{-N}, \; t \geq 0; \tag{6.17}$$

and similarly for $t \leq 0$ with P_{in}.

Proof. If $\hat{\psi}$ has support in a given compact set K of momentum space then

$$|\psi(\vec{x})| \leq C(K)\|\psi\|. \tag{6.18}$$

□

A simple reformulation in terms of operators is

Corollary 6.6.

$$\|F(|\vec{x}| \leq a|t|)\, e^{-iH_0 t} P_{out} F(\|\vec{v}\| \geq 3a)\| \leq$$

$$\leq C_N (1 + a|t|)^{-N} \text{ for } t \geq 0, \tag{6.19}$$

and with P_{in} for $t \leq 0$.

Proof. For small times observe that all operators are bounded. For large times the rapid decay of the wave function (6.14) more than compensates the power-increase of the volume of the integration region.

□

Remark. If a state is localized far out it may be useful to restrict the summation over i in the definition of $P_{in/out}$ to those i's with $\|i\| \geq i_0 > 0$. Denote the corresponding operators by $P_{out}(i_0)$ etc.. Then the classically forbidden region contains $\|\vec{x}\| \leq i_0 + at$. In the proof of Theorem 6.4 one can replace (6.15) by:
For $\|\vec{x}\| \leq at + i_0/2$ we have with $\|i\| \geq i_0$

$$\|\vec{x} - i - jt\| \geq \begin{cases} \|i\| - i_o/2 \\ |t|\,(\|j\| - a) \end{cases} \tag{6.15'}$$

and in (6.16) we get a term

$$\sum_{\|i\| \geq i_o} (1 + \|i\| - i_o/2)^{-\nu-1}$$

which decays for growing i_o. Thus one can choose decaying constants, as an example we give the modification of Corollary 6.6:

$$\|F(|\vec{x}| \leq a|t| + i_o/2)e^{-iH_ot} P_{out}(i_o) F(\|\vec{v}\| \geq 3a)\|$$

$$\leq C_N(i_o) (1 + a|t|)^{-N} \text{ for } t \geq 0, \tag{6.20}$$

where $C_N(i_o) \to 0$ as $i_o \to \infty$; similarly with $P_{in}(i_o)$ for $t \leq 0$.

Using the spectral projections of the dilation generator D to characterize incoming and outgoing states (5.12), (5.13) the proper statement reads:

Theorem 6.7. Let $g \in C_o^\infty(\mathbb{R})$ obey $g(p) = 0$ if $p = |\vec{p}| \geq 3m\,v_o$. Denote $E_o = (m/2)v_o^2$, then for any Ψ and $d \in \mathbb{R}$

a) $$\sup_{|\vec{x}| \leq v_o|t|} \left|(e^{-iH_ot} g(p) F(D \geq d)\Psi)(\vec{x})\right| \leq$$

$$\leq \|\Psi\| C'(N,g) (1 + E_o t)^{-N} \text{ for } t \geq \max(0, -d/E_o) \tag{6.21}$$

and similarly with $F(D \leq d)$ for $t \leq \min(0, d/E_o)$.

b) $$\|F(|\vec{x}| \leq v_o t)\, e^{-iH_ot} g(p) F(D \geq d)\| \leq$$

$$\leq C(N,g)(1 + E_o t)^{-N} \text{ for } t \geq \max(0, -d/E_o) \tag{6.22}$$

and similarly as $t \to -\infty$.

As a preparation we show

Lemma 6.8. For the state $\psi_{\vec{y}}(\vec{x}) = \psi(\vec{x} - \vec{y})$ with $\hat{\psi} \in C_o^\infty(\mathbb{R}^\nu)$, $\hat{\psi}(\vec{p}) = 0$ if $|\vec{p}| \leq 3m\,v_o$

$$\sup_{|\vec{y}| \leq v_o t} \| F(D \leq E_o t)\, e^{-iH_o t} \psi_{\vec{y}} \| \leq C_N (1 + E_o t)^{-N} \quad \text{for } t \geq 0 \tag{6.23}$$

$$\sup_{|\vec{y}| \leq v_o t} \| F(D \geq E_o t) e^{-iH_o t} \psi_{\vec{y}} \| \leq C_N (1 - E_o t)^{-N} \quad \text{for } t \leq 0 \tag{6.24}$$

where $E_o = (m/2) v_o^2$.

Remark. We give a heuristic argument to explain why the lemma is true and to develop our intuition of free propagation in D-space. A simple calculation shows formally

$$e^{iH_o t} D e^{-iH_o t} = D + 2H_o t; \tag{6.25}$$

a precise justification is given in the next section. Thus a state travels in D-space with a speed $2H_o$. For large t all states $\psi_{\vec{y}}$ with $|\vec{y}| \leq v_o t$ lie in D-space above $\approx - p\, v_o t$. After time t they should lie above

$$-p\, v_o t + 2(1/2m) p^2\, t = p(v - v_o) t > 2m\, v_o^2\, t = 4E_o t .$$

Thus the lemma gives an estimate of the parts of the tails propagating far out into the classically forbidden region.

Proof. We use the Mellin transform representation (5.10), (5.11).

$$(e^{-iH_o t} \psi_{\vec{y}})^o (\lambda, \omega) = (2\pi)^{-1/2} \int dp\, p^{(\nu-2)/2}\, p^{i\lambda} \times$$

$$\times \exp(-itp^2/2m - i\, p\, \omega \cdot \vec{y}) \hat{\psi}(p, \omega)$$

$$= (2\pi)^{-1/2} \int dp\, e^{-if(p)}\, p^{(\nu-2)/2}\, \hat{\psi}(p, \omega) \tag{6.26}$$

where

$$f(p) = t\, p^2/2m + p\, \omega \cdot \vec{y} - \lambda \ln p. \tag{6.27}$$

Using stationary phase estimates we will show for $t > 0$, $\lambda < 2E_o t$ the stronger estimate

$$|(e^{-iH_o t} \psi_{\vec{y}})^o (\lambda, \omega)| \leq C'_N (1 - (\lambda - 2E_o t))^{-N} \tag{6.28}$$

which implies (6.23) and similarly for (6.24). For $-(\lambda - 2E_o t) \geq t$ we use

$$f(p) = (-\lambda + 2E_o t)\,\phi(p). \tag{6.29}$$

The phase function

$$\phi(p) = \ln p + [tp^2/2m + p\,\omega \cdot \vec{y} - 2E_o t \ln p](-\lambda + 2E_o t)^{-1} \tag{6.30}$$

has bounded derivatives in a neighborhood of the support of $\hat{\psi}$ and for the allowed parameter values. Moreover

$$\frac{d\phi(p)}{dp} = \frac{1}{p} + [t\,p/m + \omega \cdot \vec{y} - 2E_o t/p](-\lambda + 2E_o t)^{-1} \tag{6.31}$$

is strictly positive because $(1/p) > (m/v_o)$ and

$$t(p/m - v_o - 2E_o/p) > v_o\, t \geq 0. \tag{6.32}$$

Thus Lemma 6.2 gives (6.28). If $t \geq (-\lambda + 2E_o t)$ we split $f(p) = t\,\phi'(p)$. The phase function

$$\phi'(p) = p^2/2m + p\,\omega \cdot \vec{y}/t - 2E_o \ln p + \ln p(-\lambda + 2E_o t)/t \tag{6.33}$$

also has bounded derivatives in the relevant region and its first derivative is strictly positive by (6.32). Thus (6.28) holds throughout and Lemma 6.8 is proved.

□

We now follow Perry [51] for the

<u>Proof of Theorem 6.7.</u> Obviously b) follows from a). To show a)

$$\begin{aligned} &|(e^{-iH_o t} g(p)\,F(D \geq -E_o t)\,\Psi)(\vec{x})| = \\ &(2\pi)^{-\nu/2} \int d^\nu p \;\exp(i\vec{p}\,\vec{x} - i\,t\,p^2/2m)\,g(p)\,(F(D \geq -E_o t)\Psi)^\wedge(\vec{p}) \\ &= (2\pi)^{-\nu/2}\,(e^{+iH_o t} g_{\vec{x}}\,,\,F(D \geq -E_o t)\Psi) \\ &\leq (2\pi)^{-\nu/2}\,\|F(D \geq E_o(-t))\,e^{-iH_o(-t)} g_{\vec{x}}\| \cdot \|\Psi\|. \end{aligned} \tag{6.34}$$

Here g is the vector with momentum space wave function $g(p)$ and $g_{\vec{x}}$ is as in Lemma 6.8, $(\cdot,\cdot)$ the inner product. Now (6.24) implies (6.21) if $t \geq \max(0, -d/E_o)$.

□

Remark. From the proof it is obvious that one could have used in (6.21), (6.22) instead of the sharp cutoff function $F(D \geq d)$ any bounded cutoff function $\phi(D)$ with rapid decay to the left. The contribution from $\phi(D)\, F(D \geq -E_o t)$ decays rapidly and the same is true for $\|\phi(D)\, F(D \leq -E_o t)\|$.

The high energy cutoff was convenient for the proof because then g is square integrable. On the other hand there is no physical reason to exclude high energies, these states leave the cone $|\vec{x}| \leq v_o |t|$ even faster. Denote by $\phi(p)$ a lower cutoff function with $\phi(p) = 0$ for $p \leq 3m\, v_o$ and $\phi(p) = 1$ for $p \geq 4m\, v_o$, and decompose it into pieces as considered above. Define

$$g_o(p) = \phi(p) - \phi(p - 3m\, v_o) \tag{6.35}$$

and generally

$$\begin{aligned} g_l(p) &= \phi(p - 3l\, m\, v_o) - \phi(p - 3(l + 1)m\, v_o) \\ &= g_o(p - 3l\, m\, v_o) \end{aligned} \tag{6.36}$$

then

$$\phi(p) = \sum_{l=o}^{\infty} g_l(p) \tag{6.37}$$

We have $g_l(p) = 0$ if $p \leq 3m\, v_l$ where $v_l = (l + 1)\, v_o$. Now we easily get

Theorem 6.9. Let the smooth function ϕ obey $\phi(p) = 0$ (resp. 1) for $p \leq 3m\, v_o$ (resp. $\geq 4m\, v_o$). Then

a)
$$\begin{aligned} \sup_{|\vec{x}| \leq v_o t} &\left|(e^{-iH_o t}\, \phi(p)\, F(D \geq d)\, \Psi)(\vec{x})\right| \leq \\ &\leq \|\Psi\|\, C'(N,\phi)\, (E_o t)^{-N}, \text{ and} \end{aligned} \tag{6.38}$$

b)
$$\begin{aligned} \|F\, |\vec{x}| \leq v_o\, t)\, e^{-iH_o t}\, &\phi(p)\, F(D \geq d)\| \leq \\ &\leq C(N,\phi)(1 + E_o t)^{-N} \end{aligned} \tag{6.39}$$

for $t \geq \max(0, -d/E_o)$. Similarly with $F(D \leq d)$ for $t \leq \min(0, d/E_o)$. $E_o = (m/2)v_o^2$.

Proof. In part b) the operators are bounded, therefore b) follows from a). The decomposition (6.37) $\phi = \Sigma\, g_l$ allows to use for each term the estimate (6.21) of Theorem 6.7. with

$E_1 = (m/2)v_1^2 = (1+1)^2 E_o$. Thus

$$\sup_{|\vec{x}| \leq v_o t} \left| (e^{-iH_o t} \phi(p) F(D \geq d)\Psi)(\vec{x}) \right| \leq \tag{6.40}$$

$$\leq \|\Psi\| (E_o t)^{-N} \sum_l C'(N,g_l)(l+1)^{-2N}$$

Comparing (6.8) and (6.26) where g_l has to be used as $\hat{\psi}$ we see that $C'(N,g_l)$ has polynomial increase from the factor $p^{(\nu-2)/2}$ but the derivatives of g_l are bounded independent of l, and the size of the integration region is fixed. Therefore the sum over l on the R.H.S. of (6.40) converges for big enough N. □

Let us now turn to the formal time operator T. We consider the subspace $F(H > E_o)\mathcal{H}$, $E_o > 0$, then f(T) is well defined for a suitable class of functions f as discussed in the previous section following (5.20) and (5.21). If f is admissible for a given E_o then also $f(\cdot - t)$ is admissible, and one immediately gets

$$e^{iH_o t} f(T) e^{-iH_o t} = f(T+t). \tag{6.41}$$

There is a very close connection between the Lax Phillips scattering theory [44,52] and our geometric approach. They are almost identical in spirit but quite different in the technical details. The main difference is the semiboundedness of the Hamiltonian. If characteristic functions of a half line were admissible functions of T (they are not in quantum mechanics), then $\chi_{[0,\infty)}(T)\mathcal{H}$ would be an outgoing subspace in the sense of Lax and Phillips by (6.41). A good approximation thereof are the smooth cutoff functions $f_{\pm}$ defined in the paragraph preceding (5.22) which have rapidly decaying tails. Thus we have only approximately absorbing subspaces and the translation representation, a crucial tool of the Lax Phillips approach, does not exist.

Before proving a theorem let us give the intuitive argument why we expect the desired propagation properties with $f_+(T)$ as a characterization of outgoing states. A state which is at time 0 essentially localized in the positive part of T-space will be localized at time t above t in T-space. The formal expression (5.16) $T = (4H_o)^{-1}\vec{p}\cdot\vec{x} + \vec{x}\cdot\vec{p}(4H_o)^{-1} > t$ implies $|\vec{x}| > (2E_o/m)^{1/2}\, t$ because $2p(4H_o)^{-1} \leq (2E_o/m)^{-1/2}$. The main task of the proof is to control the tails. As a preparation we estimate how the translation in $\vec{x}$-space propagates a state in T-space. The state is given by its energy wave function $g(E) \in C_o^\infty$ and the translation by $\vec{x}$ is multiplication by $\exp(-i\vec{p}\cdot\vec{x}) = \exp(-i\, p\, \omega\cdot\vec{x}) =$
$= \exp(-i\, p\, y) = \exp(-i\sqrt{2mE}\, y)$.

<u>Lemma 6.10.</u> For $g \in C_o^\infty(\mathbb{R})$, $g(E) = 0$ for $E \leq E_o/2 > 0$.

$$\left|(2\pi)^{-1/2} \int dE\, e^{iE\tau}\, e^{-i\sqrt{2mE}\,y}\, g(E)\right| \le$$

$$\le \begin{cases} C(g) & \text{for all } \tau, y \\ C(N,g)\,(1+|\tau|)^{-N} & \text{for } |y| < |\tau|\sqrt{E_o/4m}\,. \end{cases} \tag{6.42}$$

Proof. The first inequality holds with $C(g) = (2\pi)^{-1/2} \int dE\, |g(E)|$. For the second we write the exponent as $i\,\tau\,\phi(E)$ with phase function

$$\phi(E) = (E - \sqrt{2m\,E}\;y\,/\,\tau). \tag{6.43}$$

Its derivative is bounded below:

$$d\phi(E)/dE = 1 - y/(\tau\; 2\sqrt{E/2m}\;) > 1/2 \tag{6.44}$$

if $|y| < |\tau|\sqrt{E_o/4m} \le |\tau|\sqrt{E/2m}$ on supp g. Moreover all higher derivatives are bounded uniformly in the paramter range. Thus Lemma 6.2 implies the second estimate.

□

Theorem 6.11. Let $f_\pm$ (T) be as defined in the paragraph preceding (5.22). For any $\Psi \in F(2E_o \le H_o \le b)\mathcal{H}$, $b < \infty$ and $v_o = \sqrt{E_o/m}/4$

$$\sup_{|\vec{x}| \le v_o|t|} \left|(e^{-iH_ot} f_+(T)\Psi)(\vec{x})\right| \le \|\Psi\|\, C'_N (1+|t|)^{-N}, \tag{6.45}$$

$$\|F(|\vec{x}| \le v_o|t|)\; e^{-iH_ot} f_+(T) F(2E_o \le H_o \le b)\| \le C_N (1+|t|)^{-N},$$

and (6.46)

$$\|F(|\vec{x}| \le v_o|t|) e^{-iH_ot} g(H_o) f_+(T)\| \le C_N (1+|t|)^{-N} \tag{6.46'}$$

for $t > 0$ if $g \in C_o^\infty$, $g(E) = 0$ for $E \le E_o/2$. Similarly with $f_-(T)$ for $t < 0$.

Proof. $f_+(T)\; F(2\,E_o \le H_o \le b) = g(H_o)\; f_+(T)\; F(2E_o \le H_o \le b)$ for a function g as given above.

$$\left|(e^{-H_ot}\, g(H_o)\; f_+(T)\;\Psi)(\vec{x})\right|$$

$$= (2\pi)^{-\nu/2}\left|\int d^\nu p\, \exp(i\,\vec{p}\,\vec{x} - i\,t\,p^2/2m) g(p^2/2m)\,(f_+(T)\Psi)^\wedge(\vec{p})\right|$$

$$\le (2\pi)^{-\nu/2}\, \|f_+(T) e^{+iH_ot}\, g_{\vec{x}}\| \cdot \|\Psi\|\,. \tag{6.47}$$

With (6.41) this simplifies to:

$$\| f_{\pm}(T) e^{+iH_o t} g_{\vec{x}} \|^2 = \| f_{\pm} (T - t) g_{\vec{x}} \|^2$$

$$\leq (\int d\Omega) \sup_{\omega} \int d\tau |f_{+}(\tau - t)|^2 \times$$

$$\times (2\pi)^{-1} |\int dE \exp(i E \tau - i\sqrt{2mE}\, \omega \cdot \vec{x}) \bar{g}(E)|^2$$

$$\leq \text{const.} \int_{|\tau| \leq |t|/2} d\tau \, |f_{\pm} (\tau - t)|^2 \, |C(\bar{g})|^2 +$$

$$+ \text{const.} |C(N,\bar{g})|^2 \int_{|\tau| \geq |t|/2} d\tau \, (1 + |\tau|)^{-2N} . \tag{6.48}$$

Here $\bar{g}(E) = \sqrt{m}\,(2m\,E)^{(\nu-2)/4} g(E)$ is the energy space wave function. The second summand decays rapidly in $|t|$ by (6.42) for $|\vec{x}| \leq |t| \sqrt{E_o/m}/4 = v_o |t|$. The same is true for the first summand with + (resp. −) for t > 0 (resp. < 0). (6.46) follows immediately from (6.45).

□

Remark. Using the auxiliary space definition of the outgoing subspace (5.24) one would have got in (6.47) a factor

$$\| F(H_o \geq 0)\, F(T \geq 0) F(H_o \geq 0) e^{+iH_o t} g_{\vec{x}} \| \tag{6.49}$$

$$\leq \| F(T \geq 0) e^{+iH_o t} g_{\vec{x}} \| = \| F(T \geq t) g_{\vec{x}} \| \tag{6.50}$$

$$\leq \text{const.}\; C(N,\bar{g}) \{ \int_{|\tau| \geq t} d\tau \, (1 + |\tau|)^{-2N} \}^{1/2} , \tag{6.51}$$

where the norm in (6.49) is in the physical Hilbert space and those in (6.50) are in the extended T-space. Thus (6.46') remains true with this splitting.

Again there is an extension to subspaces without high energy cutoff. We won't give it here because our proof is more complicated than the two proofs of this fact which we have given above. Also Yafaev has announced this result [67] for his splitting (5.24), (5.25). We have discussed this question here because (i) the restriction to a bounded energy interval is unphysical, (ii) the question arises when treating time dependent potentials [63,67]. But in this report we won't need it because a restriction to a dense set of states will be sufficient. There are other methods than stationary phase estimates to control the propagation properties of outgoing states. One can use an expansion of the explicitly known convolution kernel of exp $(-iH_o t)$ in $\vec{x}$-space [18], or commutator properties of H_o and D [46], or spectral properties of $\vec{x} + \vec{v}\, t$ on suitable subspaces for large t [19].

Simon observed [61] that one can treat more general free Hamiltonians $H_o(\vec{p})$ with velocity operator $\vec{v}(\vec{p}) = \vec{\nabla}_p H_o(\vec{p})$. Below we will use the following physical condition: the states with compact kinetic energy support away from certain energy values form a dense set of states with strictly positive velocity. (or weaker: these states approximate any vector in $\mathcal{H}^{cont}$.) For the estimates above one needs in addition that $H_o(\vec{p})$ is sufficiently smooth on this dense set and that $H_o(\vec{p}) \to \infty$ as $p \to \infty$. The latter condition implies that states with bounded kinetic energy have compact support in momentum space. Decompositions into incoming and outgoing states were indicated at the end of the last section. Then the propagation estimates for states with energy support in a compact interval remain true, e.g. Corollary 6.5 and Theorem 6.11. See [28,61] for details. The results for an energy range which is unbounded above like Theorem 6.4 and Corollary 6.6 are true if the velocity increases at least like a small power of p, because then we can sum up the contributions from infinitely many "velocity-cells". If $H_o(\vec{p}) = p$ a simple direct proof with the splitting (5.24), (5.25) can be given.

7. ASYMPTOTIC EVOLUTION OF SELECTED OBSERVABLES

We resume the discussion of dynamical questions treated in Section 4. Now we impose some decay requirements at infinity on the potential and we gain from this better control of the asymptotic behavior of scattering states. We obtain our results by studying the asymptotic time evolution of suitable observables like x^2 and D on $\mathcal{H}^{cont}$.

For a self-adjoint operator A denote the time translated one with *interacting* time evolution

$$A(t) = e^{iHt} A e^{-iHt} \tag{7.1}$$

We will show e.g. that $x^2(t)$ and $D(t)$ grow asymptotically as $t \to \infty$, thus a state in $\mathcal{H}^{cont}$ eventually will be localized far from the scatterer and will be outgoing. Mathematically we study the time evolution as an automorphism group on the observables (the self-adjoint operators) $A \to A(t)$ and we investigate the limits for some of them.

We give the theorem and its proof for a smaller class of interactions to avoid some technicalities. Further details and extensions can be found in [23]. Assume that V is an H_o-compact multiplication operator in $\vec{x}$-space and it can be decomposed

$$V(\vec{x}) = V_s(\vec{x}) + V_l(\vec{x}) \tag{7.2}$$

where the short-range part obeys

$$|\vec{x}|\ V_s(\vec{x})(H_o + 1)^{-1} \text{ is compact} \tag{7.3}$$

and the distributional derivative of the long-range part $V_l(\vec{x})$ obeys

$$\vec{x}\cdot\vec{\nabla}\,V_l(\vec{x})\ (H_o+1)^{-1} \text{ is compact.} \tag{7.4}$$

In (7.3) and (7.4) one can replace the free resolvent by that of H by (2.4).

These conditions say roughly that V_s and the gradient of V_l are of short range in the sense that they decay faster than $|\vec{x}|^{-1}$. With respect to decay at infinity this condition is slightly weaker than the short-range condition introduced in Section 2. Potentials like

$$V(r) = r^{-\varepsilon} \text{ for } r \geq r_o,\ \varepsilon > 0; \tag{7.5}$$

$$V(r) = (\ln r)^{-2} \text{ for } r \geq r_o; \tag{7.6}$$

are examples for V_l. Oszillating potentials like

$$V(r) = r^{-\beta} \sin r^{\alpha},\ r \geq r_o, \tag{7.7}$$

satisfy (7.3) for $\beta > 1$ and (7.4) for $\alpha < \beta$. (Note that for some of the very long-range potentials which satisfy (7.4) one does not even know existence of modified wave operators, e.g. for (7.6), and a potential $V_s(\vec{x}) = (r \ln r)^{-1}$ requires modified wave operators.)

We state without proof a few domain properties of various unbounded self-adjoint operators which we will need below. Let V be a Kato bounded perturbation of H_o (this is implied by the H_o-compactness which we assumed above), then $\mathcal{D}(H) = \mathcal{D}(H_o)$. The operator

$$N = H_o + x^2 \tag{7.8}$$

is self-adjoint on

$$\mathcal{D}(N) = \mathcal{D}(H_o) \cap \mathcal{D}(x^2). \tag{7.9}$$

$\mathcal{D}(N)$ is a core for the self-adjoint operators H, H_o, x^2, and $D = (1/2)\ (\vec{p}\cdot\vec{x} + \vec{x}\cdot\vec{p})$; and it is time translation invariant

$$e^{-iHt}\,\mathcal{D}(N) = \mathcal{D}(N). \tag{7.10}$$

Let A be essentially self-adjoint on $\mathcal{D}(N)$ then $\mathcal{D}(N)$ is a core for all A(t) by (7.10). The commutator $i[H,A]$ is a symmetric quadratic form on $\mathcal{D}(N) \times \mathcal{D}(N)$; it is the time derivative of the quadratic form A(t) on that domain. If it is the quadratic form of a self-adjoint

operator with core $\mathcal{D}(N)$, denoted again by $i[H,A]$, then the following operator identity holds on the common core $\mathcal{D}(N)$

$$A(t_2) = A(t_1) + \int_{t_1}^{t_2} e^{iHt}\, i[H,A]\, e^{-iHt}\, dt, \tag{7.11}$$

the strong integral is well defined on $\mathcal{D}(N)$. For a proof of (7.10) and (7.11) see e.g. [27,37].

We are ready to state the main result of this section. For strong resolvent convergence and the notation see Section 2, $\mathcal{H}^{cont} = P^{cont}\mathcal{H}$ is the continuous spectral subspace of the full Hamiltonian H.

Theorem 7.1. For $H_o = -(1/2m)\,\Delta$ let $V(\vec{x})$ be an H_o-compact potential satisfying (7.2) - (7.4). Then in the sense of strong resolvent convergence

a) $$\lim_{|t|\to\infty} \frac{m}{2}\frac{x^2(t)}{t^2} = H\,P^{cont} \tag{7.12}$$

b) $$\lim_{|t|\to\infty} D(t)/t = 2\,H\,P^{cont} \tag{7.13}$$

c) $$\lim_{|t|\to\infty} H_o(t) = H \quad \text{on } \mathcal{H}^{cont} \tag{7.14}$$

Before proving the theorem we derive a corollary and discuss its implications.

Corollary 7.2. Let the assymptions of Theorem 7.1 hold, then for $\psi \in \mathcal{H}^{cont}$, any R and d:

$$\lim_{|t|\to\infty} \|F(|\vec{x}| \leq R)\, e^{-iHt}\,\psi\| = 0, \tag{7.15}$$

$$\lim_{t\to\infty} \|F(D \leq d)\, e^{-iHt}\,\psi\| = 0, \tag{7.16}$$

$$\lim_{t\to-\infty} \|F(D \geq d)\, e^{-iHt}\,\psi\| = 0, \tag{7.17}$$

$$w-\lim_{|t|\to\infty} e^{-iHt}\,\psi = 0. \tag{7.18}$$

If moreover $F(H < E_o)\psi = 0$ for some $E_o = (m/2)v_o^2 > 0$, then

$$\lim_{|t|\to\infty} \|F(|\vec{x}| \leq v_o|t|)\, e^{-iHt}\,\psi\| = 0, \tag{7.19}$$

$$\lim_{t \to \infty} \| F(D \leq 2E_o t) \, e^{-iHt} \Psi \| = 0, \tag{7.20}$$

$$\lim_{t \to -\infty} \| F(D \geq 2E_o t) \, e^{-iHt} \Psi \| = 0. \tag{7.21}$$

<u>Proof.</u> For any $\varepsilon > 0$ there is an $E_\varepsilon = (m/2)v_\varepsilon^2 > 0$ such that $\| F(H \leq E_\varepsilon) \Psi \| < \varepsilon$. Then for any $R < \infty$

$$\lim_{|t| \to \infty} \| F(|\vec{x}| \leq R) \, e^{-iHt} \Psi \|$$

$$< \varepsilon + \lim_{|t| \to \infty} \| F(|\vec{x}| \leq v_\varepsilon |t|) \, e^{-iHt} F(H \geq E_\varepsilon) \Psi \| .$$

Thus (7.19) implies (7.15), and similarly (7.20) and (7.21) imply (7.16) and (7.17). To show (7.19)

$$e^{iHt} F(|\vec{x}| \leq v_o |t|) \, e^{-iHt}$$

$$= F\left(\frac{m}{2} \frac{x^2(t)}{t^2} \leq E_o \right) \tag{7.22}$$

$$\to F(H P^{cont} \leq E_o) \tag{7.23}$$

by (7.12) and the fact that the operators involved have continuous spectrum. By assymption the limit (7.23) annihilates Ψ and (7.19) follows. For $t > 0$ (and similarly for $t < 0$)

$$e^{iHt} F(D \leq 2E_o t) \, e^{-iHt} \tag{7.24}$$

$$\to F(H P^{cont} \leq E_o)$$

by (7.13), thus (7.20) and (7.21) follow. Finally (7.18) follows from (7.16) for $t > 0$ because $F(D \geq d)$ tends strongly to zero as $d \to \infty$. Similarly for $t < 0$ with (7.17). Note that we have not used (7.14) of the theorem.

□

By imposing requirements on the decay at infinity of the potential we have shown that a state from $\mathcal{H}^{cont}$ will eventually leave any bounded region in the time limit and not only in the time mean as was shown in Theorem 4.1. States with strictly positive energy will have asymptotically strictly positive kinetic energy and they leave the origin with the corresponding minimal speed v_o. Dollard [15] proposed to define the set of scattering states as

the closure of the set of states which satisfy (7.19) for arbirarily small but positive v_o's. We have shown that $\mathcal{H}^{cont}$ consists of scattering states in this sense. Moreover we have shown that any state in $\mathcal{H}^{cont}$ will be outgoing in the far future and incoming in the remote past if we use the definition (5.12), (5.13) with D. Thus the subspaces of incoming and outgoing states are asymptotically absorbing for the interacting time evolution too, and not only for the free one as discussed in the previous section. The limits (7.19)-(7.21) are essentially estimates of the propagation into the classically forbidden region, in the previous section we proved rapid decay of these quantities for the free time evolution on suitable subsets of states.

Proof of Theorem 7.1. As quadratic forms on $\mathcal{D}(N) \times \mathcal{D}(N)$

$$i[H, (m/2)\, x^2] = (1/4)\, i[p^2, x^2] = D \tag{7.25}$$

$$\begin{aligned} i[H,D] &= i[H_o, D] + i[V,D] \\ &= 2H_o - \vec{x}\cdot\vec{\nabla} V(x) =: 2H - I\,. \end{aligned} \tag{7.26}$$

Let us assume for simplicity that the potential satisfies (7.4); for the short-range part an extra technical argument is necessary which we will omit. Then by our assumptions the interaction term

$$I = 2V + \vec{x}\cdot\vec{\nabla} V(\vec{x}) \text{ is H-compact,} \tag{7.27}$$

and both commutators (7.25) and (7.26) are essentially self-adjoint on $\mathcal{D}(N)$. The integral representation (7.11) applies for D(t) and $x^2(t)$ and we get on $\mathcal{D}(N)$

$$\begin{aligned} D(t)/t &= D(0)/t + t^{-1} \int_o^t d\tau\, e^{iH\tau} \{2H - I\}\, e^{-iH\tau} \\ &= D(0)/t + 2H - t^{-1} \int_o^t d\tau\, I(\tau)\,. \end{aligned} \tag{7.28}$$

If $\mathcal{D}(N) \cap \mathcal{H}^{cont}$ is a core for $H \restriction \mathcal{H}^{cont}$ it is sufficient to show that

$$(D(t)/t)\Psi \to 2H\,\Psi \text{ for } \Psi \in \mathcal{D}(N) \cap \mathcal{H}^{cont}. \tag{7.29}$$

We expect that this is true in general but we can prove it only if all bound states lie in $\mathcal{D}(N)$. Assuming this

$$\|(D(0)/t)\Psi\| = |t|^{-1}\, \|D\Psi\| \to 0. \tag{7.30}$$

$$\|t^{-1} \int_o^t d\tau\, I(\tau)\,(H-z)^{-1}\,(H-z)\Psi\| \le$$

$$\leq \| t^{-1} \int_o^t d\tau \, C(\tau) \, P^{cont} \| \cdot \| (H - z)\Psi \| \to 0 \tag{7.31}$$

where we have used the compactness (7.27) and Lemma 4.2 about the time average of compact operators.

If the core assumption does not hold one uses $\mathcal{D}(H) \cap \mathcal{H}^{cont}$ as a core for $H \upharpoonright \mathcal{H}^{cont}$ and approximates Ψ's from this set by

$$\Psi_\lambda = (1 + \lambda x^2)^{-1} \Psi \in \mathcal{D}(N). \tag{7.32}$$

With $\lambda(t) \to 0$ slowly as $|t| \to \infty$ the same results follow.

That the R.H.S. of (7.28) converges to zero on eigenstates follows from the virial theorem, but a direct proof is simpler. Let $H\Psi = E\Psi$, it is sufficient to show that for any s

$$\| [\exp(i\, s\, D(t)/t) - 1]\Psi \|$$

$$= \| \exp(i(H - E)t)[\exp(i\, D\, s/t) - 1]\Psi \| \to 0. \tag{7.33}$$

But $\exp(i\, D\, \lambda)$ is strongly continuous in λ. This finishes the proof of (7.13).

Similarly we get on $\mathcal{D}(N)$

$$\frac{m}{2} \frac{x^2(t)}{t^2} = \frac{m}{2} \frac{x^2(0)}{t^2} + \frac{D(0)}{t} + H +$$

$$+ t^{-2} \int_o^t dt' \int_o^{t'} d\tau \, I(\tau) \tag{7.34}$$

and we conclude (7.12) by the analogous argument.

Finally observe that by assumption $(H - z)^{-1} V (H_o - z)^{-1}$ is compact, thus for (7.14)

$$\| e^{iHt} [(H_o - z)^{-1} - (H - z)^{-1}] e^{-iHt} \Psi \| \to 0 \tag{7.35}$$

for any $\Psi \in \mathcal{H}^{cont}$ by the weak convergence (7.18).

□

For the omitted details of the proof, inclusion of stronger singularities and velocity dependent potentials see [23], also for references to related work like the algebraic approach to long range scattering.

8. THE WAVE OPERATORS

In the previous section we studied the interacting time evolution as an automorphism group on some observables. Asymptotically the limits coincide with those obtained with the free time evolution. Now we turn to the strongest notion of asymptotically free motion where one requires that the unitary groups asymptotically do not differ any more on states. There are two physically different asymptotic conditions which we discuss next.

Suppose a vector $\Psi \in \mathcal{H}$ satisfies

$$\lim_{\tau \to \infty} \sup_{t \geq 0} \| (e^{-iHt} - e^{-iH_o t}) e^{-iH\tau} \Psi \| = 0. \tag{8.1}$$

We call such a state a scattering state. If one waits long enough until the scattering is over (τ big) the future time evolution from τ on ($t \geq 0$) will not depend any more on the potential, it is asymptotically free. One may also ask whether for a vector $\Phi \in \mathcal{H}$

$$\lim_{\tau \to \infty} \sup_{t \geq 0} \| (e^{-iHt} - e^{-iH_o t}) e^{-iH_o \tau} \Phi \| = 0 \tag{8.2}$$

holds. We call these vectors asymptotic configurations because Φ represents a "boundary condition at $t = +\infty$" for which there exists a solution of the interacting Schrödinger equation. The easiest way to see this is to use the wave operator (= Møller operator [45])

$$\operatorname*{s-lim}_{t \to \infty} e^{iHt} e^{-iH_o t} = \Omega_- \tag{8.3}$$

defined on those vectors where the limit exists.

<u>Lemma 8.1.</u>

$$\lim_{t \to \infty} e^{iHt} e^{-iH_o t} \Phi \tag{8.4}$$

converges if and only if (8.2) holds. The range of the wave operator is exactly the subspace of those Ψ's for which (8.1) holds.

<u>Proof.</u>

$$\| (e^{iH(t+\tau)} e^{-iH_o(t+\tau)} - e^{iH\tau} e^{-iH_o \tau}) \Phi \|$$

$$= \| (e^{-iH_o t} - e^{-iHt}) e^{-iH_o \tau} \Phi \| . \tag{8.5}$$

Thus the Cauchy criterion for (8.4) is equivalent to (8.2). By interchanging H and H_o (8.1) is equivalent to convergence of

$$\lim_{t \to \infty} e^{iH_0 t} e^{-iHt} \Psi = \Phi , \tag{8.6}$$

and this is equivalent to

$$\lim_{t \to \infty} e^{iHt} e^{-iH_0 t} \Phi = \Psi , \tag{8.7}$$

i.e. $\Psi \in \text{Ran } \Omega_-$.

□

In particular we have seen

$$(\Omega_-)^* \Psi = \begin{cases} \lim\limits_{t \to \infty} e^{iH_0 t} e^{-iHt} \Psi & \text{if } \Psi \text{ satisfies (8.1)} \\ 0 & \text{otherwise.} \end{cases} \tag{8.8}$$

For any Φ which fulfills (8.2), and $\Psi = \Omega_- \Phi$ the following relation holds

$$\lim_{t \to +\infty} \| e^{-iHt} \Psi - e^{-iH_0 t} \Phi \| = 0. \tag{8.9}$$

In this sense Φ is an asymptotic (here outgoing) configuration. Showing (8.2) amounts to proving that there exists a solution of the interacting Schrödinger equation $\exp(-iHt)\Psi$ which fulfills the "boundary condition at $t = \infty$" (8.9). Therefore we call the verification of (8.2) or (8.4) the *existence problem*. If the potential decays sufficiently fast at infinity then one expects that the wave operators exist for any configuration $\Phi \in L^2(\mathbb{R}^\nu)$. We will shortly prove this.

Certainly (8.1) should not hold for bound states, they are influenced by the potential forever. But physical experience tells us that we should expect (8.1) to hold for all states orthogonal to the bound states. This is the *completeness problem*.

In the next section we will give various common notions of asymptotic completeness and prove them.

Similarly one considers the limits in the negative time direction with analogous results. The incoming wave operator is

$$\underset{t \to -\infty}{\text{s-lim}} \; e^{iHt} e^{-iH_0 t} = \Omega_+ , \tag{8.10}$$

the silly sign convention is a tradition of the physics literature which can be understood historically.

Let us now prove existence of the wave operators in the time dependent approach. (Alternatively one can use eigenfunction expansions [33].) The method is due to Cook who gave the first mathematically rigorous proof for potential scattering. We also use later variations, extensions, and refinements [9,42,43,32,58,59,38] adapted for our special case. Related ideas are used in Hepp's version [31] of the Haag-Ruelle scattering theory [29]. We remind the reader of Lemma 3.1 where we proved that local compactness (3.2) and the weak assumption of decay at infinity (2.14) are equivalent to compactness of the difference of the free and interacting resolvents (8.11). Condition (8.12) is the same as (2.15) discussed above.

Theorem 8.2. (Existence of the wave operators) Let

$$(H - z)^{-1} - (H_o - z)^{-1} \tag{8.11}$$

be compact and let the potential be of short range

$$\| g(H)\ V\ g(H_o)\ F(|\vec{x}| \geq R) \| = h(R) \in L^1(\mathbb{R}_+, dR) \tag{8.12}$$

for all $g \in C_o^\infty(\mathbb{R})$. Then the wave operators

$$\Omega_\pm = \text{s-lim}_{t \to \pm\infty}\ e^{iHt}\ e^{-iH_o t} \tag{8.13}$$

exist on the whole Hilbert space.

Proof. The approximating operators for Ω_- are bounded uniformly in t, therefore it is sufficient to verify (8.13) on a total set. Consider the functions $g \in C_o^\infty(\mathbb{R})$ which are one on some interval and which have compact support which does not include 0. For a suitably chosen set of such g's the set of vectors $\Phi = g(H_o)\Phi$, with $\phi(\vec{x})$ in the Schwartz space $S(\mathbb{R}^\nu)$ is a total set in $L^2(\mathbb{R}^\nu)$.

$$\lim_{t \to \infty} g(H) e^{iHt}\ e^{-iH_o t}\ \Phi$$

$$= g(H)\ e^{iH\tau}\ e^{-iH_o\tau}\ \Phi$$

$$+ \lim_{t \to \infty} \int_\tau^t dt' \frac{d}{dt'} [g(H) e^{iHt'}\ e^{-iH_o t'}\ g(H_o)]\ \Phi . \tag{8.14}$$

The derivative in the second term is the bounded continuous operator valued function

$$i\ e^{iHt'}\ g(H)\ V\ g(H_o)\ e^{-iH_o t'}, \tag{8.15}$$

thus we can estimate the integral by

$$\int_\tau^\infty dt' \| g(H)\, V\, g(H_o)\, e^{-iH_o t'}\, g(H_o)\Phi \|$$

$$\leq \int_\tau^\infty dt' \| g(H)\, V\, g(H_o)\, F(|\vec{x}| \geq v_o t') \| \tag{8.16}$$

$$+ \| g(H)\, V\, g(H_o) \| \int_\tau^\infty dt' \| F(|\vec{x}| \leq v_o t')\, e^{-iH_o t'}\, g(H_o)\Phi \| .$$

The first integral converges because of assumption (8.12) for any $v_o > 0$. It can be made arbitrarily small by choosing τ large enough. The same applies to the second summand if v_o is chosen small enough such that $g(E) = 0$ for $E < (m/2)\, v_o^2$ by Lemma 6.1.

The compactness (8.11) implies compactness of

$$(1 - g)(H) - (1 - g)\,(H_o) \tag{8.17}$$

$$= g(H_o) - g(H).$$

Since $(1 - g)(H_o)\Phi = 0$ we have

$$\| (1 - g)(H)\, e^{iHt}\, e^{-iH_o t}\, \Phi \|$$

$$= \| [\, (1 - g)(H) - (1 - g)(H_o)]\, e^{-iH_o t}\, \Phi \| . \tag{8.18}$$

This tends to zero for $|t| \to \infty$ because $\exp(-i\, H_o t)\Phi$ converges weakly to zero [c.f.(2.26)]. Thus Ω_- exists and similarly for Ω_+.

□

We have used the short range condition (8.12) because it is particularly useful for the analysis of scattering states. It is close to optimal in the following sense: consider a potential which outside a ball of radius R_o is spherically symmetric, monotone decreasing (or increasing) to zero at infinity, and smooth. Then the wave operators exist if and only if the integrability condition (8.12) holds, i.e. iff

$$\int_{R_o}^\infty dr\, |V(r)| < \infty . \tag{8.19}$$

Otherwise the wave operator differs from the long range modified one (Section 10) by an "infinite phase factor."

On the other hand in the anisotropic case one may have less decay in special directions. Or one may split a square integrable

potential into an infinite sequence of pieces and spread them widely over the $\nu \geq 3$ dimensional space. Then Cook's original existence proof works but the decay of the norm in (8.12) may be arbitrarily slow. Moreover the wave operators exist for a wide class of oszillating potentials with slow decay. Roughly speaking the physical condition for the existence of the wave operators is the finiteness of the two fold time-integral of the gradient of the potential (the force) evaluated along almost all classical free trajectories.

The relation

$$\lim_{t \to \infty} e^{iHt} e^{-iH_o t} e^{iH_o \tau}$$

$$= \lim_{t' \to \infty} e^{iH\tau} e^{iHt'} e^{-iH_o t'}$$

implies the *intertwining property*

$$f(H)\, \Omega_{\pm} = \Omega_{\pm}\, f(H_o) \tag{8.20}$$

for a wide class of functions including those which have an integrable Fourier transform. As a strong limit of unitary operators the wave operators are isometric, they are unitary operators from the Hilbert space $\mathcal{H}$ onto their ranges, Ran Ω_- (resp. Ran Ω_+). Thus the intertwining relation implies

$$H \restriction \text{Ran}\, \Omega_- = \Omega_-\, H_o\, \Omega_-^*, \tag{8.21}$$

$$H_o = \Omega_-^*\, H\, \Omega_- \tag{8.22}$$

and similarly for Ω_+. Since H_o has purely absolutely continuous spectrum the same is true for H restricted to either of the ranges:

$$\text{Ran}\, \Omega_{\mp} \subset \mathcal{H}^{ac} \tag{8.23}$$

because for any $\Psi = \Omega_-\, \Phi \in \text{Ran}\, \Omega_-$

$$(\Psi, F(H \leq E)\Psi) = (\Phi, \Omega_-^*\, F(H \leq E)\, \Omega_-\, \Phi)$$

$$= (\Phi, F(H_o \leq E)\Phi)$$

which is absolutely continuous in E for any $\Phi \in \mathcal{H}$. Moreover the ranges are left invariant under the time evolution

$$e^{-iHt}\, \text{Ran}\, \Omega_{\mp} = \text{Ran}\, \Omega_{\mp}\,. \tag{8.24}$$

We will study the ranges further in the next section.

Although the wave operators are a convenient mathematical tool

they are hard to measure in an experiment, because it is practically impossible to determine an interacting state at time $t = 0$. A better accessable quantity is the S-operator (or scattering operator)

$$S = (\Omega_-)^* \, \Omega_+ \tag{8.25}$$

which maps incoming asymptotic configurations to outgoing ones. It commutes with the free time evolution.

In some heuristic arguments below we will say that $(\Omega_- -1)$ measures the influence of the interaction in the future because

$$\lim_{t \to \infty} \| (e^{-iHt} - e^{-iH_o t}) \Phi \| = \| (\Omega_- -1) \Phi \| . \tag{8.26}$$

A stronger estimate uses the integral representation of $(\Omega_- - 1)$

$$\sup_{t \geq o} \| (e^{-iHt} - e^{-iH_o t}) \Phi \|$$

$$\leq \int_o^\infty d\tau \| V e^{-iH_o t} \Phi \|$$

which applies to Kato bounded potentials.

9. ASYMPTOTIC COMPLETENESS

Different statements about the ranges of wave operators are called asymptotic completeness by various authors. We adopt the following terminology: A pair of Hamiltonians H_o and H which describes a quantum mechanical scattering system is called *asymptotically complete* if

$$\text{Ran } \Omega_{\mp} = \mathcal{H}^{cont} , \tag{9.1}$$

we call it *weakly asymptotically complete* if

$$\text{Ran } \Omega_{\mp} = \mathcal{H}^{ac} , \tag{9.2}$$

and *unitarity of the S-operator* holds if

$$\text{Ran } \Omega_- = \text{Ran } \Omega_+ . \tag{9.3}$$

Obviously (9.1) ⇒ (9.2) ⇒ (9.3). The strongest statement (9.1) is suggested by experience from experiments: any state orthogonal to a bound state should be asymptotically free in the future in the sense of (8.1)

$$\lim_{\tau \to \infty} \sup_{t \geq 0} \| (e^{-iHt} - e^{-iH_0 t}) e^{-iH\tau} \psi \| = 0, \tag{8.1}$$

and similarly in the past. We will prove this shortly for suitable potentials. From (8.23) it follows that asymptotic completeness (9.1) is equivalent to weak asymptotic completeness (9.2) together with absence of a singular continuous spectrum of H

$$\mathcal{H}^{sing} = \emptyset . \tag{9.4}$$

The reason for the splitting into (9.2) and (9.4) is mainly technical because in the time independent approach it is usually very hard to prove (9.4).

It follows easily from the definition of the S-operator (8.25) that the unitarity of S is equivalent to (9.3) which is necessary and sufficient for S and S* to be isometric. Physically it says that any state which was free in the remote past will be free again in the far future and vice versa. It excludes that freely incoming particles are trapped.

It is clear mathematically that one can construct counterexamples to all of these conditions (9.1) - (9.4), but it was a long standing prejudice, mainly among physicists, that these counterexamples would be mathematical pathologies which cannot occur for physically reasonable forces. This belief has to be modified due to the counterexamples constructed by Pearson, see [50] and references therein. E.g. (9.3) is violated by a model for adsorption at a severe local singularity, or (9.4) is violated by a potential which decays so slowly that the states can never become free. The important fact is that the forces are reasonable and their effects physically understandable, although they do not occur in nature. The modified (and correct) expectation is that (9.1) holds for physically realistic potentials which have sufficient decay at infinity and are not too singular locally.

For the Coulomb force in three dimensions and other long-range forces the wave operators have to be modified. We will treat this case in the next section.

Recall the discussion in Section 3 that compactness of the difference of the free and interacting resolvents (9.6) is equivalent to local compactness and decay at infinity. We use the short-range condition discussed in Sections 2 and 8: for any $g \in C_0^\infty(\mathbb{R})$

$$\| g(H)\, V\, g(H_0)\, F(|\vec{x}| \geq R) \| =: h(R) \in L^1(\mathbb{R}_+, dR). \tag{9.5}$$

In Theorem 8.2 we have proved existence of the wave operators on the whole Hilbert space under the same assumptions.

Theorem 9.1. Assume that

$$(H - z)^{-1} - (H_o - z)^{-1} \text{ is compact} \tag{9.6}$$

and that the short-range condition (9.5) is satisfied. Then asymptotic completeness holds, i.e.

$$\text{Ran } \Omega_{\mp} = \mathcal{H}^{cont} = \mathcal{H}^{ac}, \tag{9.7}$$

and in particular $\mathcal{H}^{sing} = \emptyset$.

We will give two different proofs, but we start with some arguments common to both of them.

Proof. We consider the outgoing wave operator Ω_-, the proof for Ω_+ is analogous. Since Ran Ω_- is closed it is sufficient to show that a set of vectors dense in $\mathcal{H}^{cont}$ lies in the range of Ω_-. A convenient dense set consists of the states with compact energy support which does not include zero. For any such state $\Psi \in \mathcal{H}^{cont}$ there is a function $g \in C_o^\infty$ with $0 \notin \text{supp } g$, $0 \leq g \leq 1$, such that

$$g(H)\Psi = \Psi. \tag{9.8}$$

Since Ran Ω_- is time translation invariant it is sufficient to show that $\exp(-iHt)\Psi$ is arbitrarily close to Ran Ω_- for some time t. This is implied by the stronger looking

$$\| (\Omega_- - 1) e^{-iH\tau} \Psi \| < \varepsilon, \quad \tau = \tau(\varepsilon). \tag{9.9}$$

The physical reason why (9.9) should hold is that $(\Omega_- - 1)$ measures how much the free and interacting time evolutions differ in the future. If one waits until the scattering is over, τ big, (9.9) should be small. The intertwining relation (8.20) and (9.8) allow to rewrite (9.9):

$$\| (\Omega_- - 1) e^{-iH\tau} \Psi \|$$

$$= \| (\Omega_- - 1)\, g(H_o)\, e^{-iH\tau} \Psi$$

$$+ (\Omega_- - 1)\, [g(H) - g(H_o)]\, e^{-iH\tau} \Psi \|$$

$$\leq \| (\Omega_- - 1) \quad g(H_o) \quad e^{-iH\tau} \Psi \| \tag{9.10}$$

$$+ 2 \| [g(H_o) - g(H)] e^{-iH\tau} \Psi \| . \tag{9.11}$$

Since $g(H) - g(H_o)$ is compact by (9.6) the norm (9.11) will tend to zero for any sequence of times τ_n such that $\exp(-iH\tau_n)\Psi$ tends weakly to zero. To estimate (9.10) we first prove a lemma which slightly extends an observation of Mourre [46].

Lemma 9.2. Let the interaction satisfy (9.5) and (9.6). For any $g \in C_o^\infty(\mathbb{R})$ with $g(E) = 0$ for $E < E_1 > 0$:

$$(\Omega_- -1)\, g(H_o)\, P_{out} \text{ is compact,} \tag{9.12}$$

$$(\Omega_- -1)\, g(H)\, P_{out} \text{ is compact,} \tag{9.12'}$$

for every P_{out} as defined in Section 5: (5.12), (5.24), (5.5) with any constant a such that $a^2 < E_1/5m\nu$, or (5.22) with $E_o \leq E_1$. Similarly with Ω_+ for the corresponding $P_{in's}$.

Proof. For $\bar{g} \in C_o^\infty(\mathbb{R})$ with $\bar{g}(E) = 1$ on supp g we have that (9.12) equals

$$\begin{aligned} &\bar{g}(H)(\Omega_- -1)\, \bar{g}(H_o)\, g(H_o)\, P_{out} \\ &+ [\bar{g}(H) - \bar{g}(H_o)]\, g(H_o)\, P_{out}\,. \end{aligned} \tag{9.13}$$

The second summand in (9.13) is compact by (9.6), the first summand has the integral representation

$$i \int_0^\infty dt\, e^{-iHt}\, \bar{g}(H)\, V\, \bar{g}(H_o)\, g(H_o)\, e^{-iH_o t}\, P_{out}. \tag{9.14}$$

The integrand is a norm continuous compact operator valued function ot t, thus any finite integral is compact. It is sufficient to show that the integral for $t \geq T$ is small in norm for large T. By Cook's estimate this is bounded by

$$\int_T^\infty dt\, \| \bar{g}(H)\, V\, \bar{g}(H_o)\, e^{-iH_o t}\, g(H_o)\, P_{out} \|$$

$$\leq \int_T^\infty dt\, \| \bar{g}(H)\, V\, \bar{g}(H_o)\, F(|\vec{x}| \geq v_o t) \|\, \| g(H_o) P_{out} \| \tag{9.15}$$

$$+ \| \bar{g}(H)\, V\, \bar{g}(H_o) \| \int_T^\infty dt \| F(|\vec{x}| \leq v_o t) e^{-iH_o t}\, g(H_o) P_{out} \|. \tag{9.16}$$

The integral (9.15) converges for any $v_o > 0$ by (9.5) and it is small for large T. If v_o is small enough depending on E_1 also (9.16) converges. This follows from the estimates of the propagation into the classically forbidden regions: Theorem 6.7 for the definition (5.12), Theorem 6.11 for (5.22), and the remark following it for (5.24). For (5.5) the j-summation can be restricted to $\|j\| \geq 3a$ and Corollary 6.6 applies. Obviously (9.12) and (9.12') are equivalent.

□

Note that for all P_{out}: $g(H_o)\, P_{out}\, F(H_o > E) \to 0$ as $E \to \infty$, and for any R: $F(|\vec{x}| > R')\, P_{out}\, F(|\vec{x}| \leq R) \to 0$ as $R' \to \infty$. Local compactness together with either of these properties implies that

$$g(H_{(o)})\, P_{out}\, F(|\vec{x}| \leq R) \tag{9.17}$$

is compact for any R. Thus it is necessary and sufficient for the compactness (9.12), (9.12') that

$$\lim_{R \to \infty} \| (\Omega_- - 1) g(H_{(o)})\, P_{out}\, F(|\vec{x}| \geq R) \| = 0. \tag{9.18}$$

This explains best the physical content of Lemma 9.2. Among the outgoing states with strictly positive, bounded energy only finitely many will have significant interaction in the future. namely those which are localized where the potential is strong. The states which are in addition far away from the scatterer won't be influenced: (9.18).

One can give a simple direct proof of (9.18). We use (9.13) to add a $\bar{g}(H)$ on the left of the operator in (9.18) and we apply the remark following Corollary 6.6. With $P_{out}(i_o)$ as given there and the j-summation restricted to $\|j\| \geq 3a$ we get

$$\begin{aligned} &\|\bar{g}(H)(\Omega_- - 1)\, g(H_o)\, P_{out}(i_o)\| \\ &\leq \int_o^\infty \|\bar{g}(H)\, V\, g(H_o)\, F(|\vec{x}| \geq a\, t + i_o/2)\| dt\, \|P_{out}(i_o)\| \\ &\quad + \|\bar{g}(H)\, V\, g(H_o)\| \int_o^\infty dt \|F(|\vec{x}| \leq a\, t + i_o/2) e^{-iH_o t}\, P_{out}(i_o)\|. \end{aligned} \tag{9.19}$$

Both terms are arbitrarily small for i_o large enough, the first by (9.5), the second by (6.20). Finally $P_{out} - P_{out}(i_o)$ is localized in $\|\vec{x}\| \leq i_o$ up to rapidly decaying tails, thus

$$\lim_{R \to \infty} \| [P_{out} - P_{out}(i_o)]\, F(|\vec{x}| \geq R) \| = 0. \tag{9.20}$$

This finishes the direct proof of (9.18).

<u>Remark.</u> If the potential is bounded and of short-range, i.e.

$$\|V\, F(|\vec{x}| \geq R)\| \in L^1(\mathbb{R}_+, dR), \tag{9.21}$$

then also very fast particles won't be influenced by the potential and we obtain the stronger

$$(\Omega_- -1) P_{out} F(H_o \geq E_o) \text{ is compact} \tag{9.22}$$

for the definition (5.5), $a^2 \leq E_o/5\,m\nu$ by Corollary 6.6, or

$$(\Omega_- -1)\ \phi(\sqrt{2m\ H_o})\ P_{out} \text{ is compact} \tag{9.23}$$

for the definition (5.12) by Theorem 6.9, ϕ as given there. The compactness is clear for any bounded energy interval, but the high energy contribution, (9.22) for large E_o, is arbitrarily small in norm. Similarly for (9.23). See [63] for applications.

Now we will finish the proof of Theorem 9.1 in two ways. The first is very short and elegant but it requires stronger hypotheses; the second is more elementary and it has wider applicability.

<u>First version:</u> Assume that in addition the assumptions of Theorem 7.1 hold, or slightly weaker: assume that (7.13) holds which implies (7.16) - (7.18). From the compactness (9.12) it follows that for any $\varepsilon > 0$ there is a $d_\varepsilon > 0$ such that

$$\| (\Omega_- - 1)\ g(H_o)\ \ F(D \geq d_\varepsilon) \| < \varepsilon \tag{9.24}$$

where we have used (5.12): $P_{out} = F(D \geq 0)$. By (7.16)

$$\lim_{\tau \to \infty} \| F(D \leq d_\varepsilon)\ e^{-iH\tau}\ \psi \| = 0 \tag{9.25}$$

for any ε. Thus (9.10) vanishes as $\tau \to \infty$. The same applies for (9.11) because of (7.18).

□

<u>Second version:</u> This time we use the sequence of times $\tau_n \to \infty$ which has been constructed in Corollary 4.3 using only local compactness. Then (9.11) vanishes as $n \to \infty$ by (4.15) and also

$$\lim_{n \to \infty} \| (\Omega_- - 1)\ g(H_o)\ \ P_{out}\ e^{-iH\tau_n}\psi \| = 0.$$

To estimate (9.10) and thus to finish the proof it remains to show that

$$\lim_{n \to \infty} \| P_{in}\ e^{-iH\tau_n}\ \psi \ \| = 0. \tag{9.26}$$

In the first version this followed from Theorem 7.1. The operators P_{in} are self-adjoint with the exception of (5.6) where P_{in} and P_{in}^* are essentially equivalent, see the remark after Theorem 6.4. By taking adjoints the counterpart of (9.12) is:

$$P_{in}\ g(H_o)\ (\Omega_+^* - 1) \qquad \text{is compact.} \tag{9.27}$$

Then we can rewrite (9.26) using weak convergence to zero

$$\begin{aligned}
&\lim_{n\to\infty} P_{in}\, e^{-iH\tau_n}\,\Psi \\
&= \lim P_{in}\ g(H_o)\ e^{-iH\tau_n}\Psi \\
&= \lim P_{in}\ g(H_o)\ \Omega_+^*\ e^{-iH\tau_n}\Psi \\
&\quad - \lim P_{in}\ g(H_o)\ (\Omega_+^* -1)\ e^{-iH\tau_n}\ \Psi \\
&= \lim P_{in}\ g(H_o)\ e^{-iH_o\tau_n}\ \Omega_+^*\ \Psi .
\end{aligned} \tag{9.28}$$

In Section 6 we have shown that

$$\lim_{t\to -\infty} \| F(|\vec{x}| \le v_o|t|)e^{-iH_ot}\ g(H_o)\ P_{in}^{(*)} \| = 0$$

for all P_{in} (resp. P_{in}^*). By taking adjoints

$$\text{s-}\lim_{\tau\to +\infty}\ P_{in}\ g(H_o)\ e^{-iH_o\tau} = 0. \tag{9.29}$$

Thus (9.28) vanishes and the proof is complete. The physical reason for (9.26) and (9.29) ist that a state which is incoming at a very late time τ_n must have been incoming and localized extremely far out at time zero. But at that time it was localized somewhere.

<u>Variant of the second version:</u> Below following Theorem 11.1 we will prove directly that the singular continuous spectrum is absent if Lemma 9.2 holds. Then one may use Ψ from the absolutely continuous spectral subspace and the Riemann-Lebesgue-lemma implies

$$\text{w-}\lim_{|\tau|\to\infty}\ e^{-iH\tau}\ \Psi = 0. \tag{9.30}$$

Thus one can avoid to use the results of Section 4 and continue as above with the limit $\tau \to \infty$ instead of $\tau_n \to \infty$.

□

The main ingredient of this proof is patience. If one waits long enough then a continuum state is localized far out, then its kinetic and total energy coincide and zero velocities can be ex-

cluded. Moreover the state is outgoing and the future time evolution is approximately free. We thus avoid to treat the interacting dynamics explicitly, the deepest dynamical statement we used is Lemma 9.2. It was proved by applying propagation properties of the *free* dynamics. Our "hardest" analytical tools were the stationary phase method, compactness arguments, and a bit of abstract nonsense to include singularities etc. This was possible because we were led by our physical intuition and each step of our proof has a direct physical interpretation. Moreover the proof is much shorter than time independent ones. There are several different ways to assemble a complete proof by picking a small fraction of the material presented here.

Short–range interactions are particularly simple because the set of states, on which the future interaction is weak, is very big. Therefore very simple properties of the interacting dynamics were sufficient to show that a state will eventually lie in this set. Long-range interactions are more difficult in this respect.

10. THE COULOMB PROBLEM

In the last two sections we restricted ourselves to short range forces and thereby excluded the physically important forces between charged particles. For these long-range forces the observables studied in Section 7 have asymptotic limits but the stronger asymptotic conditions (8.1) and (8.2) which compare the free and interacting unitary evolution groups do not hold any more. Even at large distances the tail of the long-range force has some effect on the evolution of the states, $\exp(-i\, H_o t)$ is not a good approximation, but the asymptotic influence of the potential is so weak and of a special kind that one still may talk about a "modified free" time evolution. For long-range forces the decay of the force, i.e. the gradient of the potential is important. Our geometric methods apply if for the long-range part of the potential V_1

$$|\vec{\nabla}\, V_1(\vec{x})| \leq \text{const.}\,(1 + |\vec{x}|)^{-1-\gamma}, \tag{10.1}$$

$$1 \geq \gamma > (2\nu + 2)/(2\nu + 3), \tag{10.2}$$

ν the space dimension. Since the only physically relevant long-range force is the Coulomb- (or gravitational) force with $\gamma = 1$ the interesting cases for applications are covered. On the other hand the condition (10.2) is an artefact of our estimates and there is no other mathematical or physical reason for it. We give it to stress that the special scaling, symmetry, or positivity properties which are often so useful for Coulomb problems, are irrelevant here. There are much stronger results, e.g. [32] for existence of modified wave operators and [1, 34, 40] for completeness, or see references given in [55] for earlier work. We present our approach nevertheless because

it is the only method which extends to two cluster scattering even if both clusters are charged (ion - ion - scattering) at arbitrary energies, see Section 12.

Of the two advantages of the geometric method, its conceptual and technical simplicity, the second one is partially lost in the treatment of long range forces. We expect, however, that some of the lengthy estimates can be replaced by simpler arguments. Therefore we will give here only the main strategy and refer to [19,20] for complete details.

The basic ideas of long-range scattering are due to Dollard [13,14]: The position of a particle $\vec{x}$ can be approximated at large times by $\vec{v}\cdot t$, $\vec{v}$ the velocity. More precisely

$$\vec{x}(t) - \vec{v}(t) \cdot t \tag{10.3}$$

shold grow only slowly, e.g. logarithmically in t for Coulomb forces. Then at time t for the long-range part of the potential V_1

$$V_1(\vec{x}) \approx V_1(\vec{v} \cdot t) \tag{10.4}$$

is a better (and in our case good enough) approximation of $V_1(\vec{x})$ than to set it equal to zero. Moreover the unitary time evolution generated by the time dependent "modified free" Hamiltonian

$$H_1(t) = H_o + V_1(\vec{v} \cdot t) \tag{10.5}$$

can easily be given explicitly

$$U(t, t') = \exp[-i\, H_o(t - t') -i \int_{t'}^{t} d\tau\, V_1(\vec{v}\cdot\tau)]\,. \tag{10.6}$$

Both summands in the exponent commute and one can easily estimate propagation properties, e.g. with the stationary phase method.

The modified asymptotic conditions are then:

$$\lim_{\tau \to \infty} \sup_{t \geq o} \| [e^{-iHt} - U(t + \tau,\tau)]\, U(\tau,0)\, \Phi \| = 0 \tag{10.7}$$

for all Φ, which is equivalent to existence of the *modified wave operator*

$$\Omega_-^c = s - \lim_{t \to \infty} e^{iHt}\, U(t, 0) \tag{10.8}$$

by the same argument as in Lemma 8.1. Similarly the range of Ω_-^c consists of those vectors Ψ with

$$\lim_{\tau \to \infty} \sup_{t \geq 0} \| [e^{-iHt} - U(t + \tau, \tau)] \, e^{-iH\tau} \Psi \| = 0. \tag{10.9}$$

If (10.9) holds for all $\Psi \in \mathcal{H}^{cont}$ then we say that asymptotic completeness holds. The analogous statements for negative times correspond to the incoming modified wave operator Ω_+^c .

The crucial tool to prove existence and completeness is again Cook's estimate. Our main result is

Theorem 10.1. Let $H_o = -(2m)^{-1} \Delta$ and let $H = H_o + V$ satisfy

$$(H - z)^{-1} - (H_o - z)^{-1} \text{ is compact.} \tag{10.10}$$

The potential can be split

$$V = V_s + V_1, \tag{10.11}$$

where V_s is a short-range potential (2.15) and V_1 is a multiplication operator with the bounded function $V_1(\vec{x})$ which fulfills

$$| (\vec{\nabla} V_1)(\vec{x}) | \leq C(1 + |\vec{x}|)^{-1-\gamma}, \tag{10.12}$$

$$1 \geq \gamma > (2\nu + 2) / (2\nu + 3). \tag{10.13}$$

Then the modified wave operators (10.8) exist and are complete.

Proof of Existence. One considers a dense set of states Φ with $\hat{\phi}(\vec{p}) \in C_o^\infty(\mathbb{R}^\nu)$, $\vec{0} \notin \text{supp}\, \hat{\phi}$. Let $g \in C_o^\infty(\mathbb{R})$, $|g| \leq 1$, $g(E) = 0$ for $E < (m/2)v_o^2$,be such that $g(H_o)\, \Phi = \Phi$. By the same arguments as in Section 8 it is sufficient to show that

$$\int_o^\infty \| \frac{d}{dt} [g(H) \, e^{iHt} \, U(t,0) \, \Phi \| \, dt < \infty. \tag{10.14}$$

This can be bounded by

$$\int_o^\infty dt \| g(H) \, V_s \, g(H_o) \, F(|\vec{x}| \geq v_o t) \|$$

$$+ \| g(H) \, V_s \, g(H_o) \| \int_o^\infty dt \, \| F(|\vec{x}| \leq v_o t) \, U(t,0) \Phi \|$$

$$+ \int_o^\infty dt \, \| [V_1(\vec{x}) - V_1(\vec{v}.t)] \, U(t, 0) \, \Phi \| \tag{10.15}$$

The first integral converges due to (2.15). Below we will indicate

finer estimates which imply finiteness of the third term and rapid decay of the integrand in the second summand. The latter also implies that $U(t, 0)\,\Phi$ tends weakly to zero.

□

The splitting of the potential into a long-range and a short-range part is highly arbitrary and therefore the modified free time evolution U and the modified wave operators are non unique. If V_1 and V_1' differ by a short-range multiplication operator then the corresponding modified wave operators differ in momentum space by the unitary multiplication operator with phase

$$\int dt[V_1(\vec{v}\cdot t) - V_1'(\vec{v}\cdot t)] \tag{10.16}$$

which is finite for $\vec{v} \neq \vec{0}$. Recall that the wave operators map asymptotic configurations to interacting states and thus the unitary change is a relabelling of the asymptotic configurations which is physically irrelevant. In particular this covers the arbitrariness of the choice of the origin of time and in $\vec{x}$-space which affects $\Omega_\pm^c$. The same ambiguity is present in the short-range case as well and one can use it to improve the rate of convergence of the wave operators. Usually it does not appear because there is a unique "natural" choice of the asymptotic free motion.

In the short-range case the future time evolution is approximately free for states with strictly positive energy which are outgoing and localized far from the scatterer. Equivalently for functions g as above

$$(\Omega_- -1)\, g(H_o)\, P_{out} \tag{10.17}$$

is compact. The same cannot be true for the modified wave operators in the long-range case. Let such a state be localized near $\vec{y}$ at time 0. Then the tail of the long-range potential will be well approximated at time t by $V_1\ (\vec{y} + \vec{v}\cdot t)$, which differs from $V_1(\vec{v}\cdot t)$ at small and intermediate times. The modified wave operator will be well approximated by the phase factor (convergent for $\vec{v} \neq \vec{o}$

$$\exp\{i \int[V_1\ (\vec{v}\cdot t) - V_1(\vec{y} + \vec{v}\cdot t)]\,dt\}, \tag{10.18}$$

but only if the state is well localized near $\vec{y}$. For spread out states we have a complicated dependence on position and momentum. To handle this we use the phase-space cell decomposition introduced in Section 5

We introduce an auxiliary modified free time evolution $\underline{U}$ which depends on the initial position of the part of the state and the corresponding auxiliary outgoing wave operator $\underline{\Omega}_-$. (For the negative time direction everything holds with the obvious changes.) Recall

the phase-space decomposition of Section 5.1. $f_i\,\Psi$ is the part of Ψ well localized near $i \in \mathbb{Z}^\nu$, $\Sigma f_i = 1$. Set

$$\underline{U}\;(t, t') = \sum_i U(i;\, t,t')\, f_i \tag{10.19}$$

$$U(i;\, t,t') = \exp[\,-i\, H_o(t - t') - i \int_{t'}^{t} d\tau\; V_1(i + \vec{v}\cdot\tau)]$$

$$\underline{\Omega}_- = \underset{t \to \infty}{\text{s-lim}}\; e^{iHt}\, \underline{U}(t, 0)\;. \tag{10.20}$$

One can show that for any fixed j with $\|j\| > 4a$

$$\sup_{t \geq o} \| \sum_{\substack{i\cdot j \geq 0 \\ |i| > R}} g(H)[\,e^{-iHt} - U(i;t,0)]\,\chi_j\, f_i \| \to 0 \tag{10.21}$$

as $R \to \infty$ which implies with P_{out} defined in (5.5)

$$(\underline{\Omega}_- - 1)\; g(H_o)\; P_{out} \;\text{ is compact} \tag{10.22}$$

for g as given above and properly chosen parameter a. Given (10.21) it is now easy to go back to the modified wave operator and to finish the

Proof of Completeness. For simplicity we will assume that Theorem 7.1 holds, this is an additional restriction on V_s. In the general case a slightly longer argument applies as in Section 9. Pick a vector $\Psi \in \mathcal{H}^{cont}$ with strictly positive, bounded energy, i.e. $g(H)\;\Psi = \Psi$. By Theorem 7.1

$$\lim_{t \to \infty}\; [\,1 - g(H_o)]\, e^{-iHt}\;\; \Psi = 0. \tag{10.23}$$

Let $J' \subset a\,\mathbb{Z}^\nu$ be a finite set suci that

$$\sum_{j \in J'} \chi_j\; g(H_o) = g(H_o)\,. \tag{10.24}$$

Choosing a small enough one can arrange that $J' \subset \{j \in a\,\mathbb{Z}^\nu : \|j\| \geq 5a\}$. Since the velocity transfer of the f_i is smaller than a there is a finite set $J \subset \{j \in a\,\mathbb{Z}^\nu : \|j\| \geq 4a\}$ such that

$$\|\sum_{j \notin J} \chi_j \sum_i{}' f_i\;\, e^{-iHt}\;\; \Psi\| < \varepsilon \tag{10.25}$$

for $t > T_1$, where Σ' is any partial sum over $i \in \mathbb{Z}^\nu$. In what follows

we always understand that the j-summation runs over the finitely many elements of J.

Moreover by Corollary 7.2 the state is asymptotically outgoing and localized far from the scatterer. Therefore we can find a big enough time $t' = t'(R) > T_1$ such that for arbitrarily big fixed R

$$\| e^{-iHt'} \Psi - \sum_{\substack{i\cdot j \geq 0 \\ |i| > R}} \chi_j f_i e^{-iHt'} \Psi \| < 2\varepsilon. \tag{10.26}$$

Now we use (10.21) and determine R such that the norm in (10.21) is smaller than $\varepsilon/|J|$ for the finitely many $j \in J$, and we use this R in (10.26) and the corresponding t'. We then get

$$\| e^{-iH(t + t')} \Psi - \sum_{\substack{i\cdot j \geq 0 \\ |i| > R}} U(i;\, t,\, 0)\, \chi_j f_i e^{-iHt'} \Psi \| < 3\varepsilon \tag{10.27}$$

for all $t \geq 0$. We have chosen the first waiting time t' big enough to ensure that within a prescribed error the state is outgoing, far from the scatterer and has bounded kinetic energy away from zero. From now on we keep t' fixed. Then we can restrict the i-summation to a finite sum as well, i.e. for some R'

$$\sup_{t \geq 0} \| e^{-iH(t + t')} \Psi - \sum_{\substack{i\cdot j \geq 0 \\ R < i < R'}} U(i;\, t,\, 0)\chi_j f_i e^{-iHt'} \Psi \| < 4\varepsilon. \tag{10.28}$$

For the remaining finitely many terms (denote their number by N) we use the time t to wait again until each of them is in the range of the modified wave operator within an error ε/N. Then $\exp[-i\, H(t+t')]\Psi$ lies in the range of the modified wave operator within an error of 5ε and completeness holds.

One easily sees convergence of

$$\operatorname*{s-lim}_{t \to \infty} U^*(t,0)\, U(i;\, t,\, 0) =: U_i \tag{10.29}$$

for any i, U_i is a finite phase factor in momentum space on states with non-zero velocities. The existence of the modified wave operators implies in particular that for any state Φ

$$\lim_{t \to \infty} \|(\Omega_-^c - 1)\, U(t, 0)\, \Phi\| = 0, \tag{10.30}$$

i.e. the distance of $U(t, 0)\,\Phi$ from the range of Ω_-^c becomes arbitrarily small. Inserting for Φ the finitely many vectors

$$U_i\, \chi_j\, f_i\, e^{-iHt'}\, \Psi$$

we see that the distance of each of the N summands in (10.28) from the range of the modified wave operator actually can be made smaller than ε/N by choosing t large enough. This completes the proof of Theorem 10.1 provided (10.21) is valid.

□

Let us briefly rephrase in words our strategy. The first waiting time t' had the same effect as in the short-range case. The approximations will be good for t' large, therefore the state will be spread out over a large region. But in the presence of long-range forces the future time evolution from t' onwards is not yet approximately free but has to be modified depending on the position and momentum of the state. It is desirable to have an approximate time evolution which depends on the momentum and time only, because Fourier analysis then provides powerful tools for estimates.To achieve this we split the spread out state $\exp(-i\,Ht')\,\Psi$ into pieces which are well localized near i. For each localized piece we use a different approximate time evolution $U(i;\, t, 0)$ which depends on the momentum only. The rapid decay of the pieces of the state away from i, expressed e.g. in (5.3), (5.4), allows to suppress the $\vec{x}$-dependence. This auxiliary time evolution $\underline{U}$ (10.19) should well approximate the true time evolution if the state is outgoing and far out, i.e. if only the weak and smooth tails of the long range forces contribute. This is the content of condition (10.21). It is then easy to see that the trajectories under the auxiliary time evolution $\underline{U}$ come arbitrarily close to the range of the modified wave operators. Except for the proof of (10.21) much weaker decay assumptions on the force would be sufficient.

It is easy to see that for each individual outgoing part $U(i;\, t, 0)$ is a good approximate time evolution for large $|i|$. The technically unpleasant problem which requires subtle estimates in the proof of (10.21) is the infinite sum over i. Here we will only indicate the argument, the full proof can be found in [19,20]. As usual (10.21) is bounded by

$$\int_0^\infty dt\, \Big\| \sum_{\substack{i\cdot j \geq 0 \\ |i| > R}} g(H)[V_s + V_l(\vec{x}) - V_l(i + \vec{v}\cdot t)]\, U(i;\, t, 0)\, \chi_j\, f_i \Big\|. \tag{10.31}$$

Without changing anything we can insert between V_s and U a $\bar{g}(H_o)$ where $\bar{g} \in C_o^\infty$ is one on the energy support of χ_j. Then the short range contribution is bounded by (the sums are as in (10.31))

$$\|g(H)V_s \bar{g}(H_o)\| \times \int_0^\infty dt \|F(|\vec{x}| \leq a\,t + R/2)\,\Sigma\, U(i;\,t,\,0)\chi_j\, f_i\| \tag{10.32}$$
$$+ \sup_{t \geq 0} \|\Sigma\, U(i;\,t,\,0)\chi_j\, f_i\| \int_0^\infty dt \| g(H)V_s\, \bar{g}(H_o)F(|\vec{x}| \geq at + R/2)\|.$$

The first summand is arbitrarily small for large R because for $\|j\| \geq 4a$ and any N

$$\|F(|\vec{x}| \leq a\,t + R/2)\,\Sigma\, U(i;\,t,\,0)\chi_j\, f_i\| \leq C_N(1 + t + R)^{-N} \tag{10.33}$$

as can be shown with a stationary phase argument similar to those in Section 6. A simpler version of it has been used in the existence proof in (10.15). The integral in the second summand is small for large R by assumption. The supremum in front of the integral involves the long-range part of the potential only. If we can prove (10.21) for a Hamiltonian $H' = H_o + V_1$ then the supremum is an approximation of the uniformly bounded $\exp(-i\,H't)\,\Sigma\,\chi_j\, f_i$ and thus is finite. Consequently it is sufficient to prove (10.21) for pure long-range forces. Its validity when any short-range force is added follows from (10.32).

Note that due to the outgoing condition one has

$$|i + \vec{v}\cdot t| > at + |i|/2 \tag{10.34}$$

and the estimate (10.33) allows to consider $V_1(\vec{x})$ only for $|\vec{x}| \geq at + |i|/2$. Thus in (10.31) on the i-th component at time t only that part of the long-range potential contributes with argument bigger than $(at + |i|/2)$. Now it is clear how the freedom to split the potential into its long- and short-range part can be used most efficiently.We have to control the overlap of the contributions from different cells, e.g. by estimating their tails in $\vec{x}$-space. The effective range of propagation of the terms involving $V_1(i + \vec{v}\cdot t)$ is governed by the decay of the Fourier transform of V_1. One can arrange that the part of the long-range potential lying farther out has Fourier transform with faster decay. A construction with this effect has been given by Hörmander (Lemma 3.3 in [32] or see [19]). With this particular V_1 one then can carry out a lengthy estimate to show

$$\lim_{R\to\infty} \int_0^\infty dt \| \sum_{\substack{i\cdot j\geq 0\\ |i|>R}} [V_1(\vec{x}) - V_1(i + \vec{v}\, t)] U(i;\, t,0) \chi_j f_i \| = 0 \tag{10.35}$$

to finish the proof of (10.21). This is "the second summand in (44)" of [19], its estimate has been improved in [20].

To conclude this section we mention that the intertwining property

$$f(H)\, \Omega^c_{\pm} = \Omega^c_{\pm}\, f(H_o) \tag{10.36}$$

holds for the modified wave operators as well. Therefore Ran $\Omega^c_{\pm} \subset \mathcal{H}^{ac}$ and Theorem 10.1 implies absence of a singular continuous spectrum of the Hamiltonian. Similarly one gets

$$\mathrm{Ran}\; \Omega_{\pm}\, g(H_o)\, P_{in/out} \subset \mathcal{H}^{ac} \;. \tag{10.37}$$

11. MORE ABOUT BOUND STATES

We have shown in Section 4 that bound states can be characterized equivalently by their localization which is uniform in time or by being eigenvectors of the Hamiltonian. And in Section 3 we have seen that conditions (2.14) and (3.2) imply that the negative spectrum of H is discrete, i.e. the negative eigenvalues are at most finitely degenerate and they cannot accumulate at negative values. As observed by Simon [61] one can use the results from scattering theory to show the same properties for positive energies.

Theorem 11.1. Let the potential V obey the conditions of Theorem 9.1 or in the presence of long-range forces those of Theorem 10.1, then all non-zero eigenvalues are at most finitely degenerate and zero is the only possible finite accumulation point of eigenvalues.

Remark. For general free Hamiltonians $H_o(\vec{p})$ the energy values which could be infinitely degenerate or which could be accumulation points of eigenvalues are those which correspond to $\vec{p}$-values where the velocity $\vec{\nabla} H_o(\vec{p})$ is zero or not defined.

Proof. Assume there is an orthonormal sequence of bound states Φ_n with eigenvalues in a closed interval which does not contain zero. Then there is a function $g \in C_o^\infty$, $0 \notin$ supp g with

$$g(H)\, \Phi_n = \Phi_n \;. \tag{11.1}$$

Denote by Q the projector onto $(\mathrm{Ran}\, \Omega_+)^\perp \cap (\mathrm{Ran}\, \Omega_-)^\perp$ (in the short-range case) then also $Q\, \Phi_n = \Phi_n$. By compactness of $g(H) - g(H_o)$ and

the fact that Φ_n converges weakly to zero we get

$$\|\Phi_n\| = \|Q\, g(H)\, \Phi_n\| \tag{11.2}$$

$$\to \|Q\, g(H_o)\, \Phi_n\| = \|Q\, g(H_o)\, (P_{in} + P_{out})\, \Phi_n\|$$

as $n \to \infty$. By definition of Q

$$- Q\, g(H_o)\, P_{in} = Q\, (\Omega_+ -1)\, g(H_o)\, P_{in}, \tag{11.3}$$

which is compact by Lemma 9.2, and similarly for the outgoing part. Thus for large enough n

$$\|\Phi_n\| \leq \varepsilon + \|Q\, (\Omega_+ -1)\, g(H_o)\, P_{in}\, \Phi_n\| + \|Q(\Omega_- -1)\, g(H_o)\, P_{out}\, \Phi_n\|. \tag{11.4}$$

All terms on the RHS are arbitrarily small for big enough n which gives a contradiction.

In the long range case the same argument works if Q is chosen as the projector onto

$$(\text{Ran}\ \underline{\Omega}_+\ g(H_o)\ P_{in})^{\perp} \cap (\text{Ran}\ \underline{\Omega}_-\ g(H_o)\ P_{out})^{\perp}$$

by using (10.22) and (10.37). □

<u>Remarks:</u> 1. The preceding proof provides a direct proof of absence of a singular continuous spectrum in H if Lemma 9.2 or (10.22) is known. If a singular continuous spectrum in H were present one could construct a sequence Φ_n with the same properties as used above.

2. For bounded potentials the eigenvalues are also bounded above as follows from (9.22), (9.23).

Another observation made in [61] relates the decay of eigenfunctions with negative eigenvalue to the decay of the potential.

<u>Theorem 11.2.</u> Let V be a form-bounded perturbation of H_o and let Φ be a bound state for $H = H_o + V$

$$H\, \Phi = E\, \Phi,\ E < 0. \tag{11.5}$$

Let $g \in C_o^\infty$ be a real function with $g(E) = 1$, i.e. $g(H)\, \Phi = \Phi$. Then

$$\|F(|\vec{x}| \geq R)\, \Phi\| \leq \|g(H)\, V\, (H_o - E)^{-1}\, F(|\vec{x}| \geq R)\|. \tag{11.6}$$

Proof. The form-boundedness and (11.5) imply that

$$\Phi = -(H_o - E)^{-1} V \Phi = -(H_o - E)^{-1} V g(H) \Phi. \tag{11.7}$$

Consequently

$$\|F(|\vec{x}| \geq R)\Phi\| \leq \|F(|\vec{x}| \geq R) (H_o - E)^{-1} V g(H)\| \cdot \| \Phi \|.$$

□

For local potentials one knows exponential decay of these states, thus the result is of interest for pseudodifferential operators V describing velocity dependent forces. For applications to multiparticle systems see [61].

12. REFINEMENTS AND EXTENSIONS

All results except those in Section 7 have a simple extension to more general free Hamiltonians [28,61]. They arise e.g. in the Dirac equation, classical wave equations, or in the description of impurity scattering in crystals. The underlying space may be curved [10]. For a direct geometric treatment of classical waves see [52].

The potential need not be a symmetric operator but it may have a negative imaginary part to describe absorption as in the optical models of nuclear physics [61]. External constant electric and magnetic fields [61] or Yang-Mills fields [10] may be added. For oszillating potentials the decay may be slower [8]. By adding the Kupsch-Sandhas trick [41,61] one can include e.g. hard cores, or one can use the two Hilbert space scattering theory with a more general identification operator [28]. By adjusting the subspaces of scattering states one can include adsorption by strong local singularities [4]. Time dependent potentials have been treated in [63,67].

Besides these refinements for simple scattering systems the main challenge is to develop the geometric time dependent method to the study of multiparticle systems. A first step is based on the observation that two-cluster scattering (like e.g. atom-atom scattering) should not be much harder than two-particle scattering. Indeed one can define geometrically "asymptotic two-cluster states" in the same spirit as Ruelle's geometric characterization of bound states and scattering states discussed in Section 4. Then one proves that the direct sum of the ranges of the two-cluster channel wave operators coincides with the subspace of the asymptotic two-cluster states [21]. An important new result was the full inclusion of Coulomb forces. Moreover the usual restriction to energies below the 3-cluster threshold can be avoided.

For three and more particles a hard additional problem arises: the possibility of multiple scattering before the particles become free. This requires more information about the interacting time evolution. There are preliminary results like an extension of Theorem 7.1 to three particle systems [23] but they are not yet sufficient.

13. REFERENCES

1. *S. Agmon*: Lectures at AMS-meeting, Salt Lake City, 1978 and private communication.
2. *W.O. Amrein, J.M. Jauch, K.B. Sinha*: Scattering Theory in Quantum Mechanics, Benjamin, Reading 1977.
3. *W.O. Amrein, V. Georgescu*: On the characterization of bound states and scattering states in quantum mechanics, Helv. Phys. Acta 46, 635 - 658(1973).
4. *W.O. Amrein, D.B. Pearson, M. Wollenberg*: Evanescence of states and asymptotic completeness, preprint Univ. of Genève UGVA-DPT 1980/05 - 242, 1980.
5. *M. Birman*: A local criterion for the existence of wave operators, Izv. Akad.Nauk Mat. 32, 914 - 942(1968), eng. trans. Math. USSR-Izv. 2, 879 - 906(1968).
6. *W. Brenig, R. Haag*: Allgemeine Quantentheorie der Stoßprozesse, Fortschr. Phys. 7, 183 - 242(1959).
7. *J.M. Combes*: These proceedings.
8. *M. Combescure*: Spectral and scattering theory for a class of strongly oszillating potentials, Commun. Math. Phys. 73, 43 - 62(1980).
9. *J.M. Cook*: Convergence to the Møller wave-matrix, J. Math. and Phys. 36, 82 - 87(1957).
10. *P. Cotta-Ramusino, W. Krüger, R. Schrader*: Quantum scattering by external metrics and Yang-Mills potentials, Ann. Inst. Henri Poincaré A 31, 43 - 71(1979).
11. *E.B. Davies*: Quantum Theory of Open Systems, Academic Press, New York 1976.
12. *E.B. Davies*: On Enss' approach to scattering theory, Duke Math. J. 47, 171 - 185(1980).
13. *J.D. Dollard*: Asymptotic convergence and the Coulomb interaction, J. Math. Phys. 5, 729 - 738(1964).
14. --: Quantum mechanical scattering theory for short-range and Coulomb interactions, Rocky Mt. J. Math. 1, 5-88(1971).

15. --: On the definition of scattering subspaces in nonrelativistic quantum mechanics, J. Math. Phys. 18, 229 - 232(1977).

16. *V. Enss*: Characterization of particles by means of local observables, Commun. Math. Phys. 45, 35 - 52(1975).

17. --: A note on Hunziker's theorem, Commun. Math. Phys. 52, 233-238(1977).

18. --: Asymptotic completeness for quantum mechanical potential scattering, I. Short range potentials, Commun. Math. Phys. 61, 285-291(1978).

19. --:--, II. Singular and long-range potentials, Ann. Phys. (N.Y.) 119, 117-132(1979).

20. --: Addendum to [19], preprint Univ. Bielefeld BI-TP 79/26, 4p., 1979; unpublished.

21. --: Two-cluster scattering of N charged particles, Commun. Math. Phys. 65, 151-165(1979).

22. --: A new method for asymptotic completeness, in: Mathematical Problems in Theoretical Physics, K. Osterwalder ed., Lec re Notes in Physics 116, Springer, Berlin 1980.

23. --: Asymptotic observables on scattering states, in preparation.

24. *V. Enss, B. Simon*: Bounds on total cross sections in atom-atom and atom-ion collisions by geometric methods, Phys. Rev. Lett. 44, 319-322 and 764(1980).

25. --: Finite total cross sections in nonrelativistic quantum mechanics, Commun. Math. Phys. 76, 177-209(1980).

26. --: Total cross sections in non-relativistic scattering theory, to appear in: Classical, Semiclassical, and Quantum Mechanical Problems in Mathematics, Chemistry, and Physics, K. Gustafson and W.P. Reinhardt eds., Plenum, New York 1980/81.

27. *J. Fröhlich*: Application of commutator theorems to the integration of representations of Lie algebras and commutation relations, Commun. Math. Phys. 54, 135-150(1977).

28. *J. Ginibre*: La méthode "dépendant du temps" dans le problème de la complétude asymptotique, preprint Univ. Paris-Sud, LPTHE 80/10, 1980.

29. *R. Haag*: Quantum field theories with composite particles and asymptotic completeness, Phys. Rev. 112, 669-673(1958).

30. *R. Haag, J.A. Swieca*: When does a quantum field theory describe particles? Commun. Math. Phys. 1, 308-320(1965).

31. *K. Hepp*: On the connection between the LSZ and Wightman quantum field theory, Commun. Math. Phys. 1, 95-111(1965).

32. *L. Hörmander*: The existence of wave operators in scattering theory, Math. Z. 146, 69-91(1976).

33. *T. Ikebe*: Eigenfunction expansions associated with the Schrödinger operators and their application to scattering theory, Arch. Rational Mech. Anal. 5, 1-34(1960).

34. *T. Ikebe, H. Isozaki*: A stationary approach to the existence and completeness of long-range wave operators, preprint Kyoto Univ., 1980.

35. *K. Jörgens, J. Weidmann*: Spectral Properties of Hamiltonian Operators, Lecture Notes in Mathematics 313, Springer, Berlin 1973.

36. *T. Kato*: Perturbation Theory for Linear Operators, Springer, Berlin 1966.

37. --: Linear evolution equations of "hyperbolic" type, J. Fac. Sci. Univ. Tokyo 17, 241-258(1970).

38. --: On the Cook-Kuroda criterion in scattering theory, Commun. Math. Phys. 67, 85-90(1979).

39. --: Private communication.

40. *H. Kitada*: Scattering theory for Schrödinger operators with long-range potentials II. J. Math. Soc. Japan 30, 603-632(1978).

41. *J. Kupsch, W. Sandhas*: Møller operators for scattering on singular potentials, Commun. Math. Phys. 2, 147-154(1966).

42. *S.T. Kuroda*: On the existence and the unitary property of the scattering operator, Nuovo Cimento 12, 431-454(1959).

43. *R. Lavine*: Absolute continuity of positve spectrum for Schrödinger operators with long range potentials, J. Func. Anal. 12, 30-54(1973).

44. *P. D. Lax, R.S. Phillips*, Scattering Theory, Academic Press, New York 1967.

45. *C. Møller*: General properties of the characteristic matrix in the theory of elementary particles I, Danske Vid. Selsk. Mat.-Fys. Medd. 23, 1-48(1945).

46. *E. Mourre*: Link between the geometrical and the spectral transformation approaches in scattering theory, Commun. Math. Phys. 68, 91-94(1979).

47. *D.B. Pearson*: An example in potential scattering illustrating the breakdown of asymptotic completeness, Commun. Math. Phys. 40, 125-146(1975).

48. --: General theory of potential scattering with absorption at local singularities, Helv. Phys. Acta 48, 639-653(1975).

49. --: Singular continuous measures in scattering theory, Commun. Math. Phys. 60, 13-36(1978).

50. --: These proceedings.

51. *P.A. Perry*: Mellin transforms and scattering theory, I. short-range potentials, Duke Math. J. 47, 187-193(1980).

52. *R.S. Phillips*: Scattering theory for the wave equation with a short-range perturbation, to appear in the volume in honor of J. Segal, V. Guillemin ed.; and paper in preparation.

53. *M. Reed, B. Simon*: Methods of Modern Mathematical Physics, I. Functional Analysis, Academic Press, New York 1972.

54. --:--, II. Fourier Analysis, Self-Adjointness, Academic Press, New York 1975.

55. --:--, III. Scattering Theory, Academic Press, New York 1979.

56. --:--, IV. Analysis of Operators, Academic Press, New York 1978.

57. *D. Ruelle*: A remark on bound states in potential scattering theory, Nuovo Cimento 61 A, 655-662(1969).

58. *M. Schechter*: A new criterion for scattering theory, Duke Math. J. 44, 863-872(1977).

59. *B. Simon*: Scattering theory and quadratic forms: on a theorem of Schechter, Commun. Math. Phys. 53, 151-153(1977).

60. --: Geometric methods in multiparticle quantum systems, Commun. Math. Phys. 55, 259-274(1977).

61. --: Phase space analysis of simple scattering systems: extensions of some work of Enss, Duke Math. J. 46, 119-168(1979).

62. *B.Sz.-Nagy*: Extensions of Linear Transformations in Hilbert Space Which Extend Beyond This Space (Appendix to F. Riesz, B.Sz.-Nagy: Functional Analysis) Ungar, New York 1960.

63. *K. Veselić, V. Enss*: On the characterization of bound states and scattering states for time-dependent Hamiltonians, in preparation.

64. *E. Vock, W. Hunziker*: Weyl's criterion and perturbation theory, preprint ETH Zürich, 1980.

65. *C. R. Wilcox*: Scattering states and wave operators in the abstract theory of scattering, J. Func. Anal. 12, 257-274(1973).

66. *D.R. Yafaev*: On the proof of Enss of asymptotic completeness in potential scattering theory, preprint Leningrad Branch Mathematical Institute E-2-79, 1979.

67. --: Asymptotic completeness for the multidimensional non-stationary Schrödinger equation (in russian) Doklady Akad. Nauk 251, 812-816(1980).

STATIONARY SCATTERING THEORY

J.M. COMBES

Université de Toulon et du Var
and
Centre de Physique Théorique, Section 2
C.N.R.S - Luminy - Case 907
Centre de Physique théorique
F-13288 MARSEILLE CEDEX 2

INTRODUCTION

We present a complementary approach to the time dependent scattering theory described by V. Enss for one-body Schrödinger operators. Roughly speaking the stationary theory is concerned with those objects you read about in textbooks on quantum theory like scattering waves and amplitudes. The starting point of the old theory is not here an asymptotic condition for large times, as for wave-operators of V. Enss lecture, but rather for large distances. It is known as "Sommerfeld radiation condition" and leads, when incorporated in the Schrödinger equation as a "Cauchy condition at infinity", to the celebrated Lippman-Schwinger equation. In the more recent abstract stationary theory some generalized form of the Lippman-Schwinger equation plays the basic role ; solving this equation leads to a linear map between generalized eigenfunctions of the perturbed and unperturbed operators. This map is the "section" at fixed energy of the wave-operator from the time dependent theory. Although the radiation condition does not appear explicitly in this formulation it can be shown to hold a posteriori in a variety of situations, thus restoring the link with physical theories. A general approach to the radiation condition for a large class of Partial Differential operators is described in the work of S. Agmon and L. Hörmander[1] ; we will mention some of their results here.

In these lectures I will describe an abstract framework for the stationary theory. It is strongly inspired from the Kato-Kuroda theory[2] and also related to the more recent two Hilbert space theories of T. Kato[3], M. Schecter[4], or I. Segal[5]. As we will see it allows to incorporate many of the technical progress of these last years. Among them are S. Agmon's "elliptic" a priori estimate method and the geometric methods ; these last have appeared as particularly useful for many investigations on the bound state problem for N-body Hamiltonians[6,7,8]. They also seem to be promising for the scattering problem.

THE STANDARD FORMULATION OF STATIONARY SCATTERING THEORY FOR THE ONE-BODY QUANTUM PROBLEMS

The basic ansatz of orthodox textbooks on quantum scattering in R^n by a local potential $V(X)$ vanishing at infinity is the existence of a family of solutions $U_+^{(k)}(X)$ of the reduced Schrödinger equation :

$$(-\Delta + V)\, u_+^{(k)}(x) = 0$$

which can be decomposed as

$$u_+^{(k)}(x) = u_0^{(k)}(x) + v_+^{(k)}(x) \tag{1}$$

where $v_+^{(k)}(x)$ is a "scattered wave" and $u_0^{(k)}(x) = \exp(ikx)$
By this one means essentially that $v^{(k)}(x)$ has the asymptotic form

$$v_+^{(k)}(x) \approx |x|^{\frac{1-n}{2}} \exp(i|k||x|)\, f(|k|, w) \tag{2}$$

where w is the angular variable of the particle. This is known as "Sommerfeld Radiation condition" which can be more rigorously stated as

$$\lim_{R\to\infty} \int_{|X|=R} \left| \frac{\partial v_+^{(k)}}{\partial |X|} - i|k|\, v_+^{(k)} \right|^2 d\sigma = 0 \tag{3}$$

The function f is called the "Scattering amplitude" and from it one gets the scattering cross-section

$$\sigma(k) = \int_S |f(k,\omega)|^2\, d\omega$$

where S is the unit sphere in R^n.

It has been known for a long time that solutions in class

C^2 of the Helmoltz equation $(-\Delta - k^2)u = 0$ outside a bounded obstacle can be decomposed in a unique way into a solution of $(-\Delta - k^2)u_0 = 0$ in whole space and a function satisfying radiation condition (3). For scattering by a potential with non-compact support the work of T. Ikebe[9] represents one of the most complete investigation of (1) which can be accomplished by integral equation methods (see also Reed-Simon[10]). Using Fredholm theory in a suitable Banach space containing plane waves Ikebe shows existence of a complete set of solutions $u_+^{(k)}(x)$ satisfying (1) and (2) provided the potential V decays faster than $|x|^{-2-\varepsilon}$, $\varepsilon > 0$, at infinity. Such solutions satisfy the integral equation and the Cauchy conditions (1) and (2) at infinity ; it is known as the "Lippman-Schwinger equation" and reads :

(4) $$u_+^{(k)}(x) = u_0^{(k)}(x) - \int G_k^+(x-y)\, V(y)\, u_+^{(k)}(y)\, dy$$

Here $G_k^+(x)$ is the fundamental solution of $(-\Delta - k^2)G_k^+ = \delta$ satisfying (3) ; e.g. for N = 3 one has $G_k^+(x) = (4\pi|x|)^{-1}\exp i|k||x|$

It appears that the limitation on the allowed decay of the potential in comparison to the $|x|^{-1-\varepsilon}$ decay allowed by Enss' analysis is linked to the inhomogeneity $u_0^{(k)}(x) = \exp(ikx)$ in (4). In order that the kernel $G_k^+(x-y)V(y)$ defines a compact operator between suitable Banach spaces containing plane waves it is necessary to impose such restrictions. One can expect that with different inhomogeneities $U_0^{(k)}(x)$ having some decay at infinity, for example spherical waves, one should be able to enlarge the class of potentials for which solutions of (4) exist. This possibility will be described in these lectures.

As a preparation for the next sections let us notice that (4) can be rewritten in a more abstract operator formalism. In fact $G_k^+(x-y)$ is the boundary value as $\varepsilon \to 0^+$ of the resolvent kernel $\langle x, (-\Delta - (k^2 + i\varepsilon))^{-1} y\rangle$; so at least formally (5) reads

(5) $$u_+^{(k)} = u_0^{(k)} - \lim_{\varepsilon \downarrow 0^+} (-\Delta - (k^2 + i\varepsilon))^{-1} V u_+^{(k)}$$

The second resolvent equation gives the formal solution :

(6) $$u_+^{(k)} = u_0^{(k)} - \lim_{\varepsilon \downarrow 0^+} (-\Delta + V - (k^2 + i\varepsilon))^{-1} V u_0^{(k)}$$

In order to give a meaning to (5) and (6) one has to specify in which type of topology the $\varepsilon = 0$ limits are taken. Even if $VU_0^{(k)}$ is in L^2 these limits certainly don't exist in L^2 since in general for potentials tending to zero at infinity k^2 is in the essential spectrum of $-\Delta + V$. The next paragraph will be concerned with a general analysis of these boundary values of resolvents.

THE LIMITING ABSORPTION PRINCIPLE

Let A be a self-adjoint operator on a separable Hilbert space $\mathcal{H}$. If λ is in $\sigma(A)$ it is well-known that the resolvents $(A-(\lambda \pm i\varepsilon))^{-1}$ $\varepsilon \in \mathbb{R}\setminus\{0\}$, have no weak limit in $\mathcal{L}(\mathcal{H})$ the algebra of bounded linear operators on $\mathcal{H}$. This follows from the uniform boundedness theorem and the estimate $\|(A-(\lambda \pm i\varepsilon))^{-1}\|_{\mathcal{L}(\mathcal{H})} = \varepsilon^{-1}$. Nevertheless the following question can make sense :

Problem.

Let $\lambda \in \mathbb{R}$; is it possible that for some Banach space $\mathfrak{X}$, continuously and densely contained in $\mathcal{H}$ (which we will abbreviate by $\mathfrak{X} \hookrightarrow \mathcal{H}$), the limits

$$R^{\pm}(\lambda) = \lim_{\varepsilon \downarrow 0^+} (A-(\lambda \pm i\varepsilon))^{-1} \tag{7}$$

exist in some topology (weak, strong or norm) of $\mathcal{L}(\mathfrak{X}, \mathfrak{X}^*)$, the algebra of bounded linear maps from $\mathfrak{X}$ to $\mathfrak{X}^*$?

Notice that since

$$\mathfrak{X} \hookrightarrow \mathcal{H} \hookrightarrow \mathfrak{X}^* \tag{8}$$

and $(A-(\lambda \pm i\varepsilon))^{-1} \in \mathcal{L}(\mathcal{H})$, $\varepsilon > 0$, one has a fortiori $(A-(\lambda \pm i\varepsilon))^{-1} \in \mathcal{L}(\mathfrak{X}, \mathfrak{X}^*)$; so it makes sense to talk about limits in $\mathcal{L}(\mathfrak{X}, \mathfrak{X}^*)$.

In the following we will denote by the same symbol $\langle \cdot , \cdot \rangle$ the scalar product in $\mathcal{H}$ and the duality between $\mathfrak{X}$ and $\mathfrak{X}^*$

Definition 1

If (7) holds for some Banach space $\mathfrak{X}$ we will say that the Limiting Absorption Principle (L.A.P.) holds for A at λ in $\mathfrak{X}$.

Remarks

1) If $\lambda \in \rho(A)$, then $(A-(\lambda \pm i\varepsilon))^{-1}$ converges in norm in $\mathcal{L}(\mathcal{H})$ hence a fortiori in $\mathcal{L}(\mathfrak{X}, \mathfrak{X}^*)$ for an arbitrary $\mathfrak{X}$ satisfying (8).

2) If $\lambda \in \sigma_p(A)$, the point spectrum of A, then the L.A.P. holds for no $\mathfrak{X}$ at λ.

This follows from :

(8') $$\lim_{\varepsilon\downarrow 0^+} \pm i\varepsilon \langle \varphi, (A-(\lambda\pm i\varepsilon))^{-1}\psi\rangle = \langle \varphi, E(\{\lambda\})\psi\rangle$$

where $E(.)$ denotes the spectral family of A. If the L.A.P. holds at λ in $\mathfrak{X}$ then $\langle \varphi, E(\{\lambda\})\psi\rangle = 0$, $\forall\, \varphi, \psi \in \mathfrak{X}$. By the density of $\mathfrak{X}$ in $\mathcal{H}$ this would imply $E(\{\lambda\}) = 0$ contradicting $\lambda \in \sigma_p(A)$.

From this last remark follows that if the L.A.P. holds in $\mathfrak{X}$ for all $\lambda \in I$, where I is some open interval in R, then the spectral measure $E((-\infty,\lambda) = E_\lambda$ is continuous on I. This can be improved as follows :

Proposition 1

Let $I \subset \sigma(A)$ be a bounded open interval such that the L.A.P. holds for A in $\mathfrak{X}$ on I. Then $E(I)\mathcal{H} \subset \mathcal{H}_{ac}(A)$ where $\mathcal{H}_{ac}(A)$ is the subspace of absolute continuity for A.

Proof : Let $\varphi \in \mathfrak{X}$; then by the spectral theorem one has

$$F(z) = \langle \varphi, (A-z)^{-1}\varphi\rangle = \int \frac{1}{\lambda - z}\, d\langle \varphi, E_\lambda \varphi\rangle$$

By a classical result of de La Vallée-Poussin the set

$$B = \left\{ \lambda \in \mathbb{R},\ F^+(\lambda) = \lim_{\varepsilon\downarrow 0^+} F(\lambda + i\varepsilon) < \infty \right\}$$

is such that the restriction of $d\langle \varphi, E_\lambda \varphi\rangle$ to B is absolutely continuous with respect to the Lebesgue measure. So in particular $E(I)\varphi \in \mathcal{H}_{ac}(A)$, for all $\varphi \in \mathfrak{X}$. Since $\mathfrak{X}$ is dense in $\mathcal{H}$ the conclusion follows.

For the following it is essential to recall the connection between the boundary values $F^+(\lambda)$, $\lambda \in I$, and the Radon-Nikodym derivative of the spectral measure :

(9) $$\frac{d}{d\lambda}\langle \varphi, E_\lambda \varphi\rangle = \frac{1}{\pi}\, \mathrm{Im} \langle \varphi, R^+(\lambda)\varphi\rangle \quad , \varphi \in \mathfrak{X}.$$

This suggests the

Definition 2

Let $\varphi \in \mathfrak{X}$ and assume the L.A.P. holds for A at λ in $\mathfrak{X}$ with $\lambda \in \sigma_c(A)$ (the continuous spectrum of A). The element of $\mathfrak{X}^*$ given by

(10) $$\varphi(\lambda) = (R^+(\lambda) - R^-(\lambda))\varphi$$

will be called the spectral trace of φ at λ with respect to A.

Notice the mapping $\varphi \in \mathcal{H} \to \varphi(\lambda) \in \mathcal{H}^*$ is continuous. Now let

$$N_\lambda = \{\varphi \in \mathcal{H}, \operatorname{Im} \langle \varphi, R^+(\lambda)\varphi \rangle = 0\}$$

Then

Proposition 2

N_λ is a closed subspace of $\mathcal{H}$ consisting of these $\varphi \in \mathcal{H}$ such that $\varphi(\lambda) = 0$

Proof : Let us consider the sesquilinear form on $\mathcal{H} \times \mathcal{H}$

(11) $a(\lambda; \psi, \varphi) = i^{-1} \langle \psi, (R^+(\lambda) - R^-(\lambda))\varphi \rangle$

Notice that sesquilinearity follows from self-adjointness of A. Now a is continuous on $\mathcal{H} \times \mathcal{H}$ since $R^\pm(\lambda) \in \mathcal{L}(\mathcal{H}, \mathcal{H}^*)$ it is also positive since

(12) $$a(\lambda; \varphi, \varphi) = \lim_{\varepsilon \downarrow 0^+} i^{-1} \langle \varphi, [(A-(\lambda+i\varepsilon))^{-1} - (A-(\lambda-i\varepsilon))^{-1}]\varphi \rangle$$

$$= \lim_{\varepsilon \downarrow 0^+} \langle \varphi, \frac{2\varepsilon}{(A-\lambda)^2+\varepsilon^2}\varphi \rangle$$

So the kernel of a is a closed subspace of $\mathcal{H}$; since this kernel is precisely N_λ the first assertion of the Proposition 2 follows. Now by Cauchy-Schwartz inequality one has $\langle \psi, [R^+(\lambda) - R^-(\lambda)]\varphi \rangle = 0$ for all $\psi \in \mathcal{H}$ if and only if $\varphi(\lambda) = 0$ which concludes the proof.

The left annihilator of N_λ , denoted by $\mathcal{E}_\lambda$, is defined as

$$\mathcal{E}_\lambda = \{u \in \mathcal{H}^*, \langle u, \varphi \rangle = 0 \text{ for all } \varphi \in N_\lambda\}$$

Proposition 3

Assume L.A.P. holds for A at λ in $\mathcal{H}$. Then

i) $\mathcal{E}_\lambda$ is a closed subspace of $\mathcal{H}^*$

ii) If $\varphi \in \mathcal{H}$ then $\varphi(\lambda) \in \mathcal{E}_\lambda$

Proof : i) is a standard result and ii) follows from

$$\langle \psi, [R^+(\lambda) - R^-(\lambda)]\varphi \rangle = -\langle [R^+(\lambda) - R^-(\lambda)]\psi, \varphi \rangle$$

and Proposition 2.

In general $\mathcal{E}_\lambda$ contains strictly the set of traces. Notice however the following

Proposition 4

Assume $\mathfrak{X}$ is reflexive. Then $\varphi \in \mathfrak{X} \longrightarrow \varphi(\lambda) \in \mathcal{E}_\lambda$ has a dense image in $\mathcal{E}_\lambda$.

Proof : Let $F_\lambda = \{\varphi(\lambda), \varphi \in \mathfrak{X}\}$ and assume the closure of F_λ for the $\mathfrak{X}^*$ topology is not dense in $\mathcal{E}_\lambda$. By Hahn-Banach theorem there exists $\psi \in \mathfrak{X}^{**} = \mathfrak{X}$ such that $\langle u, \psi\rangle \neq 0$ and $\langle \varphi(\lambda), \psi\rangle = 0 \;\forall \varphi \in \mathfrak{X}$. In particular one has $\langle \psi(\lambda), \psi\rangle = \operatorname{Im} \langle \psi, R^+(\lambda)\psi\rangle = 0$ which implies $\psi \in N_\lambda$ and accordingly $\langle u, \psi\rangle = 0$, a contradiction.

Remark : It is easy to show that if $u \in \mathcal{E}_\lambda$ then

$$\langle u, (A-\lambda)v\rangle = 0$$

for all $v \in \mathcal{D}(A)$ such that $(A-\lambda)v \in \mathfrak{X}$. When A is a self-adjoint realisation of a differential operator P(D,x) this allows to interpret $\mathcal{E}_\lambda$ as a space of weak solutions in $\mathfrak{X}^*$ of $(P(D,x)-\lambda)u = 0$. For example, when A is the Laplacian on $L^2(R^n)$ and $B(R^n)$ is the Besov space introduced below, a result of S. Agmon and L. Hörmander states that for $\lambda > 0$, $\mathcal{E}_\lambda$ is exactly the set of distributions u whose Fourier transforms $\tilde{u}(\xi)$ have support on $\xi^2 = \lambda$ and are square-integrable on this sphere. In this case they also show that $\mathcal{E}_\lambda$ is exactly the set of traces $\varphi(\lambda)$, $\varphi \in B(R^n)$, thus improving the general result of Proposition 4.

THE L.A.P. AND SPECTRAL THEORY

Let us look more closely at the connection between the notion of spectral trace $\varphi(\lambda)$, $\varphi \in \mathfrak{X}$, and the spectral decomposition for A. For this we introduce $h(\lambda)$ as the closure of $\{\varphi(\lambda), \varphi \in \mathfrak{X}\}$ for the norm associated to the scalar product :

$$\langle \varphi(\lambda), \psi(\lambda)\rangle_{h(\lambda)} = (2i\pi)^{-1} \langle \varphi, [R^+(\lambda) - R^-(\lambda)]\psi\rangle \tag{13}$$

By the arguments used in the proof of Proposition 2 it is easy to check that the above set of traces is a pre-Hilbert space.

We want to construct a unitary operator U from $\mathcal{H}_{ac}(A)$ to

an Hilbert integral

$$\mathfrak{h} = \int^{\oplus} h(\lambda)\, d\lambda$$

mapping A onto the diagonal operator of multiplication by λ in $\mathfrak{h}$. For this we make the following assumption :

There exists a family $(I_n)_{n\in\mathbb{N}}$ of disjoint open intervals such that

i) $\overline{\bigcup_n I_n} = \mathbb{R}$

ii) The L.A.P. holds on I_n, $\forall n \in \mathbb{N}$, in $\mathfrak{X}$.

This implies that $\mathcal{H} = \mathcal{H}_{ac}(A) \oplus \mathcal{H}_p(A)$ where $\mathcal{H}_p(A)$ is the closed subspace spanned by eigenvectors of A . This follows in particular from Proposition 1 which implies that the singular spectrum of A is denumerable and is contained in the union of the end-points of the I_n's. Without restricting generality we can assume $\mathcal{H}_p(A) = \{0\}$ and $\sigma(A)$ consists of the closure of one single (possibly infinite) open interval I.

We refer to J. Dixmier[12] for the theory of Hilbert integrals. What is needed is a denumerable set of vector fields $\lambda \in \bar{I} \to \varphi_n(\lambda) \in h(\lambda)$, $n \in \mathbb{N}$, such that :

(14i) The mappings $\lambda \in \bar{I} \to \langle \varphi_n(\lambda), \varphi_m(\lambda)\rangle_{h(\lambda)}$ are measurable $\forall\, n,m$.

(14ii) $\int_{\bar{I}} \|\varphi_n(\lambda)\|^2_{h(\lambda)}\, d\lambda < \infty$

(14iii) $\{\varphi_n(\lambda),\ n \in \mathbb{N}\}$ is a total set in $h(\lambda)$.

It is enough to define $\varphi_n(\cdot)$ on I. For this we consider a basis $(\varphi_n)_n$ of $\mathfrak{X}$. Since $\langle \varphi_n(\lambda), \varphi_m(\lambda)\rangle_{h(\lambda)} = \frac{d}{d\lambda}\langle \varphi_n, E_\lambda \varphi_m\rangle$ by (9) and (13) properties (14i) and (14ii) hold. The continuity of the mapping $\varphi \in \mathfrak{X} \to \varphi(\lambda) \in h(\lambda)$ implies in addition (14iii).

Now the mapping $\varphi \in \mathfrak{X} \to \int^{\oplus}_{\sigma(A)} \varphi(\lambda)\, d\lambda$ is isometric for the norm induced by $\mathcal{H}$ on $\mathfrak{X}$ and the norm on $\mathfrak{h}$:

$$\left\| \int^{\oplus}_{\sigma(A)} \varphi(\lambda)\, d\lambda \right\|^2 = \int_{\sigma(A)} \|\varphi(\lambda)\|^2_{h(\lambda)}\, d\lambda$$

So it can be continued to a unitary mapping U from $\mathcal{H}$ to h . Obviously U diagonalises A. Then the spectral trace defined by

(10) on $\mathfrak{X}$ can be extended to all of $\mathcal{H}_{ac}(A)$ in the sense that to $\varphi \in \mathcal{H}_{ac}(A)$ there corresponds by U a square integrable vector field $\lambda \in \bigcup_n I_n \longrightarrow \varphi(\lambda) \in h(\lambda)$. In the following we will still call it the spectral trace of φ (although for a general $\varphi \in \mathcal{H}$ it is only defined almost everywhere).

To conclude this section let us establish a useful relation between $\mathcal{E}_\lambda$ and $h(\lambda)$:

Proposition 5

There exists a positive finite constant C such that $\forall \varphi \in \mathfrak{X}$

(15) $\|\varphi(\lambda)\|_{h(\lambda)} \geqslant C\|\varphi(\lambda)\|_{\mathfrak{X}^\times}$

Accordingly

(16) $h(\lambda) \subset \mathcal{E}_\lambda$

Proof : One has

$$\begin{aligned}\|\varphi(\lambda)\|_{h(\lambda)} &= (2\pi)^{-1} \sup_{\psi \in \mathfrak{X}} \left< \frac{\psi(\lambda)}{\|\psi(\lambda)\|_{h(\lambda)}}, \varphi(\lambda) \right>_{h(\lambda)} \\ &= \sup_{\psi \in \mathfrak{X}} \left| \left< \frac{\psi}{\|\psi(\lambda)\|_{h(\lambda)}}, \varphi(\lambda) \right> \right| \\ &\geqslant C \sup_{\psi \in \mathfrak{X}} \left| \left< \frac{\psi}{\|\psi\|_{\mathfrak{X}}}, \varphi(\lambda) \right> \right| \\ &= C\|\varphi(\lambda)\|_{\mathfrak{X}^*}\end{aligned}$$

In the last equality we have used the boundedness of the map $\psi \in \mathfrak{X} \longrightarrow \psi(\lambda) \in h(\lambda)$ due to (13).

EXAMPLES

As the first and simplest example let us consider the Laplacian $A_0 = -\Delta$ on $L^2(R^n)$ with its natural domain of self-adjointness $\mathcal{H}_2(\mathbb{R}^n) = \{u \in L^2(\mathbb{R}^n), D^\alpha u \in L^2(\mathbb{R}^n), |\alpha| \leqslant 2\}$. The spectral decomposition of A_0 is well-known ; A_0 has absolutely continuous spectrum consisting of R^+. An explicit spectral representation is easy to construct using Fourier transform $\mathcal{F}$. For $\lambda > 0$ and $\varphi_0 \in C^\infty(\mathbb{R}^n)$ define $\varphi_0(\lambda) \in L^2(S)$, with S the unit sphere in R^n, by

(17) $\varphi_0(\lambda, \omega) = \lambda^{\frac{n-1}{2}} (\mathcal{F}\varphi)(\sqrt{\lambda}\,\omega)$

By Parseval equality one has

$$\|\varphi\|^2_{L^2(\mathbb{R}^n)} = \int \|\varphi_o(\lambda)\|^2_{L^2(S)} \, d\lambda$$

On the other hand it is well-known (see e.g. Lions-Magenes[13]) that the mapping $\varphi \in C_o^\infty(\mathbb{R}^n) \longrightarrow \varphi_o(\lambda) \in L^2(S)$ is continuous for the norm :

$$\|\varphi\|^2_{L^2_s} = \int (1+|x|^2)^s \, |\varphi(x)|^2 dx$$

provided $s > \frac{1}{2}$. This property suggests the choice $\mathcal{H} = L^2_s(\mathbb{R}^n)$, $s > \frac{1}{2}$. This is done by S. Agmon[14] with the following result :

Theorem 1

For all $\lambda \in \mathbb{R}\setminus\{0\}$ the limits

$$R_o^\pm(\lambda) = \lim_{\varepsilon \downarrow 0^+} (-\Delta - (\lambda \pm i\varepsilon))^{-1}$$

exist in the norm topology of $\mathcal{L}(L^2_s, L^2_{-s})$, $s > \frac{1}{2}$. Furthermore the family of operators

$$R_o^\pm(z) = \begin{cases} (-\Delta - z)^{-1} & , \ \mathrm{Im}\, z \gtrless 0 \\ R^\pm(\lambda) & , \ z = \lambda \pm i0 \ , \ \lambda \in \mathbb{R} \end{cases}$$

is norm continuous in $\mathcal{L}(L^2_s, L^2_{-s})$. The details of the proof can be found in Reed-Simon[10]

In the Agmon's L^2_s theory[14] the mappings $\varphi \in L^2_s(\mathbb{R}^n) \rightarrow \varphi_o(\lambda) \in \mathcal{E}_\lambda$ are not surjective and the inclusion (17) is strict. This follows from the fact (see Lions-Magenes[13]) that the traces $\varphi_o(\lambda)$, $\varphi \in L^2_s(\mathbb{R}^n)$ are in the Sobolev-space $\mathcal{H}_{s-\frac{1}{2}}(S)$ which is strictly contained in $L^2(S)$ hence in $\mathcal{E}_\lambda$ by Proposition 5. Furthermore[1] the L^2_{-s} topology is strictly weaker than the $L^2(S)$ topology on the set of distributions whose Fourier transform have support on S.

One has a feeling that the "best" intermediate space $\mathcal{H}$ should optimize Propositions 4 and 5. This optimal space should be some inductive limit of spaces $L^2_s(\mathbb{R}^n)$, $s > \frac{1}{2}$ when s tends to $\frac{1}{2}$. It is not $L^2_{1/2}(\mathbb{R})$ for which the spectral trace is not continuous but something very close to it namely the Besov type space B defined as follows :

$$\text{(18)} \quad B(\mathbb{R}^n) = \left\{ u \ , \ \|u\|^2_{B(\mathbb{R}^n)} = \sum_{j=1}^\infty 2^{j-1} \int_{2^{j-2} < |x| < 2^{j-1}} |u(x)|^2 dx < \infty \right\}$$

Although this definition is not very intuitive from a physical point of view, the dual space norm is a lot more appealing if one likes to think of scattering theory from the point of view of the radiation condition (2) :

$$\|u\|^2_{B^*(\mathbb{R}^n)} = \sup_{R>1} \left(\frac{1}{R} \int_{|x|<R} |u(x)|^2 \, dx \right)$$

The following results are proved by S. Agmon and L. Hörmander[1].

Theorem 2

i) The L.A.P. holds for A_0 on $R \setminus \{0\}$.

ii) For any $\lambda > 0$ one has

$$L^2(S) = \{ \varphi_0(\lambda) , \varphi \in B(\mathbb{R}^n) \}$$

and

$$\| \varphi_0(\lambda) \|_{L^2(S)} = C \| \varphi_0(\lambda) \|_{B^*(\mathbb{R}^n)}$$

for some finite constant C .

That $B(R^n)$ certainly is the best choice not only mathematicaly but also from a physical point of view follows from the fact that the radiation condition finds in this framework the following natural formulation[1] :

Theorem 3

Let $u \in B^*(R^n)$ be such that $u = R_0^{\pm}(\lambda) f$ for some $\in B(\mathbb{R}^n)$
Then

$$\lim_{R\to\infty} \frac{1}{R} \int_{|x|<R} \left| u(x) - |x|^{\frac{1-n}{2}} f^{\pm}\left(\sqrt{\lambda}, \frac{x}{|x|}\right) \exp(\pm i \sqrt{\lambda}|x|) \right|^2 dx = 0$$

where

$$f^{\pm}(\sqrt{\lambda}, \omega) = c \lambda^{-1} \varphi_0(\lambda, \pm \omega)$$

with $\varphi_0(\lambda)$ given by (17) and

$$c = (\pi/2)^{1/2} \exp(\pm i \pi (n-3)/4)$$

Taking (5) into account this last result obviously shows that for a large class of potentials V and for free waves $u_0^{(k)}$, the scattered waves $v_+^{(k)}(x)$ in (1) will satisfy Sommerfeld radiation condition. This concludes this presentation of the L.A.P. for the Laplacian.

In view of our later discussion of the three-body problem let us state without proof the following general result[15] :

Proposition 6

Let A_1 and A_2 be self-adjoint operators on Hilbert spaces $\mathcal{H}_1$ and $\mathcal{H}_2$. Assume the L.A.P. holds for A_1 (resp. A_2) on $R \setminus \Sigma_1$ (resp. $R \setminus \Sigma_2$) on $\mathcal{X}_1$. (resp. $\mathcal{X}_2$) where Σ_1 (resp. Σ_2) is a closed set. Then L.A.P. holds for

$$A = A_1 \otimes I_{\mathcal{H}_2} + I_{\mathcal{H}_1} \otimes A_2$$

in $\mathcal{X} = \mathcal{X}_1 \otimes \mathcal{X}_2$ on $\mathbb{R} \setminus (\Sigma_1 + \Sigma_2)$.

PERTURBATIVE APPROACH TO THE LIMITING ABSORPTION PRINCIPLE

We now develop a perturbative framework allowing to prove the L.A.P. for a self-adjoint operator A obtained by a perturbation of another operator A for which the L.A.P. is known to hold. This approach uses a natural abstract generalization of the Lippman-Schwinger equation and leads in a very natural way to the stationary theory of wave-operators.

So let A_0 and A be self-adjoint operators on separable Hilbert spaces $\mathcal{H}_0$ and $\mathcal{H}$ respectively. In order to compare states in these spaces we need an "identification operator" J $\in \mathcal{L}(\mathcal{H}_0, \mathcal{H})$, the space of bounded linear mappings from $\mathcal{H}_0$ to $\mathcal{H}$. We want the identification of states in $\mathcal{H}$ to be unique ; this is satisfied under the condition :

$(H_1\text{-i})$ $JJ^* = 1_{\mathcal{H}}$

To avoid unessential domain problems we require furthermore

$(H_1\text{-ii})$ $J\mathcal{D}(A_0) \subset \mathcal{D}(A)$

The perturbation V is then defined as

$$V = AJ - JA$$

so that under condition $(H_1\text{-ii})$ one has $\mathcal{D}(V) \supset \mathcal{D}(A_0)$. Notice that under condition (19) the "effective perturbation" VJ^* is symmetric.

Now let $\mathcal{X}_0 \hookrightarrow \mathcal{H}_0$ and $\mathcal{X} \hookrightarrow \mathcal{H}$ be reflexive Banach

spaces. If $z \in \rho(A_0)$ then the restriction of $(A_0-z)^{-1}$ to $\mathcal{H}_0$ considered as a mapping from $\mathcal{H}_0$ to $\mathcal{H}_0^*$ is a bounded operator which we will denote by $R_0(z)$; we define the operator $R(z) \in \mathcal{L}(\mathcal{H}, \mathcal{H}^*)$ in the same way. Our basic assumptions will be :

(H_2-i) $J\mathcal{H}_0 = \mathcal{H}$

(H_2-ii) J extends to a bounded linear map from $\mathcal{H}_0^*$ to $\mathcal{H}^*$ (still denoted by J).

(H_3) Let $K_0(z)$, $z \in \rho(A_0)$, be the linear map from $\mathcal{H}_0$ to $\mathcal{H}$ defined by

$$\begin{cases} \mathcal{D}(K_0(z)) = \{\varphi_0 \in \mathcal{H}_0 , V(A_0-z)^{-1}\varphi_0 \in \mathcal{H}\} \\ K_0(z)\varphi_0 = V(A_0-z)^{-1}\varphi_0 \end{cases}$$

Then $K_0(z) \in \mathcal{K}(\mathcal{H}_0, \mathcal{H})$, the space of compact mappings from $\mathcal{H}_0$ to $\mathcal{H}$.

(H_4) Let $\lambda \in \sigma(A_0)$ be such that the L.A.P. is true for A_0 at λ in $\mathcal{H}_0$; then there exist $K_0^{\pm}(\lambda) \in \mathcal{K}(\mathcal{H}_0, \mathcal{H})$ such that

$$K_0^{\pm}(\lambda) = \text{norm} \lim_{\varepsilon \downarrow 0^+} K_0(\lambda \pm i\varepsilon)$$

(H_5) $u \in \mathcal{H}^*$ and $(J^* + K_0^{\pm}(\lambda)^*)\, u = 0$ imply $u \in \mathcal{H}$ and $(A-\lambda)\, u = 0$

Remarks :

Under assumptions (H_2-i) J considered as a map from $\mathcal{H}_0$ to $\mathcal{H}$ is a semi-Fredholm operator [11] with zero deficiency. Since $K_0^{\pm}(\lambda)$ is compact, $J + K_0^{\pm}(\lambda)$ also is semi-Fredholm with finite deficiency. Assumption (H_5) allows to control the situations where this deficiency is non zero ; this happens only if λ is an eigenvalue of A .

The starting point of this perturbative analysis is the following generalization of the resolvent equation

$$(A-z)^{-1} J = J(A_0-z)^{-1} - (A-z)^{-1} V (A_0-z)^{-1} \tag{20}$$

leading to

$$R(z) J = J R_0(z) - R(z) K_0(z) \tag{21}$$

The first main consequence of our assumptions is :

Theorem 4

Let the L.A.P. hold for A_0 at λ in $\mathcal{X}_0$. Then under assumptions (H_1) --- (H_5) it holds for A at λ in $\mathcal{X}$ unless $\lambda \in \sigma_p(A)$. In this case λ is an eigenvalue of finite multiplicity.

Proof : Assume first that $J+K_0^{\pm}(\lambda)$ has zero deficiency, i.e. the range of $J+K_0^{\pm}(\lambda)$ is all of $\mathcal{X}$. Then by (H_4) there exists a family $\Gamma(\lambda \pm i\varepsilon)$ of left pseudo-inverses for $J+K_0(\lambda \pm i\varepsilon)$ satisfying

$$(J+K_0(\lambda \pm i\varepsilon))\,\Gamma(\lambda \pm i\varepsilon) = \mathbb{1}_{\mathcal{X}}$$

and converging in norm to a left pseudo-inverse $\Gamma^{\pm}(\lambda)$ for $J+K_0^{\pm}(\lambda)$. Then

$$R(\lambda \pm i\varepsilon) = J R_0(\lambda \pm i\varepsilon)\,\Gamma(\lambda \pm i\varepsilon)$$

converges as ε tends to zero in the same topology as $R_0(\lambda \pm i\varepsilon)$ so that L.A.P. holds for A at λ.

Assume now that $J+K_0^{\pm}(\lambda)$ has non zero deficiency ; then by (H_5) λ is an eigenvalue of A. Let $\Gamma^{\pm}(\lambda)$ be a left pseudo-inverse for $J+K_0^{\pm}(\lambda)$ such that

$$(J+K_0^{\pm}(\lambda))\,\Gamma^{\pm}(\lambda) = P^{\pm}(\lambda)$$

where $P^{\pm}(\lambda)$ is a projection operator on the range of $J+K_0^{\pm}(\lambda)$; since this range has finite codimension it follows that $I-P^{\pm}(\lambda)$ is a finite rank operator. As before consider a sequence $\Gamma(\lambda \pm i\varepsilon) \in \mathcal{L}(\mathcal{X}, \mathcal{X}_0)$ of left pseudo-inverses for $J+K_0(\lambda \pm i\varepsilon)$ such that $\|\Gamma(\lambda \pm i\varepsilon) - \Gamma^{\pm}(\lambda)\| \to 0$ as $\varepsilon \to 0^+$. If $P(\lambda \pm i\varepsilon)$ is defined by

$$(J+K_0(\lambda \pm i\varepsilon))\,\Gamma(\lambda \pm i\varepsilon) = P(\lambda \pm i\varepsilon) \tag{22}$$

then one gets multiplying (21) by $\Gamma(\lambda \pm i\varepsilon)$ on the left:

$$R(\lambda \pm i\varepsilon)\,P(\lambda \pm i\varepsilon) = J R_0(\lambda \pm i\varepsilon)\,\Gamma(\lambda \pm i\varepsilon) \tag{23}$$

Now let $\psi \in \mathcal{X}$ belong to the range of $P^{\pm}(\lambda)$. Then $\forall \varphi \in \mathcal{X}$:

$$\lim_{\varepsilon \downarrow 0^+} \langle \varepsilon R(\lambda \pm i\varepsilon)\varphi, \psi \rangle = \lim_{\varepsilon \downarrow 0^+} \langle \varepsilon R(\lambda \pm i\varepsilon)\varphi, P(\lambda \pm i\varepsilon)\psi \rangle$$

In this equality we used the fact that $\varepsilon R(\lambda \pm i\varepsilon)\varphi$ is bounded in $\mathcal{H}$ hence in $\mathcal{X}^*$ and that by (22) $\|P(\lambda \pm i\varepsilon) - P^{\pm}(\lambda)\|$ tends to zero. Finally it follows from (8) and (23)

that

$$(24) \quad \langle \psi, E(\{\lambda\})\psi \rangle = \pm \lim_{\varepsilon \downarrow 0^+} \varepsilon \langle R_0(\lambda \pm i\varepsilon) J^* \psi, \Gamma(\lambda \pm i\varepsilon)\psi \rangle$$

$$= 0$$

since $J^*\psi \in \mathcal{H}_0$ by (H_2-i) and $\Gamma(\lambda \pm i\varepsilon)\psi$ converges in $\mathcal{H}_0$ to $\Gamma^{\pm}(\lambda)\psi$.
Since ψ belongs to a subspace of finite codimension in $\mathcal{H}$ then by a density argument it follows that $E\{\lambda\}$ is a finite rank projection operator ; this concludes the proof.

Remark

The operators $(I - P^{\pm}(\lambda)^*)$ are projections onto $N(J^* + K_0^{\pm}(\lambda)^*)$ the null space of $J^* + K_0^{\pm}(\lambda)^*$. Now (24) and a density argument show that $P^{\pm}(\lambda)^* E(\{\lambda\}) = 0$; hence $(I - P^{\pm}(\lambda)^*)\varphi = \varphi$ if $\varphi \in E(\{\lambda\})\mathcal{H}$. This implies that $E(\{\lambda\})\mathcal{H} \subset N(J^* + K_0^{\pm}(\lambda)^*)$
Since assumption (H_5) gives the converse inclusion one gets

$$(25) \quad N(J^* + K_0^{\pm}(\lambda)^*) = E(\{\lambda\})\mathcal{H}$$

Assuming now that λ is not an eigenvalue of A we want to construct mappings between some spaces of generalized eigenfunctions for A_0 and A with eigenvalue λ. Remember that for potential scattering such a mapping is given formally by (6) which provides solutions of the Lippman-Schwinger equation (5). The next result shows that any generalized solution of $(A - \lambda)u = 0$ in $\mathcal{E}_\lambda$ can be identified to the sum of a solution of $(A_0 - \lambda)\, u_0 = 0$, $u_0 \in \mathcal{E}_\lambda^0$, and an "ingoing (or outgoing) state".

We define N_λ^0 and $\mathcal{E}_\lambda^0$ by (11) and (12) with A_0 instead of A .

Proposition 7

Under assumptions (H_1) --- (H_5) and if $\lambda \notin \sigma_p(A)$ one has

$$(J^* + K_0^{\pm}(\lambda)^*)\,\mathcal{E}_\lambda \subset \mathcal{E}_\lambda^0$$

Proof : By Proposition 4 it is enough to show that $(J^* + K_0^{\pm}(\lambda)^*)\varphi(\lambda) \in \mathcal{E}_\lambda^0$ if $\varphi \in \mathcal{H}$. This holds by (12) if $\langle \varphi_0, (J^* + K_0^{\pm}(\lambda)^* \varphi(\lambda) \rangle = 0 \ \forall \varphi_0 \in N_\lambda^0$. Now

$$\langle \varphi_0, J^* + K_0^{\pm}(\lambda)^* \varphi(\lambda) \rangle = \pm \lim_{\varepsilon \downarrow 0^+} 2i\varepsilon \langle \varphi_0, (J^* + (A_0 - (\lambda \pm i\varepsilon))^{-1} V^*)$$

$$(A - (\lambda \pm i\varepsilon))^{-1}(A - (\lambda \mp i\varepsilon))^{-1}\varphi \rangle$$

By the second resolvent equation the right member of this equality is also equal to $\pm \lim_{\varepsilon \downarrow 0^+} 2i\varepsilon \langle R_0(\lambda \pm i\varepsilon)\varphi_0, J^* R(\lambda \pm i\varepsilon)\varphi \rangle$ Then it follows from Schwartz inequality that

(26) $$|\langle \varphi_0, (J^* + K_0^{\pm}(\lambda))\varphi(\lambda)\rangle| \leq \lim_{\varepsilon \downarrow 0^+} (2\varepsilon \| R_0(\lambda \pm i\varepsilon)\varphi_0\| \, \|R(\lambda \pm i\varepsilon)\varphi\|$$
$$= 2\pi \|\varphi_0(\lambda)\|_{h_0(\lambda)} \|\varphi(\lambda)\|_{h(\lambda)}$$

This completes the proof since $\varphi_0(\lambda) = 0$.

One can formulate a more intrinsic version of Proposition 7 which is a posteriori independent of the choice of auxiliary Banach spaces $\mathfrak{X}_0$ and $\mathfrak{X}$:

Proposition 8

Assume L.A.P. holds for A_0 at λ in $\mathfrak{X}_0$. Then under assumptions (H_1)---(H_5) and if $\lambda \notin \sigma_p(A)$ there exist $Z_{\pm}(\lambda) \in \mathcal{L}(h(\lambda), h_0(\lambda))$ and $W_{\pm}(\lambda) \in \mathcal{L}(h_0(\lambda), h(\lambda))$ such that $\forall \varphi \in \mathfrak{X}$:

i) $Z_{\pm}(\lambda)\varphi(\lambda) = -(J^* + K_0^{\pm}(\lambda)^*)\varphi(\lambda)$

ii) $Z_{\pm}(\lambda) = W_{\pm}(\lambda)^*$

iii) $W_{\pm}(\lambda)$ has dense range in $h(\lambda)$.

Proof : We define $W_{\pm}(\lambda)$ on the set of traces $\{\varphi_0(\lambda), \varphi_0 \in \mathfrak{X}_0\}$ by :

(27) $$W_{\pm}(\lambda)\varphi_0(\lambda) = (R^+(\lambda) - R^-(\lambda))(J + K_0^{\mp}(\lambda))\varphi_0$$

Then if $\varphi \in \mathfrak{X}$ one has

(28) $$2i\pi \langle \varphi(\lambda), W_{\pm}(\lambda)\varphi_0(\lambda)\rangle_{h(\lambda)} = -\langle (J^* + K_0^{\pm}(\lambda)^*)\varphi(\lambda), \varphi_0\rangle$$

Then it follows from (26) that $W_{\pm}(\lambda)$ extends to a bounded operator from $h_0(\lambda)$ to $h(\lambda)$ such that $\|W_{\pm}(\lambda)\| \leq 1$. Properties i) and ii) follow from (28). Finally by Definition (27), $W_{\pm}(\lambda)\varphi(\lambda)$ is the spectral trace at λ of $(J + K_0^{\mp}(\lambda))\varphi_0$; since the null space of $J^* + K_0^{\pm}(\lambda)^*$ is zero the range of the semi-Fredholm operator $J + K_0^{\mp}(\lambda)$ is all of $\mathfrak{X}$. Since the set of spectral traces is dense in $h(\lambda)$ by construction property iii) follows.

Remarks

1) One does not expect in general $W_{\pm}(\lambda)$ to be isometric. In fact an easy calculation shows that $\|W_{\pm}(\lambda)\varphi_0(\lambda)\|^2_{h(\lambda)} =$

$$\lim_{\varepsilon\downarrow 0^+} \varepsilon\pi^{-1} \|J(A_0-(\lambda\pm i\varepsilon))^{-1}\varphi_0\|^2 \leq \lim_{\varepsilon\downarrow 0^+} \varepsilon\pi^{-1} \|(A_0-(\lambda\pm i\varepsilon))^{-1}\varphi_0\|^2 = \|\varphi_0(\lambda)\|^2_{h_0(\lambda)}$$

On the other hand we will see later (see last remark) that $Z_\pm(\lambda)$ are isometric. Accordingly by ii) $W_\pm(\lambda)$ are in fact partially isometric.

2) The projection operators

$$P_\pm(\lambda) = Z_\pm(\lambda) W_\pm(\lambda)$$

satisfy

$$\|W_\pm(\lambda) P_\pm(\lambda)\varphi_0(\lambda)\|_{h(\lambda)} = \|P_\pm(\lambda)\varphi_0(\lambda)\|_{h_0(\lambda)}$$

By the preceding remark one has

$$\|P^\pm(\lambda)\varphi_0(\lambda)\|^2_{h(\lambda)} = \lim_{\varepsilon\downarrow 0^+} \varepsilon\pi^{-1} \|J(A_0-(\lambda\pm i\varepsilon))^{-1}\varphi_0\|^2$$

One can show that $P_\pm(\lambda)$ are the traces on the spectral subspace $h_0(\lambda)$ of the limits (if they exist !) :

$$P_\pm = s\text{-}\lim_{t\to\pm\infty} e^{iA_0t} J^* J e^{-iA_0t}$$

Notice that $P_\pm$ commute with A_0 and that by $(H_1\text{-i})$ $P_\pm$ are projection operators.

CONNECTION WITH TIME DEPENDENT SCATTERING THEORY

We now establish the connection between this abstract stationary construction and the time dependent approach to scattering. In the two Hilbert space formalism the wave-operators are defined in the following way, extending the definition given by V. Enss in his lectures for potential scattering :

$$W_\pm(A_0, A; J) = s\text{-}\lim_{t\to\pm\infty} e^{iAt} J e^{-iA_0t} P_{ac}(A_0)$$

where $P_{ac}(A_0)$ denotes the orthogonal projection operator on the absolutely continuous part of A_0.

It is not difficult to show using Abel's limit that if $W_\pm(A_0,A;J)$ exist they admit the following integral representation :

(29) $$\langle\varphi, W_\pm(A_0,A;J)\varphi_0\rangle = \langle\varphi, J\varphi_0\rangle \pm i \lim_{\varepsilon\downarrow 0^+} \int_0^{\pm\infty} dt\, e^{-\varepsilon|t|} \langle e^{-iAt}\varphi, V e^{-iA_0t}\varphi_0\rangle$$

for all $\varphi\in\mathcal{H}$ and $\varphi_0\in\mathcal{H}_{ac}(A_0)\cap\mathcal{D}(A_0)$. Notice that since by $(H_1\text{-ii})$ V is A_0-bounded the integrand in (29) is bounded by $Ce^{-\varepsilon|t|}$ for some constant C depending only

on $\|\varphi\|_{\mathcal{H}}$ and the A_0-graph norm of φ_0 .

The representation (29) is essential for the proof of the main result of this section. Before stating it we need the following :

Definition 3

Let $I \subset R$ be some closed interval. We will say that I is Regular (with respect to $A_0, A; \mathfrak{X}_0, \mathfrak{X}$) if the following properties hold :

i) L.A.P. holds for A_0 in $\mathfrak{X}_0$ on I .

ii) Let $I_\pm = \{z \in \mathbb{C}, z = \lambda \pm i\gamma, \gamma > 0, \lambda \in I\}$; then $R_0(z)$ is uniformly bounded in $\mathcal{L}(\mathfrak{X}_0, \mathfrak{X}_0^*)$ on I .

iii) $K_0(z)$ is continuous for the norm topology on I with continuous boundary values on both sides of I .

We will need the following :

Lemma 1

Assume (H_1) --- (H_5) are satisfied and let I be a bounded regular interval. If $\psi_0 \in \mathfrak{X}_0$ and $\varphi_0 = E_0(I)\psi_0$ one has:

i) $\varphi_0(\lambda) = \psi_0(\lambda) \quad \forall \lambda \in I$

ii) $V e^{-iA_0 t}\varphi_0 \in \mathfrak{X} \quad \forall t \in \mathbb{R}$

iii) $\langle u, V e^{-iA_0 t}\varphi_0\rangle = \int d\lambda \exp(-i\lambda t)\langle u, (K_0^+(\lambda) - K_0^-(\lambda))\psi_0\rangle$
$\forall u \in \mathfrak{X}^*$

Proof : By the functional calculus one has $\varphi_0 = \int_I^{\oplus} \psi_0(\lambda)\, d\lambda$ which implies i). Consider now the $\mathfrak{X}$-valued integral $\Psi_t = \int_I d\lambda \exp(-i\lambda t)(K_0^+(\lambda) - K_0^-(\lambda))\psi_0$. By iii) of Def. 3 one has $\|\Psi_t\|_{\mathfrak{X}} \leq \int_I d\lambda \,\|(K_0^+(\lambda) - K_0^-(\lambda))\psi_0\|_{\mathfrak{X}}$. The proof of iii) is an elementary exercise in functional calculus when u belongs to the dense domain of V^* i.e. $V^* u \in \mathfrak{X}_0$; for general $u \in \mathfrak{X}^*$ it follows by a continuity argument. Obviously iii) implies $\Psi_t = V e^{-iA_0 t}\varphi_0$ hence ii).

We can now state the

Theorem 5

Assume $W_\pm(A_0, A; J)$ exist and (H_1)---(H_5) are satisfied. Then for any bounded regular interval I such that $I \cap \sigma_p(A) = \emptyset$ one has $\forall \varphi \in \mathcal{H}$ and $\forall \varphi_0 \in E_0(I)\mathcal{H}_0$:

(30) $$\langle \varphi, W_{\pm}(A_0, A; J)\varphi_0\rangle = \int_I d\lambda \, \langle \varphi(\lambda), W_{\pm}(\lambda)\varphi_0(\lambda)\rangle_{h(\lambda)}$$

Proof : Since $E_0(I)\mathfrak{X}_0$ is dense in $E_0(I)\mathcal{H}_0$ and $W_{\pm}(\lambda)$, $W_{\pm}(A_0, A; J)$ are bounded, it is enough to show (30) with $\varphi_0 = E_0(I)\psi_0$ for some $\psi_0 \in \mathfrak{X}_0$ and with $\varphi \in \mathfrak{X}$. By the lemma above one has

$$\int_0^{\pm\infty} e^{-\varepsilon|t|} \langle e^{-iAt}\varphi, V e^{-iA_0t}\varphi_0\rangle$$
$$= (2i\pi)^{-1} \int_0^{\pm\infty} dt\, d\lambda \langle e^{-i(A-(\lambda\pm i\varepsilon))t}\varphi, (K_0^+(\lambda) - K_0^-(\lambda))\psi_0\rangle$$

The integrand on the r.h.s. of this equality is in $L^1(\mathbb{R}^{\pm}\times I)$; in fact it is bounded by $Ce^{-\varepsilon|t|}\|\varphi\|\,\|\psi_0\|_{\mathfrak{X}_0}$ with $C = \sup_{\lambda\in I}\|K_0^{\pm}(\lambda)\|$. So by Fubini's theorem we can integrate first on λ obtaining by the functional calculus:

$$\int_0^{\pm\infty} dt\, e^{-\varepsilon|t|}\langle e^{-iAt}\varphi, V e^{-iA_0t}\varphi_0\rangle$$
$$= \mp(2i\pi)^{-1}\int_I d\lambda \langle R(\lambda\pm i\varepsilon)\varphi, (K_0^+(\lambda) - K_0^-(\lambda))\psi_0\rangle$$

Now since I is regular it follows from (21) and the arguments used in the proof of Theorem 4 that R(z) also is uniformly bounded on I . Since I is compact one obtains by Riemann-Lebesgue Lemma :

$$\lim_{\varepsilon\downarrow 0^+}\int_I d\lambda \langle R(\lambda\pm i\varepsilon)\varphi, (K_0^+(\lambda) - K_0^-(\lambda))\psi_0\rangle$$
$$= \int_I d\lambda \langle R^{\pm}(\lambda)\varphi, (K_0^+(\lambda) - K_0^-(\lambda))\psi_0\rangle$$

whence finally by (29) :

$$\langle \varphi, W_{\pm}(A_0, A; J)\varphi_0\rangle$$
$$= -(2i\pi)^{-1}\int_I d\lambda \langle [(R_0^+(\lambda) - R_0^-(\lambda))J^* - (K_0^+(\lambda)^* - K_0^-(\lambda)^*)R^{\pm}(\lambda)]\varphi, \psi_0\rangle$$

It is easy to show using (21) in the limit $z = \lambda \pm i0$ that

(31) $$(R_0^+(\lambda) - R_0^-(\lambda))J^* - (K_0^+(\lambda) - K_0^-(\lambda))^* R^{\pm}(\lambda)$$
$$= [J^* + K_0^{\pm}(\lambda)]^*(R^+(\lambda) - R^-(\lambda))$$

so that the proof is completed with the help of Proposition 8.

ASYMPTOTIC COMPLETENESS

The Theorem 5 combined with Proposition 8 provides a very powerful way to prove asymptotic completeness of wave-operators. We will need the following extra assumptions:

(H_6-i) There exists a denumerable family of disjoint open intervals $(I_n)_{n\in N}$ such $\bigcup I_n = R$ and the L.A.P. holds for A_0 on each I_n in $\mathfrak{X}_0^n$.

(H_6-ii) Any closed interval I such that $I \subset I_n$ for some n is regular.

Let us show first

Proposition 9

Under assumptions (H_1) --- (H_5)(H_6) the L.A.P. holds for A on each I_n except possibly at some isolated points which can accumulate only at ∂I_n and are eigenvalues of A with finite multiplicity.

Proof : By Theorem 4 it is enough to show that $\sigma_p(A) \cap I_n$ consists of isolated eigenvalues. Now $\lambda \in \sigma_p(A) \cap I_n$ if and only if the deficiency of $J + K_0^{\pm}(\lambda)$ is non-zero. By the continuity of $K_0^{\pm}(\lambda)$ on I_n following from (H_6-ii) and Fredholm operator theory[11,16], this happens either on all of I or at some isolated points of I. The first possibility being obviously excluded one gets the stated property.

Theorem 6

If $W_{\pm}(A_0, A; J)$ exist and (H_1) --- (H_6) hold one has

$$W_{\pm}(A_0, A; J)\mathcal{H}_{ac}(A_0) = \mathcal{H}_{ac}(A)$$

Proof : Assume φ is in the orthogonal complement of $W_{\pm}(A_0, A; J)\mathcal{H}_{ac}(A_0)$; then $\forall \varphi_0 \in \mathcal{H}_0$:

$$\langle \varphi, W_{\pm}(A_0, A; J)\, E_0(I)\varphi_0 \rangle = 0$$

for any regular interval I since then $E_0(I)\mathcal{H}_0 \subset \mathcal{H}_{ac}(A_0)$ by Proposition 1. In particular if $I \cap \sigma_p(A) = \emptyset$ one has by (30) :

$$\langle \varphi, W_{\pm}(A_0, A; J)\, f(A_0)\, E_0(I)\varphi_0 \rangle$$

$$= \int_I d\lambda\, f(\lambda) \langle \varphi(\lambda), W_{\pm}(\lambda)\, \varphi_0(\lambda) \rangle_{h(\lambda)} = 0$$

for all $f \in L^{\infty}(I)$ and $\varphi_0 \in E_0(I)\mathcal{H}_0$. Then almost everywhere on I :

$$\langle \varphi(\lambda), W_{\pm}(\lambda)\varphi_0(\lambda)\rangle_{h(\lambda)} = 0$$

Taking a basis $(\varphi_{0,n})_n$ of $\mathcal{H}_0$ one has $\langle \varphi(\lambda), W_{\pm}(\lambda)\varphi_{0,n}(\lambda)\rangle_{h(\lambda)} = 0$ for all n almost everywhere. Since $(\varphi_{0,n}(\lambda))_n$ is a total set in $h_0(\lambda)$ and by Proposition 8 iii) it follows that $\varphi(\lambda) = 0$ almost everywhere on I and accordingly $E(I)\varphi = 0$ since the spectral measure of A is absolutely continuous on I by Proposition 1.

Now by assumption (H_6 ii) this implies $E(I_n \setminus \sigma_p(A))\varphi = 0$ for all n ; so $E(\cdot)\varphi$ has support on $(\bigcup_n \partial I_n) \cup \sigma_p(A)$ and accordingly $\varphi \in \mathcal{H}_{ac}^{\perp}(A)$

Remarks

1) In general the assumptions (H_1) --- (H_6) don't imply existence of wave-operators but only of the weak Abel limits

$$w \lim_{\varepsilon \to 0^+} \int_0^{\pm\infty} dt\, e^{-\varepsilon|t|}\, e^{iAt}\, J e^{-iA_0 t}\, P_{ac}(A_0)$$

2) Let I be a regular interval such that $I \cap \sigma_p(A) = \emptyset$. Then one can show that asymptotic completeness implies existence of the strong limits :

$$W_{\pm}^*(A_0, A; J)\, E(I) = s\cdot\lim_{t \to \pm\infty} e^{iA_0 t} J e^{-iAt} E(I)$$

Accordingly if $\varphi \in E(I)\mathcal{H}$ one has

(32) $$\| W_{\pm}^*(A_0, A; J)\varphi \| = \lim_{t \to \pm\infty} \| J^* e^{-iAt}\varphi \| = \|\varphi\|$$

On the other hand Theorem 5 and Proposition 8 ii) imply :

$$\| W_{\pm}^*(A_0, A; J)\varphi\|^2 = \int_I d\lambda \langle Z_{\pm}(\lambda)\varphi(\lambda), W_{\pm}^*(A_0, A; J)\varphi\rangle$$

$$= \int_I d\lambda\, \| Z_{\pm}(\lambda)\varphi(\lambda)\|^2_{h_0(\lambda)} = \int_I d\lambda \|\varphi(\lambda)\|^2_{h(\lambda)}$$

Since this remains true when one replaces φ by $f(A)\varphi$, $f \in L_{\infty}(I)$, one obtains finally by (32)

$$\| Z_{\pm}(\lambda)\varphi(\lambda)\|^2_{h_0(\lambda)} = \|\varphi(\lambda)\|^2_{h(\lambda)}$$

for all $\varphi \in \mathcal{H}$. This shows that $Z_{\pm}(\lambda)$ is isometric for all $\lambda \in I$

EXAMPLES

Potential Scattering

Let us come back to the situation described by V. Enss ; it corresponds to $A_0 = -\Delta$ on $\mathcal{H} = L^2(R^n)$ and $A = -\Delta + V$ where V is a real multiplicative operator on $\mathcal{H} = \mathcal{H}_0 = L^2(\mathbb{R}^n)$; here we take J = 1. We assume for simplicity $\mathcal{D}(A) = \mathcal{D}(A_0)$ in accordance with (H_1-ii). Following S. Agmon[6] we define :

Definition 4

V is a short range potentialof $(1+|X|^2)^{\frac{1+\varepsilon}{2}} V(X)$ is a compact mapping from $\mathcal{H}_2(\mathbb{R}^n)$ to $L^2(R^n)$ for some $\varepsilon > 0$.

Now the mapping $u(x) \longrightarrow (1+|X|^2)^{-\frac{1+\varepsilon}{2}} u(x)$ is bounded from $\mathcal{H}_{m,s}(\mathbb{R}^n)$ to $\mathcal{H}_{m,s+1+\varepsilon}(\mathbb{R}^n)$ for all $m \in N$, $s \in \mathbb{R}$ where

$$\mathcal{H}_{m,s}(\mathbb{R}^n) = \left\{ u \in L^2_{loc}(\mathbb{R}^n), (1+|x|^2)^{s/2} D^\alpha u \in L^2(\mathbb{R}^n), |\alpha| \leq \right.$$

Then for all $s \in R$ multiplication by V also is a compact mapping from $\mathcal{H}_{2,s}(\mathbb{R}^n)$ to $L^2_{1+\varepsilon+s}(\mathbb{R}^n)$.

The short range class contains potentials bounded at large distances by $C(1+|X|^2)^{\frac{-1-\varepsilon}{2}}$ for some $C, \varepsilon > 0$ and satisfying locally a regularity assumption $V \in L^p_{loc}(\mathbb{R}^n)$ with p = 2 for $n \leq 3$ and $p > \frac{n}{2}$ for $n \geq 4$. This regularity assumption guarantees $\mathcal{D}(A) = \mathcal{D}(A_0)$.

Now with the choice

$$\mathcal{X}_0 = \mathcal{X} = L^2_{\frac{1+\varepsilon}{2}}(\mathbb{R}^n)$$

assumptions (H_2) and (H_3) are satisfied. To control (H_4) notice that

$$V(A_0 - (\lambda \pm i\varepsilon))^{-1} = V(A_0+1)^{-1} + (\lambda - 1 \pm i\varepsilon) V(A_0+1)^{-1}(A_0 - (\lambda \pm i\varepsilon))^{-1}$$

By Theorem 1 and compactness of $V(A_0+1)^{-1}$ as a mapping from $\mathcal{X}_0^*$ to $\mathcal{X}_0$ one gets easily in view of (H_4) and (H_6) :

Proposition 10

Any closed interval $I \subset R \setminus \{0\}$ is regular.

It remains to verify (H_5) ; for this we need the following results :

Lemma 2

Let $\lambda \neq 0$; then for all $s > \frac{1}{2}$ there exists a finite positive constant C such that

$$\| R_0^{\pm}(\lambda) f \|_{\mathcal{H}_{2,s-1}} \leq C \| f \|_{L^2_s}$$

for all

$$f \in N^0_{\lambda,s} = \left\{ f \in L^2_s(\mathbb{R}^n) \,,\, \operatorname{Im} \langle f, R_0^{\pm}(\lambda) f \rangle = 0 \right\}$$

Lemma 3

Assume $(I + K_0^{\pm}(\lambda)) f = 0$ for some $f \in L^2_s(\mathbb{R}^n)$ with $s \geq \frac{1+\varepsilon}{2}$; then $f \in N^0_{\lambda,s}$

Proof : Assume $(I + K_0^{\pm}(\lambda)) f = 0$, $f \in L^2_s(\mathbb{R}^n)$ and let $u = R_0^{\pm}(\lambda) f$, $u \in \mathcal{H}_{2,-s}(\mathbb{R}^n)$.

Then f = -Vu and

$$\begin{aligned} \operatorname{Im} \langle f, R_0^{\pm}(\lambda) f \rangle &= \operatorname{Im} \langle f, u \rangle \\ &= -\operatorname{Im} \langle Vu, u \rangle \\ &= \lim_{\varepsilon \downarrow 0^+} \langle V(A_0 - (\lambda \pm i\varepsilon))^{-1} f, (A_0 - (\lambda \pm i\varepsilon))^{-1} f \rangle \\ &= 0 \text{ since } V \text{ is symmetric.} \end{aligned}$$

Then we have :

Proposition 11

The null space of $I + K_0^{\pm}(\lambda)^*$ is contained in $\mathcal{H}_{2,s}(\mathbb{R}^n)$ for all $s \in R$.

Proof : By the Fredholm alternative the null spaces of $I + K_0^{\pm}(\lambda)$ and $I + K_0^{\pm}(\lambda)^*$ have the same dimension. One can easily construct explicitly a bijective map between these spaces by showing that if $u = R_0^{\pm}(\lambda) f$ one has :

$$(33) \quad (I + K_0^{\pm}(\lambda)) f = 0 \iff (I + K_0^{\pm}(\lambda)^*) u = 0$$

Then the Proposition will follow from Lemmas 2 and 3 if we show that the null space of $I + K_0^{\pm}(\lambda)$ is contained in $L^2_s(\mathbb{R}^n)$ for all s . Now by Lemma 2 $f \in L^2_{\frac{1+\varepsilon}{2}}$ implies $u \in \mathcal{H}_{2,-\frac{1}{2}+\varepsilon}$ hence $f = -Vu \in L^2_{\frac{1+2\varepsilon}{2}}$. Repeating this argument n times gives $f \in L^2_{\frac{1+n\varepsilon}{2}}$ for all n ; this concludes the proof.

Corollary :

If $(I+K_0^{\pm}(\lambda)^*)\,u = 0$ for some $\lambda \neq 0$; then $u \in \mathcal{H}_{2,s}(\mathbb{R}^n)$ for all $s \in R$ and $(A-\lambda)u = 0$.

Proof : It remains to show that u is an eigenvector of A . For all $g \in C_0^\infty(\mathbb{R}^n)$ one has :

$$\langle(-\Delta-\lambda)g,u\rangle = \lim_{\varepsilon\downarrow 0^+} \langle(-\Delta-\lambda)g,(-\Delta-(\lambda\pm i\varepsilon))^{-1}f\rangle$$
$$= \langle g,f\rangle$$

so that $(-\Delta-\lambda)u = f$ in the sense of distributions. But this equality also holds in $L^2(R^n)$ since $u \in \mathcal{H}_{2,s}(\mathbb{R}^n)$ for all s . Finally f = -Vu concludes the proof.

Now all assumptions (H_1) --- (H_6) are satisfied so that we have :

Theorem 7

Let V be a short-range potential. Then :

i) $W_\pm(A_0,A;J)$ exist and are complete.

ii) The continuous singular part of A is empty.

iii) The point spectrum of A outside zero consists of isolated eigenvalues with finite multiplicity. The corresponding eigenfunctions are in $\mathcal{H}_{2,s}(\mathbb{R}^n)$ for all $s \in R$.

Proof : We refer to V. Enss' lectures for the proof that wave-operators exist. Completeness follow then from Theorem 6. Assertion ii) and the first part of iii) follow from Propositions 1 and 9 ; (in fact $(H_6\text{-}i)$ holds with the two sets $I_1 = \mathbb{R}^-\setminus\{0\}$ and $I_2 = \mathbb{R}^+\setminus\{0\}$). Decay properties follow from (25) and the above Corollary.

The Three-Body Problem.

Let us consider now a three-particle quantum system with local pair interactions $V_{ij}(X_i - X_j)$, i = j , where X_i denotes the position of particle i. Assuming that these interactions vanish at infinity there are four different types of "asymptotic freedom" corresponding respectively to three infinitely separated particles not interacting anymore or one cluster $\alpha = (ij)$ of two particles not interacting with k , (ijk) = (123). We introduce an associated partition of unity with the help of C^∞ functions J_D, where D labels the above four cluster decompositions. We require the following properties :

(35.i) J_D is homogeneous of degree 0.

(35.ii) $\sum_D J_D^2 = 1$

(35.iii) For some positive constant a, $\operatorname{Supp} J_D \subset \{|x_i - x_j| > a|X|\}$ if i,j belong to different clusters of D. Here $|X|$ denotes the total length of the vector X representing the three-particle configuration in the center of mass system.

The unperturbed operator A_0 is chosen as

$$A_0 = \oplus H_D$$

acting on $\mathcal{H}_0 = \oplus_D L^2(\mathbb{R}^6)$ where H_D is obtained from the Hamiltonian

$$A = H_0 + \sum_{1 \le i < j \le 3} V_{ij}$$

by removing interactions between particles belonging to different clusters of D. The kinetic energy operator H_0 (in the center of mass system) is a six-dimensional Laplacian with coefficients depending on the masses of the particles. Using Jacobi coordinates one can show that each two cluster Hamiltonian H_D has the structure described in Proposition 6.

(34) $$H_D = \frac{P_k^2}{2/2_k} \otimes I_{L^2(x_i - x_j)} + I_{L^2(y_k)} \otimes h_\alpha$$

where (y_k, p_k) are the relative position and momentum operators for particle k with respect to the center of mass of (i,j) with (i,j,k) = (1,2,3). The operator h_α is the two-particle Hamiltonian for the pair α with interaction V_{ij}. If we assume that two-particle interactions are short-range, h_α is known by the preceding example to satisfy the L.A.P. outside the set

$$\Sigma_\alpha = \{0\} \cup \sigma_p(h_\alpha)$$

in $L^2_{\frac{1+\varepsilon}{2}}(\mathbb{R}^3)$ for some $\varepsilon > 0$. Since $P_k^2/2/2_k$ also does outside zero it follows from Proposition 6 that H_D satisfies the L.A.P. in $L^2_{\frac{1+\varepsilon}{2}}(\mathbb{R}^6)$ outside Σ_α. Finally A_0 satisfies the L.A.P. in

$$\mathcal{X}_0 = \oplus_D L^2_{\frac{1+\varepsilon}{2}}(\mathbb{R}^6)$$

in $R \setminus \Sigma$ where

$$\Sigma = \{0\} \cup (\cup_\alpha \Sigma_\alpha)$$

Now the Hilbert space $\mathcal{H}$ is $L^2(R^6)$ and we choose

$\mathcal{X} = L^2_{\frac{1+\varepsilon}{2}}(\mathbb{R}^6)$. We construct the identification map as follows : let $\varphi_0 = \oplus_D \varphi_D$ be an arbitrary element of $\mathcal{H}_0$. Then

$$J\varphi_0 = \sum_D J_D \varphi_D$$

It is easy to show that if $\varphi \in \mathcal{H}$ then $J^*\varphi = \oplus_D J_D \varphi$

By (35.ii) one has then $JJ^* = 1$ and obviously $J\mathcal{X}_0 = \mathcal{X}$. The perturbation V is given by

$$V(\oplus_D \varphi_D) = \sum_D J_D V_D \varphi_D + \sum_D [H_0, J_D]\varphi_D \tag{36}$$

Assumption (35.iii) implies that the terms $J_D V_D$ decay at infinity like $|X|^{-1-\varepsilon}$. Homogeneity of J_D implies on the other hand that $[H_0, J_D]$ is a differential operator of order one with coefficients tending to zero at infinity like $|X|^{-1}$. Accordingly this term cannot be treated by the usual method for short-range potentials since it is not a compact mapping from $\mathcal{H}_{2,-s}(\mathbb{R}^6)$ to $L^2_s(R^6)$ for some $s > \frac{1}{2}$.

However one can expect that V falls into the class of perturbations satisfying the general assumptions (H_4) and (H_6). Let us motivate this hope by the following remarks.

1) The "effective interaction" VJ^* is short-range. This was already mentioned for the local multiplicative term in V in the r.h.s. of (36). Furthermore one has

$$\sum_D [H_0, J_D] J_D = - \sum_D J_D [H_0, J_D]$$

by (35.ii). Accordingly

$$\sum_D [H_0, J_D] J_D = \frac{1}{2} \sum_D [[H_0, J_D], J_D] \tag{37}$$

$$= \frac{1}{2} \sum_D |\nabla J_D|^2$$

where ∇ denotes the 6-dimensional gradient modulo some kinematical factors. By (37) the second term on the r.h.s. of (36) gives a contribution to the effective interaction which decays like $|X|^{-2}$. However we have chosen in these lectures to not present the abstract stationary theory in terms of the effective interaction so we will not develop this argument any longer.

2) The main reason why $K_0^{\pm}(\lambda)$ might satisfy (H_4) although V is not short-range is the radiation condition satisfied

by ingoing or outgoing states $R_0^{\pm}(\lambda) f$, $f \in B(\mathbb{R}^6)$. One can easily convince oneself by looking at Agmon-Hörmander's form[1] of the radiation condition (Theorem 3) or at Theorem 6.3 of their paper that $[H_0, J_D]$ acting on states satisfying a radiation condition improves the decay a little bit more than just $|x|^{-1}$; this suggests strongly[15] working in Besov space (18) instead of $L_s^2(R^6)$, $s > \frac{1}{2}$.

3) To control the "Fredholm alternative" with the help of (H_5) in potential scattering a very useful tool was provided by the trace result of Lemma 2 . It turns out that a trace property can be proved inductively for operators A of the type described in Proposition 6 provided it holds for A_1 and A_2. This brings another essential tool for an analysis of the N-body problem along lines identical to those of potential scattering modulo the geometrical methods.

To conclude this short presentation of the geometrical ideas in many particle scattering problems, let me mention the recent work of P. Perry, I. Sigal, and B. Simon[17]. These authors show absence of continuous singular spectrum and discreteness of eigenvalues by Mourre's method[18]. The interested reader can find there more technical details on the geometrical approach.

REFERENCES

1. S. Agmon, L. Hörmander, J. Anal. Math. 30:1 (1976).
2. T. Kato, S. Kuroda, "Rocky Mountains, J. Math. 1:127 (1971).
3. T. Kato, J. Fac. Sci. Univ. Tokyo, Sect. 1A, 24:503 (1971).
4. M. Schecter, J. Math. Pures Appl. 57:373 (1974).
5. I. Sigal, Scattering Theory for Many Particle Systems, To appear in "Lecture Notes in Mathematics".
6. V. Enss, Comm. math. Phys. 52:233 (1977).
7. B. Simon, Comm. math. Phys. 58:205 (1978).
8. P. Deift, W. Hunziker, B. Simon, and E. Vogt, Comm. math. Phys. 64, 1 (1978).
9. T. Ikebe, Arch. Rational Mech. Anal. 5:1 (1960).
10. M. Reed, B. Simon, Methods of Modern Mathematical Physics, III. Scattering Theory, Academic Press (1979).
11. T. Kato, Perturbation Theory for Linear Operators, Springer-Verlag, New York (1966).
12. J. Dixmier, Les algèbres d'opérateurs dans l'espace hilbertien, Gauthier-Villars, Paris (1969).

13. J.L. Lions, Magenes, Problèmes au limites non homogènes et applications, Dunod (1968).
14. S. Agmon, Ann. Scuol. Sup. Pisa, Ser.IV, Vol.II, 151 (1975).
15. J.M. Combes, work in preparation.
16. M. Schecter, Principles of Functional Analysis, Academic Press (1973).
17. P. Perry, I. Sigal, and B. Simon, Princeton Preprint (1980).
18. E. Mourre, Marseille Preprint, to appear in Comm. math. Phys.

SPECTRAL PROPERTIES AND ASYMPTOTIC EVOLUTION IN POTENTIAL SCATTERING

D. B. Pearson

Department of Applied Mathematics
University of Hull
Hull, England

INTRODUCTION

Consider a single non-relativistic particle moving in a potential $V(\underline{r})$. The Hamiltonian H will be some self-adjoint extension of $-\nabla^2 + V(\underline{r})$, acting in the Hilbert space $\mathcal{H} = L^2(\mathbb{R}^3)$ (We choose units such that $\hbar = 2m = 1$.)

<u>Mathematically</u>, we should be interested in the spectral properties of H, in the location of the discrete spectrum, the nature of the continuous spectrum, and so on. <u>Physically</u>, we should be interested in the time development of the quantum system and especially, in scattering theory, with the asymptotic behaviour of the system for large positive and negative times. It is one of the purposes of these lecture notes to relate these two descriptions, the mathematical and the physical, and to show how spectral properties can profoundly influence how the particle behaves.

It is customary to list the conditions which are to be imposed on the classes of potentials to be dealt with. Usually the potential locally should be not too singular, and should behave reasonably at infinity, implying that V (or more generally the derivatives of V) is of short range. One then deduces the spectral properties to be demanded of any self-respecting Hamiltonian, and shows that scattering theory works in the usual sense (existence of wave operators, unitarity of the scattering operator, and so on.) All of this is admirably covered in the notes of Volker Enss.

Here our aim is rather different, and perhaps complementary to the usual approach. We set out to explore a wider area in which our mathematical and physical intuitions may occasionally desert us. Simplifying assumptions about the potential will be kept to a minimum. Some of our potentials will locally be highly singular, or quite badly behaved at infinity. This may allow unusual spectral properties which may lead to unexpected physical consequences.

As is common to other kinds of exploration, we shall have to work quite hard to leave our usual habitat. Perhaps, after this foray into a wider and unfamiliar world, we shall wish to return to more comfortable surroundings. But we will do so with greater knowledge of what lies beyond, and our experiences should keep us, at least, from straying too close to the boundaries of our chosen domain.

The organisation of these notes are as follows.

In Section 1 we describe the various types of spectrum which can occur, and the corresponding subspaces of the Hilbert Space. We present some preliminary results relating spectral properties with the description of the time-development of the system. In particular, states in the singular continuous subspace exhibit different behaviour from those in the absolutely continuous subspace as regards localisation in finite regions.

In Section 2 we lay the groundwork for the spectral analysis of a large class of Schrödinger Hamiltonians. The basic idea is to construct the spectral measure of the differential operator from a limit of absolutely continuous measures. Each measure of this limiting sequence is given explicitly by a density function which depends in a simple way on solutions of the time-dependent Schrödinger equation.

In Section 3 we present examples of Sturm-Liouville operators on a finite interval which give rise to absolutely continuous spectrum, and discuss the consequent physical interpretation for potential scattering in three dimensions. The potentials are locally highly singular and oscillating.

Section 4 presents examples of long range non-singular potentials which give rise to singular continuous spectrum, and analyses some physical consequences of this phenomenon.

In Section 5 we set some of these examples in a more general setting, and show that they can occur quite naturally within a framework of scattering by short range but locally singular potentials. For this fairly general class of potentials, we present a systematic analysis of the relation between spectral properties and the physical phenomena which accompany them.

1. SPECTRUM AND SUBSPACES

The spectral theorem for self-adjoint operators allows us to write $H = \int \lambda dE_\lambda$, where $\{E_\lambda\}$ is the one parameter family of projection operators (resolution of the identity) which characterises the Hamiltonian H. The Hilbert space can be decomposed $\mathcal{H} = \mathcal{H}_d + \mathcal{H}_c$ where $\mathcal{H}_d$, the discrete subspace of H, is spanned by the (normalisable) eigenstates of H, and $\mathcal{H}_c$, the continuous subspace, is the orthogonal subspace to $\mathcal{H}_d$. In fact, $\mathcal{H}_c$ can itself be decomposed as follows.

Given $f \in \mathcal{H}$, define a measure μ_f on Borel subsets Σ of $\mathbb{R}$ by the formula

$$\mu_f(\Sigma) = \int_\Sigma 1 \, d\langle f, E_\lambda f\rangle \tag{1}$$

the integral being in the Lebesque-Stieltjes sense. It may be that the measure μ_f is derivable from a density function γ_f. This means that γ_f is a non-negative function satisfying $\int_{\mathbb{R}} \gamma_f(\lambda) d\lambda < \infty$ and such that

$$\mu_f(\Sigma) = \int_\Sigma \gamma_f(\lambda) \, d\lambda \tag{2}$$

We then say that the measure μ_f is <u>absolutely continuous</u> (with respect to Lebesque measure $d\lambda$). $\gamma_f(\lambda)$ is then defined for almost all λ by

$$\gamma_f(\lambda) = \frac{d}{d\lambda} \langle f, E_\lambda f\rangle \quad , \text{ a.e.} \tag{3}$$

and the set of all $f \in \mathcal{H}$ such that μ_f is absolutely continuous is called the <u>absolutely continuous subspace</u> (a.c. subspace) of H, and denoted by $\mathcal{H}_{a.c.}$ (It is not obvious that this defines a <u>subspace</u> rather than a <u>subset</u>, though this can be proved; see [1].)

For $f \in \mathcal{H}_c$, it may happen that the measure μ_f is <u>singular</u> (with respect to Lebesque measure.) This means that the real line can be split into two disjoint subsets Σ_1 and Σ_2 such that the Lebesque measure of Σ_1 is zero whereas the μ_f -measure of Σ_2 is zero. In other words, μ_f is concentrated entirely on a set Σ_1 which is very small in the sense of having Lebesque measure zero. For examples of such <u>singular continuous</u> measures see for example [2]. The set of all $f \in \mathcal{H}_c$ such that μ_f is singular is called the <u>singular continuous subspace</u> of H, and denoted by $\mathcal{H}_{s.c.}$. Again it can be shown that this defines a subspace rather than a subset. Moreover, $\mathcal{H}_{s.c.}$ is orthogonal to $\mathcal{H}_{a.c.}$ and we have the decomposition

$$\mathcal{H}_c = \mathcal{H}_{a.c.} \oplus \mathcal{H}_{s.c.}$$

Thus

$$\mathcal{H} = \mathcal{H}_d \oplus \mathcal{H}_{a.c.} \oplus \mathcal{H}_{s.c.}$$

All three of these subspaces are reduced by H. They are invariant under the operation by any function of H, and in particular by the spectral projections E_λ. Here we shall be concerned mostly with $\mathcal{H}_{a.c.}$ and with $\mathcal{H}_{s.c.}$.

For $M > 0$, let us define the subset $\mathcal{S}_M$ of $\mathcal{H}_{a.c.}$, consisting of those $f \in \mathcal{H}_{a.c.}$ such that

$$\frac{d}{d\lambda} \langle f, E_\lambda f \rangle \leq M, \ a.e. \tag{4}$$

and let $\mathcal{S}$ be the union of all $\mathcal{S}_M$ for $M > 0$. Thus, for $f \in \mathcal{S}$, the derivative of $\langle f, E_\lambda f \rangle$ is (essentially) bounded. For arbitrary $f \in \mathcal{H}_{a.c.}$ define χ_M to be the characteristic function of the set $\{\lambda ; \frac{d}{d\lambda} \langle f, E_\lambda f \rangle \leq M\}$. Then if $f_M = \chi_M(H) f$ we have

(i) $$\frac{d}{d\lambda} \langle f_M, E_\lambda f_M \rangle = \chi_M(\lambda) \frac{d}{d\lambda} \langle f, E_\lambda f \rangle \leq M$$

and (ii) $$\| f_M - f \|^2 = \int (\chi_M(\lambda) - 1)^2 \, d \langle f, E_\lambda f \rangle$$

$\to 0$ as $M \to \infty$ by the Lebesque dominated convergence theorem. From (i) it follows that $f_M \in \mathcal{S}_M$ and from (ii) that $\mathcal{S}$ is dense in $\mathcal{H}_{a.c.}$.

In view of the relation $e^{-iHt} = \int e^{-i\lambda t} dE_\lambda$, one might expect, through use of the Fourier transform, to be able to characterise the a.c. subspace in terms of the time development of states. A result which goes some way towards doing this is as follows:

THEOREM 1

$\mathcal{H}_{a.c.}$ consists of those f for which a dense subset $\mathcal{S}(f)$ of $\mathcal{H}$ exists, such that $\phi \in \mathcal{S}(f) \Rightarrow \langle \phi, e^{-iHt} f \rangle \in L^2(\mathbb{R})$, as a function of t.

PROOF

Assume such a dense subset exists, for some f, and let $\phi \in \mathcal{S}(f)$

Let $\hat{F} \in C_0^\infty(\mathbb{R})$ be the Fourier transform of the function F. Then $\int_{\mathbb{R}} F(t) \langle \phi, e^{-iHt} f \rangle dt$ exists. Using the spectral theorem for the evolution operator e^{-iHt} and inverting the orders of integration with respect to λ and t, the integral is just $\sqrt{2\pi} \langle \phi, \hat{F}(H) f \rangle$. Moreover, since $\langle \phi, e^{-iHt} f \rangle \in L^2$, the integral is bounded in absolute value by const, $\| F \|$, or equivalently by const. $\| \hat{F} \|$. Thus the Riesz representation theorem allows us to write, for some $\theta \in L^1(\mathbb{R})$, $\langle \phi, \hat{F}(H) f \rangle = \int_{\mathbb{R}} \hat{F}(s) \theta(s) ds$. Taking a sequence of such $\hat{F}$ converging to the characteristic function of a finite interval $[a, b]$ it is not difficult to show that

$$\langle \phi, (E_b - E_a) f \rangle = \int_a^b \theta(s)\, ds .$$

(The limiting case a = b confirms that a single value of λ cannot have positive measure.) Thus θ is the density function, at least for intervals, and standard measure theoretic constructions [3] show that θ is the density function for arbitrary Borel sets. In other words, the charge $d\langle \phi, E_\lambda f \rangle$ is absolutely continuous. (We refer to a charge, not a measure, since in general $\langle \phi, E_\lambda f \rangle$ need not be positive, or even real.)

Defining the projections P_d, $P_{a.c.}$ and $P_{s.c.}$ onto their respective subspaces, and decomposing the charge $d\langle \phi, E_\lambda f \rangle$ into its components, we know that the singular continuous component $d\langle \phi, E_\lambda P_{s.c.} f \rangle$ must vanish. Integrating over the entire real line implies that $\langle \phi, P_{s.c.} f \rangle$ must be zero. But the ϕ's form a dense set. Hence $P_{s.c.} f = 0$. Similarly $P_d f = 0$, so that $f \in \mathcal{H}_{a.c.}$ as required.

To prove the converse of this result, suppose $f \in \mathcal{H}_{a.c.}$ and take $\mathcal{S}$ to be defined as above. For $\phi \in \mathcal{S}$, $d\langle \phi, E_\lambda f \rangle$ is a.c. and a limiting argument, with Schwartz's inequality, shows that

$$\left| \frac{d}{d\lambda} \langle \phi, E_\lambda f \rangle \right|^2 \leq \frac{d}{d\lambda} \langle \phi, E_\lambda \phi \rangle \frac{d}{d\lambda} \langle f, E_\lambda f \rangle \quad \text{a.e.}$$

Since $\frac{d}{d\lambda} \langle \phi, E_\lambda \phi \rangle \leq M$ for some $M > 0$ and $\int_{\mathbb{R}} \frac{d}{d\lambda} \langle f, E_\lambda f \rangle = \|f\|^2 < \infty$, this implise that $\frac{d}{d\lambda} \langle \phi, E_\lambda f \rangle \in L^2$. Hence the Fourier transform is also L^2, so that

$$\langle \phi, e^{-iHt} f \rangle = \int_{-\infty}^{\infty} e^{-i\lambda t} \frac{d}{d\lambda} \langle \phi, E_\lambda f \rangle \, d\lambda$$

$\in L^2$ as a function of t. Thus the Theorem holds with $\mathcal{S}(f) \equiv \mathcal{S} \oplus \mathcal{H}_d \oplus \mathcal{H}_{s.c}$, on noting that the discrete and s.c. components of ϕ do not contribute to $\langle \phi, e^{-iHt} f \rangle$. ■

The following corollary generalises part of the statement of the Theorem.

COROLLARY

Let T be an operator of trace class and suppose $f \in \mathcal{S}$ Then

$$\int_{-\infty}^{\infty} |\langle e^{-iHt} f, T e^{-iHt} f \rangle| \, dt < \infty \tag{5}$$

PROOF

Suppose $f \in \mathcal{S}_M$, and use the representation

$$T = \sum_{k=1}^{\infty} c_k |\phi_k\rangle\langle\psi_k| ,$$

where $\sum_{k=1}^{\infty} |c_k| < \infty$ and $\|\phi_k\| = \|\psi_k\| = 1$.

Then

$$\int_{-T}^{T} |\langle e^{-iHt} f, T e^{-iHt} f\rangle| \, dt$$

$$= \lim_{n\to\infty} \int_{-T}^{T} \left| \sum_{k=1}^{n} c_k \langle e^{-iHt} f, \phi_k\rangle\langle\psi_k, e^{-iHt} f\rangle \right| dt$$

$$\leq \lim_{n\to\infty} \left\{ \int_{-T}^{T} dt \sum_{k=1}^{n} |c_k| |\langle e^{-iHt} f, \phi_k\rangle|^2 \right\}^{1/2}$$

$$\times \left\{ \int_{-T}^{T} dt \sum_{k=1}^{n} |c_k| |\langle\psi_k, e^{-iHt} f\rangle|^2 \right\}^{1/2} \qquad (6)$$

Now from the proof of the Theorem we derive the upper bound $M\sqrt{2\pi}$ for the L^2 norm of each of the functions $\langle e^{-iHt} f, \phi_k\rangle$ and $\langle\psi_k, e^{-iHt} f\rangle$. Hence the integral on the l.h.s. of eq(6) is bounded, uniformly in T, by $M\sqrt{2\pi} \sum_{k=1}^{\infty} |c_k|$, and the result follows.

We now have the following characterisation due to Sinha [4] of the a.c. subspace in the case of non-singular potentials. (The generalisation for singular potentials will follow.)

THEOREM 2

Suppose $V(\underline{r})$ is a bounded potential such that $V(\underline{r}) \to 0$ as $|\underline{r}| \to \infty$. Let $H = -\nabla^2 + V$. Denote by $E_{|\underline{r}|<a}$ the projection, in position space, onto the ball $|\underline{r}| < a$, so that, for a given state f, $\|E_{|\underline{r}|<a} f\|^2$ is the probability of finding the particle within the region $|\underline{r}| < a$. Then $\mathcal{H}_{a.c.}$ is the closure of the set of all $f \in \mathcal{H}$ satisfying

$$\int_{-\infty}^{\infty} dt \, \| E_{|\underline{r}|<a} e^{-iHt} f \|^2 < \infty \qquad (\forall a > 0) \qquad (7)$$

PROOF

Remembering that $\mathcal{S}$ is dense in $\mathcal{H}_{a.c.}$, the proof follows easily from two basic results, namely

$$\text{(i)} \quad f \in \mathcal{S} \Rightarrow \int_{-\infty}^{\infty} dt \, \| E_{|\underline{r}|<a} \, e^{-iHt} E_\Delta f \|^2 < \infty ,$$

where $E_\Delta = E_b - E_a$ is the spectral projection of H corresponding to any finite interval $\Delta = (a,b]$,

$$\text{(ii)} \quad \int_{-\infty}^{\infty} dt \, \| E_{|\underline{r}|<a} \, e^{-iHt} f \|^2 dt < \infty \quad \forall a > 0$$
$$\Rightarrow f \in \mathcal{H}_{a.c.}.$$

To prove (i), let ρ be the operator of multiplication, in position space, by a smooth function $\rho(\underline{r})$ satisfying $\rho \equiv 1$ for $|\underline{r}| < a$ and $\rho \equiv 0$ for large $|\underline{r}|$. Then $\rho \, \overline{D}(H) \subseteq D(H_0)$, so that $(H_0 + 1) \rho E_\Delta$ is defined as an operator on $\mathcal{H}$, and hence bounded (by the closed graph theorem.) On the other hand, $E_{|\underline{r}|<a} (H_0 + 1)^{-1}$ is a Hilbert-Schmidt operator. Multiplying these two operators together, we have that $E_{|\underline{r}|<a} E_\Delta$ is Hilbert-Schmidt, and hence that $E_\Delta E_{|\underline{r}|<a} E_\Delta$ is of trace class. Result (i) now follows from the corollary to Theorem 1, on setting $T = E_\Delta E_{|\underline{r}|<a} E_\Delta$. Note that, as Δ is varied, the range of E_Δ is dense in $\mathcal{H}$.

To prove (ii), suppose that f satisfies $\| E_{|\underline{r}|<a} e^{-iHt} f \| \in L^2$. Then $\langle E_{|\underline{r}|<a} \phi, e^{-iHt} f \rangle \in L^2$ for any $\phi \in \mathcal{H}$. Now apply Theorem 1, with $\mathcal{S}(f)$ being the range of $E_{|\underline{r}|<a}$ as a is varied. Then $f \in \mathcal{H}_{a.c.}$ as required. ■

Given an initial state f, the integral in the statement of the Theorem may be interpreted as the total time spent by the particle in the region $|\underline{r}| < a$. Modulo taking closures, we can then interpret $\mathcal{H}_{a.c.}$ as consisting of those initial states for which the time spent by the particle in any finite region is finite. We have exactly the same characterisation even if $V(\underline{r})$ has local singularities, except that the finite region must be located at strictly positive distance from the singularities. (<u>Singularity</u>, in this context, means, roughly, any behaviour locally as singular as $1/|\underline{r}|^2$.) This suggests that states in $\mathcal{H}_{s.c.}$ are in some sense <u>bound states</u>, and we shall be able to confirm this, at least for short range potentials, in section (5).

<u>REMARK</u>

It follows from the above results, and for arbitrary self-adjoint H, that $e^{-iHt}f$ converges weakly to zero for $f \in \mathcal{H}_{a.c.}$ This cannot, however, be used to characterise $\mathcal{H}_{a.c.}$ since it may also happen that $e^{-iHt}f$ converge weakly to zero for some f in $\mathcal{H}_{s.c.}$ There may also be f in $\mathcal{H}_{s.c.}$ such that $e^{-iHt}f$ does <u>not</u> converge weakly to zero. It seems that the only simple characterisation along these lines is a description of $\mathcal{H}_c$ in terms of <u>time-averaged</u> convergence to zero (see [5].) We shall only prove one half of

this result, namely

LEMMA 1

Suppose $f \in \mathcal{H}_c$. Then, for any $\phi \in \mathcal{H}$,

$$\lim_{T \to \infty} \frac{1}{T} \int_{-T}^{T} dt \, | < \phi, e^{-iHt} f > |^2 = 0 .$$

PROOF

Write the integral in the form

$$\frac{1}{T} \int_{-T}^{T} dt < f, e^{+iHt} \phi >< \phi, e^{-iHt} f > .$$

Write $e^{+iHt} = \int e^{i\lambda t} dE_\lambda \; ; \; e^{-iHt} = \int e^{-i\mu t} dE_\mu .$

We have, then, a triple integration in which we may carry out first of all the t integral to give

$$\iint \frac{e^{i(\lambda-\mu)T} - e^{-i(\lambda-\mu)T}}{i(\lambda-\mu)T} d_\lambda < f, E_\lambda \phi > d_\mu < \phi, E_\mu f > .$$

The integrand is bounded uniformly in λ, μ and T. The product measure $d_\lambda \, d_\mu$ is absolutely continuous, and $\lambda = \mu$ has measure 0 in $\mathbb{R} \times \mathbb{R}$. Hence the limit is zero by the Lebesgue dominated convergence theorem. ■

REMARK

By the Schwartz inequality we have

$$\left\{ \int_{-T}^{T} dt \frac{| < \phi, e^{-iHt} f > |}{T} \right\}^2 \leq \frac{2}{T} \int_{-T}^{T} dt \, | < \phi, e^{-iHt} f > |^2$$

$$\leq \frac{2}{T} \int_{-T}^{T} dt \, | < \phi, e^{-iHt} f > |$$ if both ϕ and f are normalised. Hence the power 2 may be dropped in the statement of the Theorem.

2. GENERATION OF SPECTRAL MEASURES

Let us turn now to something rather more concrete and specific. We consider the ordinary differential operator $H = -\frac{d^2}{dr^2} + V(r)$ on $[0,\infty)$ or on some subinterval of $[0,\infty)$. To keep things simple we shall suppose that V(r) is locally bounded, even at r = 0, but V is allowed to behave badly either at $r = \infty$ or at the right hand point of the subinterval. Denote by $\psi(r;\lambda)$ the solution of the time independent Schrödinger equation at energy λ, namely

$$-\frac{d^2\psi}{dr^2} + V\psi = \lambda\psi \qquad (r>0) \tag{8}$$

subject to the initial conditions $\psi(0) = 0$, $\psi'(0) = 1$.

To fix ideas, consider first the case of a finite interval $[0,L]$ and suppose that V is bounded at both end points. Let H be the self-adjoint operator $H = -d^2/dr^2 + V(r)$ acting in $L^2(0,L)$ with boundary conditions $\phi(0) = \phi(L) = 0$. Then H has purely discrete spectrum, with non-degenerate eigenvalues $\{\lambda_i\}$.

We may expand arbitrary $f \in L^2(0,L)$ as a generalised Fourier series in terms of (non-normalised) eigenfunctions $\psi(r;\lambda_i)/\|\psi(r;\lambda_i)\|^2$, the norm being in $L^2(0,L)$. Denoting the coefficients in this expansion by $g(\lambda_i)$, we then have

$$\left.\begin{aligned} f &= \sum_i g(\lambda_i)\psi(r;\lambda_i)/\|\psi(r,\lambda_i)\|^2 \\ g(\lambda_i) &= \int_0^L f(r)\psi(r;\lambda_i)\,dr \\ \int_0^L dr\,|f(r)|^2 &= \sum_i |g(\lambda_i)|^2/\|\psi(r;\lambda_i)\|^2 \end{aligned}\right\} \tag{9}$$

The spectral function ρ for the differential operator acting in $L^2(0,L)$ is defined (up to an additive constant) to be constant between any pair of adjacent eigenvalues, to be continuous from the right at each eigenvalue, and at each eigenvalue to have a jump discontinuity given by

$$[\operatorname{disc}\rho(\lambda)]_{\lambda=\lambda_i} = 1/\|\psi(r;\lambda_i)\|^2 \tag{10}$$

Eqs. (9) may be rewritten in terms of the spectral function as follows.

$$f(r) = \int_{-\infty}^{\infty} g(\lambda)\psi(r;\lambda)\,d\rho(\lambda)$$

$$\left.\begin{aligned} g(\lambda) &= \int_0^L f(r)\,\psi(r;\lambda)\,dr \\ \int_0^L |f(r)|^2 dr &= \int_{-\infty}^{\infty} |g(\lambda)|^2 d\rho(\lambda) \end{aligned}\right\} \quad (11)$$

Eqs. (11), which have been expressed in a form more readily generalisable to cases where H may have continuous spectrum, define a one-to-one correspondence between elements f of $L^2(0, L)$ and elements g of $L^2(\mathbb{R}; d\rho)$. This correspondence is both onto and norm-preserving. It is apparent, on operating by H on the equation for g(r), that H is equivalent to multiplication of g by λ. In other words, H is unitarily equivalent to multiplication by λ in the Hilbert space $L^2(\mathbb{R}; d\rho)$. A knowledge of the spectral function is therefore sufficient for a complete spectral analysis of H.

The spectral measure μ of the differential operator is simply the Lebesque-Stieltjes measure $d\rho$. Thus the measure of an interval $(a, b]$ is just $\rho(b) - \rho(a)$.

Now we want to consider the case of $H = -d^2/dr^2 + V(r)$ acting in $L^2(0, r_\infty)$, where V(r) is bounded in any subinterval $[0, r_n)$ with $r_n < r_\infty$, but may behave badly as r approaches r_∞. To this end, consider an increasing sequence of points $\{r_n\}$ (n = 1, 2, ..) such that $r_n > 0$ and $\lim_{n\to\infty} r_n = r_\infty$. (We allow the possibility $r_\infty = \infty$). Let H be the self-adjoint operator $-d^2/dr^2 + V$ acting in $L^2(0, r_\infty)$, with boundary condition $\phi(0) = 0$. (We shall consider only the case where r_∞ is the limit point ([6]), in which case no boundary condition is needed at the right hand endpoint. If r_∞ is limit circle we should arrive at a spectral matrix rather than a spectral function.) Let $H_n = -d^2/dr^2 + V$ acting in $L^2(0, r_n)$ with boundary conditions $\phi(0) = 0, \phi(r_n) = 0$. Let $d\rho_n$ be the corresponding spectral measure μ_n. We refer the reader to [6] for the proof of the following Lemma.

LEMMA 2

H has a non-decreasing spectral function ρ such that eqs. (11) hold, with L replaced by r_∞, and define a unitary correspondence between $L^2(0, r_\infty)$ and $L^2(\mathbb{R}, d\rho)$, in which H is unitarily equivalent to multiplication by λ in the second space. Moreover, ρ is the limit of ρ_n as $n\to\infty$, in the sense that $\rho(b) - \rho(a) = \lim_{n\to\infty} \cdot (\rho_n(b) - \rho_n(a))$ provided a, b are not points of discontinuity of ρ.

REMARK

The discontinuities of ρ are, as before, at the eigenvalues of H, and ρ is strictly increasing at points of the continuous spectrum.

If we are to use Lemma 2 to study the spectral function of H, we need to consider the behaviour of the solution $\psi(r;\lambda)$ of eq.(8) at $r=r_n$ for increasing values of n. Let us define, for $\lambda>0$, functions $R(r,\lambda)$, $\theta(r,\lambda)$ and $I(r,\lambda)$ by

$$\left.\begin{aligned}\psi &= \lambda^{-1/2} R\cos\theta \\ \psi' &= R\sin\theta\end{aligned}\right\} \tag{12}$$

and

$$I = \int_0^r \psi^2\,dr \tag{13}$$

In terms of ψ and ψ', this gives

$$\left.\begin{aligned}R^2 &= \psi'^2 + \lambda\psi^2 \\ \theta &= \tan^{-1}(\psi'/\sqrt{\lambda}\,\psi)\end{aligned}\right\} \tag{14}$$

We define R to be positive. Then R depends continuously on r in the interval $0\le r<r_\infty$. We take $\theta(0,\lambda)=\pi/2$; θ may also be defined as a continuous function of r in the same interval, and both functions are continuous in λ for $\lambda\geqslant 0$. Here R provides an estimate of the amplitude of the solution ψ and of its derivative, θ describes the phase and I determines the norm in the Hilbert space $L^2(0,r)$. It should be noted that there are alternative possible definitions of the functions R and θ; the choice employed here is particularly appropriate in the context of potentials which approximate to zero over most of the real line.

Accepting the principle that it is simpler to differentiate than to integrate, the function $I(r,\lambda)$ may be expressed in terms of ψ and ψ' as follows. Standard arguments with Wronskians lead to the formula

$$\psi(r;\lambda_1)\psi'(r;\lambda_2) - \psi(r;\lambda_2)\psi'(r;\lambda_1) = (\lambda_1-\lambda_2)\int_0^r \psi(r;\lambda_1)\psi(r;\lambda_2)\,dr\ .$$

Dividing by $(\lambda_1-\lambda_2)$ and letting λ_1 approach λ_2, we have

$$I(r,\lambda) = \int_0^r \psi^2(r;\lambda)\,dr = \psi'\frac{\partial\psi}{\partial\lambda} - \psi\frac{\partial\psi'}{\partial\lambda} \tag{15}$$

The distribution of eigenvalues of H_n depends on the variation of $\theta(r,\lambda)$ with λ at $r=r_n$. A simple calculation gives

$$\frac{\partial\theta}{\partial\lambda} = \frac{\sqrt{\lambda}\left(\psi\frac{\partial\psi'}{\partial\lambda} - \psi'\frac{\partial\psi}{\partial\lambda} - \psi\psi'/2\sqrt{\lambda}\right)}{\psi'^2+\lambda\psi^2}$$

which, expressed through (12) and (15) in terms of R, Θ and I, becomes

$$\frac{\partial\theta}{\partial\lambda} + (2\lambda)^{-1}\sin\theta\cos\theta = \frac{-\sqrt{\lambda}\,I(\lambda)}{R^2(\lambda)} \tag{16}$$

Since we are interested mainly in what happens at the points $r = r_n$ (and incidentally, in the subsequent applications of these ideas, the sequence of points will need to be choses with great care), let us define

$$\left.\begin{aligned} R_n(\lambda) &= R(r_n, \lambda) \\ \theta_n(\lambda) &= \theta(r_n, \lambda) \\ I_n(\lambda) &= I(r_n, \lambda) \end{aligned}\right\} \tag{17}$$

Eq. (16) then becomes

$$\frac{d\theta_n}{d\lambda} + (2\lambda)^{-1}\sin\theta_n\cos\theta_n = -\sqrt{\lambda}\,I_n/R_n^2 . \tag{18}$$

The following Lemma allows us to express, under certain conditions, the spectral function ρ in terms of the $R_n(\lambda)$ for large values of n.

LEMMA 3

Suppose that, as $n \to \infty$, we have the following uniform estimate in λ, for λ in some closed interval $[a, b]$ not containing the origin:

$$I_n(\lambda) = p(\lambda)\,F(n) + o(F(n)) \tag{19}$$

where $p(\lambda)$ is continuous and strictly positive, and F(n) is monotonic increasing without bound.

Then the spectral measure μ of H is given, for subintervals Δ of $[a, b]$ by

$$\mu(\Delta) = \lim_{n\to\infty} \int_\Delta \frac{\sqrt{\lambda}}{\pi[R_n(\lambda)]^2}\,d\lambda \tag{20}$$

PROOF

We have $\mu(\Delta) = \lim_{n\to\infty} \mu_n(\Delta)$, where μ_n is the spectral measure of H_n. Let $\lambda_1, \lambda_2, \ldots\ldots, \lambda_k$, for some fixed value of n, be the eigenvalues of H_n (in increasing order) which lie within the interval Δ, and let λ_{k+1} be the following eigenvalue. Then

$$\mu_n(\Delta) = \sum_{j=1}^{n} 1/I_n(\lambda_j) . \tag{21}$$

Define a step function $\tilde{I}_n(\lambda)$ by

$$\tilde{I}_n(\lambda) = I_n(\lambda_j) \qquad (\lambda_j \le \lambda < \lambda_{j+1}) \tag{22}$$

It is a consequence of classical Sturm-Liouville theory that the values of θ_n at consecutive eigenvalues are related by the equation

$$\theta_n(\lambda_{j+1}) - \theta_n(\lambda_j) = -\pi \tag{23}$$

Hence $(I_n(\lambda_j))^{-1} = -(1/\pi)\int_{\lambda_j}^{\lambda_{j+1}} (\tilde{I}_n(\lambda))^{-1}\theta_n'(\lambda)d\lambda$, so that (21), together with (18), imply that

$$\mu_n(\Delta) = \frac{1}{\pi}\int_{\lambda_1}^{\lambda_{k+1}} \frac{(2\lambda)^{-1}\sin\theta_n\cos\theta_n}{\tilde{I}_n(\lambda)}\,d\lambda$$

$$+ \frac{1}{\pi}\int_{\lambda_1}^{\lambda_{k+1}} \frac{\sqrt{\lambda}\, I_n(\lambda)}{\tilde{I}_n(\lambda)}\left[\frac{1}{R_n(\lambda)}\right]^2 d\lambda \tag{24}$$

The first integral on the r.h.s. vanishes in the limit as $n \to \infty$, and we have

$$\mu(\Delta) = \lim_{n\to\infty} \frac{1}{\pi}\int_{\lambda_1}^{\lambda_{k+1}} \frac{\sqrt{\lambda}\, I_n(\lambda)}{\tilde{I}_n(\lambda)}\left[\frac{1}{R_n(\lambda)}\right]^2 d\lambda \tag{25}$$

Notice that $I_n(\lambda) \to \infty$ implies that, on the r.h.s., the integral taken between any consecutive pair of eigenvalues is vanishingly small in the limit. This means that we can replace λ_1 and λ_{k+1} as integration limits by the lower and upper endpoints of the interval Δ. Moreover, for any $d > 0$, the total integral taken between all consecutive pairs λ_j, λ_{j+1} such that $(\lambda_{j+1} - \lambda_j) \geq d$ is also vanishingly small in the limit. So we can suppose without loss of generality that consecutive eigenvalues of H_n are close together, which with (19) and the continuity of $\rho(\lambda)$ implies that $I_n(\lambda)/\tilde{I}_n(\lambda)$ converges uniformly to 1. Eq. (20) now follows immediately from (25). ■

It may be worth while, before proceeding to more esoteric applications, to review the implications of Lemma 3 in a more familiar context, that of an L^1 potential on the infinite interval $[0, \infty)$. In that case $r_\infty = \infty$ and, as we shall see, the sequence $\{r_n\}$ may be chosen arbitrarily (assuming $r_n \to \infty$).

From (14) we have, on substituting from (12),

$$\frac{\partial}{\partial r} R^2 = 2\psi'(\psi'' + \lambda\psi) = 2\lambda^{-1/2}\, V R^2 \sin\theta\cos\theta$$

so that
$$\frac{\partial}{\partial r}\log R^2 = 2\lambda^{-1/2}\, V(r)\sin\theta\cos\theta \tag{26}$$

Integrating with respect to r, it follows that log R^2 converges, uniformly in our interval Δ, to a limit as $r \to \infty$ In particular, we can define

$$R(\lambda) = \lim_{n\to\infty} R_n(\lambda) \tag{27}$$

with convergence uniform in Δ.

Similarly, the variation of θ with r is given by

$$\frac{\partial\theta}{\partial r} = -\sqrt{\lambda} + \lambda^{-1/2} V \cos^2\theta \tag{28}$$

so that $\theta + \sqrt{\lambda}\, r$ converges uniformly to a limit δ. Thus we have, for large r, the asymptotic behaviour

$$\psi(r;\lambda) \sim \lambda^{-1/2} R(\lambda) \cos(\sqrt{\lambda}\, r - \delta(\lambda)).$$

The estimate (19) for $I_n(\lambda)$ with

$$\rho(\lambda) = (2\lambda)^{-1}[R(\lambda)]^2, \qquad F(n) = r_n, \text{ follows.}$$

Moreover, in this case (20) becomes

$$\mu(\Delta) = \int_\Delta \frac{\sqrt{\lambda}}{\pi[R(\lambda)]^2}\, d\lambda,$$

so that the spectral measure is absolutely continuous for positive energies. In this instance it may be convenient to make the change of variable $\lambda = k^2$, $d\rho = \frac{2}{\pi} k^2 dk/R^2$, to redefine $\tilde{\psi} = \sqrt{\frac{2}{\pi}}\, k\psi/R$ as the solution (8) and to modify the transform in (11) by taking

$$g(k) = \int_0^\infty f(r)\, \tilde{\psi}(r;k^2)\, dr,$$

$$f(r) = \int_0^\infty g(k)\, \tilde{\psi}(r;k^2)\, dk,$$

in which case

$$\int_0^\infty |f(r)|^2 dr = \int_0^\infty |g(k)|^2 dk.$$

(Actually this assumes f to have positive spectral support with respect to H; if not, contributions from negative energy eigen - functions have to be taken into account.)

L' potentials represent, of course, a very special case of the formalism developed above. In general $R_n(\lambda)$ need not converge pointwise to a limit, nor need the measure given by (20) be absolutely continuous. There is another direction in which the Lemma may be generalised, that of relaxing the conditions (19) on the behaviour of I_n for large n. Indeed, all that is needed in the proof is some estimate allowing the replacement of $I_n/\tilde{I}_n$ by 1 in the limit. This step could be justified, for example, by the assumptions

(i) that the eigenvalues of H_n cluster in the limit $n \to \infty$ (i.e. the maximum distance between consecutive eigenvalues converges to zero).

and (ii) the maximum relative variation of I_n (meaning (max I_n - min I_n)/min I_n), over intervals $[\lambda_j, \lambda_{j+1}]$ between consecutive eigenvalues, converges to zero as $n \to \infty$. In other words, between any adjacent pair of eigenvalues I_n should not oscillate too much.

DEFINITION

An increasing sequence $\{r_n\}$ tending to r_∞ is said to be an asymptotic sequence for the potential V(r) in the interval $[0, r_\infty)$ if the spectral measure μ of $-d^2/dr^2 + V(r)$ (acting in $L^2(0, r_\infty)$ with boundary condition $\phi(0) = 0$) is given, for subintervals Δ of some strictly positive closed interval $[a, b]$, by eq. (20).

Thus Lemma 3, or its generalisation in (i) and (ii) above, provide sufficient conditions for the existence of asymptotic sequences. Here are some cases, with $r_\infty = \infty$, for which asymptotic sequences may be found. In the first two instances (one of which has already been noted) every sequence $\{r_n\}$ tending to infinity is an asymptotic sequence. In the third case the $\{r_n\}$ have to be chosen more carefully.

(1) $V \in L^1(0, \infty)$.

(2) V is bounded and periodic.

(3) Let V be an infinite sequence of bumps as in Diag. 1.

Diagram 1

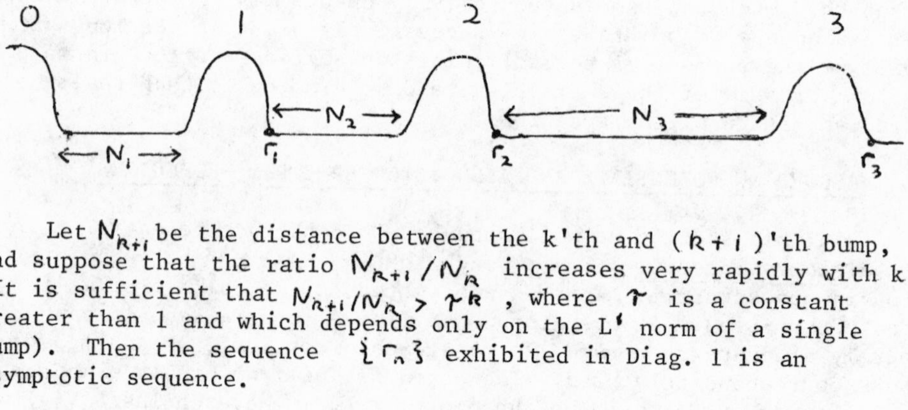

Let N_{k+1} be the distance between the k'th and $(k+1)$'th bump, and suppose that the ratio N_{k+1}/N_k increases very rapidly with k. (It is sufficient that $N_{k+1}/N_k > \tau^k$, where τ is a constant greater than 1 and which depends only on the L^1 norm of a single bump). Then the sequence $\{r_n\}$ exhibited in Diag. 1 is an asymptotic sequence.

We shall omit in this case the detailed estimates that allow us to conclude, with a given rapid increase of the N_k, that $\{r_n\}$ is an asymptotic sequence. The result depends on verifying properties (i) and (ii) above. This requires estimates of (a) the distance between consecutive eigenvalues of H_n (in fact $(\lambda_{j+1} - \lambda_j) = O(1/N_n)$) and (b) the relative variation in I_n between adjacent eigenvalues (this being $O((\lambda_{j+1} - \lambda_j) R_{n-1} \tau^n)$). It is possible, however, to convince oneself of the validity of eq. (20) in this case, without specifying the precise manner in which the $\{r_n\}$ should

increase, by the following line of argument. (Diagram 2)

Diagram 2

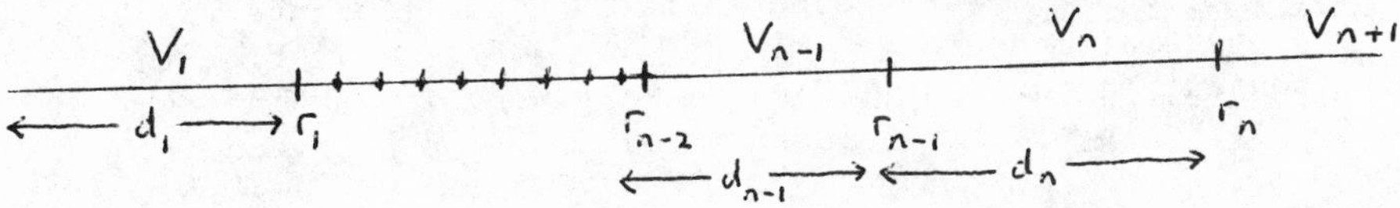

Divide the interval $[0, \infty)$ into an infinite sequence of subintervals of length d_1, d_2, ..., where d_n increases very rapidly and tends to infinity. Suppose the potential V(r) is made by patching together the sequence of L' potentials V_1, V_2, ..., where $V = V_k$ on the interval of length d_k. Having inductively chosen the points of subdivision r_1, r_2, ... r_{n-1}, choose the value of r_n sufficiently large that $I_n / \tilde{I}_n$ is very close to 1, uniformly in Δ. This involves deciding for how great a distance d_n the potential should be the same as V_n. We know this is possible, from the results already obtained for L^1 potentials. Having done this, choose the value of r_{n+1} to make $I_{n+1} / \tilde{I}_{n+1}$ uniformly closer still to 1, and so on. Then $\lim_{n\to\infty} I_n / \tilde{I}_n = 1$, uniformly in Δ, and we have verified eq. (20) in this case.

By extension of this argument, we can deal with potentials which are made up of patched together periodic potentials, or periodic potentials with L' perturbations, etc. The main point is that the potential should successively change its character at each of a sequence of points having successively wider separation. It will be the purpose of the following two sections to look in more detail at some of these examples, and to examine in each case the precise nature of the spectrum.

3. ABSORBING POTENTIALS AND ASSOCIATED SPECTRAL PROPERTIES

In this section we will show how to construct potentials V(r) on a finite interval $[0, r_\infty)$ such that $-d^2/dr^2 + V$ has absolutely continuous spectrum. This is unexpected, since folk lore has it that Sturm-Liouville operators on a finite interval should have purely discrete spectrum. To go against folk lore, we shall have to work correspondingly hard. The potential will be highly singular and oscillating in the neighbourhood of $r = r_\infty$. Nevertheless, V(r) will be bounded by some inverse power of $(r_\infty - r)$.

We first define the basic building block of our potential. For a small and positive, define

$$V_a(r) = \sum_{j=1}^{8} \lambda_j \, \delta(r - ja) \tag{29}$$

where the coefficients of the 8 δ-functions are given respectively by

$$\lambda_1, \lambda_2, \ldots, \lambda_8 = (a^{-3/2}-a^{-1}), (a^{-1/2}-a^{-1}), (a^{-1/2}-a^{-1}), (a^{-3/2}-a^{-1}),$$
$$(a^{-1/2}-a^{-1}), (a^{-3/2}-a^{-1}), (a^{-3/2}-a^{-1}), (a^{-1/2}-a^{-1}).$$

We can construct a 2 x 2 transfer matrix, which transforms $\binom{\psi}{\psi'}$ at r = 0 into $\binom{\psi}{\psi'}$ at r = 8a, for solutions ψ of the Schrödinger equation $-\psi'' + V_a\psi = \lambda\psi$, at energy $\lambda = k^2$. (We adopt the convention

$$\binom{\psi}{\psi'}_{r=ja} = \lim_{\varepsilon\to 0+} \binom{\psi}{\psi'}_{r=ja+\varepsilon}.)$$

For example, the transfer matrix from r = 0 to r = 2a is just

$$\begin{pmatrix} \cos ka, & k^{-1}\sin ka \\ \lambda_2\cos ka - k\sin ka, & \lambda_2 k^{-1}\sin ka + \cos ka \end{pmatrix}\begin{pmatrix} \cos ka, & k^{-1}\sin ka \\ \lambda_1\cos ka - k\sin ka, & \lambda_1 k^{-1}\sin ka + \cos ka \end{pmatrix}$$

which works out, for small a, as

$$\begin{pmatrix} a^{-1/2} + O(a^{3/2}) & a^{1/2} + O(a) \\ -\frac{1}{3}k^2 a^{1/2} + O(a) & a^{1/2} + O(a^{3/2}) \end{pmatrix}$$

The transfer matrix from r = 2a to r = 4a is just the transposed of this. After a lot of matrix multiplication we end up with the transfer matrix from r = 0 to r = 8a, which has the form

$$\begin{pmatrix} 1 - \frac{5}{3}k^2 + O(a^{1/2}) & 1 + O(a^{1/2}) \\ -\frac{5}{3}k^2 + O(a^{1/2}) & 1 + O(a) \end{pmatrix}$$

so we define

$$M(k^2) = \begin{pmatrix} 1 - \frac{5}{3}k^2 & 1 \\ -\frac{5}{3}k^2 & 1 \end{pmatrix} \tag{30}$$

The transfer matrix thus approximates to $M(k^2)$ for small a. A sequence $\{a_n\}$ of positive numbers is chosen, rapidly converging to zero, and the interval $[0, r_\infty)$ is divided into consecutive subintervals of length $8a_n$ (n = 1, 2, 3, ...), where $r_\infty = 8\sum_1^\infty a_j$. The points of subdivision are given by $r_n = 8\sum_1^n a_j$, and the potential V(r) is defined by

$$V(r) = V_{a_n}(r - r_{n-1}) \qquad (r_{n-1} < r \le r_n) \tag{31}$$

Thus V is made up of a sequence of basic units V_a, defined over smaller and smaller subintervals of length 8a. Since the a's are rapidly decreasing, the coefficients λ_j will get larger and larger, both positive and negative, so that V is wildly oscillating as r approaches r_∞.

Now, for solutions of eq. (8), the transfer matrix from 0 to r_n is a product of n transfer matrices, each of which approximates, as a_j decreases, to the matrix $M(k^2)$ defined by eq. (30). So we expect the transfer matrix from 0 to r_n to behave, for large n, something like $(M(k^2))^n$. The most convenient way of quantifying

this asymptotic behaviour is to define a sort of discrete analogue of a wave operator, namely

$$W(k^2) = \lim_{n\to\infty} [M(k^2)]^{-n} M(k^2; (r=0) \to (r=r_n))$$

This limit may be shown to exist, provided the a_n tend to zero sufficiently rapidly in a sense which we shall leave here unspecified, and to define a 2 x 2 matrix depending continuously on energy. These and other details will be found in [7].

The matrix $M(k^2)$, in the interval $0 < k^2 < 12/5$, has complex eigenvalues $\exp(\pm i\alpha(k^2))$, in terms of which we can calculate powers of $M(k^2)$, to give

$$\sin\alpha\, [M(k^2)]^n = \begin{pmatrix} (1-\frac{5}{3}k^2)\sin n\alpha - \sin(n-1)\alpha & \sin n\alpha \\ -\frac{5}{3}k^2 \sin n\alpha & \sin n\alpha - \sin(n-1)\alpha \end{pmatrix}$$

This immediately determines the asymptotic behaviour, near r_∞, of ψ and ψ', since

$$\begin{pmatrix} \psi \\ \psi' \end{pmatrix}_{r=r_n} \sim [M(k^2)]^n W(k^2) \begin{pmatrix} 0 \\ 1 \end{pmatrix}$$

The asymptotic behaviour of I_n may now be determined, for example by application of eq. (15), to give

$$\lim_{n\to\infty} \frac{1}{n} I_n(\lambda) = \frac{5}{36\sin^2\alpha} \left\{ \left[W(\lambda)\begin{pmatrix}0\\1\end{pmatrix}\right]^T \begin{pmatrix} 10\lambda, & -5\lambda \\ -5\lambda, & 6 \end{pmatrix} \left[W(\lambda)\begin{pmatrix}0\\1\end{pmatrix}\right] \right\} \tag{32}$$

This allows us to apply Lemma 3 by setting, in eq. (19), $F(n) = n$, and $p(\lambda)$ equal to the r.h.s. of (32). All of the conditions of the Lemma are met, so that the spectral measure of $H = -d^2/dr^2 + V$ in $L^2(0, r_\infty)$, with boundary condition $\phi(0) = 0$, is given for closed subintervals Δ of $(0, 12/5)$ by eq. (20). Although $R_n(\lambda)$ does not converge pointwise to a limit as $n \to \infty$, it follows from the asymptotic formula for ψ, ψ' at $r = r_n$, together with the above expression for the n'th power of $M(k^2)$, that $R_n(\lambda)$ is bounded both above and below, uniformly in both n and λ for $\lambda \in \Delta$. Hence the spectral measure given by eq. (20) in this case is absolutely continuous in the interval $(0, 12/5)$. The density function $d\rho/d\lambda$ is given by the limit of $\sqrt{\lambda}/\pi[R_n(\lambda)]^2$, in the sense of distributions. The spectral properties of H outside the interval $(0, 12/5)$ do not concern us here, but a detailed analysis shows that outside this interval the spectrum is purely discrete. Moreover, this discrete spectrum is bounded from below, so that H is semi-bounded. This in itself is measure of the precision with which the potential must be constructed in order to manifest a.c. spectrum. For it may be shown (see [8] for example) that for semi-bounded Hamiltonians of this type the spectrum of $-d^2/dr^2 + gV$ must be purely discrete for all values of g in the interval $0 \le g < 1$. In other words a change in the coupling constant, however small, may lead to complete disappearance of the a.c. spectrum! Thus a.c. spectrum coming from

a differential operator in a finite interval tends to be a rather unstable feature. It should be stressed that the singular nature of the spectrum is nothing to do with the fact that we have used δ-functions rather than a potential which is locally bounded away from r_∞. It would be perfectly possible to replace the δ's by appropriate δ-approximating functions, and indeed this is carried out in [7] and the modified estimates made. So the same general features described here apply even to examples of potentials which are C^∞ away from the singularity. Such potentials may, as already described, be bounded by inverse powers of $r_\infty - r$.

What of the physical consequences of the spectral properties in this example? Let $\varphi(r)$ be a smooth function, defined on the interval $[0, r_\infty)$, such that $\varphi(r) \equiv 1$ near $r = 0$ and $\varphi(r) \equiv 0$ near $r = r_\infty$. An application of an argument used already in the proof of Theorem 2 implies that $\varphi E_\Delta(H)$ is compact (even Hilbert-Schmidt). Hence also $E_{r<c} E_\Delta(H)$ is compact for $0<c<r_\infty$. Now we have seen, at the end of Section 1, that, for $f \in \mathcal{H}_{a.c.}$, $e^{-iHt}f$ converges weakly to zero. By a well known consequence of compactness, $E_{r<c}\, e^{-iHt} E_\Delta(H)f$ converges strongly to zero as $t \to \pm\infty$. In other words, for initial states in $\mathcal{H}_{a.c.} \cap \text{range}\,(E_\Delta(H))$, and hence for all initial states in $\mathcal{H}_{a.c.}$, the probability of finding the particle near $r = 0$ tends to zero as $|t| \to \infty$, and the particle approaches the singularity $r = r_\infty$, in the sense that the probability of finding the particle near r_∞ converges to 1. We shall say that such a state is asymptotically absorbed.

We can translate this phenomenon into its proper quantum-mechanical setting of a particle moving in 3-dimensions as follows. Define

$$\begin{aligned} V_1(|\underline{r}|) &= V(r_\infty - |\underline{r}|) \qquad (0 < |\underline{r}| \le r_\infty) \\ &= 0 \qquad\qquad\qquad (|\underline{r}| > r_\infty) \end{aligned}$$

In the partial wave subspace $l = m = 0$, $H_1 \equiv -\Delta + V_1$ is unitarily equivalent to the ordinary differential operator

$$-\frac{d^2}{dr^2} + V_1(r) \quad \text{in} \quad L^2(0, \infty).$$

This operator has absolutely continuous spectrum in the interval $[0, \infty)$ which is doubly degenerate for $0 < \lambda < 12/5$. The interpretation of this degeneracy is that there exist states in $\mathcal{H}_{a.c.}$ which are asymptotically absorbed into the origin as $t \to +\infty$, and states which are absorbed as $t \to -\infty$. In general, an incoming state from $|\underline{r}| = \infty$ will have non-zero probability of absorption into the singularity. Although the usual wave operators $\Omega_\pm(H, H_0)$ exist and are isometries, asymptotic completeness will not hold, and the scattering operator will not be unitary. All of this comes from a local singularity - in fact the potential here is even of finite range. (It is perhaps of interest, in view of this somewhat bizarre behaviour, to recollect that the energy spectrum is bounded from

below and to note that there is a finite total scattering cross-section!) Notice that in this case the states absorbed at $t = +\infty$ are not identical to those absorbed at $t = -\infty$, due to coupling between local behaviour and behaviour at $|\underline{r}| = \infty$

4. SINGULAR CONTINUOUS SPECTRUM

Consider again the potential of Example 3 at the end of section 2, consisting of an infinite sequence of bumps of increasingly wide separation. Part of the potential is represented schematically in the diagram below.

We will take $r_n = b_n$, and assume that N_{n+1}/N_n increases sufficiently rapidly that we have, repeating eq (20),

$$\mu(\Delta) = \lim_{n\to\infty} \int_\Delta \frac{\sqrt{\lambda}}{\pi [R_n(\lambda)]^2} d\lambda \tag{33}$$

Probably the reader will have realised by now that these potentials provide the canonical examples of singular continuous spectrum. Where does this singular spectrum come from? We first of all attempt to answer this question in mathematical terms. Later we shall look more closely at the physical implications.

From eqs (26) and (28) we see that, in any of the intervals $[b_{n-1}, a_n]$ in which $V(r) \equiv 0$, the function $R(r,\lambda)$ remains independent of r, whereas the Phase $\theta(r,\lambda)$ decreases linearly with distance. In particular, we have

$$\left.\begin{aligned} R(a_n,\lambda) &= R(b_{n-1},\lambda) = R_{n-1}(\lambda) \\ \theta(a_n,\lambda) &= \theta_{n-1}(\lambda) - N_n\sqrt{\lambda} \end{aligned}\right\} \tag{34}$$

On the other hand, we see from (26) that $R(r,\lambda)$ changes across each bump, for example

Diagram 3

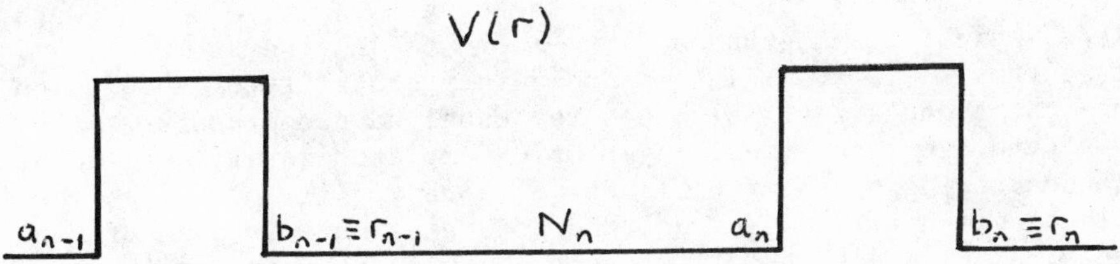

$$\left[\frac{R(b_{n-1},\lambda)}{R(a_{n-1},\lambda)}\right]^2 = \exp\left\{2\lambda^{-1/2}\int_{a_{n-1}}^{b_{n-1}} dr\, V(r)\sin\theta(r,\lambda)\cos\theta(r,\lambda)\right\}. \tag{35}$$

For a given range of energies, there will be some constant $\gamma > 1$ (depending on the L^1 norm of V across a single bump) so that R cannot change by more than a multiplicative constant γ across a single bump. Similarly, if R decreases across a bump, the value at b_{n-1} cannot be smaller than γ^{-1} times the value at a_{n-1}. (This comes from the change of variables $r \leftrightarrow -r$.) So, across a single bump, $R(r,\lambda)$ may increase or decrease, but by not too much in either case. Whether we have increase or decrease of R depends on the sign of the integral within the exponential on the r.h.s. of eq (35). This in turn depends critically on the phase $\theta(r,\lambda)$ across the bump in question. By the time we arrive at the following bump $[a_n, b_n]$ this phase, because of the large factor N_n in the second of eqs(34), will be varying extremely with energy λ. In fact we can think of $\sin 2\theta(r,\lambda)$ in that case as very much analogous to $\sin 2\Theta$ where Θ is a random variable, uniformly distributed in the interval $[0,\pi]$. This implies that, in eq (33) as n increases, $R_n(\lambda)$ will vary irregularly with λ and n, much as though, for fixed λ and as n is increased, the value of the function is obtained through some random process.

It turns out that $R_n(\lambda)$ tends to infinity <u>in measure</u>. (This means that, for any given $N > 0$, the Lebesgue measure of the set of points λ, such that $R_n(\lambda) < N$, converges to zero.) However, for given n, there will always be a set of points (having small Lebesgue measure), such that $R_n(\lambda)$ is very small. These will be the points for which $R(r,\lambda)$ has <u>decreased</u>, at successive bumps of the sequence, an inordinately large number of times. Though relatively insignificant as far as Lebesgue measure is concerned, these points are those on which, in the limit $n \to \infty$, the μ-measure is concentrated. We have, in fact, that $R_n(\lambda)$ actually converges to <u>zero</u> in μ-measure. This, then, is how singular continuous measure is generated. (The fact that we get continuous rather than discrete measure comes from the <u>uniform</u> convergence of μ_n to μ over subintervals of a given closed interval not containing $\lambda = 0$.) We get a set of points having Lebesgue measure zero, on which the μ-measure is concentrated.

A simple example illustrating some of these ideas is as follows. Let $\{N_j\}$ be a sequence of positive numbers increasing very rapidly to infinity, and define a measure μ by

$$\mu(\Delta) = \lim_{n\to\infty} \int_\Delta d\lambda \prod_{j=1}^{n} \left(1 + \tfrac{1}{2}\sin N_j \lambda\right) \tag{36}$$

for subintervals Δ of a given finite fixed interval. Then μ extends to a singular continuous measure on Borel subsets of the given interval. The intuition behind this example is that, if $\{\theta_j\}$ is a sequence of independent random variables, randomly distributed over the interval $(0,2\pi)$, then with probability 1 we have

$$\prod_{j=1}^{\infty} (1 + \tfrac{1}{2} \sin \theta_j) = 0$$

This comes from the law of large numbers in probability theory, since the average value of $\log(1 + \frac{1}{2}\sin\theta)$ is negative, and the infinite product corresponds to an infinite series of logarithms. Hence, again in (36), the integrand converges almost everywhere to zero, and the μ-measure is concentrated on the complement of this set of points. The following Theorem generalises this example. The proof is to be found in [8].

THEOREM 3

Let $\{f_n(k,y)\}$ $(\alpha \le k \le \beta, -\infty < y < \infty, n = 1, 2, \ldots)$ be periodic in y with period c, continuously differentiable, and satisfy

i) $f_n(k,y) \geq \text{const} > 0$,

ii) $\bar{f}_n(k) \equiv \frac{1}{c}\int_0^c f_n(k,y)\,dy = 1$,

iii) $\sum_{n=1}^{\infty} -m_n(k) = \infty \quad , \quad \alpha \le k \le \beta$,

where

$$m_n(k) = \overline{\log(f_n(k,y))} = \frac{1}{c}\int_0^c \log f_n(k,y)\,dy ,$$

iv) For N sufficiently large, $f_n(k, Nk)$ is an analytic function of k, $\alpha \le k \le \beta$.

Given a sequence $\{N_i\}$ (i = 1, 2, 3, ...) of increasing positive numbers with $\lim_{i\to\infty} N_i = \infty$, define the measures μ_n by

$$\mu_n(\Delta) = \int_\Delta dk \prod_{i=1}^{n} f_i(k, N_i k) ,$$

for subintervals Δ of $[\alpha, \beta]$.

Then the sequence $\{N_i\}$ may be chosen such that $\lim_{n\to\infty} \mu_n(\Delta) = \mu(\Delta)$ exists for each subinterval of $[\alpha, \beta]$ and defines a singular continuous measure on Borel subsets of $[\alpha, \beta]$.

How is this result applied to the measure μ defined by (33)? (Actually a slight generalisation, given in [8] as a Corollary to the main Theorem, is needed.) Comparing with eq (33), we see that the role of the f_i in this case is taken by $[R_{i-1}/R_i]^2$. The change of integration variable to k $(\lambda = k^2)$ is inessential, but a useful notational convenience.

We need, then, to calculate R_n/R_{n-1}. This can be done by solving the Schrodinger equation between the points r_{n-1} and r_n (see Diagram 3). From eq (34), using also eq (12) for ψ and ψ' we have, at $r = a_n$,

$$\left.\begin{aligned} \psi &= k^{-1} R_{n-1}(k^2) \cos(\theta_{n-1} - N_n k) \\ \psi' &= R_{n-1}(k^2) \sin(\theta_{n-1} - N_n k) \end{aligned}\right\} \tag{37}$$

Let M denote the transfer matrix across a single bump. Then we can use M, with (37), to calculate ψ and ψ' at $r = r_n$. Having done so, we need only evaluate $\psi'^2 + k^2\psi^2$ at b_n to determine the value of R_n^2. This calculation leads to

$$\left(\frac{R_n}{R_{n-1}}\right)^2 = \frac{1}{f_n(k, N_n k, \theta_{n-1}(k^2))}$$ where, setting

$y = N_n k$, we have

$$[f_n(k, y, \theta)]^{-1} = A(k) + B(k)\cos 2(\theta - y) + C(k)\sin 2(\theta - y)$$

with

$$\left.\begin{aligned} A &= \tfrac{1}{2}(M_{11}^2 + M_{22}^2 + k^{-2}M_{21}^2 + k^2 M_{12}^2) \\ B &= \tfrac{1}{2}(M_{11}^2 + k^{-2}M_{21}^2 - M_{22}^2 - k^2 M_{12}^2) \\ C &= (k\, M_{11}M_{12} - k^{-1}M_{21}M_{22}) \end{aligned}\right\}$$

The average values of f_n over y are derivable from standard integrals. In fact

$$\overline{f_n} = (A^2 - B^2 - C^2)^{-1/2} = (\det M)^{-1} = 1, \quad \text{and}$$

$$\overline{m_n} = \overline{\log f_n} = \log(2/A+1) < 0.$$

(In this application, the only dependence of f_n on n is through the argument $\theta_{n-1}(k^2)$.) Thus all of the conditions (i) to (iv) of Theorem 3 can be met, so that the spectrum of H = $-d^2/dr^2 + V(r)$ in this case is singular continuous along the entire positive real line.

Let us take a while now to consider some possible generalisations of this example. To start with, the shape of individual bumps can be purely arbitrary. In particular, $V(r)$ at the bumps may be either positive or negative, the sign will only affect the nature of the spectrum for negative energies. The bumps do not have to be identical; they can even be of dimishing height g_n, where g_n tends to zero as $n \to \infty$. In that case, condition (iii) of Theorem 3 imposes a constraint on the rate at which the g_n are allowed to decrease if we are to have singular continuous spectrum. We expect this, since for example $\Sigma |g_n| < \infty$ would imply $V \in L^1$, in which case H is known to be spectrally absolutely continuous. In fact, perhaps rather surprisingly, the borderline is at L^2 potentials in that (provided always that the separation distance N_n increases sufficiently rapidly) we have

$$\Sigma |g_n|^2 < \infty \implies \text{spectrum a.c. for } \lambda > 0\text{, but}$$

$$\Sigma |g_n|^2 = \infty \implies \text{spectrum s.c. for } \lambda > 0.$$

This result gives a clue to the physical intepretation of singular continuous spectrum. Provided $g_n \to 0$, we are in a position to apply Theorem 2 of Section 1. For initial states in $\mathcal{H}_{s.c.}$, the total time spent in any finite region will be infinite. To look at the problem in a different way, consider a particle moving to the right (in Diagram 3) and encountering a sequence of obstacles or bumps. At each successive obstacle there will be probability p_n of transmission and q_n of reflection. In fact, if we are prepared to discount interference effects and treat each obstacle separately, we can carry out the quantum mechanical calculation of q_n and p_n. Let us compare this quantum phenomenon with the classical process of a particle encountering a series of obstacles, again with probabilities p_n and q_n of transmission/reflection. The process is then said to be recurrent if the particle returns, with probability 1, to the origin. And the condition for recurrence in the classical problem, namely $\Sigma q_n/p_n = \infty$, corresponds exactly, for the potentials considered here, to the condition $\Sigma |g_n|^2 = \infty$ for singular continuous spectrum. This result is in striking confirmation of the interpretation of Theorem 2 in this case.

Further insight into the nature of singular continuous spectrum may be derived from a consideration of the Fourier transform of the spectral measure with respect to k. Returning to the potential sketched in Diagram 3, with identical bumps, detailed estimates

show that $\int e^{-ikr_n} d\mu$ does not converge to zero as $n \to \infty$. Since $k = \sqrt{\lambda}$, this implies that $e^{-i\sqrt{H}t}$ does not converge weakly to zero, as would be the case if the spectrum of H (and therefore of $\sqrt{H}$) were purely absolutely continuous. It is expected, however, though as yet unproved, that due to "spreading of the wave-packet", e^{-iHt} will converge weakly to zero, at least for the class of Hamiltonians considered here. For a description, in terms of wave packets, of the evolution of states in $\mathcal{H}_{s.c.}$, see [8]. Let us conclude this section with a brief mention of another line of development, viz. the generation of singular continuous spectrum via local singularities. Examples of short range (even finite range) potentials such that $-d^2/dr^2 + V$ has singular continuous spectrum may be constructed effectively by combining the main example of Section 3 with the main example of the present section.

Define V(r) to be the sum of the potential defined by eq (31) and the potential $\sum_{i=1}^{\infty} \delta(r - r_{n_i})$, where $n_i = \sum_{j=1}^{i} N_j$ and the $\{N_j\}$ are rapidly increasing and tend to infinity as $j \to \infty$. In other words, place an addition δ-function, for each i, at the n_i'th point of the asymptotic sequence $\{r_n\}$ for the original singular potential. Then N_i, being a measure of the difference between the locations of consecutive δ's, will play a precisely analogous role to the N_i in Theorem 3 and in the examples of the present section. The same arguments as were employed at the end of section 2 show that $\{r_{n_i}\}$ forms an asymptotic sequence of the modified potential, and one cango on to show that $-d^2/dr^2 + V$, acting in $L^2(0, r_\infty)$, has singular continuous spectrum in this case for $0 < \lambda < 12/5$. (In order to compute the F_n of Theorem 3, it is technically simpler here to modify the definition of R_n and θ_n, in which case Proposition (3) of [7] may be applied directly.) We assume always that the N_i increase sufficiently rapidly as $i \to \infty$

To define a Schrödinger operator in $L^2(\mathbb{R}^3)$, now set

$$V_1(|\underline{r}|) = V(r_\infty - |\underline{r}|) - 1, \quad 0 < |\underline{r}| \leq r_\infty$$
$$= 0, \quad |\underline{r}| > r_\infty$$

The subtraction of +1 from the potential in the region $|\underline{r}| \leq r_\infty$ shifts the location of the singular spectrum, with the result that $H_1 = -\nabla^2 + V_1$ has singular continuous spectrum for $-1 < \lambda < 0$. (But not for $\lambda > 0$; the singular spectrum for positive energies gets "swallowed up" by the absolutely continuous component coming from the short-range nature of the potential. See [9].)

5. SPECTRAL ANALYSIS AND ASYMPTOTIC BEHAVIOUR FOR SHORT-RANGE SINGULAR POTENTIALS

I hope in these notes to have shown that some insight into spectral properties and evolution of states in potential scattering

can be gained from the examination of suitable constructed examples. We have met two distinct types of phenomena in these examples, the first (the wave trap) associated with absorption at local singularities (perhaps, as pointed out by V. Enss, more accurately described as adsorption) and the second (singular continuous spectrum) a feature of long-range but non singular potentials which succeed, through multiple transmissions and reflections, in confining a particle locally for inordinately long periods of time. Perhaps, after all, these two phenomena are not really distinct, since we saw at the end of the last chapter that they may occur in combination.

Singular continuous spectrum, unlike local absorption, seems to be a relatively stable phenomenon. (Perhaps because to work it relies on the absence of interference between waves, whereas the wave trap depends crucially on very carefully arranged interference.) At least in the examples presented here, s.c. spectrum manifests itself, independently of the magnitude of coupling constant, as a rather drastic alteration of the nature of the spectrum for potentials subjected at large distances to a series of localised perturbations of rapidly increasing spearation. Mathematically, the limiting measure on the r.h.s. of (33) fails to distribute itself sufficiently uniformly over a given interval, and one might even regard the more familiar situation of absolutely continuous spectrum as a rather restricted special case!

It is time now to set all this within a more general context, in which we attempt to characterise possible types of asymptotic behaviour for large t, and link these with associated spectral properties of H. Knowing that both absorption and s.c. spectrum may occur already for short range potentials, it will be enough to consider this case. So we assume

$$V(\underline{r}) = O(|\underline{r}|^{-1-\varepsilon})$$

for some $\varepsilon > 0$ as $|\underline{r}| \to \infty$

(This is also much the easiest case to treat, though probably Coulomb-like behaviour at infinity could also be handled.)

Take $V(\underline{r})$ to be locally bounded except for a singularity at $\underline{r} = 0$. We place no restriction on the nature of this singularity. (Our treatment could easily be extended to more general singular sets, for example to arbitrary closed, bounded subsets of $\mathbb{R}^3$ having Lebesgue measure zero.) Let ρ be the operator of multiplication, in position space, by some smooth function $\rho(\underline{r})$ satisfying $\rho(\underline{r}) \equiv 0$ near $\underline{r} = 0$ and $\rho(\underline{r}) \equiv 1$ for large $|\underline{r}|$. Rather standard arguments, based on Cook-type estimates, imply the existence of the limits $\underset{t\to\pm\infty}{s\text{-}\lim}\, e^{iHt} \rho\, e^{-iH_0t}$ on the entire Hilbert space $\mathcal{H} = L^2(\mathbb{R}^3)$. Since $(1-\rho)e^{-iH_0t}$ converges strongly to zero, these limits are identical to the usual wave operators

$$\Omega_{\pm} = \underset{t \to \pm\infty}{s\text{-}\lim}\, e^{iHt} e^{-iH_0 t} \tag{38}$$

Let $P_{\pm}$ denote the projection operators onto the positive/negative spectral subspace of the generator of the dilation group. These operators and their properties are discussed in the notes of V. Enss, in which it is shown that, for non-singular potentials and for any smooth function having compact support not containing the origin, the two operators $(\Omega_{\pm} - 1)\phi(H_0) P_{\pm}$ are compact. In the present context, the local singularity at $\underline{r} = 0$ prevents this result from holding. However, the operator ϱ acts as a smooth cut-off near $\underline{r} = 0$, so that one can use the bound

$$\int_0^{\pm\infty} dt \, \| (H\varrho - \varrho H_0)\, \phi(H_0)\, e^{-iH_0 t} P_{\pm} \| < \infty$$

to deduce that in this case it is the operators

$(\Omega_{\pm} - \varrho)\, \phi(H_0)\, P_{\pm}$ which are compact.

The only other property of the projections $P_{\pm}$ which we shall need is that

$$\underset{t \to \pm\infty}{s\text{-}\lim}\; P_{\mp}\, e^{-iH_0 t} = 0 \tag{39}$$

We have the identity, for arbitrary $f \in \mathcal{H}$,

$$\begin{aligned} &\varrho\, e^{-iHt} \phi(H) f \\ &= [\varrho\, \phi(H) - \phi(H_0)\varrho\,]\, e^{-iHt} f \\ &\quad + P_+\, \phi(H_0)(\varrho - \Omega_+^*)\, e^{-iHt} f \\ &\quad + P_-\, \phi(H_0)(\varrho - \Omega_-^*)\, e^{-iHt} f \\ &\quad + P_+\, \phi(H_0)\, e^{-iH_0 t}\, \Omega_+^* f + P_-\, \phi(H_0)\, e^{-iH_0 t}\, \Omega_-^* f. \end{aligned} \tag{40}$$

In order to describe asymptotic behaviour for large t, we shall need the notion of a bound state and of a scattering state (see [5]).

DEFINITION

The vector $f \in \mathcal{H}$ is said to be a bound state, as $t \to \infty$, if, given any $\varepsilon > 0$, there exists $a > 0$ (depending on ε), such that

$$\| E_{|\underline{r}| > a}\, e^{-iHt} f \| < \varepsilon$$

for all $t > 0$ (equivalently for all t sufficiently large.)

The vector $f \in \mathcal{H}$ is said to be a scattering state, as $t \to \infty$, if

$$\lim_{t\to\infty} \| E_{|\underline{r}|<a}\, e^{-iHt} f \| = 0$$

for all $a > 0$.

In other words, bound states remain localised whereas scattering states move off to infinity. It is not difficult to show that the set of bound states form a subspace of $\mathcal{H}$, which we denote by $\mathcal{H}_0^+$, and the set of scattering states forms a subspace $\mathcal{H}_\infty^+$. Care should be taken to distinguish between $\mathcal{H}_0^+$ and $\mathcal{H}_d$. In general $\mathcal{H}_d$ is a proper subspace of $\mathcal{H}_0^+$. Note also that $\mathcal{H}_0^+$ and $\mathcal{H}_\infty^+$ are mutually orthogonal.

The subspaces $\mathcal{H}_0^-$ and $\mathcal{H}_\infty^-$, defined for $t \to -\infty$, are in general distinct from $\mathcal{H}_0^+$ and $\mathcal{H}_\infty^+$.

THEOREM 4

$$\mathcal{H} = \mathcal{H}_0^+ \oplus \mathcal{H}_\infty^+ ,$$

where $\mathcal{H}_\infty^+$ is the range of Ω_+ and $\mathcal{H}_0^+$ is the orthogonal subspace to the range of Ω_+.

PROOF.

The result follows easily from

(i) range $\Omega_+ \subseteq \mathcal{H}_\infty^+$,

and (ii) $(\text{range}\ \Omega_+)^\perp \subseteq \mathcal{H}_0^+$.

To prove (i), note that, for $g \in$ range Ω_+,

$$\| e^{-iHt} g - e^{-iH_0t} h \| \to 0 \text{ as } t \to \infty, \text{ for some } h \in \mathcal{H} .$$

In other words, $e^{-iHt}g$ behaves for large t like a free-particle state, for which it is known that $\| E_{|\underline{r}|<a}\, e^{-iH_0t} h \| \to 0$. Hence g is a scattering state.

To prove (ii), suppose $f \perp$ range Ω_+ and that f has compact spectral support (of H) not containing the origin. Then ϕ can be found in (40) such that $\phi(H)f = f$. Using $\Omega_+^* f = 0$, eq (40) becomes

$$g\, e^{-iHt} f = C\, e^{-iHt} f + P_-\, e^{-iH_0t} \phi(H_0)\, \Omega_-^* f \tag{41}$$

where C is a compact operator. We have used here the compactness of $\rho\phi(H) - \phi(H_0)\rho$. This follows from the result that $\rho(H-i)^{-1} - (H_0-i)^{-1}\rho$ is compact by using the complex Stone-Weierstrass theorem to approximate ϕ by polynomials in $(\lambda+i)^{-1}$ and $(\lambda-i)^{-1}$. For details see Lemma (2) of [10].

Suppose that f is <u>not</u> a bound state for $t \to \infty$. Then increasing sequences $\{t_n\}$, $\{a_n\}$ can be found such that $t_n \to \infty$, $a_n \to \infty$ and

$$\| E_{|\underline{r}|>a_n} e^{-iHt_n} f \| > K > 0 , \tag{42}$$

for all n, where K is a constant.

Since norm bounded sequences have weakly convergent subsequences, we can also suppose without loss of generality that $e^{-iHt_n}f$ tends weakly to a limit h, say. Substitute now t = t_n in eq(41) and use eq (39). Then, as $n \to \infty$, we have the <u>strong</u> limit

$$\underset{n\to\infty}{\text{s-lim}}\ \rho e^{-iHt_n} f = Ch \tag{43}$$

Hence

$$\underset{n\to\infty}{\text{s-lim}}\ E_{|\underline{r}|>a_n} e^{-iHt_n} f = \underset{n\to\infty}{\text{s-lim}}\ E_{|\underline{r}|>a_n} \rho e^{-iHt_n} f$$

$$= \underset{n\to\infty}{\text{s-lim}}\ E_{|\underline{r}|>a_n} Ch = 0 . \tag{44}$$

But this contradicts eq (42) above, and we can conclude that f is indeed a bound state. The argument can be extended from f having compact spectral support to arbitrary f in (range $\Lambda_+)^\perp$ by taking strong limits and using the fact that $\mathcal{H}_0^+$ is closed. The only f in (range $\Lambda_+)^\perp$ which cannot be treated by such limiting arguments are zero energy eigenvectors of H, but they are in $\mathcal{H}_0^+$ anyway. ■

How do states in $\mathcal{H}_0^+$ behave for large t? For eigenstates of H, and linear combinations, the time development is rather simple, but what about states in $\mathcal{H}_c$? Suppose first that $e^{-iHt}f$ converges weakly to zero as $t \to \infty$. Then eq (41) implies that $\rho e^{-iHt}f$ converges strongly to zero. Hence $E_{|\underline{r}|>a} e^{-iHt}f$ goes strongly to zero for any a > 0, and the particle is found asymptotically, with probability 1, in a small neighbourhood of the origin.

DEFINITION

The state $f \in \mathcal{H}$ is said to be an absorbed state, as $t \to \infty$, if, for any $a > 0$

$$\lim_{t\to\infty} \| E_{|\underline{r}|>a}\, e^{-iHt} f \| = 0 .$$

What we have seen is that any state in $\mathcal{H}_0^+$ which converges weakly to zero is an absorbed state. This applies in particular to all states of $\mathcal{H}_0^+$ which are also in $\mathcal{H}_{a.c.}$. Conversely, any absorbed state is in $\mathcal{H}_0^+$ and converges weakly to zero. The kinetic energy of such states tends to infinity, in the sense that

$$\lim_{t\to\infty} \| E_{H_0<M}\, e^{-iHt} f \| = 0 ,$$

for any $M > 0$.

States in $\mathcal{H}_c$ do not necessarily converge weakly to zero. However, we have already seen that for such states $\langle \phi, e^{-iHt} f \rangle$ converges to zero in time average. Hence, for $f \in \mathcal{H}_c \cap \mathcal{H}_0^+$, the particle will approach the origin on time average, in the sense that

$$\lim_{T\to\infty} \frac{1}{T} \int_{-T}^{T} dt\, \| E_{|\underline{r}|>a}\, e^{-iHt} f \| = 0 .$$

We can interpret this as saying that states in $\mathcal{H}_c \cap \mathcal{H}_0^+$ spend most of the time in a small neighbourhood of the origin. One consequence of this is that there exists an increasing sequence of times $\{t_n\}$ such that, at $t = t_n$ and in the limit $t_n \to \infty$, the particle is abosrbed. This will apply in particular to all states in the singular continuous subspace.

What does all this mean for scattering theory? Asymptotic completeness (range Ω_+ = range Ω_-) will hold if and only if $\mathcal{H}_\infty^+ = \mathcal{H}_\infty^-$, or equivalently if and only if $\mathcal{H}_0^+ = \mathcal{H}_0^-$. Since these spaces are defined geometrically, we have a geometric characterisation of asymptotic completeness. Since neither $\mathcal{H}_d$ nor $\mathcal{H}_{s.c.}$ carry a label + or −, everything will depend on a comparison of $\mathcal{H}_0^+ \cap \mathcal{H}_{a.c}$ with $\mathcal{H}_0^- \cap \mathcal{H}_{a.c.}$ (This justifies mathematical physicists in their constant preoccupation, in the question of asymptotic completeness, with the absolutely continuous subspace of H.) Asymptotic completeness will hold if and only if the subspace of states absorbed at time $t = -\infty$ is the same as the

subspace of states absorbed at $t = +\infty$. Usually one tries to prove $\mathcal{H}_0^\pm \cap \mathcal{H}_{a.c.} = \{0\}$, for example by use of relative compactness (see Enss notes). However, it is also possible for a.c. to hold even if $\mathcal{H}_0^\pm \cap \mathcal{H}_{a.c.}$ is non-trivial, for example in situations where H has negative a.c. spectrum.

To sum up, the various subspaces and types of asymptotic behaviour which we have met are as follows:

Type 1: range Ω_+

States in range Ω_+ ($\equiv \mathcal{H}_\infty^+$) behave for large t like free particle states. They are the states that every good scattering theorist likes to work with.

Type 2: $\mathcal{H}_0^+$

The subspace of bound states. The particle remains localised. Within type 2 we have

Type 2(i):

Eigenstates of H, and their linear combinations.

Type 2(ii):

The subspace of absorbed states. The particle moves in to the origin, its kinetic energy tends to infinity.

Type 2(iii):

The subspace of $\mathcal{H}_{s.c.}$ where weak convergence to zero fails to hold. The particle spends most of its time near the origin with very large kinetic energy, but takes time off to explore the surrounding region from time to time.

For various Hamiltonians, all of these types of behaviour are possible. (So far type 2(iii) has only been proved for Schrödinger Hamiltonians in cases where $\sqrt{H}$, rather than H, is the generator of the evolution group. This is connected with decay properties of the Fourier transform of the spectral measure; see Section 5). For short range potentials singular at $\underline{r} = 0$, the above list is exhaustive. No other types of asymptotic behaviour are possible, and a general state will be a linear combination of states of the above types. A similar classification holds for short range potentials with other singular sets, though in that case some of these spaces may then break up into still further components.

It would be good to be able to analyse the long range case, assuming always $V \to 0$ as $|\underline{r}| \to \infty$, with comparable generality. We do not yet have, in this case, a sufficiently detailed description of the possible types of asymptotic behaviour. Certainly there is no lack of examples, from potentials generating singular continuous spectrum to the violation of asymptotic completeness [11] by potentials having different decay rates in various directions. If past experience is anything to go by, we can expect (with high probability) that further surprises may be in store for us.

REFERENCES

1 T. Kato,"Perturbation Theory for Linear Operators" (2nd Edition); Berlin, Springer (1976).

2 P. Halmos, "Measure Theory", Princeton: Van Norstrand (1950).

3 R. Bartle, "The Elements of Integration", New York: Wiley (1966)

4 K. Sinha, Ann. Inst. Henri Poincaré 26:263 (1977).

5 W. Amrein and V. Georgescu, Helv. Phys. Acta 46:635 (1973).

6 E. Coddington and N. Levinson, "Theory of Ordinary Differential Equations", New York, McGraw-Hill (1955).

7 D. Pearson, Comm. Math. Phys. 40:125 (1975).

8 D. Pearson, Comm. Math. Phys. 60:13 (1978).

9 W. Amrein and V. Georgescu, Helv. Phys. Acta 47:517 (1974).

10 W. Amrein, D. Pearson and M. Wollenberg, Evanescence of States and Asymptotic Completenes , to be published in Helv.Phys. Acta.

11 D. Yafaev, Comm. Math. Phys. 65:167 (1979).

SCHRÖDINGER OPERATORS WITH EXTERNAL HOMOGENEOUS ELECTRIC AND MAGNETIC FIELDS

Ira W. Herbst*

Department of Mathematics
University of Virginia
Charlottesville, Virginia 22903

These lectures split into two parts, the first concerns homogeneous electric fields and the second homogeneous magnetic fields. The reader is referred to the beautiful survey of Hunziker [36] for an introduction to the subject. To complement the one-body results reviewed in [36] I have given a rather detailed analysis of the N-body Stark effect following the work in [27,31,32]. There is essentially only one new development (not present in the published literature) which is presented here: The rather complicated semigroup Weinberg-van Winter analysis used in [32] has been replaced by a simpler technique.

In writing these notes I have omitted an alternative method developed by Graffi and Grecchi for studying Stark resonances which is based on writing the Laplace operator in parabolic coordinates. This technique has yielded much fruit in the study of the Hydrogen Stark effect. The reader is referred to [11,15,19,20,21, 22,23,25,38,54].

In contrast to the electric field part of the lectures the material on magnetic fields presented here is much narrower in scope. I have concentrated on the subject of enhanced binding in homogeneous fields, and even there have omitted much which is relevant. The material is based on the work in [6,8,9]. On the subject of enhanced binding the reader should also consult [39] and for further results on Schrödinger operators with magnetic fields, [5,7,14].

*Research supported by NSF Grant No. MCS 78-00101

I would like to express my thanks to Joseph Avron and Barry Simon for an extremely enjoyable collaboration during which most of the work discussed here was completed.

I. THE STARK EFFECT

I.1. Introduction:

In this part of the lectures we discuss the theory of the operator

$$H(f) = -\Delta + f\hat{e} \cdot \vec{x} + V \tag{1.1}$$

and its N-body generalizations. Here V is multiplication by the real valued function $v(\vec{x})$. When $v(\vec{x}) = -|\vec{x}|^{-1}$, H(f) describes the Hydrogen atom in an external electric field pointing in the direction of the unit vector $\hat{e}$.

In most quantum mechanics texts, the study of this operator is at the level of perturbation theory in the parameter f, the electric field strength. A notable exception is [38] where the lifetime of Hydrogen in an electric field is estimated for small f. The decay of Hydrogen in an electric field can be understood as a barrier penetration problem. The potential energy, $-|\vec{x}|^{-1} + f\hat{e} \cdot \vec{x}$, is shown in Figure 1 as a function of $x_{||} = \hat{e} \cdot \vec{x}$ for some fixed value of $\vec{x}_{\perp} = \vec{x} - (\vec{x} \cdot \hat{e})\hat{e}$ and $f > 0$.

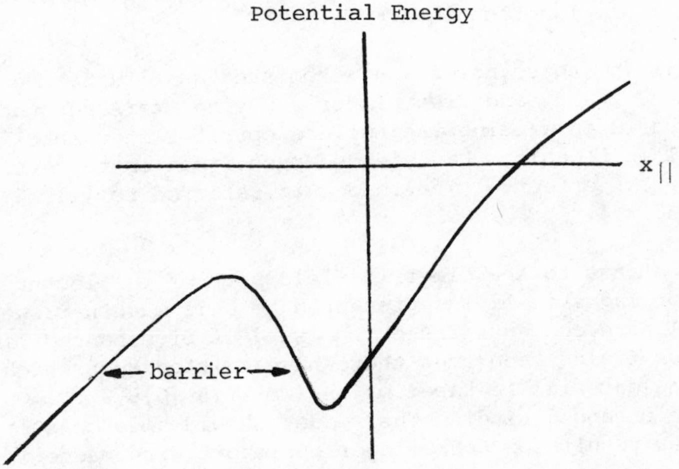

Figure 1. Stark Potential Energy Barrier.

A particle initially in the potential well will, under the action of $\exp(-itH(f))$, eventually tunnel out to the left. The corresponding lifetime of a Hydrogenic state is for $f > 0$ roughly of order of magnitude [8] $\exp(cf^{-1})$ for some $c > 0$ and thus for small f, the system acts very much as if it had bound states, at least for a very long time.

Even though there is a certain richness of structure in the system with Hamiltonian H(f), this operator itself is very simple. For example we have:

<u>Theorem 1.1 [26]:</u> <u>For</u> $f > 0$ <u>the operator</u> $H(f) = -\Delta + f\hat{e}\cdot\vec{x} - |\vec{x}|^{-1}$ <u>in</u> $L^2(\mathbb{R}^3)$ <u>is unitarily equivalent to multiplication by</u> x_1 <u>in</u> $L^2(\mathbb{R}^3)$.

We do not prove this theorem here. More general results of this type can be found in [1,4,26,47,49,52,53]. We mention it mainly to point out that besides the non-existence of bound states, an analysis of the operator H(f) <u>alone</u> does not offer much information about the system in question.

In order to extract more information than Theorem 1.1 allows, we introduce an assumption which permits us to continue certain matrix elements of the resolvent $(z-H(f))^{-1}$ from $\mathrm{Im} z > 0$ to the "second sheet". While other possibilities exist [4,22,53], we use the dilation analytic formalism of [2,10] which is suitable for Coulomb interactions. Thus we assume that with

$$u(\theta)\phi(\vec{x}) = e^{3\theta/2}\phi(e^{\theta}\vec{x}), \quad \theta \in \mathbb{R} \tag{1.2}$$

$$V(\theta) = u(\theta)Vu(-\theta) \tag{1.3}$$

we have

$$V(\theta)(-\Delta+1)^{-1} \text{ has a compact operator valued analytic continuation to } \{\theta : |\mathrm{Im}\theta| < \theta_0\} \tag{1.4}$$

for some $\theta_0 > 0$. If (1.4) is satisfied we say that V is dilation analytic in $\{\theta : |\mathrm{Im}\theta| < \theta_0\}$.

Under this assumption, it will turn out that matrix elements of $(z-H(f))^{-1}$ between dilation analytic vectors (i.e. vectors ψ such that $u(\theta)\psi$ has an analytic continuation to $\{\theta : |\mathrm{Im}\theta| < \phi_0\}$ for some $\phi_0 > 0$) have meromorphic continuations from $\mathrm{Im} z > 0$ to all of $\mathbb{C}$ (at least for the atomic case.) We call the poles which occur <u>resonances</u> and denote the set of all such poles by $\mathcal{R}_f$. These poles have direct

physical relevance as can be seen by the following theorem.

Theorem 1.2 [28]: *Define $V_{\vec{a}}$ as multiplication by $V(\vec{x}+\vec{a})$ and suppose in addition to (1.4) the potential V is such that*

a). *$V(\vec{x})$ is bounded*

b). *$V_{\vec{a}}(-\Delta+1)^{-1}$ has an analytic compact operator valued continuation to $\mathbb{C}^3$ in the variable $\vec{a}$.*

Then the set of resonances $\mathcal{R}_f$ is a discrete set $\{E_1,\cdots,E_n,\cdots\}$ in $\{z:\text{Im}z<0\}$ which if infinite satisfies $\lim_{n\to\infty}\text{Im}E_n=-\infty$. Given any eigenvalue and any $\alpha>0$ there are complex numbers $c_n(\psi,f)$ so that for $t\to\infty$

$$(\psi,e^{-itH(f)}\psi)=\sum_{\substack{n\\ \text{Im}E_n>-\alpha}} c_n(\psi,f)e^{-itE_n}+O(e^{-\alpha t}) \tag{1.5}$$

The reader is referred to [28] for a more general result and a proof which does not use the condition (1.4). Unfortunately theorem 1.2 does not apply to Hydrogen although if the delta function nuclear charge density is replaced with a Gaussian density ρ, the resulting potential $V(\vec{x})=-\int|\vec{x}-\vec{y}|^{-1}\rho(\vec{y})d^3y$ meets the requirements of Theorem 1.2.

It is the resonances $\mathcal{R}_f$ which will be the subject of study in the electric field part of these lectures. Our major goals are as follows.

Section I.2: Here we extend the dilation analytic formalism of [2,10] to cover N-body Schrödinger operators with external electric fields. This turns out to be highly non-trivial but the payoff is a framework in which one can talk about resonances.

Section I.3: A stability result is proved here. Discrete eigenvalues of $H(0)$ turn into resonances of $H(f)$ which are continuous at $f=0$. The proof given here is new. It avoids the rather involved semigroup Weinberg-van Winter analysis of [32].

Section I.4: The asymptotic nature of the Rayleigh-Schrödinger perturbation series for the resonances is proved along with certain analyticity properties which allow a proof of a kind of Borel summability.

Section I.5: Here some open problems are discussed.

I.2. DILATION ANALYTIC FORMALISM

In the Hilbert space $L^2(\mathbb{R}^{3(N+1)})$ let

$$\tilde{H}_0(f,\theta) = -e^{-2\theta} \sum_{i=1}^{N+1} (2m_i)^{-1}\Delta_i + e^{\theta} f \sum_{i=1}^{N+1} q_i \hat{e} \cdot \vec{x}_i \tag{2.1}$$

be the scaled Hamiltonian of N+1 non-interacting particles of charges q_i and masses m_i, $i=1,\cdots,N+1$ in an electric field $-f\hat{e}$. We remove the center of mass motion from (2.1) by writing

$$\tilde{H}_0(f,\theta) = H_0(f,\theta) \otimes I + I \otimes [-e^{-2\theta}(2M)^{-1}\Delta_{\vec{R}} + e^{\theta} fQ\hat{e} \cdot \vec{R}] \tag{2.2}$$

where

$$M = \sum_{i=1}^{N+1} m_i, \quad Q = \sum_{i=1}^{N+1} q_i, \quad \vec{R} = (\sum_{i=1}^{N+1} m_i \vec{x}_i) M^{-1}.$$

Here $H_0(f,\theta)$ is an operator in $L^2(\mathbb{R}^{3N}, d^3r_1 \cdots d^3r_N)$ which can be written

$$\begin{aligned} &H_0(f,\theta) = e^{-2\theta} K + e^{\theta} fX \\ &K = -\sum_{i=1}^{N} (2\mu_i)^{-1}\Delta_i, \quad X = \sum_{i=1}^{N} \lambda_i \hat{e} \cdot \vec{r}_i \end{aligned} \tag{2.3}$$

The numbers $\mu_i > 0$ are the usual reduced masses while the λ_i are certain functions of m_i and q_i. It is important to understand that if all particles are uniformly accelerated by the electric field then the operator X is identically zero. In fact this condition is necessary and sufficient for the vanishing of X.

<u>Proposition 2.1:</u> $X \equiv 0$ <u>if and only if</u>

$$q_i/m_i = q_j/m_j \quad \text{all } i,j.$$

<u>Proof:</u> $X = (\sum_{i=1}^{N+1} q_i \vec{x}_i - Q\vec{R}) \cdot \hat{e}$. Using the definitions of $\vec{R}$ and Q above the result easily follows. □

Until now we have been rather formal. We now specify the domain of $H_0(f,\theta)$ as $\mathcal{D}(K) \cap \mathcal{D}(X)$, and since we will eventually want to discuss analyticity in the variable f, we allow both f and θ to be complex.

The following quadratic estimate is fundamental in our development:

<u>Proposition 2.2 [27]:</u> <u>There is a constant</u> $c > 0$ <u>independent of</u> (f,θ) <u>so that for all</u> $f \neq 0$ <u>and</u> $\psi \in \mathcal{D}(H_0(f,\theta)) \equiv \mathcal{D}(K) \cap \mathcal{D}(X)$

$$\|H_0(f,\theta)\psi\|^2 + c|f|^{4/3}\|\psi\| \geq d(f,\theta)(\|K\psi\|^2 + |\alpha|^2\|X\psi\|^2) \quad (2.4)$$

where

$$d(f,\theta) = \frac{1}{4}\,|e^{-2\theta}\alpha^{-1}\mathrm{Im}\alpha|^2; \quad \alpha = fe^{3\theta} \quad (2.5)$$

<u>Remark:</u> There is a small error in the analogous estimate in [27], Proposition II.4, but this error does not affect the developments of that paper.

<u>Proof:</u> By a simple approximation argument we need only prove (2.4) for $\psi \in \mathcal{S}(\mathbb{R}^{3N})$. Thus the following computations are to be interpreted as taking place on vectors in $\mathcal{S}(\mathbb{R}^{3N})$ and all operator inequalities as quadratic form inequalities on $\mathcal{S}(\mathbb{R}^{3N} \times \mathcal{S}(\mathbb{R}^{3N})$:

Writing $H_0(f,\theta) = e^{-2\theta}(K + \alpha X)$ we have

$$H_0(f,\theta)^* H_0(f,\theta) = e^{-4\mathrm{Re}\theta}(K^2 + |\alpha|^2X^2 + \bar{\alpha}XK + \alpha KX)$$

$$= e^{-4\mathrm{Re}\theta}(K^2 + |\alpha|^2X^2 + \mathrm{Re}\alpha(XK + KX) - i[X,K]\mathrm{Im}\alpha) \quad (2.6)$$

We use $(X|\alpha|^{1/2} \pm K|\alpha|^{-1/2})^2 \geq 0$ to derive

$$\pm (XK + KX) \leq |\alpha|X^2 + |\alpha|^{-1}K$$

and multiply by $|\mathrm{Re}\alpha|$ to obtain

$$\mathrm{Re}\alpha(XK + KX) \geq -|\alpha^{-1}\mathrm{Re}\alpha|(K^2 + |\alpha|^2X^2) \quad (2.7)$$

Since $-i[X,K] = \sum_{j=1}^{N} a_j\,\hat{e}\cdot(-i\vec{\nabla}_j)$ for some real numbers a_j, we have for some $\beta > 0$

$$-i[X,K]\mathrm{Im}\alpha \geq -|\mathrm{Im}\alpha|\beta K^{1/2} \quad (2.8)$$

Combining (2.6) - (2.8) we have

$$e^{4\mathrm{Re}\theta}H_0(f,\theta)^*H_0(f,\theta) \geq (1-|\alpha^{-1}\mathrm{Re}\alpha|)(K^2+|\alpha|^2X^2) - |\mathrm{Im}\alpha|\beta K^{1/2} \quad (2.9)$$

so that using $\frac{1}{2}K^2$ from $K^2 + |\alpha|^2 X^2$ to bound the last term we find

$$H_0(f,\theta)^* H_0(f,\theta) \geq \frac{1}{2} e^{-4\mathrm{Re}\theta}(1 - |\alpha^{-1}\mathrm{Re}\alpha|)(K^2 + |\alpha|^2 X^2) + e^{-4\mathrm{Re}\theta} A \tag{2.10}$$

with

$$A = (1 - |\alpha^{-1}\mathrm{Re}\alpha|)\frac{1}{2}K^2 - |\mathrm{Im}\alpha|\beta K^{1/2}$$

Since $1 - |\alpha^{-1}\mathrm{Re}\alpha| \geq \frac{1}{2}|\alpha^{-1}\mathrm{Im}\alpha|^2$, (2.10) gives

$$\|H_0(f,\theta)\psi\|^2 \geq d(f,\theta)(\|K\psi\|^2 + |\alpha|^2\|X\psi\|^2) + e^{-4\mathrm{Re}\theta}(\psi, A\psi) \tag{2.11}$$

If $\mathrm{Im}\alpha = 0$ then $A = 0$ while otherwise defining

$$\gamma = |\mathrm{Im}\alpha|/(1 - |\alpha^{-1}\mathrm{Re}\alpha|),$$

we have

$$\begin{aligned} A &\geq (1 - |\alpha^{-1}\mathrm{Re}\alpha|)\inf_{x\geq 0}(\tfrac{1}{2}x^2 - \gamma\beta x^{1/2}) \\ &= (1 - |\alpha^{-1}\mathrm{Re}\alpha|)\gamma^{4/3}\inf_{x\geq 0}(\tfrac{1}{2}x^2 - \beta x^{1/2}) \\ &= -\tilde{c}(1 - |\alpha^{-1}\mathrm{Re}\alpha|)\gamma^{4/3} \end{aligned} \tag{2.12}$$

It is easy to show that $(1 - |\alpha^{-1}\mathrm{Re}\alpha|)\gamma^{4/3} \leq 2^{1/3}|\alpha|^{4/3}$ so that noting $e^{-4\mathrm{Re}\theta}|\alpha|^{4/3} = |f|^{4/3}$, (2.4) follows from (2.11) and (2.12). □

<u>Corollary 2.3:</u> <u>Suppose</u> $\mathrm{Im}(fe^{3\theta}) \neq 0$. <u>Then</u>

(i) $H_0(f,\theta)$ <u>is a closed operator</u>

(ii) $H_0(f,\theta)^* = H_0(\bar{f},\bar{\theta})$.

<u>Proof</u>: (i) If $\psi_n \in \mathcal{D}(H_0(f,\theta))$ and

$$\|H_0(f,\theta)\psi_n - \phi\| + \|\psi_n - \psi\| \to 0, \tag{2.13}$$

we must show $\psi \in \mathcal{D}(H_0(f,\theta))$ and $\phi = H_0(f,\theta)\psi$. From (2.13), $\|H_0(f,\theta)(\psi_n - \psi_m)\| \to 0$ as $n,m \to \infty$ and thus by the quadratic estimate

$$||X(\psi_n - \psi_m)|| + ||K(\psi_n - \psi_m)|| \to 0. \tag{2.14}$$

Since X and K are closed (2.14) implies $\psi \in \mathcal{D}(X) \cap \mathcal{D}(K) = \mathcal{D}(H_0(f,\theta))$ and

$$X\psi_n \to X\psi, \quad K\psi_n \to K\psi \tag{2.15}$$

But (2.15) imples $\phi = \lim_{n\to\infty} H_0(f,\theta)\psi_n = H_0(f,\theta)\psi$.

(ii) It is clear that $H_0(f,\theta)^* \supseteq H_0(\bar{f},\bar{\theta})$. Suppose $\phi \in \mathcal{D}(H_0(f,\theta)^*)$. We need only show that $\phi \in \mathcal{D}(H_0(\bar{f},\bar{\theta}))$. Let $\phi_t = e^{-tK}e^{-tr^2}\phi$ $r^2 = \sum_{i=1}^{N} |\vec{r}_i|^2$. We claim that

$$||H_0(\bar{f},\bar{\theta})\phi_t|| \leq \text{const} \tag{2.16}$$

for all $t \in (0,1]$. (Note $\phi_t \in \mathcal{S}(\mathbb{R}^{3N}) \subseteq \mathcal{D}(H_0(\bar{f},\bar{\theta}))$.) Once (2.16) is proved the quadratic estimate shows

$$||X\phi_r|| + ||K\phi_t|| \leq \text{const} \tag{2.17}$$

which using the spectral theorem and $\lim_{t\downarrow 0} \phi_t = \phi$ implies $\phi \in \mathcal{D}(X) \cap \mathcal{D}(K) = \mathcal{D}(H_0(\bar{f},\bar{\theta}))$. To show (2.16) we note that for all $\psi \in \mathcal{D}(H_0(f,\theta))$ with $||\psi|| = 1$ and $t > 0$

$$\begin{aligned}
&|(H_0(\bar{f},\bar{\theta})\phi_t,\psi)| = |(\phi_t, H_0(f,\theta)\psi)| = \\
&|(e^{-tr^2}\phi, H_0(f,\theta)e^{-tK}\psi) - fe^{\theta}(e^{-tr^2}\phi,[X,e^{-tK}]\psi)| \\
&\leq |(\phi, H_0(f,\theta)e^{-tr^2}e^{-tK}\psi)| + c(||[X,e^{-tK}]|| \\
&+ ||[e^{-tr^2},K]e^{-tK}||)||\phi|| \\
&\leq ||H_0(f,\theta)^*\phi|| + \text{const}
\end{aligned} \tag{2.18}$$

The estimate $||[X,e^{-tK}]|| + ||[e^{-tr^2},K]e^{-tK}|| \leq$ const is straightforward and will not be given. Since (2.18) says $|(H_0(\bar{f},\bar{\theta})\phi_t,\psi)| \leq \text{const}||\psi||$ for all $\psi \in \mathcal{D}(H_0(\bar{f},\bar{\theta}))$ and the latter is dense, (2.16)

follows. □

We now analyze $H_0(f,\theta)$ from another point of view. It is easy to see that the numerical range of $H_0(f,\theta)$ is contained in a half-plane. For $f \neq 0$ we introduce the operators

$$L_0(f,\theta) = ie^{-\theta}f^{-1}|f|H_0(f,\theta);\ \operatorname{Im}(fe^{3\theta}) > 0 \tag{2.19}$$

$$L_0(f,\theta) = L_0(\bar{f},\bar{\theta})^*;\quad \operatorname{Im}(fe^{3\theta}) < 0 \tag{2.20}$$

Let $\alpha = fe^{3\theta}$. If $\operatorname{Im}\alpha > 0$ then

$$L_0(f,\theta) = |f|(i\alpha^{-1}K + iX) \tag{2.21}$$

Since $\operatorname{Re} i\alpha^{-1} = |\alpha|^{-2}\operatorname{Im}\alpha > 0$, it is clear from (2.21) that the numerical range of $L_0(f,\theta)$ is contained in $\{z : \operatorname{Re}z \geq 0\}$. This suggests that $L_0(f,\theta)$ may generate a contraction semigroup. A formal computation based on the identity [4,27,49]

$$\frac{-d^2}{dx^2} + \lambda x = \exp([-\frac{d}{dx}]^3/3\lambda)\lambda x \exp([\frac{d}{dx}]^3/3\lambda)$$

suggests that for $\operatorname{Im}\alpha > 0$ this semigroup will be given by

$$P_t \equiv \exp(-it|f|X/2)\exp(-t|f|i\alpha^{-1}K)\exp(-it|f|X/2)\exp(-t^3D(f,\theta)) \tag{2.22}$$

$$\alpha = fe^{3\theta},D(f,\theta) = i\alpha^{-1}|f|^2 \sum_{j=1}^{N} \lambda_j^2/24\mu_j \tag{2.23}$$

If $\operatorname{Im}\alpha < 0$ then P_t is defined by changing t to $-t$ on the right side of (2.22). We proceed to analyze this situation further:

<u>Proposition 2.4 [27]:</u> <u>Suppose</u> $\operatorname{Im}\alpha \neq 0$. <u>Then</u>

(i) $L_0(f,\theta)$ <u>generates a strongly continuous contraction semigroup</u>:

$$e^{-tL_0(f,\theta)} = P_t,\ \|P_t\| = \exp(-t^3|\operatorname{Re}D(f,\theta)|) \tag{2.24}$$

(ii) <u>If</u> $X \neq 0$, <u>the spectrum of</u> $H_0(f,\theta)$ <u>is empty and the resolvent</u> $(z-H_0(f,\theta))^{-1}$ <u>is jointly analytic in</u> (z,f,θ) <u>for</u> $z \in \mathbb{C}$, $\operatorname{Im}(fe^{3\theta}) \neq 0$.

<u>Proof:</u> (i) It is only slightly tedious to verify that

$$P_t P_s = P_{t+s}; \; t \geq 0, \; s \geq 0$$

and that for $\phi \in \mathcal{S}(\mathbb{R}^{3N})$

$$\frac{d}{dt} P_t\phi = -L_0(f,\theta)P_t\phi \tag{2.25}$$

Note that each factor in (2.22) maps $\mathcal{S} \to \mathcal{S}$ so that $P_t\phi \in \mathcal{S}$ and (2.25) makes sense. Since P_t obviously is a strongly continuous contraction semigroup, it has a generator B which is maximal accretive: $P^t = e^{-tB}$. From (2.25) at t = 0 we find $B\phi = L_0(f,\theta)\phi$ for $\phi \in \mathcal{S}(\mathbb{R}^{3N})$. From the quadratic estimate it is easy to show that $\mathcal{S}$ is a core for $L_0(f,\theta)$ so that $B \supseteq L_0(f,\theta)$. To show that in fact $B = L_0(f,\theta)$ we need only show that $L_0(f,\theta)$ is maximal accretive, for such operators have no proper accretive extensions. Suppose ψ is orthogonal to $\mathrm{Ran}(L_0(f,\theta) + 1)$ and consider $f(t) = (\psi,P_t\phi)$ for $\phi \in \mathcal{S}(\mathbb{R}^{3N})$. By (2.25), $f'(t) = -(\psi,L_0(f,\theta)P_t\phi) = f(t)$ since ψ is orthogonal to $(L_0(f,\theta)+1)P_t\phi$. Thus $f(t) = f(0)e^t$. Since f(t) is bounded, $f(0) = (\psi,\phi) = 0$. Thus ψ is orthogonal to $\mathcal{S}(\mathbb{R}^{3N})$ and hence $\psi = 0$. Since the norm of P_t is readily calculated from (2.22) with the result as given in (2.24), the proof is complete.

(ii) If $X \neq 0$, then $|\mathrm{Re}D(f,\theta)| > 0$ so that the integral

$$\int_0^\infty e^{-tL_0(f,\theta)} e^{zt}\, dt \equiv C$$

is convergent for any $z \in \mathbb{C}$. It is readily verified that $C = (L_0(f,\theta)-z)^{-1}$. Thus $H_0(f,\theta)$ has empty spectrum. We now write

$z-H_0(f,\theta) = (1+M_0(z,f,\theta))(z_1 - H_0(f_1,\theta_1))$ for some point (z_1,f_1,θ_1) with $\mathrm{Im}(f_1e^{3\theta_1}) \neq 0$. Here

$$M_0(z,f,\theta) \equiv (z-z_1+H_0(f_1,\theta_1) - H_0(f,\theta))(z_1 - H_0(f_1,\theta_1))^{-1}$$

is a bounded analytic operator valued function of (z,f,θ) in the region $z \in \mathbb{C}$, $\mathrm{Im}(fe^{3\theta}) \neq 0$. Since $M_0(z_1,f_1,\theta_1) = 0$, $(1+M_0(z,f,\theta))^{-1}$

is analytic in a neighborhood of the point (z_1, f_1, θ_1). Thus so is $(z-H_0(f,\theta))^{-1} = (z_1-H_0(f_1,\theta_1))^{-1}(1+M_0(z,f,\theta))^{-1}$. □

We are now ready to introduce interactions. Assume we are given real valued measurable functions $v_{ij}(\vec{x})$, $1 \le i < j \le N+1$, with each v_{ij} dilation analytic in $\{\theta: |\mathrm{Im}| < \theta_0\}$. We denote by V_{ij} the operator of multiplication by $v_{ij}(\vec{x}_i - \vec{x}_j)$. The dilation group on $L^2(\mathbb{R}^{3N})$ is given by

$$U(\theta)\phi(\vec{r}_1,\cdots,\vec{r}_N) = \exp(3N\theta/2)\phi(e^{\theta}\vec{r}_1,\cdots,e^{\theta}\vec{r}_N).$$

We write $V_{ij}(\theta) = U(\theta)V_{ij}U(-\theta)$. The dilation analyticity of v_{ij} implies that $V_{ij}(\theta)(1+K)^{-1}$ has an analytic continuation from real θ to $|\mathrm{Im}\theta| < \theta_0$. Let

$$\mathcal{R} = \{(f,\theta): \mathrm{Im}(fe^{3\theta}) > 0, \ |\mathrm{Im}\theta| < \theta_0\} \tag{2.26}$$

and $\bar{\mathcal{R}} = \{(f,\theta): (\bar{f},\bar{\theta}) \in \mathcal{R}\}$.

$$V(\theta) = \sum_{1 \le i < j \le N+1} V_{ij}(\theta) \tag{2.27}$$

For $(f,\theta) \in \mathcal{R} \cup \bar{\mathcal{R}}$ we define

$$H(f,\theta) = H_0(f,\theta) + V(\theta); \quad \mathcal{D}(H(f,\theta)) = \mathcal{D}(H_0(f,\theta)) \tag{2.28}$$

If f and θ are real, $H(f,\theta)$ is defined as the closure of the right side of (2.28). We also define

$$L(f,\theta) = ie^{-\theta}f^{-1}|f|H(f,\theta); \quad (f,\theta) \in \mathcal{R} \tag{2.29}$$

$$L(f,\theta) = L(\bar{f},\bar{\theta})^*; \quad (f,\theta) \in \bar{\mathcal{R}} \tag{2.30}$$

To keep things simple we do not give a separate discussion of the situation where one of the particles (say the nucleus of an atom) has infinite mass (see [32]).

<u>Proposition 2.5 [32]:</u>

(i) <u>For all</u> (f,θ) <u>for which</u> $H(f,\theta)$ <u>has been defined</u>, $H(f,\theta)$ <u>is a closed operator with</u> $H(f,\theta)^* = H(\bar{f},\bar{\theta})$.

(ii) <u>Given a point</u> (z_1,f_1,θ_1) <u>with</u> $(f_1,\theta_1) \in \mathcal{R} \cup \bar{\mathcal{R}}$ <u>and</u> $z_1 \notin \sigma(H(f_1,\theta_1))$ <u>then</u> $(z-H(f,\theta))^{-1}$ <u>is jointly analytic in the</u>

variables (z,f,θ) in a neighborhood of the point (z_1,f_1,θ_1).

(iii) The operator $L(f,\theta)$ generates a strongly continuous semigroup, $\{e^{-tL(f,\theta)}:t \geq 0\}$ which is exponentially bounded:

$$\|e^{-tL(f,\theta)}\| \leq \exp(tE(e^{i\arg f},\theta)) \tag{2.31}$$

where $E(e^{i\arg f},\theta)$ is uniformly bounded on compact subsets of

$\{(e^{i\arg f},\theta):(e^{i\arg f},\theta) \in \mathcal{R} \cup \bar{\mathcal{R}}\}$

Proof: (i) If $(f,\theta) \in \mathcal{R} \cup \bar{\mathcal{R}}$ or f and θ are real but $fX = 0$ then (i) follows from Corollary 2.3 and the fact that $V(\theta)$ is $H_0(f,\theta)$-bounded with relative bound zero (which in turn follows from the quadratic estimate). If f and θ are real and $fX \neq 0$, the self-adjointness of $H(f,\theta)$ follows from a theorem of Faris and Lavine [17,43].

(ii) The proof of this analyticity result is exactly the same as the proof of (ii) in Proposition 2.4.

(iii) Given $f \neq 0$, define $\hat{f} = f/|f| = e^{i\arg f}$. It is enough to show that if $\mathcal{M}$ is a compact subset of $\{(f,\theta) \in \mathcal{R}:|f| = 1\}$ then there is an $M > 0$ so that for all $(f,\theta) \in \mathcal{R}$ with $(\hat{f},\theta) \in \mathcal{M}$, the numerical range of $L(f,\theta) + M$ is contained in the right half plane. The fact that $L(f,\theta) + M$ is maximal accretive then follows from this numerical range result, the fact that $L_0(f,\theta)$ is maximal accretive, and the result that $L(f,\theta)-L_0(f,\theta)$ is $L_0(f,\theta)$-bounded with relative bound zero. To derive a similar result for $\mathcal{M}$ a compact subset of $\bar{\mathcal{R}}$ we need only note the definition (2.3) and the fact that a closed operator is maximal accretive if and only if its adjoint is.

Thus given $\mathcal{M}$ as above, choose $\mu \in (0,\pi/2)$ so that for all $(f,\theta) \in \mathcal{M}$, $\mathrm{Im}(fe^{3\theta}e^{\pm i\mu}) > 0$ and let

$$\delta = \sup\{|e^{-6\theta}|\,\mathrm{Im}(fe^{3\theta}): (f,\theta) \in \mathcal{M}\}$$

By the compactness and analyticity of $v_{ij}(\theta)(-\Delta+1)^{-1}$, it is easy to show that $V(\theta)(\lambda+K)^{-1} \to 0$ as $\lambda \uparrow \infty$ uniformly on compact subsets of $\{\theta:|\mathrm{Im}\theta| < \theta_0\}$. It follows that we can choose $\lambda_0 > 0$ so that for all θ in $\{\theta:(f,\theta) \in \mathcal{M}$ for some $f\}$, the numerical range of $\lambda_0 + K + V(\theta)e^{2\theta}$ lies in the sector

$$S_\mu = \{t_1 e^{i\mu} + t_2 e^{-i\mu} : t_i \geq 0\}$$

Then for any $\psi \in \mathcal{D}(K) \cap \mathcal{D}(X)$ and any $(f,\theta) \in \mathcal{R}$ with $(\hat{f},\theta) \in \mathcal{M}$ there is a $z \in S_\mu$ so that

$$\mathrm{Re}(\psi, L(f,\theta)\psi) = \mathrm{Re}(\psi, i(\hat{f}e^{3\theta})^{-1}(K + e^{2\theta}V(\theta))\psi)$$

$$= \|\psi\|^2 \mathrm{Re}[i(\hat{f}e^{3\theta})^{-1}(z-\lambda_0)]$$

But $\mathrm{Re}(i(\hat{f}e^{3\theta})^{-1}e^{\pm i\mu}) = |e^{-6\theta}|\mathrm{Im}(\hat{f}e^{3\theta}e^{\mp i\mu}) > 0$ and $\mathrm{Re}(i[\hat{f}e^{3\theta}]^{-1}) = |e^{-6\theta}|\mathrm{Im}(\hat{f}e^{3\theta}) \leq \delta$ so that with $M = \delta\lambda_0$,

$$\mathrm{Re}(\psi, (L(f,\theta) + M)\psi) \geq 0. \qquad \square$$

For real f we define

$$\mathcal{R}_f = \bigcup_{0 < \mathrm{Im}\theta < \mathrm{Min}\{\theta_0, \pi/3\}} \sigma_{\mathrm{disc.}}(H(f,\theta)) \tag{2.32}$$

as the set of resonances of $H(f) \equiv H(f,0)$. Here $\sigma_{\mathrm{disc.}}(A)$ is the set of all isolated eigenvalues of A having finite algebraic multiplicity. The remainder of this section is devoted to elucidating the structure of $\mathcal{R}_f$ and showing that the points of $\mathcal{R}_f$ are nothing more than poles in analytically continued matrix elements of $(z-H(f))^{-1}$. In order to accomplish this we will also find $\sigma_{\mathrm{ess.}}(H(f,\theta))$. (Here $\sigma_{\mathrm{ess.}}(A) = \sigma(A) \setminus \sigma_{\mathrm{disc.}}(A)$.)

The problem is more difficult than in the case f = 0 in part because in general if f is real, $(z-H(f,\theta))^{-1}$ is not analytic in the variable θ at $\theta = 0$. However, we have the following strong convergence result as a substitute:

Lemma 2.6 [32]: Suppose Imf = 0. Then if Imz > 0,

$$\operatorname*{s-lim}_{\theta\to 0, \mathrm{Im}\theta > 0} (z-H(f,\theta))^{-1} = (z-H(f))^{-1} \tag{2.33}$$

while if Imz < 0

$$\operatorname*{s-lim}_{\theta\to 0, \mathrm{Im}\theta < 0} (z-H(f,\theta))^{-1} = (z-H(f))^{-1} \tag{2.34}$$

Proof: We need only show that if z is fixed with Imz > 0 and

and $\mathrm{Im}\theta > 0$, then for θ small enough, $\|(z-H(f,\theta))^{-1}\|$ is uniformly bounded. For then by taking the adjoint the same is true for fixed z with $\mathrm{Im}z < 0$ if $\mathrm{Im}\theta < 0$ with θ small enough and we have the following well known [37] convergence argument. If $\phi \in \mathcal{D}(K) \cap \mathcal{D}(X)$

$$[(z-H(f,\theta))^{-1}-(z-H(f))^{-1}](z-H(f))\phi = (z-H(f,\theta))^{-1}(e^{\theta}-1)fX\phi \to 0 \tag{2.35}$$

as $\theta \to 0$. But $\{(z-H(f))\phi:\phi \in \mathcal{D}(K) \cap \mathcal{D}(X)\}$ is dense since $\mathcal{D}(K) \cap \mathcal{D}(X)$ is a core for $H(f)$ [17,43] and thus we have strong convergence on a dense set. Since $\|(z-H(f,\theta))^{-1}\|$ is uniformly bounded this is easily seen to imply (2.33) and (2.34).

To derive the uniform bound we show that for $\mathrm{Im}\theta > 0$ and small, the numerical range of $H(f,\theta)$ is contained in the half plane below the line $z = -E + e^{\theta}t$, $-\infty < t < \infty$, for some fixed $E > 0$. The uniform bound easily follows from this. Because of the unitarity of $U(\phi)$ and the equation

$$U(\phi)H(f,\theta)U(-\phi) = H(f,\theta+\phi) \tag{2.36}$$

if suffices to prove that there exists an $E > 0$ so that for all small $\beta > 0$

$$(\psi,H(f,i\beta)\psi) \text{ lies below } \{-E+e^{i\beta}t:t \in \mathbb{R}\} \tag{2.37}$$

In the proof of Proposition 2.5 it was shown that given a compact subset $\tilde{M}$ of $\{\theta:|\mathrm{Im}\theta| < \theta_0\}$ and $\mu \in (0,\pi/2)$, the numerical range of $K + V(\theta)e^{2\theta} + \lambda_0$ lies in S_μ for all $\theta \in \tilde{M}$ if λ_0 is sufficiently large. We strengthen this result to show that μ can be taken to shrink as $\mathrm{Im}\theta \to 0$. Thus given $\phi_0 < \mathrm{Min}\{\theta_0,\pi/3\}$ with $\phi_0 > 0$ choose ε in $(0,1)$ and $\lambda_0 = \lambda_0(\varepsilon)$ so that

$$\|V(i\beta)(K+\lambda_0)^{-1}\| \le \varepsilon; \quad -\phi_0 \le \beta \le \phi_0 \tag{2.38}$$

In addition since $\frac{d}{d\theta}v_{ij}(\theta)e^{2\theta}(-\Delta+1)^{-1}$ is compact and analytic we can choose λ_0 to satisfy

$$\left\|\frac{d}{d\theta}[V(\theta)e^{2\theta}](K+\lambda_0)^{-1}\right\| \le \varepsilon(1-\varepsilon); \quad |\theta| \le \phi_0 \tag{2.39}$$

From (2.39) it follows that

$$\|(V(i\beta)e^{2i\beta}-V(0))(K+\lambda_0)^{-1}\| \leq \varepsilon(1-\varepsilon)|\beta|; \quad -\phi_0 \leq \beta \leq \phi_0$$

so that by interpolation

$$\|(K+\lambda_0)^{-1/2}(V(i\beta)e^{2i\beta}-V(0))(K+\lambda_0)^{-1/2}\| \leq \varepsilon(1-\varepsilon)|\beta| ;$$

$$-\phi_0 \leq \beta \leq \phi_0 \tag{2.40}$$

Hence if $\psi \in \mathcal{D}(K)$

$$|\mathrm{Im}(\psi,[K+\lambda_0+V(i\beta)e^{2i\beta}]\psi)| =$$

$$|\mathrm{Im}((K+\lambda_0)^{1/2}\psi,(K+\lambda_0)^{-1/2}(V(i\beta)e^{2i\beta}-V(0))(K+\lambda_0)^{-1/2}(K+\lambda_0)^{1/2}\psi)|$$

$$\leq \varepsilon(1-\varepsilon)|\beta|(\psi,(K+\lambda_0)\psi); \quad -\phi_0 \leq \beta \leq \phi_0 \tag{2.41}$$

Similarly from (2.38)

$$\mathrm{Re}(\psi,[K+\lambda_0+V(i\beta)e^{2i\beta}]\psi) \geq (1-\varepsilon)(\psi,(K+\lambda_0)\psi); \ -\phi_0 \leq \beta \leq \phi_0 \tag{2.42}$$

so that if $z = (\psi,[K+\lambda_0+V(i\beta)e^{2i\beta}]\psi)$, $|\mathrm{Im}z|/\mathrm{Re}z \leq \varepsilon|\beta|$. This implies the numerical range of $K + V(i\beta)e^{2i\beta} + \lambda_0$ lies in the sector $S_{\varepsilon\beta} = \{t_1e^{i\varepsilon\beta} + t_2e^{-i\varepsilon\beta} : t_i \geq 0\}$.

We now show that (2.37) holds for $0 < \beta \leq \phi_0$ if E is chosen appropriately. This is equivalent to

$$\mathrm{Im}(\psi,e^{-i\beta}(H(f,i\beta) + E)\psi) \leq 0; \quad 0 < \beta \leq \phi_0 \tag{2.43}$$

or

$$\mathrm{Im}(\psi,e^{-3i\beta}(K+V(i\beta)e^{2i\beta} + \lambda_0)\psi)$$
$$+ \mathrm{Im}(e^{-i\beta}E - \lambda_0e^{-3i\beta})\|\psi\|^2 \leq 0; \quad 0 < \beta \leq \phi_0 \tag{2.44}$$

The first term in (2.44) is ≤ 0 if ε is chosen small enough and if E is chosen large enough so is the second. □

We now introduce some notation to enable us to talk about subsystem Hamiltonians and eventually essential spectrum. Thus if $C \subseteq \{1,2,\cdots,N+1\}$, the operator $H^C(f,\theta)$ in $L^2(\mathbb{R}^{3(|C|-1)})$ is defined to be the result of removing the center of mass from the operator

$$\tilde{H}^C(f,\theta) = -e^{-2\theta}\sum_{i\in C}\Delta_i/2m_i + fe^{\theta}\sum_{i\in C} q_i\vec{x}_i\cdot\hat{e} + \sum_{i,j\in C} V_{ij}(\theta) \quad (2.45)$$

as in (2.2). ($|A|$ is the number of elements in the set A.) If $D = \{C_1,\cdots,C_k\}$ is a cluster decomposition we write $I_D(f,\theta)$ as the sum of all $V_{ij}(\theta)$ with i and j in different clusters of D. (Note $\bigcup_{i=1}^{k} C_i = \{1,\cdots,N+1\}$ and $C_i \cap C_j = \emptyset$ for $i \neq j$. The C_i are called clusters of D.) We define as is customary,

$$H_D(f,\theta) = H(f,\theta) - I_D(f,\theta) \quad (2.46)$$

and note that if $|D| \geq 2$ then

$$H_D(f,\theta) = H^{C_1}(f,\theta)\otimes I\otimes\cdots\otimes I + \cdots + I\otimes\cdots\otimes H^{C_k}(f,\theta)\otimes I$$
$$+ I\otimes\cdots\otimes I\otimes h_D(f,\theta) \quad (2.47)$$

where $h_D(f,\theta)$ is the operator which results from

$$-e^{-2\theta}\sum_{i=1}^{k}\Delta_{\vec{R}_{C_i}}/2M_{C_i} + e^{\theta}f\sum_{i=1}^{k} Q_{C_i}\vec{R}_{C_i}\cdot\hat{e}$$

when the center of mass is removed. Here

$$M_{C_i} = \sum_{j\in C_i} m_j,\quad Q_{C_i} = \sum_{j\in C_i} q_j,\quad \vec{R}_{C_i} = M_{C_i}^{-1}\sum_{j\in C_i} m_j\vec{x}_j. \quad (2.48)$$

We write

$$h_D(f,\theta) = -e^{-2\theta}k + e^{\theta}fx \quad (2.49)$$

and note that by Proposition 2.1, $x = 0$ if and only if $Q_{C_i}/M_{C_i} = Q_{C_j}/M_{C_j}$ for all i and j. In this case we say that D has equally accelerated clusters. If $x \neq 0$ (i.e. if $Q_{C_i}/M_{C_i} \neq Q_{C_j}/M_{C_j}$ for some i and j) we say that D has unequally accelerated clusters.

A system of N+1 charges and masses is said to be <u>ineffective</u> if there is a cluster decomposition D with $|D| \geq 2$ having equally accelerated clusters. Otherwise the system is said to be <u>effective.</u>

Theorem 2.7 [32]: *Suppose the system described by* $H(f,\theta)$ *is effective. Then if* $(f,\theta) \in \mathcal{R} \cup \bar{\mathcal{R}}$, $\sigma_{ess.}(H(f,\theta)) = \emptyset$. *If f is real and* $(f,\theta) \in \mathcal{R}$ *then*

$$\sigma_{disc.}(H(f,\theta)) \subseteq \{z : \text{Im} z \leq 0\}$$

Proof: It $(f,\theta) \in \mathcal{R} \cup \bar{\mathcal{R}}$ we define operators $L^C(f,\theta)$, $L_D(f,\theta)$, $\ell_D(f,\theta)$ corresponding to $H^C(f,\theta)$, $H_D(f,\theta)$, $h_D(f,\theta)$ respectively, by equations analogous to (2.29) and (2.30). Clearly an equation for $L_D(f,\theta)$ exactly like (2.47) holds with all H's changed to L's. This gives for $D = \{C_1,\cdots,C_k\}$ and $k \geq 2$,

$$e^{-tL_D(f,\theta)} = e^{-tL^{C_1}(f,\theta)} \otimes \cdots \otimes e^{-tL^{C_k}(f,\theta)} \otimes e^{-t\ell_D(f,\theta)} \tag{2.50}$$

By (iii) of Proposition 2.5, $\|e^{-tL^{C_j}(f,\theta)}\| \leq e^{t\lambda_j}$ for some λ_j while by the effectiveness hypothesis and (2.24), $\|e^{-t\ell_D(f,\theta)}\| = e^{-\lambda t^3}$ for some $\lambda > 0$. Thus using $\|A \otimes B\| = \|A\| \, \|B\|$ we have

$$\|e^{-tL_D(f,\theta)}\| \leq \exp\left(\sum_{j=1}^{k} \lambda_j t - \lambda t^3\right)$$

so that the integral $\int_0^\infty \|e^{-t(L_D(f,\theta) - z)}\| dt$ converges for all $z \in \mathbb{C}$. Hence $(z - L_D(f,\theta))^{-1}$ is entire in the variable z, i.e. the spectrum of $L_D(f,\theta)$ (or $H_D(f,\theta)$) is empty for all D with $|D| \geq 2$.

To prove $\sigma_{ess.}(H(f,\theta)) = \emptyset$ we use the Weinberg-van Winter expansion [44,48,51] for the resolvent of $L(f,\theta)$:

$$(z-L(f,\theta))^{-1} = \tag{2.51}$$

$$\sum_{S=(D_{N+1},\cdots,D_k)} (z-L_{D_{N+1}}(f,\theta))^{-1}\beta_N^S(z-L_{D_N}(f,\theta))^{-1}\beta_{N-1}^S\cdots\beta_k^S(z-L_{D_k}(f,\theta))^{-1}$$

Here S denotes a "string" of cluster decompositions and the sum is over all strings (so that the index k is also summed over). The operators β_ℓ^S are given by

$$\beta_\ell^S = \pm ie^{-\theta}|f|f^{-1} \sum' V_{ij}(\theta) \tag{2.52}$$

where in accordance with (2.29) and (2.30) the plus sign is taken if $(f,\theta) \in R$ and the minus sign if $(f,\theta) \in \bar{R}$. $\sum'$ is the sum over all pairs (i,j) with $i < j$ which connect the two clusters in $D_{\ell+1}$ which are joined together in D_ℓ. (2.51) is an algebraic identity [41] which holds whenever z is not in the spectrum of any $L_D(f,\theta)$ (including the trivial $|D| = 1$ where $L_D(f,\theta) = L(f,\theta)$).

A string $(D_{N+1},\ldots,D_k)$ is called connected if $k = 1$ and otherwise disconnected. We denote by $D(f,\theta;z)$ the part of the sum in (2.51) extending over disconnected S. We multiply the remaining part of the sum by $z - L(f,\theta) = z - L_{D_1}(f,\theta)$ on the right and call the result $I(f,\theta;z)$. We thus have [44,48,51]

$$(z-L(f,\theta))^{-1} = D(f,\theta;z) + I(f,\theta;z)(z-L(f,\theta))^{-1} \tag{2.53}$$

For $|D| \geq 2$ the operators $V_{ij}(\theta)\ (z-L_D(f,\theta))^{-1}$ appearing in (2.51) are

(a) bounded,
(b) convergent to zero as $z \to -\infty$,
(c) entire functions of z

The fact that all $V_{ij}(\theta)$ are $L_0(f,\theta)$-bounded with relative bound zero implies the same is true with $L_0(f,\theta)$ replaced with $L_D(f,\theta)$. This gives (a) and coupled with the estimate (see Proposition 2.5) $\|(z+L_D(f,\theta)+E)^{-1}\| \leq (\mathrm{Re}z)^{-1}$ for $\mathrm{Re}z > 0$ and some $E > 0$ also gives (b). (c) follows from (a) and the identity

$$\begin{aligned} &V_{ij}(\theta)(z-L_D(f,\theta))^{-1} = \\ &V_{ij}(\theta)(z_0-L_D(f,\theta))^{-1}[1+(z_0-z)(z-L_D(f,\theta))^{-1}] \end{aligned} \tag{2.54}$$

where the factor in square brackets is entire. (a), (b), and (c)

also hold true for operators of the form $(z-L_D(f,\theta))^{-1}V_{ij}(\theta)$ (which is an abuse of notation for the bounded extension of this operator from $\mathcal{D}(K)$ to $L^2(\mathbb{R}^{3N})$).

From the above statements it follows that $D(f,\theta;z)$ and $I(f,\theta;z)$ are entire function of z and that $\|I(f,\theta;z)\|$ converges to zero as $z \to -\infty$ so that $1-I(f,\theta;z)$ is invertible on an open set $\mathcal{O}$ containing all very negative z. Choose E so that $L_D(f,\theta) + E$ is accretive for all D. Then

$$(z-L(f,\theta))^{-1} = (1-I(f,\theta;z))^{-1}D(f,\theta;z) \tag{2.55}$$

for all $z \in \mathcal{O} \cap \{z:\text{Re}z < -E\}$, a non-empty open set. We claim that $I(f,\theta;z)$ is compact. The proof is given in Lemma 2.8. Assuming this, the analytic Fredholm theorem [42] implies that the right hand side of (2.55) is meromorphic in $\mathbb{C}$ with finite rank residues at its poles. By analytically continuing (2.55) we see that $\sigma_{\text{ess.}}(L(f,\theta)) = \emptyset$.

To see that $\sigma(H(f,\theta)) \subseteq \{z:\text{Im}z \leq 0\}$ if $(f,\theta) \in \mathcal{R}$ and f is real, first note that by (2.36) and the analyticity result (ii) of Proposition 2.5, the spectrum of $H(f,\theta)$ is independent of θ for $0 < \text{Im}\theta < \text{Min}\{\theta_0,\pi/3\}$. However, Lemma 2.6 says in part that given z_0 with $\text{Im}z_0 > 0$, $z_0 - H(f,\theta)$ is invertible for $\text{Im}\theta > 0$ and small enough. Thus z_0 cannot be in $\sigma(H(f,\theta))$ for any θ with $0 < \text{Im}\theta < \text{Min}\{\theta_0,\pi/3\}$. □

We now prove the compactness of $I(f,\theta;z)$ along with another result which will be helpful in finding the essential spectrum of $H(f,\theta)$ in the ineffective case. We do not make the effectiveness hypothesis in either of the following lemmas.

Lemma 2.8: *Suppose $(f,\theta) \in \mathcal{R} \cup \bar{\mathcal{R}}$. Then for each $z \notin \bigcup_{|D|\geq 2} \sigma(L_D(f,\theta))$, $I(f,\theta;z)$ is compact.*

Lemma 2.9 [32]: *Suppose $(f,\theta) \in \mathcal{R} \cup \bar{\mathcal{R}}$ and that for $|D| \geq 2$ each operator $L_D(f,\theta)$ satisfies:*

Condition C: Given any $\lambda \in \mathbb{R}$ there is a $y_\lambda > 0$ so that $\sigma(L_D(f,\theta)) \cap \{z:\text{Re}z \leq \lambda\} \subseteq \{z:|\text{Im}z| < y_\lambda\}$ and $\sup\{\|(z-L_D(f,\theta))^{-1}\|:\text{Re}z \leq \lambda,\ |\text{Im}z| \geq y_\lambda\} < \infty$.

Then $L(f,\theta)$ also satisfies condition C.

Proof of Lemma 2.8: We claim that it is enough to prove the compactness of an operator of the form

$$B' = (z-L_{D_1}(f,\theta))^{-1}V_1'(z-L_{D_2}(f,\theta))^{-1}V_2'\cdots(z-L_{D_n}(f,\theta))^{-1}V_n' \quad (2.56)$$

where z is as in the lemma, $|D_i| \geq 2$, V_i' is multiplication by a function $v'_{j_ik_i}(\vec{x}_{j_i}-\vec{x}_{k_i})$ with $v'_{j_ik_i} \in C_0^\infty(\mathbb{R}^3)$ and where the set of pairs (j_i,k_i): $i=1,\cdots,n$ is such that for any (ℓ,m), $\vec{x}_\ell - \vec{x}_m = \sum_{i=1}^{n} a_{\ell mi}(\vec{x}_{j_i}-\vec{x}_{k_i})$. From the compactness of the operators $v_{ij}(\theta)(-\Delta+1)^{-1}$ it follows (see for example [33]) that $(-\Delta+1)^{-1}v_{j_ik_i}(\theta)$ can be approximated in norm by operators of the form $(-\Delta+1)^{-1}v'_{j_ik_i}$ and thus the operator B which results from B' by replacing each V_i' by $V_{j_ik_i}(\theta)$ can be approximated in norm by operators of the form B'. Thus if such operators B' are compact so is B. Since $I(f,\theta;z)$ is a linear combination of operators of the form B, $I(f,\theta;z)$ is compact.

To show that B' is compact, first note that KB' is bounded. Secondly, if $\hat{a}$ is a unit vector we can prove that $\hat{a}\cdot(\vec{x}_{j_i}-\vec{x}_{k_i})B'$ is bounded by commuting $\hat{a}\cdot(\vec{x}_{j_i}-\vec{x}_{k_i})$ past resolvents $(z-L_{D_j}(f,\theta))^{-1}$ until it hits V_i'. The commutators are easily shown to be bounded. By using the formula $\vec{x}_\ell - \vec{x}_m = \sum_{i=1}^{n} a_{\ell mi}(\vec{x}_{j_i}-\vec{x}_{k_i})$ we easily see that with $\rho = \sum_{i<j}|\vec{x}_i - \vec{x}_j|$, $\rho B'$ is bounded. Thus

$$B' = (K+\rho+1)^{-1}[(\rho+K+1)B']$$

expresses B' as the product of a compact operator $(K+\rho+1)^{-1}$ and a bounded operator. ∎

Before proving Lemma 2.9 we state another result (without proof). It is easily proved using the maximum principle.

Lemma 2.10 [32]: Suppose $F(z)$ is an operator valued function, continuous on $\{z: 0 \leq \mathrm{Re}z \leq 1, |\mathrm{Im}z| \geq 0\} \equiv A$ and analytic in the interior of A. Suppose $\sup\{\|F(z)\| : z \in A\} < \infty$ and $\lim_{y\to+\infty} \|F(iy)\| = 0$. Then $\sup\{\|F(x+iy)\| : 0 \leq x \leq 1-\delta\} \to 0$ as $y \uparrow \infty$ for any $\delta \in (0,1)$.

Proof of Lemma 2.9: Given $E \in \mathbb{R}$ we must find $\beta > 0$ so that

$$\sigma(L(f,\theta)) \cap \{z: \mathrm{Re}z \leq E\} \subseteq \{z: |\mathrm{Im}z| < \beta\} \tag{2.57}$$

$$\sup\{\|(z-L(f,\theta))^{-1}\| : \mathrm{Re}z \leq E, |\mathrm{Im}z| \geq \beta\} < \infty \tag{2.58}$$

In the following we treat only the non-trivial case $X \neq 0$. Let $\lambda = E + 1$ and suppose y_λ is the number guaranteed to exist by our hypothesis. Define $W = \{z: \mathrm{Re}z \leq \lambda, |\mathrm{Im}z| \geq y_\lambda\}$. By hypothesis $\|(z-L_D(f,\theta))^{-1}\|$ is bounded for $z \in W$ and $|D| \geq 2$. We want to show that

$$\sup_{z \in W} \|V_{ij}(\theta)(z-L_D(f,\theta))^{-1}\| < \infty \tag{2.59}$$

First note that if in (2.54) we set $z_0 = z - C$ with $C > 0$ and large, it is sufficient to show that for some $E_0 \in \mathbb{R}$

$$\sup\{\|V_{ij}(\theta)(z+L_D(f,\theta))^{-1}\| : \mathrm{Re}z \geq E_0\} < \infty \tag{2.60}$$

We know that for any $\varepsilon > 0$ and all (ℓ,m)

$$\|V_{\ell m}(\theta)\psi\| \leq \varepsilon\|K\psi\| + C_\varepsilon\|\psi\|; \quad \psi \in \mathcal{D}(K) \tag{2.61}$$

for some C_ε and by the quadratic estimate

$$\|K\psi\| \leq C_1(\|L_0(f,\theta)\psi\| + \|\psi\|); \ \psi \in \mathcal{D}(L_0(f,\theta)) \tag{2.62}$$

(2.62) implies that for $\mu > 0$

$$\|K(L_0(f,\theta)+\mu)^{-1}\| \leq C_1(2+\mu^{-1}) \tag{2.63}$$

but by performing a translation which commutes with K and takes $L_0(f,\theta)$ to $L_0(f,\theta) + i\gamma$, (2.63) implies

$$\|K(L_0(f,\theta)+z)^{-1}\| \leq C_1(2+(\mathrm{Re}z)^{-1}); \ \mathrm{Re}z > 0 \tag{2.64}$$

Thus from (2.57), for Rez > 0 and any $\varepsilon > 0$

$$\| V_{\ell m}(\theta)(L_0(f,\theta)+z)^{-1}\| \leq \varepsilon C_1(2+(\mathrm{Rez})^{-1}) + C_\varepsilon(\mathrm{Rez})^{-1}$$

so that

$$\sup\{\| V_{\ell m}(\theta)(L_0(f,\theta)+z)^{-1}\| : \mathrm{Rez} \geq 1\} < \infty \tag{2.65}$$

$$\lim_{\mathrm{Rez}\to+\infty} \| V_{\ell m}(\theta)(L_0(f,\theta)+z)^{-1}\| = 0 \tag{2.66}$$

Since $L_D(f,\theta) - L_0(f,\theta)$ is a linear combination of $V_{\ell m}(\theta)$'s, for Rez large enough $\| [L_D(f,\theta) - L_0(f,\theta)](L_0(f,\theta)+z)^{-1}\| < 1$ and

$$V_{ij}(\theta)(z+L_D(f,\theta))^{-1} =$$
$$V_{ij}(\theta)(L_0(f,\theta)+z)^{-1}(1+[L_D(f,\theta)-L_0(f,\theta)](L_0(f,\theta)+z)^{-1})^{-1}$$

so that (2.60) holds for E_0 large enough. In addition (2.66) holds with $L_0(f,\theta)$ replaced by $L_D(f,\theta)$. Similarly these results hold for $(z-L_D(f,\theta))^{-1}V_{ij}(\theta)$ so that

$$\sup\{\| I(f,\theta;z)\| + \| D(f,\theta;z)\| ;\ z \in W\} < \infty \tag{2.67}$$

$$\lim_{\mathrm{Rez}\to-\infty} \| I(f,\theta;z)\| = 0 \tag{2.68}$$

We will show that for some $\beta > y_\lambda$

$$\sup\{\| I(f,\theta;z)\| : \mathrm{Rez} \leq E, |\mathrm{Imz}| \geq \beta\} \leq 1/2 \tag{2.69}$$

from which (2.57) and (2.58) easily follow from the Weinberg-van Winter equation (2.55), and (2.67). Because of (2.68) and Lemma 2.10, it suffices in proving (2.69) to show that for some large negative λ_1 chosen so that $\| I(f,\theta;x+iy)\| \leq 1/2$ for all $y \in \mathbb{R}$ if $x \leq \lambda_1$, we have $\lim_{y\to\pm\infty} \| I(f,\theta;\lambda_1+iy)\| = 0$. We choose λ_1 so that the expansion of $I(f,\theta;\lambda_1+iy)$ in powers of the $V_{ij}(\theta)$ converges uniformly in y (this can be accomplished because of (2.66)) and thus it sufficies to show that each term in this

expansion $\to 0$ as $y \to \pm\infty$. A typical term is a multiple of

$$G(y) = (L_0(f,\theta) - \lambda_1 - iy)^{-1} V_1 \cdots (L_0(f,\theta) - \lambda_1 - iy)^{-1} V_n \qquad (2.70)$$

where $V_i = V_{j_i k_i}(\theta)$ and the set $\{(j_i, k_i): i=1,\cdots,n\}$ makes (2.70) compact. By a simple approximation argument it is sufficient to prove $G(y) \to 0$ when each V_i is multiplication by a function $f_i(\vec{x}_{j_i} - \vec{x}_{k_i})$ with $f_i \in C_0^\infty(\mathbb{R}^3)$. In this case we write

$$G(y) = \int_{t_i > 0} g(t_1,\cdots,t_n) e^{i(t_1+\cdots+t_n)y} \prod_{i=1}^{n} dt_i$$

with $g(t_1,\cdots,t_n) = e^{-t_1 L_0(f,\theta)} V_1 \cdots e^{-t_n L_0(f,\theta)} V_n e^{\lambda_1(t_1+\cdots+t_n)}$. By the procedure used in the proof of Lemma 2.8 it is easy to see that g is compact if $\prod_{i=1}^{n} t_i > 0$. Here the representation (2.22) is use- in computing the relevant commutators. g is easily seen to be weakly measurable and since the compact operators is a separable space, g is strongly measurable. Since $g \in L^1(\prod_{i=1}^{n} dt_i)$, the Riemann-Lebesgue lemma implies $\lim_{y\to\pm\infty} \|G(y)\| = 0$. □

We make use of the following extension of Ichinose's lemma [43] to find $\sigma_{ess.}(L(f,\theta))$. It's proof it omitted.

<u>Proposition 2.11 [29]:</u> <u>Suppose</u> $A_i, i=1,2,\cdots,m$ <u>generate exponentially bounded semigroups</u> $\{e^{-tA_i}: t \geq 0\}$ <u>on separable Hilbert spaces</u> H_i. (<u>We assume</u> $\|e^{-tA_i}\| \leq Ce^{Kt}$ <u>with</u> $C = 1$.) <u>Suppose each</u> A_i <u>satisfies condition C. Then the operator A defined on</u> $H_1 \otimes \cdots \otimes H_m$ <u>by</u> $e^{-tA} = e^{-tA_1} \otimes \cdots \otimes e^{-tA_m}$ <u>also satisfies condition C and</u>

$$\sigma(A) = \sigma(A_1) + \cdots + \sigma(A_m)$$

<u>Theorem 2.12 [32]:</u> Suppose $(f,\theta) \in R \cup \bar{R}$. <u>Define</u> I <u>to be the set of all cluster decompositions</u>, D, <u>with equally accelerated clusters. Then</u>

$$\sigma_{ess.}(H(f,\theta)) = \bigcup_{D=\{C_1,\cdots,C_k\}\in I} \{\lambda e^{-2\theta}+\mu_1+\cdots+\mu_k : \lambda \geq 0, \mu_i \in \sigma_{disc.}(H^{C_i}(f,\theta))\} \tag{2.71}$$

In addition if f is real, then the discrete spectrum of $H(f,\theta)$ and all $H^{C_i}(f,\theta)$ is contained in $\{z:(\mathrm{Im}z)(\mathrm{Im}\theta) \leq 0\}$.

Proof: We assume $X \neq 0$ otherwise the proof is standard. We proceed by induction on $|C|$ to prove that $L^C(f,\theta)$ satisfies condition C. This is certainly true if $|C| = 1$ where there is no interaction. If it is true for all $|C| \leq N$, then since the operators $L_D(f,\theta)$ with $|D| \geq 2$ are given by (2.47) with all $|C_i| \leq N$, Proposition 2.11 shows that $L_D(f,\theta)$ satisfies condition C and thus by Lemma 2.9 so does $L(f,\theta) = L^{\{1,2,\cdots,N+1\}}(f,\theta)$.

Let us now suppose inductively that (2.71) holds with $H^C(f,\theta)$ ($|C| \leq N$) replacing $H(f,\theta)$. We then want to prove it for $H(f,\theta)$. By Proposition 2.11

$$\sigma(H_D(f,\theta)) = \sum_{i=1}^{k} \sigma(H^{C_i}(f,\theta)) + \sigma(h_D(f,\theta)) \tag{2.72}$$

If $D \notin I$ then $\sigma(h_D(f,\theta)) = \emptyset$ and thus $\sigma(H_D(f,\theta)) = \emptyset$. Otherwise $\sigma(h_D(f,\theta)) = \{\lambda e^{-2\theta} : \lambda \geq 0\}$. Calling the right side of (2.71) E we claim that $E = \bigcup_{|D|\geq 2} \sigma(H_D(f,\theta))$. The inclusion $E \subseteq \bigcup_{|D|\geq 2} \sigma(H_D(f,\theta))$ follows from (2.72) while if $z \in \sigma(H_D(f,\theta))$, by (2.72) $D = \{C_1,\cdots,C_k\} \in I$ and $z = \lambda e^{-2\theta} + \sum_{i=1}^{k} \mu_i$ with $\lambda \geq 0$ and $\mu_i \in \sigma(H^{C_i}(f,\theta))$. By the induction hypothesis either $\mu_i \in \sigma_{disc.}(H^{C_i}(f,\theta))$ or there is a cluster decomposition $\{C_{i1},\cdots,C_{i\ell_i}\}$ of C_i with uniformly accelerated clusters so that $\mu_i = \lambda_i e^{-2\theta} + \sum_{j=1}^{\ell_i} \mu_{ij}$ and $\mu_{ij} \in \sigma_{disc.}(H^{C_{ij}}(f,\theta))$. Note $Q_{C_{ij}}/M_{C_{ij}}$

$= Q_{C_i}/M_{C_i}$. Putting these facts together we obtain a refinement $D' = \{C'_1, \cdots, C'_\ell\}$ of D and numbers λ', μ'_i with $\lambda' \geq 0$, $\mu'_i \in \sigma_{\text{disc.}}(H^{C'_i}(f,\theta))$ so that $z = \lambda' e^{-2\theta} + \sum_{i=1}^{\ell} \mu'_i$. Thus

$E = \bigcup_{|D| \geq 2} \sigma(H_D(f,\theta))$. In a standard way the Weinberg-vanWinter equation implies $(z-H(f,\theta))^{-1}$ is meromorphic outside E with finite rank residues at its poles and thus $E \supseteq \sigma_{\text{ess.}}(H(f,\theta))$. The opposite inclusion can be proved using Hunziker's method [35] of translating the clusters far from each other.

To prove the last statement of the theorem, we proceed by induction on $|C|$ using (2.72) and the same proof used to prove the analogous result in Theorem 2.7. ∎

Although there are many other results one can now state based on the machinery we have developed (see [2,10,44] for X = 0), we complete our discussion of the dilation analytic formalism by giving a result which shows that the set of resonances R_f defined by (2.32) has some connection with the self-adjoint operator $H(f) \equiv H(f,0)$. We restrict ourselves to the effective case where the result is easier to state:

Theorem 2.13 [32]: Suppose ψ_1 and ψ_2 are dilation analytic vectors (i.e. $U(\theta)\psi_j$ has an $L^2(\mathbb{R}^{3N})$-valued analytic continuation to $\{\theta: |\text{Im}| < \phi_0\}$ for some $\phi_0 > 0$). Then for f real

$$(\psi_1, (z-H(f))^{-1}\psi_2) \tag{2.73}$$

has a meromorphic continuation from Imz > 0 to $\mathbb{C}$ with poles possible only at the points of R_f. If $z_0 \in R_f$ then for some dilation analytic ψ_j, (2.73) will have a pole at z_0.

Proof: We will prove only that for Imz > 0

$$(\psi_1, (z-H(f))^{-1}\psi_2) = (U(\bar{\theta})\psi_1, (z-H(f,\theta))^{-1}U(\theta)\psi_2) \tag{2.74}$$

for any θ with $0 < \text{Im}\theta < \text{Min}\{\theta_0, \phi_0, \pi/3\}$. The rest is standard.

The right side of (2.74) is analytic in θ in this region

(Theorem 2.7 and (ii) of Proposition 2.5), independent of Reθ (Equation (2.36)), and thus constant. But by Lemma 2.6 the right side converges to the left as $\theta \to 0$ with Im$\theta > 0$. □

I.3 STABILITY

Consider a family of operators $\{A(f):f \geq 0\}$ where in some sense $A(f)$ is close to $A = A(0)$ for small f. If A has an eigenvalue $E \in \sigma_{disc.}(A)$, of algebraic multiplicity m, a reasonable definition of stability of this eigenvalue should at least demand that in any sufficiently small neighborhood of E, $A(f)$ should have eigenvalues of combined multiplicity m for all sufficiently small $f > 0$. (This idea is formalized in [37]). In this sense the eigenvalues of Hydrogen are certainly not stable when $H = -\Delta-|\vec{x}|^{-1}$ is perturbed to $H(f) = H + f\hat{e} \cdot \vec{x}$ since for $f > 0$ $H(f)$ has no eigenvalues. Instead as we show these eigenvalues move off into the lower half complex plane and become eigenvalues of $H(f,\theta)$ for $0 < \text{Im}\theta < \pi/3$. If we realize that for these θ a discrete eigenvalue of H is also a discrete eigenvalue of $H(0,\theta)$ with the same multiplicity then it becomes clear that this eigenvalue could be stable under the perturbation of $H(0,\theta)$ to $H(f,\theta)$. This is in fact what we will show for general n-body systems. Alternatively if $\{H(f):f \geq 0\}$ is replaced by $\{H(if):f \geq 0\}$, for example, the discrete eigenvalues of $H = H(0)$ are again stable:

Theorem 3.1 [32]. *Suppose $H = H(0,0)$ (where $H(f,\theta)$ is defined as in (2.28)) has an eigenvalue $E_0 \in \sigma_{disc.}(H)$ of multiplicity m. Fix a compact subset K of $\{(\omega,\theta):(\omega,\theta) \in R,\ \text{Im}(\omega e^{\theta}) > 0,\ |\omega| = 1\}$. Then there is an $\varepsilon > 0$ and an $F_0 > 0$ so that for $|f| < F_0$ and $(\omega,\theta) \in K \cup \bar{K}$, $H(|f|\omega,\theta)$ has at most m spectral points within the disk $\{z:|z-E_0| \leq \varepsilon\}$ each is an eigenvalue of finite algebraic multiplicity and the sum of their multiplicities is exactly m. Furthermore as $|f| \downarrow 0$ each of these eigenvalues converges to E_0, uniformly for $(\omega,\theta) \in K \cup \bar{K}$.*

The proof of this Theorem will occupy the remainder of this section. One of the hardest facts to prove is given in the following proposition whose proof is postponed until the end of this section:

Proposition 3.2 [32]: *Let $\Sigma = \inf \sigma_{ess.}(H(0,0))$ and suppose K is as in Theorem 3.1. Define $\Sigma(\omega,\theta) = \text{Re}(i\omega^{-1}e^{-\theta}\Sigma)$ for $(\omega,\theta) \in K$. Then given $\varepsilon > 0$ there is an $f_\varepsilon > 0$ so that for $|D| \geq 2$*

$$\sup\{\|(z-L_D(f,\theta))^{-1}\| : \mathrm{Re}\, z \le \Sigma(\hat{f},\theta)-\varepsilon, (\hat{f},\theta) \in K,\ 0 < |f| \le f_\varepsilon\} < \infty. \quad (3.1)$$

Here $\hat{f} = e^{i\arg f}$.

Before stating the next result we need some notation: Let

$$L(\omega,\theta) = i\omega^{-1} e^{-\theta} H(0,\theta); \quad (\omega,\theta) \in R$$

and corresponding to $H^C(0,\theta)$ and $H_D(0,\theta)$ we define operators $L^C(\omega,\theta)$ and $L_D(\omega,\theta)$ similarly. The operators appearing in the Weinberg-van Winter equation for $(z-L(\omega,\theta))^{-1}$ are denoted by $\mathcal{D}(\omega,\theta;z)$ and $\mathcal{I}(\omega,\theta;z)$. ($\mathcal{D}$ corresponds to D and $\mathcal{I}$ to I).

Proposition 3.3: _Let K be as in Theorem 3.1. Given $\varepsilon > 0$ let f_ε be as in Proposition 3.2. The operators $D(|f|\omega,\theta;z)$, $I(|f|\omega,\theta;z)$, $\mathcal{D}(\omega,\theta;z)$, $\mathcal{I}(\omega,\theta;z)$ are uniformly bounded for $\mathrm{Re}\, z \le \Sigma(\omega,\theta)-\varepsilon$, $(\omega,\theta) \in K$, and $0 < |f| \le f_\varepsilon$. In addition the following hold uniformly for (z,ω,θ) as above:_

$$\operatorname*{s-lim}_{|f|\downarrow 0} D(|f|\omega,\theta;z) = \mathcal{D}(\omega,\theta;z) \quad (3.2)$$

$$\operatorname*{s-lim}_{|f|\downarrow 0} D(|f|\omega,\theta;z)^* = \mathcal{D}(\omega,\theta;z)^* \quad (3.3)$$

$$\lim_{|f|\downarrow 0} \|I(|f|\omega,\theta;z) - \mathcal{I}(\omega,\theta;z)\| = 0 \quad (3.4)$$

Proof: We omit the proof that the operators $V_{ij}(\theta)(z-L_D(|f|\omega,\theta))^{-1}$ and $(z-L_D(|f|\omega,\theta))^{-1}V_{ij}(\theta)$ are uniformly bounded in the above region. This follows from Proposition 3.2 by the methods used to prove Lemma 2.9. The arguments in Lemma 2.9 are easily seen to be uniform for $z,\omega,\theta,|f|$ in the above region. Thus the uniform bounds on $D(|f|\omega,\theta;z)$ and $I(|f|\omega,\theta;z)$ follow. To handle $\mathcal{D}$ and $\mathcal{I}$ we first note that if $(\omega,\theta) \in K$ then $|\mathrm{Im}\,\theta| \le \pi/2-\delta$ for some $\delta > 0$. Thus

$$\sigma(H_D(0,\theta)) \subseteq \{\Sigma + t_1 e^{-2\theta} + t_2 : t_i \ge 0\} \quad (3.5)$$

and therefore

$$\sigma(L_D(\omega,\theta)) \subseteq \{i\omega^{-1}e^{-\theta}\Sigma + t_1 i\omega^{-1}e^{-3\theta} + t_2 i\omega^{-1}e^{-\theta} : t_i \geq 0\} \tag{3.6}$$

Because $(\omega,\theta) \in K$ this means

$$\sigma(L_D(\omega,\theta)) \subseteq \{z : \mathrm{Re} z \leq \Sigma(\omega,\theta)\} \tag{3.7}$$

If $d(z,A)$ is the distance from z to the set A it is easy to see that if B' is a compact subset of $\{(z,\theta) : z \in \mathbb{C}, |\mathrm{Im}\theta| < \theta_0\}$ and $\varepsilon' > 0$ then

$$\sup\{\|(z-H_D(0,\theta))^{-1}\| : d(z,\sigma(H_D(0,\theta))) \geq \varepsilon', (z,\theta) \in B'\} < \infty \tag{3.8}$$

Since if $\mathrm{Re} z \leq \Sigma(\omega,\theta)-\varepsilon$, $d(z,\sigma(L_D(\omega,\theta))) \geq \varepsilon$ we have for any compact subset B of $\mathbb{C}$

$$\sup\{\|(z-L_D(\omega,\theta))^{-1}\| : \mathrm{Re} z \leq \Sigma(\omega,\theta)-\varepsilon, (\omega,\theta) \in K, z \in B\} < \infty. \tag{3.9}$$

By sectoriality considerations, $\|(z-L_D(\omega,\theta))^{-1}\|$ is uniformly bounded for z large with $\mathrm{Re} z \leq \Sigma(\omega,\theta)-\varepsilon$ and $(\omega,\theta) \in K$. Hence

$$\sup\{\|(z-L_D(\omega,\theta))^{-1}\| : \mathrm{Re} z \leq \Sigma(\omega,\theta)-\varepsilon, (\omega,\theta) \in K\} < \infty. \tag{3.10}$$

This is easily seen to give the required uniform bounds for $\mathcal{D}(\omega,\theta;z)$ and $\mathcal{I}(\omega,\theta;z)$.

The easiest way to prove the strong convergence results (3.2) and (3.3) is to realize that $D(|f|\omega,\theta;z)$ and $\mathcal{D}(|f|\omega,\theta;z)$ are linear combinations of resolvents $(z-L_D(|f|\omega,\theta;z))^{-1}$ and $(z-L_D(\omega,\theta;z))^{-1}$ with $|D| \geq 2$ [41] so that it suffices to prove strong convergence of these resolvents and their adjoints with the required uniformity. We compute

$$(z-L_D(|f|\omega,\theta;z))^{-1} - (z-L_D(\omega,\theta;z))^{-1}$$

$$= (z-L_D(|f|\omega,\theta;z))^{-1} i|f|X(z-L_D(\omega,\theta;z))^{-1} \tag{3.11}$$

and

$$X(z-L_D(\omega,\theta;z))^{-1} = (z-L_D(\omega,\theta;z))^{-1}X +$$

$$(z-L_D(\omega,\theta;z))^{-1}[X,K]ie^{-3\theta}\omega^{-1}(z-L_D(\omega,\theta;z))^{-1} \tag{3.12}$$

The last term in (3.12) is easily seen to be uniformly bounded so that by (3.11), (3.12) and the uniform bounds proved above we have uniform convergence of (3.11) on $\mathcal{D}(X)$. The uniform bounds convert this to uniform convergence on all of $L^2(\mathbb{R}^{3N})$. This proves (3.2). The proof of (3.3) is similar.

To prove (3.4) we follow the argument used to prove compactness of $I(f,\theta;z)$ in Lemma 2.8. Since the arguments used in that proof are uniform in $(z,\omega,\theta,|f|)$ in the above region, it is enough to prove convergence of an operator of the form B' in (2.56) (with f replaced by $|f|\omega$) to the operator B" with each $L_{D_i}(|f|\omega,\theta;z)$ replaced with $\mathcal{L}_{D_i}(\omega,\theta;z)$. We replace $(z-L_{D_i})^{-1}$ with $(z-\mathcal{L}_{D_i})^{-1}$ one at a time using (3.11). We then note $X = \sum_{i=1}^{n} \beta_i \hat{e} \cdot (\vec{x}_{j_i} - \vec{x}_{k_i})$ and commute $\hat{e} \cdot (\vec{x}_{j_i} - \vec{x}_{k_i})$ past resolvents until it reaches V_i' giving a bounded operator. The commutators generated in this procedure are computed using a formula similar to (3.12) and are thus seen to be uniformly bounded. Since this procedure gives an explicit factor of $|f|$ in the difference of B' and B" the proof of (3.4) is complete. ∎

<u>Proof of Theorem 3.1:</u> Suppose $E_0 \in \sigma_{disc.}(H)$, $H = H(0,0)$. If $\Sigma = \inf \sigma_{ess.}(H)$ then, of course, $E_0 < \Sigma$. We note that the projection

$$P(\theta) = (2\pi i)^{-1} \oint_{|z-E_0|=\delta} (z-H(0,\theta))^{-1} dz \tag{3.13}$$

defined for $\delta > 0$ and sufficiently small is analytic in θ for $|\mathrm{Im}\theta| < \mathrm{Min}\{\pi/2,\theta_0\}$ and thus $\mathrm{Ran}P(\theta)$ has constant dimension equal to m. In addition by analyticity $(H(0,\theta)-E_0)P(\theta) = 0$ so that E_0 is an eigenvalue of $H(0,\theta)$ of multiplicity m. Let $\xi = i\omega^{-1}e^{-\theta}$ and note $\mathrm{Re}\xi > 0$ for $(\omega,\theta) \in K$ and that

$$z - H(|f|\omega,\theta) = \xi^{-1}(z\xi - L(|f|\omega,\theta)) \tag{3.14}$$

Choose $\varepsilon' > 0$ so that $\mathrm{Re}\,\xi(E_0-\Sigma) \leq -2\varepsilon'$ for all $(\omega,\theta) \in K$ and choose $\varepsilon^* > 0$ so that $\mathrm{Re}[\xi(z-\Sigma)] \leq -\varepsilon'$ for all z in $\{z: |z-E_0| \leq \varepsilon^*\}$ and all $(\omega,\theta) \in K$. Noting that $\mathrm{Re}\,\xi\Sigma = \Sigma(\omega,\theta)$ we see from Proposition 3.2 that for $|D| \geq 2$ and $0 < |f| \leq f_{\varepsilon'}$, $(\xi z - L_D(|f|\omega,\theta))^{-1}$ is analytic in $\{z: |z-E_0| < \varepsilon^*\}$ for all $(\omega,\theta) \in K$. It then follows that $(1-I(|f|\omega,\theta;\xi z))^{-1}$ is meromorphic in this disk for $|f|$ and (ω,θ) in the same range. Thus using the Weinberg-van Winter equation

$$(\xi z - L(|f|\omega,\theta))^{-1} = (1-I(|f|\omega,\theta;\xi z))^{-1} D(|f|\omega,\theta;\xi z) \tag{3.15}$$

and noting (3.14) we see that $(z-H(|f|\omega,\theta))^{-1}$ is meromorphic in $\{z: |z-E_0| < \varepsilon^*\}$ for all $(\omega,\theta) \in K$ and $0 < |f| < f_{\varepsilon'}$. From Proposition 3.3 it also follows that $(1-\mathcal{I}(\omega,\theta;\xi z))^{-1}$ is meromorphic in this disk for $(\omega,\theta) \in K$. It is clear from the norm continuity of $1-\mathcal{I}(\omega,\theta;z)$ in (ω,θ,z) and the mermorphic nature of its inverse that given any circle $\Gamma_\lambda = \{z: |z-E_0| = \lambda\}$ with $0 < \lambda < \varepsilon^*$ we can find a $\rho < \varepsilon^*$ arbitrarily close to λ so that $1-\mathcal{I}(\omega,\theta;\xi z)$ is invertible for all $z \in \Gamma_\rho$ and all $(\omega,\theta) \in K$. If Γ_ρ has the latter property it is clear by Proposition 3.3 that if $|f|$ is small enough $1-I(|f|\omega,\theta;\xi z)$ will also be invertible for $z \in \Gamma_\rho$ and $(\omega,\theta) \in K$ and thus by (3.15) and (3.14) so will $\xi z - L(|f|\omega,\theta)$ and $z-H(|f|\omega,\theta)$. We can then define the spectral projection.

$$\begin{aligned} P_\rho(|f|\omega,\theta) &= (2\pi i)^{-1} \oint_{\Gamma_\rho} (z-H(|f|\omega,\theta))^{-1} dz \\ &= \xi(2\pi i)^{-1} \oint_{\Gamma_\rho} (z\xi - L(|f|\,\omega,\theta))^{-1} dz \end{aligned} \tag{3.16}$$

We have

$$P_\rho(|f|\omega,\theta) - P(\theta) = \xi(2\pi i)^{-1} \oint_{\Gamma_\rho} [(z\xi - L(|f|\omega,\theta))^{-1} - (z\xi - \mathcal{L}(\omega,\theta))^{-1}] dz \tag{3.17}$$

Since $D(|f|\omega,\theta;\xi z)$ and $\mathcal{D}(\omega,\theta;\xi z)$ are analytic inside $\{z: |z-E_0| < \varepsilon^*\}$ we can replace $(z\xi - L)^{-1}$ by $(z\xi - L)^{-1} - D$ and similarly $(z\xi - \mathcal{L})^{-1}$ by $(z\xi - \mathcal{L})^{-1} - \mathcal{D}$ without changing (3.17). Using (3.15) we thus have

using obvious abbreviations

$$P_\rho(|f|\omega,\theta)-P(\theta) = \xi(2\pi i)^{-1}\oint_{\Gamma_\rho} [(1-I)^{-1}ID-(1-\mathcal{I})^{-1}\mathcal{I}\mathcal{D}] \tag{3.18}$$

We want to show that (3.18) converges to zero in norm as $|f| \downarrow 0$ uniformly for $(\omega,\theta) \in K$. To see this note from Proposition 3.3 that $(1-I)^{-1}I \xrightarrow{\|\cdot\|} (I-\mathcal{I})^{-1}\mathcal{I}$ uniformly for $z \in \Gamma_\rho$ and $(\omega,\theta) \in K$ so we need only show that $\|(1-\mathcal{I})^{-1}\mathcal{I}(D-\mathcal{D})\| = \|(D^*-\mathcal{D}^*)(1-\mathcal{I})^{-1*}\mathcal{I}^*\| \to 0$ with the same uniformity. This follows from the following observations: (1) The set

$$Q = \{(1-\mathcal{I}(\omega,\theta;\xi z))^{-1*}\mathcal{I}(\omega,\theta;\xi z)^*\phi : \|\phi\| \le 1, (\omega,\theta)\in K, z\in\Gamma_\rho\}$$

is compact. (This is a consequence of the compactness of $(1-\mathcal{I})^{-1}\mathcal{I}$ and its norm continuity in the variables (ω,θ,z).)

(2) By Proposition 3.3 the strong convergence of $(D(|f|\omega,\theta;\xi z)-\mathcal{D}(\omega,\theta;\xi z))^*$ to zero is uniform for $(\omega,\theta)\in K, z\in\Gamma_\rho$ and thus for any compact set Q we automatically have

$$\sup\{\|(D(|f|\omega,\theta;\xi z)-\mathcal{D}(\omega,\theta;\xi z))^*\phi\| : \phi\in Q, (\omega,\theta)\in K, z\in\Gamma_\rho\} \to 0$$

as $|f| \to 0$.

We find the ε and F_0 of the theorem as follows: Choose a ρ with $0 < \rho < \varepsilon^*$ and $1-\mathcal{I}(\omega,\theta;z\xi)$ invertible if $z \in \Gamma_\rho$ and $(\omega,\theta) \in K$. Set $\varepsilon = \rho$. Choose F_0 so that if $0 < |f| \le F_0$ and $(\omega,\theta)\in K$, $\|P_\rho(|f|\omega,\theta) - P(\theta)\| < 1$. Then $\dim \operatorname{Ran}P_\rho(|f|\omega,\theta) = \dim \operatorname{Ran}P(\theta) = m$ so that counting multiplicity there are exactly m eigenvalues of $H(|f|\omega,\theta)$ in $\{z: |z-E_0| \le \varepsilon\}$ for all $|f|,\omega,\theta$ with $0 < |f| \le F_0$ and $(\omega,\theta)\in K$. The convergence of these eigenvalues to E_0 as $|f| \to 0$ (uniformly for $(\omega,\theta)\in K$) follows from the fact that ε can be made as small as desired in the above argument. The statment of the theorem for $(\omega,\theta)\in \bar{K}$ follows from $H(f,\theta)^* = H(\bar{f},\bar{\theta})$. □

We finish this section with a proof of Proposition 3.2. The argument we give, based on results of Gearhart in [18], replaces the semigroup Weinberg-van Winter analysis of [32]. We begin with the following general result.

Proposition 3.4: *Suppose $\{e^{-tA}:t \geq 0\}$ is a strongly continuous semigroup on a separable Hilbert space $\mathcal{H}$ with $\|e^{-tA}\| \leq e^{Kt}$. Suppose in addition that for some $\mu > -K$, $\{z:\text{Re}z \leq \mu\} \cap \sigma(A) = \emptyset$ and*

$$\sup_{x\leq\mu}\sup_{y\in\mathbb{R}} \|(x+iy-A)^{-1}\| \leq M < \infty \tag{3.19}$$

Then there is a constant C depending only on μ, K, and M (but not otherwise on A) such that

$$\|e^{-tA}\| \leq Ce^{-t\mu} \tag{3.20}$$

Proof: Our proof relies on a result of Gearhart [18] which says that if $\{e^{-tC}:t \geq 0\}$ is a strongly continuous contraction semigroup on $\mathcal{H}$ such that for some w with $\text{Re}w > 0$

$$\sup_{w\in\mathbb{Z}} \|(w+2\pi in-C)^{-1}\| = M_w < \infty$$

then $e^{-w} \notin \sigma(e^{-C})$ and

$$\|(e^{-C}-e^{-w})^{-1}\| \leq (1-e^{-\text{Re}w})^{-1}$$

$$+2(1+\text{Re}wM_w)e^{\text{Re}w}(1-e^{-\text{Re}w})^{-3/2}(1+e^{-\text{Re}w})^{1/2} \tag{3.21}$$

We set $C = t(A+K)$, $w = t(\mu+K+iy)$ with $y \in \mathbb{R}$. Then

$M_w = \sup_{n\in\mathbb{Z}} \|[t(\mu+K+iy) + 2\pi in - t(A+K)]^{-1}\| \leq t^{-1}M$ so that (3.2) implies

$$\|(1-e^{-t(A-\mu-iy)})^{-1}\| \leq C(t,\mu,K,M)$$

where

$$C(t,\mu,K,M) = [e^{t(\mu+K)}-1]^{-1}$$

$$+ 2(1+(\mu+K)M)(1-e^{-t(\mu+K)})^{-3/2}(1+e^{-t(\mu+K)})^{-1/2}$$

We note by (3.19) and Gearhart's result that if $|z| \geq 1$ then $z \notin (e^{-t(A-\mu)})$ so that we can write a Dunford integral

$$e^{-t(A-\mu)} = (2\pi i)^{-1} \oint_{|z|=1} z(z-e^{-t(A-\mu)})^{-1} dz$$

which gives

$$\|e^{-t(A-\mu)}\| \le \sup_{|z|=1} \|(z-e^{-t(A-\mu)})^{-1}\| \le C(t,\mu,K,M)$$

Note $C(t,\mu,K,M) \le C(1,\mu,K,M)$ if $t \ge 1$ while $\|e^{-t(A-\mu)}\| \le e^{\mu+K}$ if $t \le 1$ so that (3.20) holds with $C = \text{Max}\{e^{\mu+K}, C(1,\mu,K,M)\}$. □

<u>Proof of Proposition 3.2:</u> Suppose we knew that for a system of n particles ($n \ge 1$, otherwise arbitrary) with corresponding operators $H(f,\theta)$ and $L(f,\theta)$ and $\Lambda = \inf\sigma(H(0,0))$ that given $\varepsilon > 0$ there exists $g_\varepsilon > 0$ so that for $0 < |f| \le g_\varepsilon$ and $(\hat{f},\theta) \in K$

$$\|e^{-tL(f,\theta)}\| \le C_\varepsilon e^{-t(\Lambda(\hat{f},\theta)-\varepsilon)} ; \Lambda(\hat{f},\theta) = \text{Re}(i\hat{f}^{-1}e^{-\theta}\Lambda) \quad (3.22)$$

This result implies Proposition 3.2 because if $D = \{C_1, \cdots, C_k\}$, $k \ge 2$ we can drop all $V_{ij}(\theta)$ with i and j in different clusters from $L(f,\theta)$ giving the operator $L_D(f,\theta)$ for which by assumption (3.22) holds with Λ replaced by $\Sigma_D = \inf\sigma(H_D(0,0))$. But $\Sigma = \inf\sigma_{\text{ess.}}(H(0,0)) \le \Sigma_D$ so that integrating $e^{-t(L_D(f,\theta)-z)}$ with $\text{Re} z \le \Sigma(\hat{f},\theta)-2\varepsilon$ we get $\|(z-L_D(f,\theta))^{-1}\| \le \varepsilon^{-1}C_\varepsilon$. We proceed to prove this result by induction on the number of particles, n, in the system. For $n = 1$, (3.22) is trivial. Suppose it is true for all $n \le N$. We will prove it for $n = N + 1$. Thus if $L(f,\theta)$ comes from an N+1 body system, we know from our induction hypothesis that if $D = \{C_1, \cdots, C_k\}$ with $k \ge 2$ we can find $f_\varepsilon > 0$ so that $\|e^{-tL^{C_i}(f,\theta)}\| \le C'_\varepsilon e^{-t(\Lambda^{C_j}(\hat{f},\theta)-\varepsilon/2k)}$ for all such D and all $(\hat{f},\theta) \in K$, $0 < |f| \le f_\varepsilon$. Using the tensor product structure of $e^{-tL_D(f,\theta)}$ we have

$$\|e^{-tL_D(f,\theta)}\| \le (C'_\varepsilon)^k e^{-t[-\varepsilon/2 + \sum_{j=1}^{k} \Lambda^{C_j}(\hat{f},\theta)]} \quad (3.23)$$

But $\Sigma = \inf\sigma_{\text{ess.}}(H(0,0)) \le \sum_{j=1}^{k} \Lambda^{C_j}(\Lambda^{C_j} = \inf\sigma(H^{C_j}(0,0)))$ and thus integrating $e^{-t(L_D(f,\theta)-z)}$ we find that $\|(z-L_D(f,\theta))^{-1}\|$ is uniformly bounded for $\text{Re}z \le \Sigma(\hat{f},\theta)-\varepsilon$, $(\hat{f},\theta) \in K$ and $0 < |f| \le f_\varepsilon$. In this region we will use the Weinberg-van Winter equation

$$(z-L(f,\theta))^{-1} = (1-I(f,\theta;z))^{-1}D(f,\theta;z) \tag{3.24}$$

to show the existence of $g_\varepsilon > 0$ so that $\|(z-L(f,\theta))^{-1}\|$ is uniformly bounded in the region $0 < |f| \le g_\varepsilon$, $\text{Re}z \le \Lambda(\hat{f},\theta)-\varepsilon$, $(\hat{f},\theta) \in K$ from which the desired result (3.22) follows from Proposition 3.4.

By sectoriality considerations (see the proof of (iii), Proposition 2.5) we can select an E with $-E < \Lambda(\omega,\theta)-\varepsilon$ for all $(\omega,\theta) \in K$ so that $\|I(\omega,\theta;z)\| \le 1/2$ if $\text{Re}z \le -E$ or if $|\text{Im}z| \ge E$ and $\text{Re}z \le \Lambda(\omega,\theta)-\varepsilon$ for all $(\omega,\theta) \in K$. Since by Proposition 3.3

$I(|f|\omega,\theta;z) \xrightarrow{\|\cdot\|} I(\omega,\theta;z)$ uniformly in this region as $|f| \to 0$ (remember $\Lambda \le \Sigma$), using (3.24) and the uniform boundedness of $D(f,\theta;z)$ (see Proposition 3.3 again) we see that there is a g'_ε so that $\|(z-L(f,\theta))^{-1}\|$ is uniformly bounded in the region with $(\hat{f},\theta) \in K$ and $\text{Re}z \le -E$ or $|\text{Im}z| \ge E$ and $\text{Re}z \le \Lambda(\hat{f},\theta)-\varepsilon$. Let T be the solid closed rectangle with sides parallel to the lines $\text{Re}z = -E$, $\text{Re}z = \Lambda(\hat{f},\theta)-\varepsilon$ and the lines $\text{Im}z = \pm E$. It remains to bound $\|(z-L(f,\theta))^{-1}\|$ for $z \in T$. The spectral projection $P = (2\pi i)^{-1}\oint_{\partial T}(z-L(|f|\omega,\theta))^{-1}dz$ converges in norm (see the proof of Theorem 3.1) to $(2\pi i)^{-1}\oint_{\partial T}(z-L(\omega,\theta))^{-1}dz$ uniformly for $(\omega,\theta) \in K$ as $|f| \to 0$. But the latter is zero since $L(\omega,\theta)$ has no spectrum in T. Thus $P = 0$ for sufficiently small $|f|$, say $0 < |f| \le g_\varepsilon \le g'_\varepsilon$, for all $(\omega,\theta) \in K$. Hence $(z-L(f,\theta))^{-1}$ is analytic in T for $0 < |f| \le g_\varepsilon$ and $(\hat{f},\theta) \in K$. By the maximum principle $\sup\{\|(z-L(f,\theta))^{-1}\| : z \in T\} = \sup\{\|(z-L(f,\theta))^{-1}\| : z \in \partial T\}$ which we have already shown to be uniformly bounded. ∎

I.4. ANALYTICITY AND BOREL SUMMABILITY OF RESONANCE EIGENVALUES

For simplicity we will consider a non-degenerate eigenvalue E_0 in $\sigma_{disc.}(H)$ where $H = H(0,0)$. A more general analysis is carried out in [32] where analogous results are proved in the situation where on each symmetry subspace any degeneracy present is removed to first order in f.

Theorem 4.1 [32]: *Suppose E_0 is an isolated non-degenerate eigenvalue of H. Then there is an $\varepsilon > 0$ and an $F_0 > 0$ so that for all θ with $0 < \mathrm{Im}\theta < \mathrm{Min}\{\theta_0, \pi/3\}$ and all f with $0 < f < F_0$, $H(f,\theta)$ has a non-degenerate eigenvalue $E(f)$ with*

$$\sigma(H(f,\theta)) \cap \{z: |z-E_0| < \varepsilon\} = \{E(f)\}$$

Given any $\delta > 0$ there is an $R_\delta > 0$ so that $E(f)$ has an analytic continuation to the region

$$G_\delta = \{f: 0 < |f| < R_\delta, -\mathrm{Min}\{\theta_0, \pi/2\} + \delta < \arg f < \pi + \mathrm{Min}\{\theta_0, \pi/2\} - \delta\}$$

which satisfies

(i) *If $0 < |f| < R_\delta$ and $\mathrm{Im} f > 0$ then $E(f)$ is a non-degenerate eigenvalue of $H(f,0)$ and*

$$\sigma(H(f,0)) \cap \{z: |z-E_0| < \varepsilon\} = \{E(f)\}$$

(ii) $\lim_{|f|\to 0} E(f) = E_0$ *uniformly for*

$$-\mathrm{Min}\{\theta_0, \pi/2\} + \delta < \arg f < \pi + \mathrm{Min}\{\theta_0, \pi/2\} - \delta$$

Theorem 4.2 [32]: *Suppose E_0, $E(f)$ and G_δ are as in Theorem 4.1. Then the formal Rayleigh-Schrödinger perturbation series, $\sum_{n=0}^{\infty} a_n f^n$, for $E(f)$ is a strong asymptotic series:*

$$\left| \sum_{n=0}^{N} a_n f^n - E(f) \right| \le C_\delta^{N+1} |f|^{N+1} (N+1)!;\ N \ge 0 \tag{4.1}$$

uniformly for f in G_δ for any $\delta > 0$. For some $R > 0$ the Borel

transform $B(z) = \sum_{n=0}^{\infty} (n!)^{-1} a_n z^n$ is analytic in the region

$$\{z: |z| < R\} \cup \{z: \pi/2 - \mathrm{Min}\{\theta_0, \pi/2\} < \arg z < \pi/2 + \mathrm{Min}\{\theta_0, \pi/2\}\}$$

and for sufficiently small $f > 0$

$$E(if) = \int_0^{\infty} e^{-t} B(ift)\,dt$$

Proof of Theorem 4.1: Choose δ with $0 < \delta < \mathrm{Min}\{\pi/3, \theta_0\}$ and let

$$K = \{(\omega, i\beta) : |\omega| = 1,\ \beta \in \mathbb{R},\ \tfrac{1}{2}\delta \le \arg\omega + \beta \le \pi - \tfrac{1}{2}\delta,\ \tfrac{1}{2}\delta \le \arg\omega + 3\beta \le \pi - \tfrac{1}{2}\delta, |\beta| \le \theta_0 - \tfrac{1}{2}\delta\}$$

The set K is shown in Figure 2 when $\theta_0 \ge (\pi - \delta)/2$. Then from Theorem 3.1 there is an $\varepsilon > 0$ and $F_0 > 0$ so that if $0 < |f| \le F_0$ and $(\hat{f}, \theta) \in K$ we have

(1) $H(f,\theta)$ has a non-degenerate eigenvalue, $E(f,\theta)$, such that $\sigma(H(f,\theta)) \cap \{z: |z - E_0| \le 2\varepsilon\} = \{E(f,\theta)\}$ while $|E(f,\theta) - E_0| < \varepsilon$

(2) As $|f| \downarrow 0$ $E(f,\theta) \to E_0$ uniformly for $(f,\theta) \in K$

By Proposition 3.2 we can also demand (by shrinking ε and F_0 if necessary) that

(3) For all $(\hat{f}, \theta) \in K$ and $0 < |f| \le F_0$, $\sigma_{\mathrm{ess.}}(H(f,\theta)) \subseteq A(\xi)$ where $A(\xi)$ is the half-plane $\{z: \mathrm{Re}\,\xi(z - \Sigma) \ge -\varepsilon\}$ and in addition $A(\xi) \cap \{z: |z - E_0| \le 2\varepsilon\} = \emptyset$ Here we take $\xi = i\hat{f}^{-1} e^{-i\mathrm{Im}\theta}$. The half-plane $A(\xi)$ is shown in Figure 3.

By the unitary equivalence of $H(f,\theta)$ and $H(f,\theta+\phi)$ for ϕ $\mathbb{R}$, all the above statements remain valid if K is replaced by K_δ with

$$K_\delta = \{(\omega, \theta) : |\omega| = 1,\ \tfrac{1}{2}\delta \le \arg\omega + \mathrm{Im}\theta \le \pi - \tfrac{1}{2}\delta,\ \tfrac{1}{2}\delta \le \arg\omega + 3\,\mathrm{Im}\theta \le \pi - \tfrac{1}{2}\delta,\ |\mathrm{Im}\theta| \le \theta_0 - \tfrac{1}{2}\delta\}$$

By the analyticity of the resolvent of $H(f,\theta)$ ((ii) of Proposition 2.5) $E(f,\theta)$ is analytic in the variables (f,θ) for (f,θ) in the region

$$\Omega_\delta = \{(f,\theta) : 0 < |f| < F_0,\ (\hat{f}, \theta) \text{ in the interior of } K_\delta\}$$

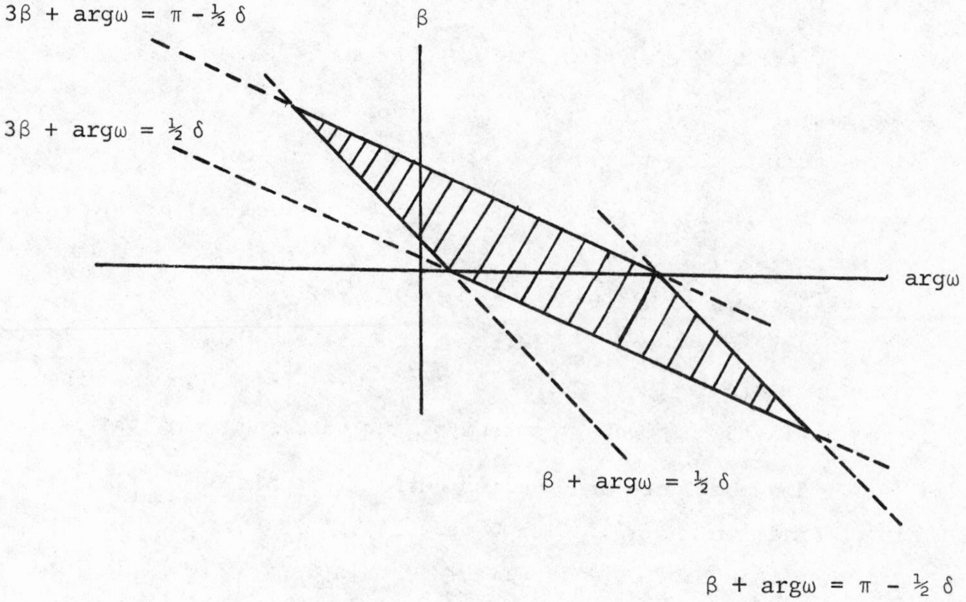

Figure 2. The set K.

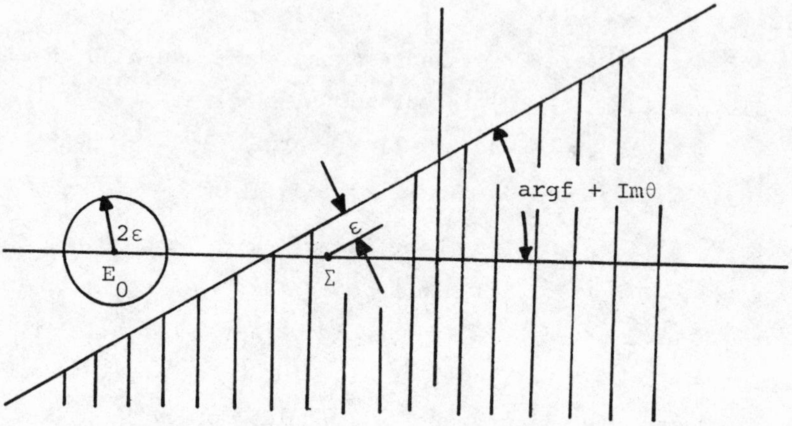

Figure 3. The half-plane $A(\xi)$.

and since $E(f,\theta) = E(f,\theta+\phi)$ for $\phi \in \mathbb{R}$, $E(f,\theta) = E(f)$ for some function $E(\cdot)$ analytic in

$$\tilde{G}_\delta \equiv \{f:(f,\theta) \in Q_\delta \text{ for some } \theta \in \mathbb{C}\}$$

If we set $R_\delta = F_0$, a bit of algebra shows that $G_\delta = \tilde{G}_\delta$ and thus (ii) and the analyticity statement of the theorem follow.

Fix f with $0 < f < F_0 = R_\delta$. Then by (1) and (3) above, if $\frac{1}{2}\delta \leq \text{Im}\theta \leq (\pi-\delta/2)/3$ and $|\text{Im}\theta| \leq \theta_0 - \frac{1}{2}\delta$, $E(f)$ is a non-degenerate eigenvalue of $H(f,\theta)$ and the disk $\{z:|z-E_0| \leq 2\varepsilon\}$ is outside $A(\xi)$ $(\xi = ie^{-i\text{Im}\theta})$ where $\sigma_{ess.}(H(f,\theta))$ is located. Let $\text{Im}\theta_1 = \frac{1}{2}\delta$. Then $A(\xi_1) \cap \{z:|z-E_0| \leq 2\varepsilon\} = \emptyset$, $(\xi = ie^{-i\text{Im}\theta_1})$. We know that any point in the discrete spectrum of any subsystem Hamiltonian $H^{C_j}(f,\theta)$ does not vary as θ varies until it is hit by $\sigma_{ess.}(H^{C_j}(f,\theta))$. This observation and (2.71) show that if $\text{Im}\theta$ is increased to anything less than $\text{Min}\{\theta_0,\pi/3,(\pi-\text{Im}\theta_1)/2\} = \text{Min}\{\theta_0,\pi/3\}$ or if $\text{Im}\theta$ is decreased to anything above zero, $\sigma_{ess.}(H(f,\theta))$ remains in the half-plane $A(\xi_1)$. Thus the points of $\sigma_{disc.}(H(f,\theta))$ in the complement of $A(\xi_1)$ remain fixed. This argument shows that $E(f)$ is an eigenvalue of $H(f,\theta)$ for $0 < \text{Im}\theta < \text{Min}\{\theta_0,\pi/3\}$ and $\sigma(H(f,\theta)) \cap \{z:|z-E_0| < \varepsilon\} = \{E(f)\}$.

Similarly fix f with $0 < |f| < R_\delta$ and $\text{Im}f > 0$. Then if $\frac{1}{2}\delta \leq \arg f \leq \pi - \frac{1}{2}\delta$, $E(f)$ is a non-degenerate eigenvalue of $H(f,0)$ and the disk $\{z:|z-E_0| \leq 2\varepsilon\}$ is outside $A(\xi)$, $\xi = i\hat{f}^{-1}$. If $0 < \arg f < \frac{1}{2}\delta$ choose θ_2 so that $\text{Im}\theta_2 + \arg f = \delta/2$. Then $(\hat{f},\theta_2) \in K_\delta$, $E(f)$ is an eigenvalue of $H(f,\theta)$ and $\sigma_{ess.}(H(f,\theta_2)) \subseteq A(\xi_2)$ with $\xi_2 = i\hat{f}^{-1}e^{-i\text{Im}\theta_2}$. Decreasing $\text{Im}\theta$ to zero we see that $\sigma_{ess.}(H(f,\theta)) \subseteq A(\xi_2)$ for all θ with $0 \leq \text{Im}\theta \leq \text{Im}\theta_2$ so that (ii) holds for $0 < \arg f < \pi-\delta/2$. If $\pi > \arg f > \pi-\delta/2$ choose θ_3 with $\text{Im}\theta_3 + \arg f = \pi-\delta/2$. We find that $(\hat{f},\theta_3) \in K_\delta$ and by the same argument (we increase $\text{Im}\theta$ to zero) (ii) holds for $0 < \arg f < \pi$. □

<u>Proof of Theorem 4.2:</u> We use the notation of the proof of Theorem 4.1 and show that the estimate (4.1) holds uniformly in G_δ. The

remainder of the theorem follows by varying δ and using Watson's theorem [24]. Let $\Gamma = \{z: |z-E_0| = 3\varepsilon/2\}$ and for $0 < |f| < R_\delta$ and $(\hat{f},\theta) \in K_\delta$ write

$$P(\theta) = (2\pi i)^{-1} \oint_\Gamma (z-H(0,\theta))^{-1} dz$$

$$P(f,\theta) = (2\pi i)^{-1} \oint_\Gamma (z-H(f,\theta))^{-1} dz;$$

Suppose $H\psi = E_0\psi$ with $||\psi|| = 1$. The following results are standard [2,10,13,44].

(1) $U(\theta)\psi = \psi(\theta)$ has an analytic continuation to $\{\theta: |\mathrm{Im}\theta| < \mathrm{Min}\{\pi/2,\theta_0\}\}$ satisfying $P(\theta)\psi(\theta) = \psi(\theta)$

(2) For some $\alpha > 0$, $||e^{\alpha\rho}\psi(\theta)||$ is uniformly bounded for θ in $M = \{i\beta: \beta \in \mathbb{R}, |\beta| \leq \mathrm{Min}\{\pi/2,\theta_0\}-\delta/2\}$. Here $\rho = \sum_{i<j} |\vec{x}_i - \vec{x}_j|$.

(3) For some $\alpha > 0$ $||e^{\lambda\rho}(z-H(0,\theta))^{-1}e^{-\lambda\rho}||$ is uniformly bounded for all $\lambda \in [0,\alpha]$, $\theta \in M$, and $z \in \Gamma$.

We write for $(\hat{f},\theta) \in K_\delta$

$$E(f) = (\psi(\bar{\theta}), H(f,\theta)P(f,\theta)\psi(\theta)) / (\psi(\bar{\theta}), P(f,\theta)\psi(\theta)) \tag{4.2}$$

and prove that both numerator and denominator have a strong asymptotic series obeying a bound like (4.1) uniformly for $0 < |f| < \tilde{R}_\delta \leq R_\delta$ and $(f,\theta) \in G_\delta$. By a standard argument [44] this suffices to prove (4.1). We note that since $P(f,\theta) \xrightarrow{||\cdot||} P(\theta)$ as $|f| \downarrow 0$ uniformly for $(\hat{f},\theta) \in K_\delta$ and $(\psi(\bar{\theta}),\psi(\theta)) = 1$, there is an $\tilde{R}_\sigma \in (0,R_\delta]$ so that $|(\psi(\bar{\theta}),P(f,\theta)\psi(\theta))| \geq 1/2$ for $0 < |f| < \tilde{R}_\delta$ and $(\hat{f},\theta) \in K_\delta$. Using Proposition 3.3 and the Weinberg-van Winter equation we can also assume

(4) $\sup\{||(z-H(f,\theta))^{-1}|| : z \in \Gamma,\ 0 < |f| < \tilde{R}_\delta,\ (\hat{f},\theta) \in R_\delta\} < \infty$

It suffices to prove the estimate for $0 < |f| < \tilde{R}_\delta$ because nothing new is learned from the estimate in the region $\tilde{R}_\delta \leq |f| \leq R_\delta$. We

consider the numerator; the denominator is similar. We thus write

$$(z-H(f,\theta))^{-1} = (z-H(0,\theta))^{-1} \sum_{n=0}^{N} [fe^{\theta}X(z-H(0,\theta))^{-1}]^{n}$$

$$+ (z-H(f,\theta))^{-1}[fe^{\theta}X(z-H(0,\theta))^{-1}]^{N+1}$$

Note that the coefficients in the expansion of the numerator and denominator of (4.2) are exactly those obtained in the formal Rayleigh-Schrödinger expansion for an eigenvalue of H(f,0) because by analyticity

$$(\psi(\bar{\theta}),(z-H(0,\theta))^{-1}(e^{\theta}X(z-H(0,\theta))^{-1})^{n}\psi(\theta))$$

$$= (\psi,(z-H)^{-1}(X(z-H)^{-1})^{n}\psi).$$

The remainder after expanding the numerator to order f^N is given by

$$(2\pi i)^{-1} \oint_{\Gamma} dzz(\psi(\bar{\theta}),(z-H(f,\theta)^{-1}(fe^{\theta}X(z-H(0,\theta))^{-1})^{N+1}\psi(\theta)) \quad (4.3)$$

Using (2) and (4) above we bound the absolute value of (4.3) by

$$\text{const}|f|^{N+1} \sup_{z\in\Gamma} ||[X(z-H(0,\theta))^{-1}]^{N+1}\psi(\theta)||$$

Let $X_N = Xe^{-\alpha\rho/N+1}$, $R_n^N(z,\theta) = \exp(\alpha\rho n/N+1)(z-H(0,\theta))^{-1}\exp(-\alpha\rho n/N+1)$.

Then

$$[X(z-H(0,\theta))^{-1}]^{N+1}\psi(\theta) = (X_N R_1^N)(X_N R_2^N)\cdots(X_N R_{N+1}^N)e^{\alpha\rho}\psi(\theta)$$

From (2) and (3) above, $||e^{\alpha\rho}\psi(\theta)||$ and $||R_n^N(z,\theta)||$ are bounded uniformly for $\theta \in M$ and $z \in \Gamma$. Clearly $||X_N||^{N+1} \leq [C(N+1)]^{N+1}$ and thus the result follows. □

I.5. OPEN PROBLEMS

We single out five areas which deserve further study:

A. Absence of Bound States

The spectral theory of the self-adjoint one-body Stark

Hamiltonian has been extensively studied with rather complete results [1,4,26,47,52,53]. In the many body case very little is known. In particular it has not been shown that a multielectron atom in the presence of a homogeneous electric field has no bound states.

B. Asymptotics of Widths

Once the non-existence of bound states is established, it follows that for small $f > 0$ the resonance eigenvalues, $E(f)$ guaranteed to exist by Theorem 4.1, have non-zero widths, $\Gamma(f) = -2\,\mathrm{Im}E(f)$. In the case of Hydrogen the asymptotics of $\Gamma(f)$ as $f \downarrow 0$ are known in great generality [11,15,25,54]. The technology for calculating $\Gamma(f)$ in more general one-body problems (not to mention multielectron atoms) does not yet exist.

C. Exponential Decay

Even if Theorem 1.2 could be generalized to include Hydrogen, it is not quite what is wanted from a physical point of view. For example, suppose ψ in Eq. (1.5) is an eigenvector of $-\Delta + V$ with nondegenerate eigenvalue $E_0 < 0$. Then one might expect that

$$\lim_{f\downarrow 0}\left|(\psi, e^{-itH(f)}\psi) - c(\psi,f)e^{-itE(f)}\right| = 0 \text{ uniformly for } t \text{ in } [0,\infty).$$

Here the resonance eigenvalue $E(f)$ (corresponding to E_0 as in Theorem 4.1) occurs in the expansion (1.5) with coefficient $(U(\bar{\theta})\psi, P(f,\theta)U(\theta)\psi) = c(\psi,f)$ which converges to $\|\psi\|^2$ as $f \downarrow 0$. But it is not even guaranteed that this is the largest coefficient appearing in (1.5) for small f. Problems of this nature deserve further study. In addition an N-body exponential decay result would be interesting.

D. Stability of Atomic Resonances

For physical applications long lived resonances are as important as bound states. If the system has a resonance in zero field (i.e. if $H(0,\theta)$ has an eigenvalue with negative imaginary part) we expect that for small $f > 0$, $H(f,\theta)$ will have an eigenvalue nearby. The techniques developed here seem incapable of showing this.

E. The Stark Ladder

Consider the operator $H(f) = \dfrac{-d^2}{dx^2} + fx + V(x)$ where V is periodic, with period 1. It has been shown [30] that if V is an entire function such that $\sup\{|V(z)| : |\mathrm{Im}z| \leq a\}$ is finite for all $a > 0$ then for $f > 0$ certain matrix elements of the resolvent of $H(f)$ have meromorphic continuations to $\mathbb{C}$. Since under translation

by an integer n, $H(f) \to H(f) + nf$, the poles of this continuation (if there are any) occur in ladders $\{E_0 + nf : n = 0, \pm 1, \cdots\}$. Although it is believed that such poles exist for small f [3,50,55], this has not been shown. It would be interesting to determine the behavior of these resonances for $f \downarrow 0$. For another approach to the problem see [12].

II. ENHANCED BINDING IN HOMOGENEOUS MAGNETIC FIELDS

In this section we study the operator

$$(-i\vec{\nabla} - \tfrac{1}{2}\vec{B}\times\vec{r})^2 + V(\vec{r}); \ \vec{B} = \text{const.}$$

and its N-body generalizations with a view to showing that binding is easier with the magnetic field $\vec{B} \neq 0$ than with $\vec{B} = 0$. Our least trivial example of this is given in Theorem 4.1 which states that any atom in a non-zero magnetic field will bind an extra electron In fact, the Hamiltonian for the negative ion has infinitely many bound states below the continuum.

The intuition behind the above phenomenon involves the fact that an electron in a homogeneous $\vec{B}$-field is already bound in a Landau orbit in the two dimensions perpendicular to $\vec{B}$. (Classically the particle moves in a circle in the plane perpendicular to $\vec{B}$ and freely in the $\vec{B}$ direction, i.e. in a spiral.) The particle thus only requires a binding force in the one dimension parallel to $\vec{B}$ to keep it from eventually moving to infinity. The situtation is further enchanced by the fact that in one dimension an arbitrarily weak attractive potential will produce a bound state.

We discuss one-dimensional weak binding in Section II.1, review the spectral analysis of the Hamiltonian describing an electron in a homogeneous magnetic field in Section II.2, and then go on to Sections II.3 and II.4 where we prove that certain systems which may not bind when $\vec{B} = 0$ have infinitely many bound states when $\vec{B} \neq 0$. Section II.3 contains one-body results and Section II.4 results on negatively charged ions.

II.1. Weak Coupling in One Dimension

It is well known that if a square well potential in 3 dimensions is sufficiently shallow, the corresponding Hamiltonian will have no bound states. The situation is very different in one dimension. Consider a potential V in $L^1(\mathbb{R})$ with $\int_{-\infty}^{\infty} V(z)dz < 0$, and the corresponding Hamiltonian $h_\lambda = \dfrac{-d^2}{dz^2} + \lambda V(z)$ with $\lambda > 0$. For purposes of illustration we have chosen to measure the strength of the potential

by the crude device of inserting a coupling constant. We want to show that for $\lambda > 0$ and small, h_λ has an eigenvalue below the continuum $[0,\infty)$ and to say something about the behavior of this eigenvalue as $\lambda \downarrow 0$.

We use a trial state $f_\alpha(z) = \alpha^{1/2} e^{-\alpha|z|}$. Note $\|f_\alpha\|_2 = 1$ and that f_α is in the form domain of h_λ. We take $\alpha = -\lambda(\int V(z)dz)/2$ and compute:

$$(f_\alpha, h_\lambda f_\alpha) = \alpha^2 + \lambda\alpha \int V(z) e^{-2\alpha|z|} dz$$

$$= -(\lambda \int V(z)dz/2)^2 (1+2(\int V(z)dz)^{-1} \int V(z)(e^{-2\alpha|z|}-1)dz)$$

$$= -(\lambda \int V(z)dz/2)^2 (1+o(\lambda))$$

where the last line follows from the dominated convergence theorem. Thus h_λ has an eigenvalue $E(\lambda)$ satisfying

$$E(\lambda) \leq -(\tfrac{1}{2} \int \lambda V(z)dz)^2 (1+o(\lambda)) \tag{1.1}$$

While we will mainly be interested in weak potentials we note that if $h_\mu \psi_\mu = e\psi_\mu$ for some small $\mu > 0$ and some $e < 0$ and if $\|\psi_\mu\|_2 = 1$, then $(\psi_\mu, h_\lambda \psi_\mu) \leq \lambda\mu^{-1} e$ for all $\lambda > \mu$ so that h_λ has an eigenvalue below zero for all $\lambda > 0$.

These results and others can all be found in [45] where in particular the inequality in (1.1) is shown to be equality.

II.2. Two Dimensional Motion in a Homogeneous Magnetic Field

We consider a charged particle moving in the xy plane under the influence of a magnetic field $\vec{B} = B\hat{z}$, $B > 0$. The Hamiltonian describing this motion is

$$h(B) = (-i\vec{\nabla} - \tfrac{1}{2}\vec{B} \times \vec{r})^2 + \partial_z^2$$
$$= -(\partial_x^2 + \partial_y^2) + \frac{B^2}{4}(x^2 + y^2) - B\ell_z$$

where $\ell_z = \hat{z} \cdot (-i\vec{r} \times \vec{\nabla}) = -i(x\partial_y - y\partial_x)$. $h(B)$ is quadratic in $\partial_x, \partial_y, x, y$ so it is not surprising that $h(B)$ is a harmonic oscillator in disguise. We introduce the independent annihilation operators

$$a_x = B^{-1/2}\partial_x + (B/4)^{1/2}x,\quad a_y = B^{-1/2}\partial_y + (B/4)^{1/2}y \tag{2.1}$$

and the related set

$$a_{\pm} = (a_x \pm ia_y)/\sqrt{2} \tag{2.2}$$

A computation shows

$$h(B) = 2B(a_+^* a_+ + 1/2),\quad \ell_z = a_-^* a_- - a_+^* a_+ \tag{2.3}$$

We use the notation $(\ell_z = m)$ for the subspace annihilated by $\ell_z - m$, $m = 0, \pm 1, \pm 2, \cdots$. From (2.3) it follows that

$$\sigma(h(B)\restriction(\ell_z = m)) = \{(|m| - m + n)B : n = 1, 2, \cdots\}$$

with each eigenvalue non-degenerate. The ground states of $h(B)$ satisfy $a_+^* a_+ \phi_m = 0$, $a_-^* a_- \phi_m = m\phi_m$, $m = 0,1,2,\cdots$. In Sections II.3 and II.4 we will need their explicit coordinate space representations:

$$\phi_m(\vec{\rho}) = (2^{m+1}\pi m!)^{-1/2} B^{1/2} (B^{1/2}\rho)^m e^{-B\rho^2/4} e^{im\phi} \tag{2.4}$$

where $\vec{\rho} = (x,y)$, $\tan\phi = y/x$. We have

$$\int |\phi_m(\vec{\rho})|^2 d^2\rho = 1,\quad \ell_z \phi_m = m\phi_m,\quad h(B)\phi_m = B\phi_m.$$

In addition we will need properties of the measure $|\phi_m(\vec{\rho})|^2 d^2\rho = \rho|\phi_m(\vec{\rho})|^2 d\rho d\phi$ for large m. We write $\rho^{2m+1} e^{-B\rho^2/2} = e^{-a(\rho)}$ and note that $a(\rho)$ has a minimum at $\rho = \rho_m$ where $\rho_m = \sqrt{(2m+1)/B}$. If we use the approximation $a(\rho) \cong a(\rho_m) + \frac{1}{2} a''(\rho_m)(\rho-\rho_m)^2$ and Stirling's formula we find

$$|\phi_m(\vec{\rho})|^2 d^2\rho \cong \frac{d\phi}{2\pi} \sqrt{B/\pi}\, e^{-B(\rho-\rho_m)^2} d\rho \tag{2.5}$$

For our purposes the estimate

$$a(\rho) \geq a(\rho_m) + \frac{1}{2}(\rho-\rho_m)^2 (\inf_{x\geq 0} a''(x))$$

suffices. It leads to the bound

$$\rho|\phi_m(\vec{\rho})|^2 \leq c\sqrt{B}\, e^{-B(\rho-\rho_m)^2/2}; \quad \rho_m = \sqrt{(2m+1)/B} \tag{2.6}$$

for some constant c.

II.3. Enhanced Binding: One Body Problem

We consider the Hamiltonian

$$H = (-i\vec{\nabla} - \tfrac{1}{2}\vec{B}\times\vec{r})^2 + V(\vec{r}); \quad \vec{B} = B\hat{z}$$

where $B > 0$ and $V(-\Delta+1)^{-1}$ is compact. Let $H_0 = (-i\vec{\nabla} - \frac{1}{2}\vec{B}\times\vec{r})^2$. The diamagnetic inequality $|(H_0+1)^{-1}\psi| \leq (-\Delta+1)^{-1}|\psi|$ [46] and an abstract result [16,40] show that the compactness of $V(-\Delta+1)^{-1}$ implies the compactness of $V(H_0+1)^{-1}$. The reader is referred to the above references for details. By a standard Weyl argument we thus have $\sigma_{ess.}(H) = \sigma_{ess.}(H_0) = [B,\infty)$ where the last equality follows from the spectral results of Section II.2. With these preliminaries we can state a typical result. Further results can be found in [8].

Theorem 3.1 [8]: *Suppose H and V are as above and in addition there is an $R > 0$ so that if χ is the characteristic function of the set $\{\vec{r}:\rho \geq R\}$ then the operator χV is not identically zero and in addition $V(\vec{r}) \leq 0$ if $|\vec{r}| \geq R$. Then*

(a) *H has at least one eigenvalue below B*

(b) *If V is azimuthally symmetric (i.e. depends only on ρ and z) then H has infinitely many eigenvalues below B.*

Proof: We can assume that for $|\vec{r}| \geq R$ $|V(\vec{r})| \leq e^{-|\vec{r}|}$ for if not set $\tilde{V}(\vec{r}) = V(\vec{r})$ if $|\vec{r}| < R$ and $\tilde{V}(\vec{r}) = \mathrm{Max}\{V(\vec{r}), -e^{-|\vec{r}|}\}$ if $|\vec{r}| \geq R$. If the result is true for $\tilde{V}(\vec{r})$ then by the min-max principle and the fact that $V(\vec{r}) \leq \tilde{V}(\vec{r})$, the result follows for V. Note that if V is azimuthally symmetric so is $\tilde{V}$. We use trial states of the form $\psi(\vec{r}) = f(z)\phi_m(\vec{\rho})$. Note that $\ell_z\psi = m\psi$. We will show that there is an $M_0 \geq 0$ with the property that for each $m \geq M_0$ there exists an f with

$$(\psi, (H-B)\psi) < 0 \tag{3.1}$$

This proves the result; for in case (b), (3.1) implies $H \restriction (\ell_z = m)$

has an eigenvalue below B. Note that

$$(\psi, (H-B)\psi) = (f, \left[\frac{-d^2}{dz^2} + V_m(z)\right] f) \text{ where}$$

$$V_m(z) = \int |\phi_m(\vec{\rho})|^2 V(\vec{r}) d^2\rho$$

By the results of Section II.1 we need only show that for large enough m, $\int V_m(z)dz < 0$. The latter integral is, up to a positive (m-dependent) multiple, given by

$$\int (\rho/R)^{2m} V(\vec{r}) e^{-B\rho^2/2} d^3r$$

$$\leq \int_{|\vec{r}| \leq R} (\rho/R)^{2m} V(\vec{r}) e^{-B\rho^2/2} d^3r + \int_{\rho \geq R} (\rho/R)^{2m} V(\vec{r}) e^{-B\rho^2/2} d^3r$$

As $m \uparrow \infty$ the second integral $\downarrow -\infty$ while the first is bounded by

$$\left(\int_{|\vec{r}| \leq R} |V(\vec{r})|^2 dr \right)^{1/2} \left(\int_{|\vec{r}| \leq R} (\rho/R)^{4m} e^{-B\rho^2} dr \right)^{1/2}$$

which converges to zero as $m \uparrow \infty$. Note that $V \in L^2_{loc.}(\mathbb{R}^3)$ since $\mathcal{D}(V) \supseteq \mathcal{D}(-\Delta)$ so that $\|V\phi\|_2 < \infty$ for all $\phi \in C_o^\infty(\mathbb{R}^3)$. ∎

II.4. Enhanced Binding: Negative Ions

In this section we consider the Hamiltonian of an ion with nuclear charge n-1 and n electrons, $n \geq 2$:

$$H(B) = \sum_{j=1}^{n} \left[(-i\vec{\nabla}_j - \tfrac{1}{2}\vec{B} \times \vec{r}_j)^2 - (n-1)|\vec{r}_j|^{-1} \right]$$

$$+ \sum_{1 \leq i < j \leq n} |\vec{r}_i - \vec{r}_j|^{-1} - 2\vec{B} \cdot \vec{S} \qquad (4.1)$$

We have included the interaction of the total election spin $\vec{S}$ with the magnetic field. The Hamiltonian of the neutral atom is given by

$$H^{n-1}(B) = \sum_{j=1}^{n-1} \left[(-i\vec{\nabla}_j - \tfrac{1}{2}\vec{B} \times \vec{r}_j)^2 - (n-1)|\vec{r}_j|^{-1} \right] + \sum_{1 \le i < j \le n-1} |\vec{r}_i - \vec{r}_j|^{-1} - 2\vec{B} \cdot \vec{S}^{n-1} \tag{4.2}$$

where $\vec{S} = \vec{S}^{n-1} + \vec{s}_n$ and $\vec{s}_n$ is the spin of the nth electron. Note that

$$H(B) = H^{n-1}(B) - \Delta_n + \tfrac{1}{4}B^2\rho_n^2 - B(\ell_{z_n} + 2s_{z_n}) + W \tag{4.3}$$

where $W(\vec{r}_1, \cdots, \vec{r}_n) = \sum_{j=1}^{n-1} (|\vec{r}_j - \vec{r}_n|^{-1} - |\vec{r}_n|^{-1})$ and we have taken $\vec{B} = B\hat{z}$.

There are two results proved in [9] concerning the bound states of $H(B)$. One says that for any $B > 0$, $H(B)$ has infinitely many bound states on the Hilbert space of functions of electron spatial and spin coordinates which are antisymmetric under simultaneous interchange of the spatial and spin coordinates of two electrons (we call this space the "physical" Hilbert space). All of the corresponding eigenvalues are below the continuum. It should be noted that there are many atoms (including all the noble gases, for example) which are thought to have no stable negative ions when $B = 0$ [34]. The other result concerns the behavior of the eigenvalues as $B \downarrow 0$. It is shown in [9] that for small B, the corresponding binding energies are at least as large as cB^3 for some $c > 0$. B^3 is a small number even with the highest laboratory fields now available and thus it is not clear that these states can be seen in the laboratory with present technology. However, these ions may exist in the neighborhood of some neutron stars.

We will state only the first result as a theorem and give a sketch of its proof.

<u>Theorem 4.1 [9]:</u> <u>Fix $B > 0$. Then the operator $H(B)$ has infinitely many bound states on the physical Hilbert space below the physical continuum.</u>

<u>Sketch of proof:</u> Let η be the ground state of $H^{n-1}(B)$ on the physical Hilbert space of $n-1$ electrons. If there is degeneracy we choose η to have a definite value of $L_z^{n-1} = -i\hat{z} \cdot (\sum_{j=1}^{n-1} \vec{r}_j \times \vec{\nabla}_j)$.

We normalize η so that $||\eta|| = 1$. The existence of η can be shown by an inductive argument which is simpler than the one which we will give below. We thus have

$$H^{n-1}(B)\eta = E\eta,\ L_z^{n-1}\eta = L_o\eta,\ ||\eta|| = 1$$

It will be important in the following that η has strong spatial decay properties:

$$|\eta(\vec{r}_1,\xi_1;\cdots\vec{r}_{n-1},\xi_{n-1})| \leq c_o \exp(-\lambda_o[\rho_1^2+\cdots+\rho_{n-1}^2 + \sqrt{z_1^2+\cdots+z_{n-1}^2}]) \quad (4.4)$$

Here $\lambda_o > 0$ and $\xi_j = \pm 1/2$ is the z-component of the spin of the jth electron. We will use a trial state Ψ constructed as follows: $\Psi = P\Phi$, where P projects onto the physical Hilbert space, and Φ is given by

$$\Phi(\vec{r}_1,\xi_1;\cdots;\vec{r}_n,\xi_n) = \quad (4.5)$$

$$\eta(\vec{r}_1,\xi_1;\cdots;\vec{r}_{n-1},\xi_{n-1})(1+\sum_{j=1}^{n-1} z_j g(\vec{r}_n))\alpha^{1/2}e^{-\alpha|z_n|}\phi_m(\vec{\rho}_n)\zeta(\xi_n)$$

$$g(\vec{r}) = -\beta(\mathrm{sgn}z)\gamma(z)(1+|\vec{r}|^2)^{-1};\ \zeta(1/2) = 1,\ \zeta(-1/2) = 0 \quad (4.6)$$

Here γ is an even C^∞ function which is 1 in a neighborhood of infinity and 0 in a neighborhood of zero. The function g gives the neutral atom a dipole moment which is felt by the extra electron in the state $\alpha^{1/2}e^{-\alpha|z|}\phi_m(\vec{\rho})\zeta(\xi)$. The parameter β regulates the size of this dipole moment. It is clear that as an operator on the physical Hilbert space, H(B) has essential spectrum $[E,\infty)$. The point E arises from moving the nth electron off to ∞ say in the z-direction, putting it in a ground state Landau orbit with zero kinetic energy in the z-direction, and in a spin state $\zeta(\xi)$. The other n-1 electrons are put in the state η. Of course the whole state must be antisymmetrized. We will show that there is an $M_o \geq 0$ so that if $m \geq M_o$ there exist $\alpha > 0$ and $\beta > 0$ so that $(\Psi,(H(B)-E)\Psi) < 0$. This will complete the proof since Ψ has total $L_z = m + L_o$. The major difficulties in the proof of this theorem which prevent it from being as simple as Theorem 3.1 stem from the fact that if $(\Phi,W\Phi) = \alpha K(\alpha,\beta,m)$ then although $K(0,0,m) < 0$, the latter decays exponentially in m. As we shall see later, this is the same order of magnitude as the exchange terms whose signs are unknown. This is the reason for taking $\beta > 0$. The large (i.e. $O(m^{-3})$) binding

then arises from the term in $(\Phi, W\Phi)$ which is linear in β and since α is not zero the sign of $K(\alpha,0,m)$ is unknown and $K(\alpha,0,m)$ must be estimated.

It turns out that the best results are obtained by taking $\beta = \text{const}$, $\alpha = (\text{const})m^{-3/2}$ and we will use these relations in our estimates without comment. The values of these constants will appear in the course of the calculation. The magnitudes of the exchange terms are small because of the estimate (based on (2.6))

$$\int e^{-\lambda\rho^2} |\phi_m(\vec{\rho})|^2 d^2\rho = O(e^{-\lambda' m}) \tag{4.7}$$

where $\lambda' > 0$ if $\lambda > 0$. This and the bound (4.4) imply $(\Psi,\Psi) = (\Phi,\Phi)/\sqrt{n} + O(e^{-\lambda m})$ for some $\lambda > 0$. A slightly more complicated argument gives $(\Psi,(H-E)\Psi) = (\Phi,(H-E)\Phi)/\sqrt{n} + O(e^{-\lambda m})$ so that it suffices to show

$$(\Phi,(H-E)\Phi) \leq -cm^{-p} \tag{4.8}$$

for some $c > 0$ and $p > 0$. We calculate the left side of (4.8) using the operator identity $-h\Delta h = -(\Delta h^2 + h^2\Delta)/2 + (\vec{\nabla}h)^2$ with $h = 1 + (\sum_{j=1}^{n-1} z_j)g(\vec{r}_n)$ and all variables except z_n included in Δ and $\vec{\nabla}$. The result is

$$\begin{aligned}(\Phi,(H-E)\Phi) &= \left\|\frac{d}{dz_n}\Phi\right\|^2 + (n-1)\|f_m g\|^2 \\ &+ \left\|\eta\left(\sum_{j=1}^{n-1} z_j\right)\right\|^2 \|f_m \vec{\nabla}_\perp g\|^2 + (\Phi,W\Phi)\end{aligned} \tag{4.9}$$

Here $\vec{\nabla}_\perp = (\partial_x, \partial_y)$ and $f_m(\vec{r}) = \alpha^{1/2}\phi_m(\vec{\rho})e^{-\alpha|z|}$. The estimate

$$\int \rho^{-k} |\phi_m(\vec{\rho})|^2 d^2\rho = O(m^{-k/2}) \tag{4.10}$$

will be very useful. A simple computation based on (4.10) shows that the first three terms of (4.9) contribute at most

$$\alpha^2 + c_1 m^{-3/2}\beta^2\alpha + O(m^{-7/2}) \tag{4.11}$$

for some $c_1 > 0$. The extra kinetic energy introduced by taking $\beta > 0$ is evident in (4.11) and will be balanced against a negative

interaction energy arising from the term linear in β which appears in $(\Phi,W\Phi)$. This term is (using $|\eta(\vec{r}_1,\cdots,\vec{r}_n)|$ to denote the norm in the spin variables)

$$D(m) = -2\beta\alpha\int dr_1\cdots dr_{n-1}dr|\eta(\vec{r}_1,\cdots,\vec{r}_n)|^2(\sum_{j=1}^{n-1} z_j)W(\vec{r}_1,\cdots,\vec{r}_{n-1},\vec{r})\cdot$$

$$e^{-2\alpha|z|}\gamma(z)\,\mathrm{sgn}z\,(1+|\vec{r}|^2)^{-1}|\phi_m(\vec{\rho})|^2$$

The dipole contribution of W is $\sum_{j=1}^{n-1}\vec{r}_j\cdot\vec{r}/|\vec{r}|^3$. The error after replacing W by this term is found to be $O(m^{-7/2})$. Similarly one can replace $\gamma(z)(1+|\vec{r}|^2)^{-1}e^{-2\alpha|z|}$ by $|\vec{r}|^{-2}$ with a small error. The result is

$$D(m) = -2\beta\alpha||\eta(\sum_{j=1}^{n-1} z_j)||^2\int|\vec{r}|^{-5}|z||\phi_m(\vec{\rho})|^2 dr + O(m^{-7/2})$$

$$\leq -2\beta\alpha m^{-3/2}c_2 + O(m^{-7/2})$$

where $c_2 > 0$. Adding this to (4.11) and minimizing with respect to β leads to $\beta = c_2/c_1$ and

$$(\Phi,(H-E)\Phi) \leq \alpha^2 - 2c_3\alpha m^{-3/2} + O(m^{-7/2}) + [(\Phi,W\Phi)-D(m)] \qquad (4.12)$$

where $c_3 > 0$. The term in $(\Phi,W\Phi)$ which does not involve β is (taking the antisymmetry of η into account)

$$I_0 = \alpha(n-1)\int dr_1 dr\,\lambda(\vec{r}_1)(|\vec{r}_1-\vec{r}|^{-1}-|\vec{r}|^{-1})|\phi_m(\vec{\rho})|^2 e^{-2\alpha|z|}$$

where $\lambda(\vec{r}_1) = \int dr_2\cdots dr_{n-1}|\eta(\vec{r}_1,\cdots,\vec{r}_{n-1})|^2$. We use a multipole expansion of $|\vec{r}-\vec{r}_1|^{-1} - |\vec{r}|^{-1}$ for $|\vec{r}_1| < |\vec{r}|/2$. (The region $|\vec{r}_1| > |\vec{r}|/2$ contributes $O(e^{-\lambda m})$ to I_0 for some $\lambda > 0$.):

$$|\vec{r}_1-\vec{r}|^{-1} - |\vec{r}|^{-1} = \sum_{\ell=1}^{3} 4\pi(2\ell+1)^{-1} \sum_{\nu=-\ell}^{\ell} \overline{Y_\ell^\nu(\hat{r}_1)} Y_\ell^\nu(\hat{r}) |\vec{r}_1|^\ell |\vec{r}|^{-\ell-1} + O(|\vec{r}_1|^4/|\vec{r}|^5) \tag{4.13}$$

The last term in (4.13) contributes $O(\alpha m^{-2}) = O(m^{-7/2})$ to I_0 so that adding back in the region $|\vec{r}_1| > |\vec{r}|/2$ we have

$$I_0 = \alpha(n-1)\int dr_1 dr \sum_{\ell=2}^{3} 4\pi(2\ell+1)^{-1} \sum_{\nu=-\ell}^{\ell} \overline{Y_\ell^\nu(\hat{r}_1)} Y_\ell^\nu(\hat{r}) |\vec{r}_1|^\ell |\vec{r}|^{-\ell-1} \cdot (e^{-2\alpha|z|}-1)\lambda(\vec{r}_1)|\phi_m(\vec{\rho})|^2 + O(m^{-7/2}) \tag{4.14}$$

Here we have used two additional facts: (a) the $\ell = 1$ term vanishes by symmetry and (b) the -1 in $(e^{-2\alpha|z|}-1)$ does not contribute at all in (4.14). The latter fact is a result of

$$\int |\phi_m(\vec{\rho})|^2 Y_\ell^\nu(\hat{r}) |\vec{r}|^{-\ell-1} dr = 0; \quad \ell \geq 1 \tag{4.15}$$

(4.15) follows from $\int P_\ell(\cos\theta)|\vec{r}|^{-\ell-1} dz = 0$ for $\ell \geq 1$. The latter can be seen by noting $f(x,y) = \int P_\ell(\cos\theta)|\vec{r}|^{-\ell-1} dz$ is a solution to Laplace's equation. However by a change of variables we see that $f = (\text{const})\rho^{-\ell}$ and $\rho^{-\ell}$ does not satisfy Laplace's equation for $\ell \geq 1$. Using $1-e^{-2\alpha|z|} \leq 2\alpha|z|$ and (4.10) gives

$$I_0 \leq (\text{const})\alpha^2 \sum_{\ell=2}^{3} \int |z||\vec{r}|^{-\ell-1} |\phi_m(\vec{\rho})|^2 d\vec{r} + O(m^{-7/2}) = O(m^{-7/2})$$

The term in $(\Phi, W\Phi)$ proportional to β^2 is handled similarly, so that from (4.12) we have

$$(\Phi,(H-E)\Phi) \leq \alpha^2 - 2c_3\alpha m^{-3/2} + O(m^{-7/2})$$

We optimize by choosing $\alpha = c_3 m^{-3/2}$ and find

$$(\Phi,(H-E)\Phi) \leq -c_3^2 m^{-3} + O(m^{-7/2})$$

which for large enough m gives (4.8). □

REFERENCES

1. S. AGMON, Proc. Tokyo Int. Conf. on Functional Analysis and Related Topics (1969).
2. J. AGUILAR and J. M. COMBES, Commun. Math. Phys. 22, 269 (1971).
3. J. E. AVRON, Phys. Rev. Lett. 37, 1568 (1976); J. Phys. A. 12, 2393 (1979).
4. J. E. AVRON and I. HERBST, Commun. Math. Phys. 52, 239 (1977).
5. J. E. AVRON, I. HERBST, and B. SIMON, Phys. Lett. 62A, 214 (1977).
6. ________, Phys. Rev. Lett. 39, 1068 (1977).
7. ________, Ann. Phys. 114, 431 (1978).
8. ________, Duke Math. J. 45, 847 (1978)
9. ________, Schrödinger operators with magnetic fields, III, Atoms in homogeneous magnetic field, submitted to Commun. Math. Phys.
10. E. BALSLEV and J. M. COMBES, Commun. Math. Phys. 22, 280 (1971).
11. L. BENASSI and V. GRECCHI, J. Phys. B. 13, 911 (1980).
12. F. BENTOSELA, Commun. Math. Phys. 68, 173 (1979).
13. J. M. COMBES and L. THOMAS, Commun. Math. Phys. 34, 251 (1973).
14. J. M. COMBES, R. SCHRADER, and R. SEILER, Ann. Phys. 111, 1 (1978).
15. R. J. DAMBURG and V. V. KOLOSOV, J. Phys. B. 11, 1921 (1978).
16. P. G. DODDS and D. H. FREMLIN, Compact operators in Banach lattices, submitted to J. Functional Analysis.
17. W. FARIS and R. LAVINE, Commun. Math. Phys. 35, 39 (1974).
18. L. GEARHART, Trans. AMS 236, 385 (1978).
19. S. GRAFFI and V. GRECCHI, Lett. Math. Phys. 2, 335 (1978).
20. ________, Commun. Math. Phys. 62, 83 (1978).
21. ________, J. Phys. B 12, 265 (1979).
22. ________, Resonances in the Stark effect of atomic systems, to appear in Commun. Math. Phys.
23. S. GRAFFI, V. GRECCHI, and B. SIMON, J. Phys. A 12, L193 (1979).
24. G. H. HARDY, Divergent series, Clarendon Press, Oxford (1949).
25. E. HARRELL and B. SIMON, The mathematical theory of resonances whose widths are exponentially small, to appear in Duke Math.J.
26. I. HERBST, Math. Zeit. 155, 55 (1977).
27. ________, Commun. Math. Phys. 64, 279 (1979).
28. ________, Commun. Math. Phys. 75, 197 (1980).
29. ________, Contraction semigroups and the spectrum of $A_1 \otimes I + I \otimes A_2$, submitted to J. Op. Th.
30. I. HERBST and J. HOWLAND, The Stark ladder and other one-dimensional external field problems, to appear in Commun. Math. Phys.
31. I. HERBST and B. SIMON, Phys. Rev. Lett. 41, 67 (1978).
32. ________, Dilation analyticity in constant electric field II, N-body problem, Borel summability, submitted to Commun. Math. Phys.
33. I. HERBST and A. SLOAN, Trans. AMS 236, 325 (1978).

34. H. HOFOP and W. C. LINEBERGER, J. Phys. Chem. Ref. Data 4, 539, (1975)
35. W. HUNZIKER, Helv. Phys. Acta 39, 451 (1966).
36. __________, Schrödinger operators with electric or magnetic fields, Proc. Int. Conf. Math. Phys., Lausanne, 1979, to appear.
37. T. KATO, Peturbation theory for linear operators, Springer, New York (1966).
38. L. D. LANDAU and E. M. LIFSHITZ, Quantum mechanics, Pergamon Press, New York (1977).
39. D. LARSEN, Phys. Rev. Lett. 42, 749 (1979).
40. L. PITT, J. Op. Th. 1, 49 (1979).
41. W. POLYZOU, J. Math. Phys. 21, 506 (1980).
42. M. REED and B. SIMON, Methods of modern mathematical physics, I, Functional analysis, Academic Press, New York (1972).
43. __________, Methods of modern mathematical physics, II, Fourier analysis, self-adjointness, Academic Press, New York (1975).
44. __________, Methods of modern mathematical physics, IV, Analysis of operators, Academic Press, New York, (1978).
45. B. SIMON, Ann. of Phys. 97, 279 (1976).
46. ________, Functional integration and quantum physics, Academic Press, New York (1979).
47. ________, Duke Math. J. 46, 119 (1979).
48. C. VAN WINTER, Mat.-Fys. Skr. Danske Vid. Selsk 2, No. 8 (1964).
49. K. VESELIC and J. WEIDMANN, Math. Zeit. 156, 93 (1977)
50. G. H. WANNIER, Phys. Rev. 117, 432 (1960); 181, 1364 (1969); Rev. Mod. Phys. 34, 645 (1962).
51. S. WEINBERG, Phys. Rev. 133, 232 (1964).
52. K. YAJIMA, J. Fac. Sci., Univ. Tokyo 26, 377 (1979).
53. K. YAJIMA, Spectral and scattering theory for Schrödinger operators with Stark-effect, II, to appear in J. Fac. Sci. Univ. Tokyo.
54. T. YAMABE, A. TACHIBANA, and H. J. SILVERSTONE, Phys. Rev. A 16, 877 (1977).
55. J. ZAK, Phys. Rev. Lett. 20, 1477 (1968); Phys. Rev. 181, 1366 (1969).

The Born-Oppenheimer Approximation

J.M.Combes, P.Duclos
Département de mathématique
Université de Toulon

R.Seiler
Institut f. theor.Physik
Freie Universität Berlin

I. INTRODUCTION

1. The Problem

In physics and chemistry the Born-Oppenheimer approximation is a very important method for analyzing the spectrum of molecules[1]. It is based on the important fact that the molecular Schrödinger operator contains one small parameter, the ratio k^4 of the electronic to the nuclear mass [2]. Perturbation theory in this parameter is however very singular.

The two main results of the article by Born and Oppenheimer (1927) are:

1. The discrete spectrum is discussed qualitatively. A formal perturbation theory for energy eigenvalues (discrete spectrum) in k is developed. To order k^6 it is enough to investigate a nuclear Schrödinger problem. The electrons are taken care of by an effective potential V (Fig.1).

2. Quantitatively the energy eigenvalues for diatomic molecules are given by the formula

$$(1)\quad E(k) = V_o + a(n+1/2)k^2 + \left(b + \frac{\ell(\ell+1)}{|R_o|^2}\right)k^4 + O(k^5)$$

n and ℓ are quantum numbers; a,b and R_o are constants which will be computed (see equ.(37)). R_o is the classical equilibrium configuration of the nuclei in the effective potential V and $V_o=V(R_o)$.

The object of these lectures is to give an outline of a proof of the two results for diatomic molecules and small energies.

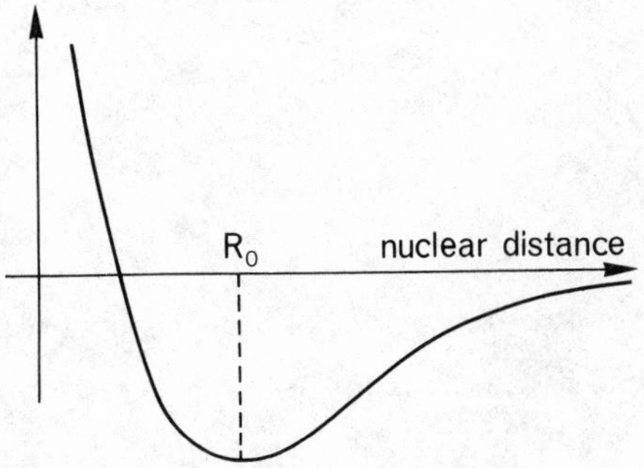

Fig. 1. Graph of a typical effective potential V_{eff} for diatomic molecules.

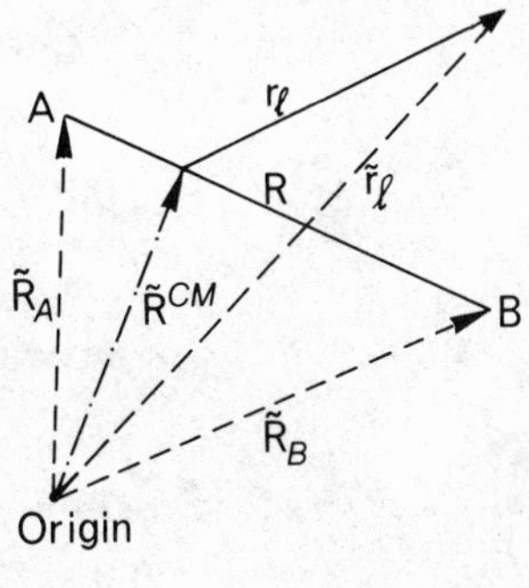

Fig. 2. The center of mass of the nucleic system (CMNS).

2. Notation and Strategy

In order to motivate formula (2) below for the Schrödinger operator of a diatomic molecule let us make a slight detour on the geometry of the system with 2 nuclei A and B (heavy particles) and L electrons (light particles).

The Hamilton function - in standard notation - is given by

$$(2)\quad \tilde{H} := \frac{1}{2M_A}\tilde{P}_A^2 + \frac{1}{2M_B}\tilde{P}_B^2 + \frac{1}{2}\sum_{\ell=1}^{L} p_\ell^2 + \underbrace{\sum_{\ell=1}^{L} V_{A,\ell}(\tilde{R}_A - \tilde{r}_\ell)}_{V_A}$$

$$+ \underbrace{\sum_{\ell=1}^{L} V_{B,\ell}(\tilde{R}_B - \tilde{r}_\ell)}_{V_B} + \underbrace{\sum_{i<k} V_{ik}(\tilde{r}_i - \tilde{r}_k)}_{V^{e\ell}} + \underbrace{V_{AB}\,(\tilde{R}_A - \tilde{R}_B)}_{V_{AB}} .$$

We shall use the abbreviation $\underline{x} = (\tilde{R}_A, \tilde{R}_B, \tilde{r}_1 \ldots \tilde{r}_L)$, $\underline{p} = (\tilde{P}_A, \tilde{P}_B, \tilde{p}_1, \ldots, \tilde{p}_L)$ and have set the electronic mass equal to one. $\tilde{H}$ generates a flow on the phase space $\Omega = R^{6(2+L)}$. In a standard fashion Ω decomposes in a direct sum of the center of mass subspace Ω^{CM} and the reduced phase space Ω^{RED}

$$\Omega^{CM} := \{(\underline{x},\underline{p}) \mid \tilde{R}_A = \tilde{R}_B = \tilde{r}_i = R_{CM} \quad \tilde{P}_A = \frac{M_A}{M_{Total}} P_{CM},\ \tilde{P}_B = \frac{M_B}{M_{Total}} P_{CM},$$

$$\tilde{p}_i = \frac{1}{M_{Total}} P_{CM} \} \cong \mathbb{R}^6$$

$$\Omega^{RED} := \{(\underline{x},\underline{p}) \mid M_A\tilde{R}_A + M_B\tilde{R}_B + \sum_{\ell=1}^{L} \tilde{r}_\ell = 0,\ \tilde{P}_A + \tilde{P}_B + \sum_{\ell=1}^{L} p_\ell = 0 \}$$

$$\cong \mathbb{R}^{6(1+L)}$$

The two subspaces are invariant under the flow generated by $\tilde{H}$. Ω^{CM} is parametrized by R_{CM}, P_{CM}. Ω^{RED} is usually parametrized with Jacobi coordinates. They are constructed so that the structure of the kinetic part is preserved. In the context of diatomic molecules it is standard to use different coordinates $R, r_1 \ldots r_L, P, p_1, \ldots p_L$. They are defined in figure 2. It is useful to introduce the following notation

$$\mu_{AB} := M_A M_B\,(M_A + M_B)^{-1}$$

$$k^4 := \mu_{AB}^{-1}$$

(3)

$$\rho := \mu_{AB}(M_A + M_B)^{-1}$$

$$\tau_{\substack{A\\B}} := \pm M_{\substack{B\\A}}(M_A + M_B)^{-1}, \quad \tau_A - \tau_B = 1$$

The restriction H(k) of $\tilde{H}$ to the reduced phase space expressed in terms of the coordinates introduced in fig.2 (CMNS) is

$$\text{(4)} \quad H(k) = \frac{1}{2} k^4 \left[P^2 + \rho(\Sigma p_\ell)^2\right] + H^{e\ell}$$

$$\text{(5)} \quad H^{e\ell} := \underbrace{\frac{1}{2}\Sigma p_\ell^2}_{H_o^{e\ell}} + \underbrace{\Sigma V_{A,\ell}(\tau_A R - r_\ell)}_{V_A(R)} + \underbrace{\Sigma V_{B,\ell}(\tau_B R - r_\ell)}_{V_B(R)} + \underbrace{\sum_{i<k} V_{ik}(r_i - r_k)}_{V^{e\ell}} + V_{AB}(R).$$

The notation V_A for the interaction term of electrons with nucleus A and $V^{e\ell}$ for electrons among themself will be used regularly . The use of these coordinates, the so called center of mass of the nuclei system (Pack and Hirschfelder, (1968)) is traditional. Certain properties of the system can be seen quite easily in this representation of the Hamiltonian. But we don't want to emphasize any particular choice of coordinates. If the k-dependence is unimportant for the problem under discussion we write H instead of H(k).

Notice that $H^{e\ell}$ is in a formal sense H(k=o) and that the $k \searrow o$ limit resembles a WKB limit. The limit is however very singular. This can easily be seen looking at examples. They show that the essential spectrum of the operator valued function H(k) is discontinuous at k=o.

The analysis of the spectrum of H(k) will proceed in three steps:

Step 1 : Analysis of spectrum and eigenfunctions of $H^{e\ell}$.

Step 2 : Reduction of the Born-Oppenheimer problem to a regular WKB problem for nuclei in an effective potential.

Step 3 : Analysis of the WKB-problem for nuclei.

Throughout we will explain the ideas using what we shall call the standard example. It is the Schrödinger operator of the diatomic molecular ion with one electron (A particular case is the H_2^+ molecule.

$$\text{(6)} \quad H(k) = \frac{1}{2} k^4 \left[p^2 + \frac{1}{4} p_1^2\right] + H^{e\ell}$$

$$(7)\quad H^{e\ell} = \frac{1}{2}p_1^2 - \frac{\alpha}{|\frac{1}{2}R - r_1|} - \frac{\alpha}{|-\frac{1}{2}R - r_1|} + \frac{\alpha}{|R|} =$$

$$= H_o^{e\ell} + V_A(R) + V_B(R) + V_{AB}(R)\ .$$

α is the fine structure constant; the nuclear charges Z_A and Z_B we have set equal to one. The nuclear masses are equal. All arguments can easily be generalized from this simple case to the more general one with arbitrary charges and masses.

As a general policy we will state theorems in full generality and proofs for the standard example only. There will be no proofs for results added in remarks.

The minimal assumption on all two body potentials in H(k) is that they are real and dilation analytic. This is abreviated in all statements by "let the potential be dilation analytic" (Aguilar and Combes, 1971). We will use the terminology Schrödinger operator and Hamiltonian synonymously.

II. THE ELECTRONIC HAMILTONIAN

1. The Clamped Nuclei Approximation

The Electronic Hamiltonian defined in equation (7) is invariant under translations of the momentum $P \to P + a$; hence $H^{e\ell}$ commutes with R and has the direct integral decomposition

$$(8)\quad H^{e\ell} = \int_{\oplus} dR\, H^{e\ell}(R).$$

$H^{e\ell}(R)$ is called the clamped nuclei Hamiltonian and has a domain of definition in the space of electronic states $\mathcal{H}^{e\ell}(R) \cong L^2(R^{3L})$. $H^{e\ell}(R)$ represents the dynamics of electrons in the external field of the nuclei in configuration R. Notice the important fact that $H^{e\ell}(R)$ acts as on operator in the electronic configuration space $L^2(R^{3L})$ which is distinct from the space of molecular configurations $L^2(R^{3(L+1)})$ which contains the domain of definition D(H(k)). For the decadent reader we might add that $L^2(R^{3(L+1)})$ can be looked at as the space of sections of a trivial fiber bundle with base R^3 and fibers $L^2(R^{3L})$, $H^{e\ell}(R)$ acting in the direction of the fibre only. The clamped nuclei Hamiltonian contains a term $V_{AB}(R)$ which is for fixed nuclear configuration just a number. Hence it is useful to introduce the notation H(R) for the nontrivial part in $H^{e\ell}(R)$,

$$(9)\quad H^{e\ell}(R) = H(R) + V_{AB}(R)$$

In the remainder of this chapter we will analyze spectrum and eigenfunctions of H(R). Notice the difference between H(k) defined by equation (4) and H(R) defined in (9).

Remark 1: The results about H(R) can be guessed by looking at analytical and numerical results computed for the standard example (Komarov, Ponomarev and Slavianov, 1976, see also fig.3).

a) Discrete eigenvalues are smooth functions of $|R|$ only.
b) The spectrum of H(R) converges to the spectrum of He^+ for $R \to o$. This is the united atom limit.
c) In the separated atom limit, $|R| \to \infty$, the spectrum of H(R) converges towards the one of Hydrogen.
d) The slope of discrete eigenvalues $E(|R|)$ at R=o is zero (Fig.3).
e) The ground state energy is monotonically increasing in $|R|$. (Lieb and Simon, 1978; M. and T.Hoffmann-Osterhoff, 1979).
f) Eigenfunctions are functions of R with values in $L^2(R^{3L})$. They are quite smooth in R.

2. General Properties of H(R)

The clamped nuclei Hamiltonian satisfies certain covariance conditions if the potentials are for instance radially symmetric (e.g. Combes and R.Seiler , 1978).

Proposition 1: Let all potentials in H(R) be radially symmetric and let $\tilde{O}$ be the unitary representation of the rotation O of $\mathbb{R}^3$ considered as a mapping from $\mathcal{H}^{el}(R)$ to $\mathcal{H}^{el}(OR)$. Then $\tilde{O}H(R)=H(OR)\tilde{O}$.

We give a formal proof only. The mathematical details can be put in easily. By definition the following equalities hold

$$(\tilde{O}\, H(R)\phi)(r_1) = \left(\frac{1}{2}\, p_1^2 - \frac{\alpha Z_A}{|\frac{1}{2}R - O^{-1}r_1|} - \frac{\alpha Z_B}{|\frac{1}{2}R + O^{-1}r_1|}\right)\phi\,(O^{-1}r_1)$$

$$= \left(\frac{1}{2}\, p_1^2 - \frac{\alpha Z_A}{|\frac{1}{2}OR - r_1|} - \frac{\alpha Z_B}{|\frac{1}{2}OR + r_1|}\right)\phi\,(O^{-1}r_1)$$

$$= (H(OR)\,\tilde{O}\phi)(r_1)$$

They imply the statement in the above proposition.

Since the spectrum of an operator is a unitary invariant we get the corollary:

$$\sigma(H(R)) = \sigma(H(OR)) \,. \tag{10}$$

Remark 2: $H(R)$ is selfadjoint on the natural domain

$$D(H_o^{e\ell}) \text{ of } H_o^{e\ell} = \sum_{\ell=1}^{L} \frac{1}{2} p_\ell^2 .$$

Eigenvalues of $H(R)$ are loosely speaking real analytic in $|R|$. The precise result is a corollary (see below) of

Theorem 2: Let the potentials be dilation analytic and let furthermore $D(\theta)$ be the unitary dilations,

$(D(\theta)\phi)(r_1 \ldots r_L) = e^{\frac{3L}{2}\theta} \phi(e^\theta r_1, \ldots e^\theta r_L)$, $\theta \in R$. Then the operator valued function $H(\theta,R) = D(\theta)\, H(e^\theta R)\, D^{-1}(\theta)$ has an analytic extension into the strip $|\text{Im}\theta| < \theta_{max}$ for a suitable θ_{max} (for a precise definition of $H(\theta,R)$ see remark 3 below). $H(\theta,R)$ takes its values in the linear space of closed operators on $L^2(R^{3L})$.

This theorem has the consequence:

Corollary 3: Let $\theta_o \in R$. Any finite system of eigenvalues $E_i(e^\theta|R_o|) \in \sigma(H(\theta,R_o))$, $i=1 \ldots N$, is analytic in θ in a neighborhood of θ_o. Hence $E_i(|R|)$ is analytic in $|R|$ in a neighborhood of $e^{\theta_o}|R_o|$ provided $R_o \neq o$ and under suitable labeling of the eigenvalues.

Remark 3: $H(\theta,R)$ is defined by

$$H(\theta,R) := \{1 + T(\tau_A R)\, D(\theta)\, V_A(0)\, (1+H_o^{e\ell})^{-1}\, D(\theta)^{-1}\, T(\tau_A R)^{-1} + (A \to B) + D(\theta) V^{e\ell} (1+H_o^{e\ell})^{-1} D(\theta)^{-1}\}\, (1+H_o^{e\ell}(\theta)) - 1 .$$

$T(a)$ is the unitary translation operator and τ_A, τ_B has been defined previously (3).

Nondegenerate eigenfunctions and eigenvalues are twice differentiable in R.
This follows from

Theorem 4: If the potentials in $H(R)$ are dilation analytic the resolvent $(H(R)-z)^{-1}$ is twice differentiable in R.

The relation between total eigenprojection $P(R)$, discrete nondegenerate eigenvalue $E(R)$ and resolvent is given by the well known formulas

$$P(R) = -\frac{1}{2\pi i} \int_\Gamma dz\, (H(R)-z)^{-1}$$

$$E(R) = \text{Trace } H(R)\, P(R) .$$

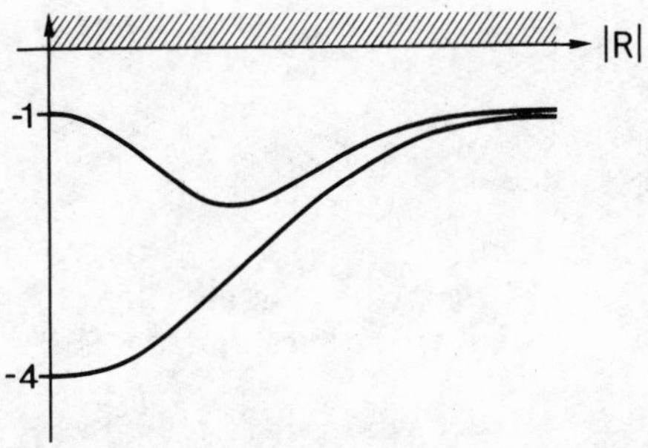

Fig. 3. Graph of the spectrum of H(R) for H_2^+ ground state, first excited state, and continuum.

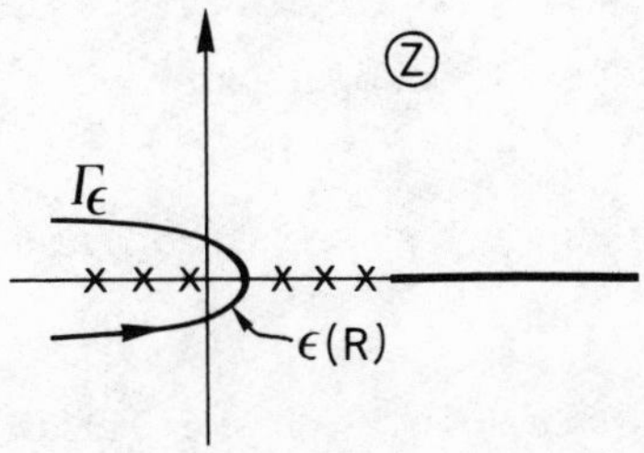

Fig. 4. Spectrum of H(R).

They imply regularity of P(R) and E(R).

Remark 4: Differentiality of eigenvalues holds even if there is degeneracy at least if the potentials are rotationally symmetric (Combes and Seiler 1980) .

We will give only a sketch of the proof of theorem 4. The first derivative in the direction e of the resolvent is given by the expression

$$(11)\quad \frac{d}{dt}\,(H(R)+te)-z)^{-1}\Big|_{t=o} = -\,(H(R)-z)^{-1}\ \operatorname{grad}_R H(R)\,(H(R)-z)^{-1}.$$

Furthermore

$$\begin{aligned}\operatorname{grad}_R H(R) &= \operatorname{grad}_R V_A(R) + \operatorname{grad}_R V_B(r)\\ &= \frac{i}{2}\left[p_1, V_A(R)\right] - \frac{i}{2}\left[p_1, V_B(R)\right].\end{aligned}$$

If inserted in (11) a typical term up to a numerical constant is $(H(R)-Z)^{-1}V_B(R)p_1(H(R)-Z)^{-1}$. It is well defined because $(H(R)-Z)^{-1}V_B(R)$ is a bounded operator and the range of the resolvent $(H(R)-Z)^{-1}$ is the domain of $H_o^{e\ell}$, hence $p_1(H(R)-Z)^{-1}$ is bounded too. This proves existence of the derivative in direction e. Since this holds for any direction and since the directional derivative is continuous in R,the derivative exists.

The second derivative can again be written in terms of resolvents and commutatators.Existence can be shown along the same lines.

Remark 5: Differentiability up to second order holds because $H_o^{e\ell}$ is a second order differential operator.

Remark 6: Eigenvectors for nondegenerate eigenvalues of H(R) are twice differentiable in R as functions with values in $L^2(\mathbb{R}^{3L})$, but not in all arguments of course. Looking for instance at the standard example the following cusp condition holds (Kato,1957 and M.Hoffmann-Ostenhoff and R.Seiler, 1980).

$$\psi_{av}(\rho, \tfrac{1}{2}R) := \int_{S^2} d\omega\, \psi\,(R, r_1 = \tfrac{1}{2}R + \rho\omega),\ \rho > o\ ,$$

$$\frac{\partial\psi_{av}}{\partial\rho}\,(0, \tfrac{1}{2}R) = \text{const.}\ \psi(R, \tfrac{1}{2}R\,).$$

S^2 is the two dimensional sphere in $\mathbb{R}^3$. The electron wave function is therefore quite singular if the electron is just on top of a nucleus, i.e. $r_1 = \pm\,\frac{1}{2}R$.

Remark 7: Theorem 4 implies the Feynman-Hellman theorem.

Remark 8: The slope of discrete eigenvalues $E(|R|)$ at $|R|=o$ is zero if the potentials are rotationally symmetric (Combes and Seiler, 1980)

Remark 9: Discrete eigenvalues $E(R)$ converge towards $E \in \sigma_d(H(R=o))$ for $R \to o$.

Remark 10: Theorem 4 can be extended to systems with many nuclei .

3. The Separated Atom Limit, $R \to \infty$

In this section we discuss the standard example only . For a full discussion we refer to the literature (Combes and Seiler, 1978; Morgan III and Simon 1979).

Let us first define the separated atom spectrum

$$\sigma_d^\infty : = \{\lim_{\lambda \to \infty} E(\lambda R) \mid E(\lambda R) \in \sigma_d(H(\lambda R))\} \quad ,$$

where σ_d denotes the discrete part of the spectrum.

Intuitively it is clear that the separated atom spectrum σ_d^∞ is very much related to the spectrum of the atomic Schrödinger operators

$$H_{\underline{A}} = \frac{1}{2} p_1^2 - \frac{\alpha}{|r_1|}$$

$$H_{\underline{B}} = \frac{1}{2} p_1^2 - \frac{\alpha}{|r_1|} \quad .$$

This was already mentioned in remark 1c of the first section and is put into a precise form in

Theorem 5: Let the two body potentials in $H(R)$ be dilation analytic. Then the separated atom spectrum is described by

i) $\sigma_d^\infty \supset \sigma_d\,(H_{\underline{A}}) \cup \sigma_d(H_{\underline{B}})$

ii) $\sigma_d^\infty \subset \sigma_d\,(H_{\underline{A}}) \cup \sigma_d(H_{\underline{B}}) \cup \Sigma_3$;

$\Sigma_3 = \{0\}$ is the three body threshold for this system.

Proof: i) The first statement follows from the following argument: Consider the identity

$$H_{\underline{A}} + T(-R)V_B(0)T(R) = T(-\tfrac{1}{2}R)H(R)T(\tfrac{1}{2}R) \, .$$

We show that the left hand side converges in the strong resolvent (s.r.) sense to H_A if $R \to \infty$.

For that it is enough to show that $T(-R)\, V_B(0) T(R) \mathfrak{f} \to 0$ for all $\mathfrak{f} \in D(H(R)) = D(H_o^{e\ell})$. Since every such $\mathfrak{f}$ is of the form $\mathfrak{f} = (H_o^{e\ell}+1)^{-1}\psi$ we have to look at

$$T(-R)V_B(0)T(R)(H_o^{e\ell}+1)^{-1}\psi = T(-R)V_B(0)(H_o^{e\ell}+1)^{-1}\; T(R)\psi$$

Since $V_B(0)(H_o^{e\ell}+1)^{-1}$ is a compact operator and since $T(R)$ is unitary and converges weakly to zero, the statement follows.

Knowing that $T(-\frac{1}{2}R)H(R)T(\frac{1}{2}R)$ converges in s.r. sense to $\underline{H_A}$ every point in $\sigma(H_A)$ is limit point of elements in $\sigma(T(-\frac{1}{2}R)H(R)\,T(\frac{1}{2}R))$ for $R \to \infty$ (see eg. T.Kato, 1966 page 431), hence also of elements in $\sigma(H(R))$. Furthermore every point in $\sigma_d(H_A)$ has to be limit point of $\sigma_d(H(R))$ due to stability of the essential spectrum by $H_o^{e\ell}$-compact perturbations. This proves the first part of the theorem.

ii) The second statement can be shown using the Weinberg-v.Winter equations or geometrical methods:
Consider the Weinberg-v.Winter equation (e.g. W.Hunziker, 1966), which is for the standard example particularly simple:

$$(12)\quad (H(R)-Z)^{-1} = D(R,Z) + I(R,Z)(H(R)-Z)^{-1}$$

$$D(R,Z) := (H_o^{e\ell}-Z)^{-1} + (H_o^{e\ell}-Z)^{-1}\; T(\tfrac{1}{2}R)\, V_A(0)(H_{\underline{A}}-Z)^{-1}\; T(-\tfrac{1}{2}R) + A \to B$$

$$I(R,Z) := (H_o^{e\ell}-Z)^{-1} T(\tfrac{1}{2}R)V_A(0)\quad (H_{\underline{A}}-Z)^{-1} T(-\tfrac{1}{2}R)T(-\tfrac{1}{2}R)V_B(0)T(\tfrac{1}{2}R) + A \leftrightarrow B$$

$$= T(\tfrac{1}{2}R)\underbrace{(H_o^{e\ell}-Z)^{-1}V_A}_{\text{compact}}(0)\underbrace{(H_A-Z)^{-1}\; T(-R)V_B(0)T(\tfrac{1}{2}R)}_{S} + A \leftrightarrow B\,.$$

The adjoint of the second term S goes strongly to zero. Since it is multiplied by a compact operator $I(R,Z)$ goes to zero in norm, $I(R,Z) \xrightarrow{n} 0$.

Because of that fact the resolvent $(H(R)-Z)^{-1}$ exists for every $Z \in \rho(H_{\underline{A}}) \cap \rho(H_B) \cap \rho(H_o^{e\ell})$ and $|R|$ sufficiently large. Hence $E_\infty \in \sigma_d$ is an element of $\sigma_d(H_{\underline{A}}) \cup \sigma_d(H_B) \cup \sigma(H_o^{e\ell})$. But since $E \leqslant 0$ the second part of the theorem is proved.

For a proof of this statement by geometrical methods we refer to the article by Morgan and Simon (1979) or also Combes and Seiler(1980).

<u>Remark 11</u>: In the case of many electrons it is important to incorporate statistics. This can be done using either of the methods of proof, see in particular the article by Morgan and Simon (1979).

Remark 12: This remark concerns a property of the standard example for the particular case $Z_A = Z_B$ and $m_A = m_B$ only . Let Π be the unitary operator mapping the electronic state space $\mathcal{H}^{e\ell}(R)$ onto itself:

$$(13) \quad (\Pi(R)\phi)(R,r_1) = \phi(R,-r_1)$$

Due to their definition H(R) and $\Pi(R)$ commute.

Using the spectral projectors $\Pi_{\pm}(R) = \frac{1}{2}(1 \pm \Pi(R))$ on even and odd wave functions we can write

$$H(R) = H_+(R) + H_-(R) \;, \; H_{\pm}(R) := \Pi_{\pm}(R)\, H(R)\, \Pi_{\pm}(R) \;.$$

The groundstate energies of $H_+(R)$ and $H_-(R)$ for $R\to\infty$ are the same up to an exponentially small term (Herring, 1962),

$$\Delta E(R) = E_-(R) - E_+(R) \sim \frac{4}{e} |R| \, e^{-|R|}$$

Assuming the existence of an asymptotic series

$$E(R) \sim \Sigma\, a_n \, |R|^{-n}$$

there is strong numerical evidence that the following relation holds (Brezin and Zinn-Justin, 1979):

$$a_n \cong \int_0^\infty d|R|\,|R|^{n-1} \left(\frac{1}{2}\Delta E(R)\right)^2 = -\frac{2}{e^2}\,\frac{(n+1)!}{2^{n+1}} \quad .$$

This is a so called Bender-Wu formula for a_n, n large (large order expansion). The proof of this relation is an, as yet unsolved, interesting problem.

III. REDUCTION TO A WKB-PROBLEM FOR NUCLEI IN AN EFFECTIVE POTENTIAL

In this chapter we will discuss a construction which allows an important simplification in the analysis of the discrete spectrum of H(k). This construction depends crucially on the upper bound ε of the energy range $(-\infty,\varepsilon)$ considerd. Furthermore the method is well suited if one looks for asymptotic expansions of discrete eigenvalues up and including 4-th order (1) in k. Beyond that it looses its simplicity.

1.Adiabatic States

In the following we will make a convenient assumption concerning the structure of the spectrum of H(R). We shall assume the existence of a smooth function $\varepsilon(R)$, real valued with the property: there exists $\delta > o$ so that

$$\text{dist}\;[\varepsilon(R),\, \sigma(H(R))] \geq \delta > o$$

It follows from our general assumptions on the potentials that the spectrum below $\varepsilon(R)$ is discrete.

Now we are ready to define the adiabatic states. Let P^{AD} be the projector

$$P^{AD} := \int_{\oplus} dR P(R)$$

where P(R) is the spectral projector of H(R) mapping on the subspace with spectrum below ε(R),

$$P(R) = -\frac{1}{2\pi i} \int_{\Gamma_\varepsilon(R)} dz \quad (H(R)-z)^{-1} .$$

$\Gamma_{\varepsilon(R)}$ is the curve in the complex plane depicted in fig.4. The range of P^{AD} is the subspace of adiabatic wave functions $\mathcal{H}^{AD}$.

Notice that due to the general assumption on ε(R) the dimensions d of range P(R) in $\mathcal{H}^{e\ell}(R)$ is R-independent.

The adiabatic wave functions $\phi \in \mathcal{H}^{AD}$ are of the following form

$$(14) \qquad \phi^{AD}(R,\underline{r}) = \sum_{\ell=1}^{d} \mathfrak{f}_i(R) e_R^i(\underline{r}),$$

where $\{e_R^i\}_{i=1}^d$ is a basis of range P(R).

Equation (14) defines a natural isomorphism between $\mathcal{H}^{AD}$ and $L^2(R^3) \oplus \ldots \oplus L^2(R^3)$ where the number of $L^2(R^2)$-spaces equals the dimensions of range P(R)=d.

For the following it is important to have the technical result which we state without proof:

Lemma 6: Let all potentials in H(R) be dilation analytic and rotationally invariant. Then there exists a choice of basis $\{e_R^i\}_1^d$ with the properties

i) $\{e_R^i\}_1^d$ is orthonormal

ii) e_R^i is twice differentiable in R.

Remark 13: The statement does not appear to depend on rotational invariance, but the available proof requires this assumption (J.M.Combes and R.Seiler, 1980).

Remark 14: The terminology adiabatic means that the wave functions $\phi^{AD} = \Sigma \mathfrak{f}_i(R) e_R^i(\underline{r})$ represent states in which the heavy nuclei move slowly enough so that the electronic motion can adapt instantaneously to the nuclear configuration. The wave function ϕ^{AD} for fixed nuclear configuration is a superposition of eigenfunctions of $H^{e\ell}(R)$.

The adiabatic Schrödinger operator is defined by

(15) $H^{AD}(k) := P^{AD} H(k) P^{AD}$,

with domain $\mathcal{D}(H^{AD}(k)) = P^{AD}\mathcal{D}(H_o) \oplus Q^{AD}\mathcal{H}$.

The definition is sensible due to the technical

Lemma 7: Let the potentials be dilation analytic. Then the adiabatic Schrödinger operator is well defined i.e.

(16) $P^{AD}\ \mathcal{D}(H_o) \subset \mathcal{D}(H_o) = \mathcal{D}(H(k))$.

Furthermore $H^{AD}(k)$ is selfadjoint.

Proof: To prove the first statement remember that P(R) is twice differentiable in R. The derivatives are uniformely bounded in R. Hence the inclusion (16) holds.

To prove selfadjointness it is enough to show that $P^{AD}H(k)Q^{AD} + Q^{AD}H(k)P^{AD}$ is H(k) bounded with bound zero in the sense of Kato. If this is the case the operator

$$P^{AD}H(k)P^{AD} + Q^{AD}H(k)Q^{AD} = H(k) - (P^{AD}H(k)Q^{AD} + Q^{AD}H(k)P^{AD})$$

is selfadjoint with domain $\mathcal{D}(H_o) = \mathcal{D}(H(k))$. It follows then that H^{AD} is selfadjoint.

To prove the zero bound consider $\phi \in \mathcal{D}(H(k))$. Then the inequality holds:

$$\begin{aligned} || P^{AD}H(k)Q^{AD}\phi||^2 &= \frac{1}{4} k^4 \int dR\, ||P(R) \left[P^2+\rho(\Sigma p_\ell)^2\right] Q(R)\phi(R)||^2 \\ &\leqslant \text{const.}\ k^4 \{\int dR||\ P(R)(\nabla_R Q(R))\ \nabla_R\phi(R)||^2 + \\ &\quad + \int dR||P(R)(\Delta_R Q(R))\phi(R)||^2 + \\ &\quad + ||P^{AD}\rho(\Sigma p\ell)^2\ Q^{AD}\phi||\ \} . \end{aligned}$$

It implies the second assertion of the lemma.

2. The Partitioning Technic

This technique, which is also called Feshbach's method, permits the substitution of the elliptic eigenvalue problem $H\phi = E\phi$ by an other problem which can be formulated in $\mathcal{H}^{AD}$ only. The prize to be paid is that the new problem is of the form

$$(H^{AD} - U(E))\ \phi^{AD} = E\ \phi^{AD}$$

where U is an nonlocal, E-dependent operator defined by

$$U(E) = P^{AD}\ H\ Q^{AD}\ (\hat{H-E})^{-1}\ Q^{AD}\ H\ P^{AD}$$

$(\hat{H-E})^{-1}$ is the resolvent of $Q^{AD} H\, Q^{AD}$ in the subspace range Q^{AD}. The new problem is not an eigenvalue problem.

Some technically important properties of U are summarized in

Lemma 8: Let $E < \varepsilon = \inf_R \varepsilon(R)$ and of course all potentials be dilation analytic. Then U(E) with $\mathcal{D}(U(E))=\mathcal{D}(H^{AD})$ is well defined symmetric, H^{AD} bounded with bound zero and positive. Furthermore - U(E) is monotonically decreasing in E and concave. H^{AD}- U(E) converges in the strong resolvant sense to H^{AD} for $E \to -\infty$.

Since the proof makes essentially only use of the methods already explained in the context of the previous lemma we omit it.

The results in lemma 7 and 8 lay the technical ground for

Theorem 9: Assume $E < \varepsilon$ and all potentials dilation analytic. Then there exists an isomorphism I between the nullspaces of H-E and H^{AD}-U(E)-E given by

$$I : \phi \to P^{AD}\phi$$

$$I^{-1}: \Psi^{AD} = P^{AD}\ \Psi^{AD} \to \Psi = \Psi^{AD} - Q^{AD}\ (\hat{H-E})^{-1}\ Q^{AD}\ H\ \Psi^{AD}.$$

We leave it as an exercise to establish that, firstly I exists in a formal sense and secondly that all formal steps can be rigorously justifies using the previously mentioned results.

Looking at theorem 9 one gets easily the impression that the complicated eigenvalue problem has been replaced by an even more complicated one. The importance of the result is however coming from the fact that the energy dependent term U(E) is in a certain sense small for small k and this uniformly in E. This will be specified in the following section.

3. Upper and Lower Bounds on Energy Eigenvalues

Theorem 9 suggests a nice geometrical characterisation of molecular energy eigenvalues:
Let $\Lambda_n(E)$ be the n-th discrete eigenvalue of H^{AD}-U(E) below ε.
Then the n-th discrete eigenvalue E_n of H is given by the intersection of the two graphs $\Lambda_n(E)$ and f(E)=E, (see fig.5).

An upper bound to E_n is easely found because of the inequality

$$H^{AD} - U(E) \leq H^{AD}$$

which holds in the sense of quadratic forms. This implies that E_n is bounded above by the n-th eigenvalue E_n^{AD} of H^{AD}. (To simplify notation we identify quadratic forms and their associated operators).

To get a lower bound is more difficult. The idea is to look for an energy dependent lower bound of the form

$$(17) \quad H^{AD} - D(E) \leqslant H^{AD} - U(E)$$

for a suitable D(E). A possible choice is suggested by the computation which has been done previously in order to prove the H^{AD}-boundedness of U(E) (Lemma 7 and lemma 8); it is

$$D(E) = \frac{1}{\varepsilon - E} \, PHQHP.$$

By construction inequality (17) holds. Notice that H^{AD}-D(E) is again concave in E and leads to a lower bound for the n-th eigenvalue E of H:

Theorem 10: Let all potentials be dilation analytic. Let furthermore E_n^{AD}, E_n, $\Lambda_n^-(E)$ be the n-th eigenvalue of H^{AD}, H, H^{AD}-D(E) respectively and assume all of them to be below $\varepsilon = \inf \varepsilon(R)$. Then the following inequalities hold :

$$\Lambda_n^- (E_n^{AD}) \leqslant E_n \leqslant E_n^{AD} .$$

The proof follows easily from the analysis of fig.5.

More explicit results are obtained if we use the isomorphism between $\mathcal{H}^{AD}$ and the direct sum of $L^2(R^3)$spaces given by the equation (14).

$$\mathcal{H}^{AD} \cong L^2(R^3) \oplus \ldots \oplus L^2(R^3) .$$

On this space H^{AD} has the following realization

$$(18) \quad H^{AD} = \frac{1}{2} k^4 (P^2 + T_{eff}) + V_{eff}$$

$$T_{eff,k\ell} := \langle e_R^k, (-\Delta_R + \rho (\Sigma p_\ell)^2 e_R^\ell \rangle - 2 \langle e_R^k, i\nabla_R e_R^\ell \rangle P$$

$V_{eff}(R)_{k\ell} = E_\ell(R)\delta_{k\ell}$, $k,\ell = 1,..,\dim P(R) = d$, where $E_\ell(R)$ is a discrete eigenvalue of $H^{e\ell}(R)$.

In this realization H^{AD} is a matrix valued one particle Schrödinger operator for the nuclear motion. The electrons are taken into account by the effective potential V_{eff} and the effective nuclear kinetic part T_{eff}. Notice that they are expressed in terms of eigenvalues and eigenfunctions of $H^{e\ell}(R)$ only.

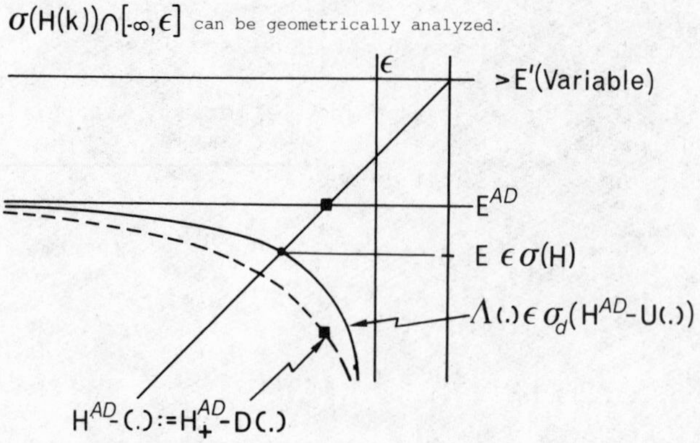

Fig. 5. Spectra of H, H^{AD}, H^{AD}-U(E), and H^{AD}-D(E).

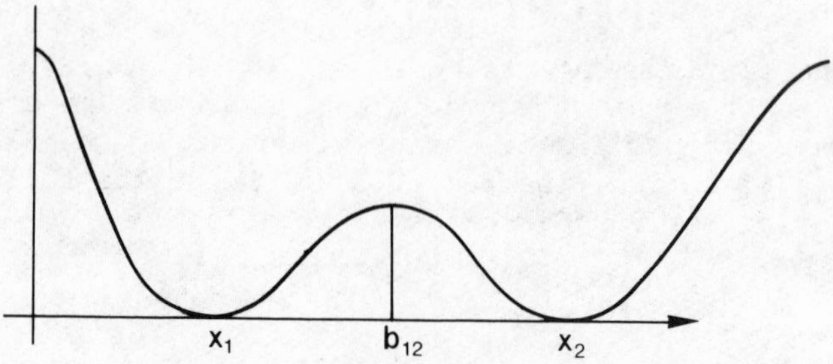

Fig. 6. Effective potential with absolute minima.

In the next and last chapter we will show - for the simplest case d=1 - that the previous inequalities determine the discrete spectrum of H(k) to lowest orders.

Remark 15: Let us define the Born-Oppenheimer Hamiltonian

$$H^{BO}(k) = \frac{1}{2} k^4 P^2 + V_{eff}$$

which coincides with H^{AD} up to the effective kinetic term. It is a remarkable result due to Brattsev (1965) and Epstein (1966) that the infimum of the spectrum of H^{BO} is a lower bound for the infimum of the spectrum of H (see Combes and Seiler (1980) for a short proof of this fact). Hence one has for the mathematical ground state of a molecule the upper and lower bounds

$$(19) \quad \inf \sigma (H^{BO}) \leqslant \inf \sigma (H) \leqslant \inf \sigma (H^{AD}).$$

IV. THE CLASSICAL APPROXIMATION

In this chapter we shall analyze the discrete spectrum of $H^{BO}(k)$ and $H^{AD}(k)$ in the limit $k \searrow o$. At the end this will be used to prove equation (1) for the energy eigenvalues of H(k) which are below ε. In all of this chapter we make the assumption d=dim P(R)=1. The generalization to the case d>1 requires more technical details.

It is to be expected that the two-nuclei system in the n-th eigenstate of $H^{BO}(k)$ behaves more and more classically as k decreases to zero. The wave function will get more localized around the minimum R_o of V_{eff} (Fig.1) and the energy eigenvalue E(k) will be absorbed in the continuum at its lower end V_o= Min $V_{eff}(R)$. Furthermore it is expected that the system behaves very much as if it would move in a harmonic oscillator potential due to the regularity of the effective potential (theorems 2,3,4). Hence the following asymptotic expression for E(k) is expected to hold:

$$(20) \qquad E(k) = V_o + (n+ {}^1/_2) \sqrt{V''(Ro)}k^2 + \ldots \quad .$$

Since all energy eigenvalues of fixed quantum number are absorbed by the same point - the minimum of the effective potential - it is importand to introduce the microscopic operator. It is defined in such a way that the spectrum is magnified by a factor k^{-2}. The justification for the choice of this factor will become clear later.

The effective potential V might have several relative and absolute minima. In the limit k=o only absolute minima are importand if one is interested in states with fixed quantum number n .

However for $k>0$, but small, all minima are of physical interest and give rise to a rich structure of eigenvalues and resonances. To analyze this we proceed along the following lines:

1. Reduce the case of several minima to the one of only one minimum by Dirichlet decoupling. Show asymptotic expansion of eigenvalues for the latter case (section 1).

2. Prove that this gives rise to asymptotic expansions of all eigenvalues for the original Schrödinger operator (section 2).

3. Remove the Dirichlet boundary conditions by Krein's formula. Prove the existence of a convergent perturbation expansion starting from the Schrödinger operator with Dirichlet boundary conditions (section 3).

We will discuss the example of a spherically symmetric potential with two absolute minima in the radial direction. This case contains already most of the problems of the general case. The proofs will only contain the main arguments. For details we refer to a forthcoming article (Combes, Duclos and Seiler, 1981).

1. Reduction to the Case with One Minimum

Consider the Schrödinger operator H^{BO} on a subspace with fixed angular momentum ℓ and effective potential $V(x=|R|)$,

$$(21)\quad L(k) := k^4\left(-\frac{d^2}{dx^2}+\frac{\ell(\ell+1)}{x^2}\right) + V(x) \text{ on } L^2\,(R^+,dx);\ \ell=0,1,\ldots\ .$$

We shall assume that $V(x)$ is a positive smooth function[4)] and has two absolute minima x_1, x_2 with asymptotic expansions around them (see assumption (26,iv) below). In this section we will reduce this problem to the case of a potential with one absolute minimum. The results will then be used in the next section to treat the case we started from (two absolute minima).

Consider now a point b_{12} between x_1 and x_2 (Fig.6) and define the Schrödinger operator

$$(22)\quad L^D(k) = L_1(k) \oplus L_2(k)\ .$$

with the same differential symbol as $L(k)$ but Dirichlet boundary conditions at b_{12}. The properties of $L^D(k)$ are given by those of $L_i(k), i=1,2$; in particular the spectrum is given by the equation

$$(23)\quad \sigma\,(L^D(k)) = \sigma\,(L_1(k)) \cup \sigma\,(L_2(k)).$$

We will later use the differential operator $L^N(k)$ with Neumann boundary conditions at b_{12}. In the remainder of this section we will concentrate on $L_1(k)$.

To $L_1(k)$ we associate the microscopic operator $\ell_1(k)$ as follows: Translation $x \to x - x_1$ followed by scaling $x \to kx$ maps $L^2([o,b_{12}],dx)$ unitarily onto $L^2([-k^{-1}x_1, k^{-1}(b_{12}-x_1)],dx)$. This map transforms $L_1(k)$ for ℓ=o into the operator $k^2\ell_1(k)+V_1$

$$(24) \quad \ell_1(k) := -\frac{d^2}{dx^2} + k^{-2} \underbrace{(V(kx+x_1)-V_1)}_{W(kx)}, \quad V_1 = \text{Min } V(x) = V(x_1).$$

By construction $\ell_1(k)$ is a differential operator with Dirichlet boundary conditions at both ends of the interval $[-k^{-1}x_1, k^{-1}(b_{12}-x_1)]$ and its spectrum is related to the one of $L_1(k)$ by

$$(25) \quad \sigma(L_1(k)) = k^2\,\sigma(\ell_1(k)) + V_1$$

The following assumptions will be made

$$(26) \quad \text{i)}\ \Omega(k) = k^{-1}[-b,b] = [-k^{-1}x_1, k^{-1}(b_{12}-x_1)]$$

$$\text{ii)}\ W(x) = \lim_{M\to\infty} W_M(x) \quad \text{on } \Omega(1).$$

$$W_M(x) = \sum_{2}^{M} a_n x^n, \quad a_2=1$$

$$\text{iii)}\ c_1x^2 \leqslant W_n(x) \leqslant c_2x^2 \text{ on } \Omega(1) \text{ for all } M \in Z_+ ;\ c_1, c_2 > o.$$

$$\text{iv)}\ W(x) - W_M(x) = O(x^{M+1}) \text{ on } \Omega(1).$$

The first assumption is introduced for convenience only and can be dropped easely. The main result of this section is summarized in

Theorem 11: Let $W(x)$ satisfy the conditions i) to iv) stated above (26). Denote by $e(k)$, $e_M(k)$ the n-th eigenvalue of

$$\ell_1(k), \ell_{1,M}(k) = -\frac{d^2}{dx^2} + k^{-2}W_M(kx)$$

respectively, where $\ell_{1,M}$ differs from ℓ_1 only through its potential term (in particular both operators have the same domain of definition). Then $e(k)$ and $e_M(k)$ have both asymptotic expansions in k

$$e(k) \sim \Sigma\, e_s k^s, \quad e_M(k) \sim \Sigma\, e_{M,s}\, k^s$$

with $e_s := e_{s+2,s}$. Furthermore the coefficients are computed by ordinary Rayleigh-Schrödinger perturbation theory of

$$\ell^{harm} = -\frac{d^2}{dx^2} + x^2 \text{ by } k^{-2}W_M(kx) - x^2.$$

The proof of this theorem is divided into several steps. The first result concerns the relation between the differential operations

$\ell_1(k)$, $\ell_1^{harm}(k) = -\frac{d^2}{dx^2} + x^2$ defined on $L^2(\Omega(k),dx)$, $\ell^{harm}(k)$ defined on $L^2(R,dx)$ with Dirichlet boundary condition on the boundary of $\Omega(k)$ and $\ell^{harm} = -\frac{d^2}{dx^2} + x^2$ with the natural domain of definition.

Lemma 12: Under the assumption of the preceeding theorem the difference of the resolvents of $\ell_1(k)$ and $\ell_1^{harm}(k)$ as well as $\ell^{harm}(k)$ and ℓ^{harm} converge to zero for $k \searrow o$ in norm.

To prove this statement we write the integral kernel of the resolvents explicitly using the Feynman-Kac formula

$$(27)\quad (\ell_1(k)+1)^{-1}(x,y) - (\ell_1^{harm}(k)+1)^{-1}(x,y) = \int_o^\infty dt\, e^{-t} \int d\mu_{xy}^t(\omega) \{\exp - k^{-2}\int_o^t ds\, W(k\omega(s)) - \exp - \int_o^t ds\, \omega(s)^2\} .$$

Clearly the integrand on the right hand side converges pointwise to zero and is uniformely bounded by the integrable function e^{-t}. Hence the difference of the integral kernels converge pointwise to zero. In fact they are integral kernels of Hilbert-Schmidt operators and can be easily dominated-uniformely in k- by an integral kernel of the resolvent of a suitably chosen harmonic oscillator. This is due to assumption (26iii).To prove the statement on $\ell^{harm}(k)$ and ℓ^{harm} we proceed analogously.

The above lemma implies the validity of the harmonic approximation for the computation of the eigenvalues of $\ell_1(k)$ respectively $L_1(k)$. The next result shows that this can be improved to higher approximations:

Lemma 13: Under the assumption of theorem 11 the following asymptotic relation holds:

$$e(k) - e_M(k) = O\,(k^{M-1}).$$

The proof of this statement follows from the identity

$$(28)\quad e(k) - e_M(k) = \text{Trace}\,(p(k)\,\ell_1(k) - p_M(k)\ell_{1,M}(k))$$

where $p(k)$, $p_M(k)$ denote the projectors on the eigenstates corresponding to $e(k)$, $e_M(k)$. They are rank one operators,

$$p(k) = |\Psi(k)\rangle\langle\Psi(k)| \ ,\ p_M(k) = |\Psi_M(k)\rangle\langle\Psi_M(k)| \quad ,$$

where $|\Psi(k)\rangle$ denotes the normalized eigenfunction and $\langle\Psi(k)|$ the corresponding linear functional in the standard notation of Dirac.

It is usefull to write $e(k) - e_M(k)$ as a sum of two terms

$$e(k)-e_M(k) = \text{Trace } p(k)(\ell_1(k)-\ell_{1,M}(k)) + \text{Trace } (p(k)-p_M(k))\ell_{1,M}(k)$$

and to estimate each one seperatly. We consider the first term only, the second one can be treated analogously modulo an argument which requires lemma 12 . Consider the equation

$$\text{Trace } p(k)\,(\ell_1(k)-\ell_{1,M}(k))=k^{-2}\int_{\Omega(k)} dx\, |\Psi(k,x)|^2\,(W(kx)- W_M(kx))\ .$$

Now the statement follows from the exponential decay of eigenfunctions $\Psi(k)$ due to assumption (26,iii) [3],

$$(29)\quad |\Psi(k,x)| \leqslant \text{const.} \exp - \alpha \frac{c_1}{2} x^2$$

and the estimate (26,iv).

In the preceeding lemma it has been stated that existence of an asymptotic series of e(k) holds if such a result is known about $e_M(k)$. The next lemma makes sure that this is indeed the case:

Lemma 14 : Under the assumption of theorem 11 the n-th eigenvalue $e_M(k)$ has an asymptotic expansion to any order in k.

The proof of this statement follows from the equations

$$(30)\quad (\ell_{1,M}(k)-z)^{-1} = (\ell_1^{harm}(k)-z)^{-1} \sum_{m=o}^{S} \left[-U_M(k)(\ell_1^{harm}(k)-z)^{-1}\right]^m + \left[-(\ell_1^{harm}(k)-z)^{-1}U_M(k)\right]^{S+1}(\ell_{1,M}(k)-z)^{-1}$$

$$(31)\quad e_M(k)= - \frac{1}{2\pi i} \text{Trace} \int_\Gamma dz\, (\ell_{1,M}(k)-z)^{-1}\, \ell_{1,M}(k)$$

where Γ is suitable chosen and $U_M(k)$ the multiplication operator by $k^{-2}W_M(kx)-x^2$. Inserting (30) into (31) one gets two expressions. To prove the lemma it is enough to show that the second one is of order k^{S+1}. This implies that all terms up to order k^S can come from the first term only. The integral can be computed by the residue theorem, so one gets a finite number of terms. A typical one is

$$\text{Trace } p^{harm}\, U_M \left[(\ell_1^{harm} \widehat{-} e^{harm})^{-1}\, U_M\right]^{S-m} \{1+(\ell_{1,M}-e^{harm})^{-1}e^{harm}\} \left[(\ell_1^{harm} \widehat{-} e^{harm})^{-1}\, U_M\right]^m$$

where we have used the notation p^{harm} for the projector on the eigenspace associated to the eigenvalue e^{harm} of ℓ_1^{harm}. To simplify

the above formula the k-dependance has been supressed. We recall that $\wedge$ denotes the reduced resolvent. The main point in the argument that all these terms are of order k^{S+1} is the exponential decay of eigenfunctions(29).

Until now we have made perturbation theory with respect to $\ell_1^{harm}(k)$. In order to eliminate the k-dependence in the non-perturbed operator we replace in the above expressions $\ell_1^{harm}(k)$ by ℓ^{harm}. Of course the potential $k^{-2}W_M(kx)-\bar{x}^2$ is considered as a polynomial on the real line. It can be shown that this change introduced only an exponentially small error which does not effect the asymptotic expansion.
The results stated in Lemmas 12,13,14 imply Theorem 11.

Remark 16: In this section we have considered differential operators with Dirichlet boundary conditions. Analogous results hold for the case of Neumann or mixed boundary conditions. They are trickier to prove (Duclos, 1981).

2. The Case with Two Minima

Now we come back to the case of a potential with two absolute minima at x_1 and x_2. For that we have to change the notation in the following way because we have to use simultaneously eigenvalues of operators with and without Dirichlet boundary conditions. We denote by $e_1^D(k)$ and $\varphi_1^D(k)$ the eigenvalue and eigenfunction of $\ell_1(k)$ and by $e_1^N(k)$, $\varphi_1^N(k)$ the corresponding objects for the Neuman case. The following result will be important.
Lemma 15: The asymptotic expansions of eigenvalues are the same for Dirichlet or Neuman boundary conditions at b_{12}.
This lemma is valid for $L^D(k)$ and $L^N(k)$ as well as for the corresponding microscopic operator. We shall do the proof in the microscopic case with i=1.

Let us consider the nth eigenvalue $e_1^D(k)$ and the corresponding eigenfunction $\varphi_1^D(k)$ of $\ell_1(k)$ as well as $e_1^N(k)$, $\varphi^N(k)$ for the Neuman case. By Green's formula or partial integration one can easily see that

$$(32)\quad |e_1^D(k)-e_1^N(k)|=|<\varphi_1^N(k),\varphi_1^D(k)>||\varphi_1^N(k,k^{-1}b)\varphi_1^{D\prime}(k,k^{-1}b)|$$

That the first term on the right hand side of (32) converges to one is a consequence of lemma 12 and its generalization to the Neuman case. The second term is exponentially small because $|\varphi_1^N(k,k^{-1}b)|$ and $|\varphi_1^{D\prime}(k,k^{-1}b)|$ are bounded by $p(k)\exp -Ak^{-2}$, $p(k)$ a polynomial in k, A a strictly positive constant, due to the exponential fall off of eigenfunctions (29). (Remember that $b=b_{12}-x_1$, see 26i)).

In the remainder of this section the spectra of $L(k)$, $L^D(k)$ and $L^N(k)$ will be related:

Theorem 16 : Let $E(k)$, $E^D(k)$, $E^N(k)$ be the nth eigenvalues of $L(k)$, $L^D(k)$, $L^N(k)$ respectively. Assume that $V(x)$ is a smooth function with two degenerate minima at x_1, x_2 such that the results of the previous section can be applied. Then $E(k)$, $E^D(k)$, $E^N(k)$ have the same assymptotic expansion for $k \searrow 0$.

Since the following inequalities hold, $E^N(k) \leqslant E(k) \leqslant E^D(k)$ it is enough to prove that $E^N(k)$ and $E^D(k)$ have the same asymptotic expansion. The previous inequalities are a well known consequence of the inequalities on quadratic forms corresponding to the operators $L^N(k)$, $L(k)$, $L^D(k)$ and Weyl's "mini-max principle".

The crucial step in the proof of the theorem is the result already stated in lemma 15. Let $E_i^D(k)$ be the nth eigenvalue of $L_i(k)$ and $E_i^N(k)$ the corresponding eigenvalue for the operator with Neuman boundary conditions. Then they have the same asymptotic expansion for $k \searrow 0$.

To finish the proof of theorem 15 it is enough to realize the following fact which is essentially a consequence of lemma :
Let $E^D(k)$ and $E^N(k)$ be the n-th eigenvalue of $L^D(k)$ and $L^N(k)$. Then there exists for sufficiently small k, numbers p and i such that $E^D(k)$ and $E^N(k)$ are also the p-th eigenvalue of $L_i(k)$ and the corresponding Neuman operator for the same i. Furthermore i and p are k-independent. For the above argument we have assumed that the potential has different asymptotic expansions around the two absolute minima. The statement can however still be proved in the case of two identical asymptotic expansion as for instance for the double well potential (Combes, Duclos and Seiler 1981, Duclos 1981).

In this section we argued that the eigenvalues of $L^N(k)$, $L(k)$, $L^D(k)$, are asymptotically all the same. In the following one we want to go beyond asymptotic statements.

3. The expansion in the tunneling Parameter

Before we explain the aim of this section, we have to introduce new operators. Let $\ell(k)$ be the microscopic operator obtained from $L(k)$ in the same manner as $\ell_1(k)$ from $L_1(k)$, i.e. translation by x and dilation around the first minimum. $\ell^D(k)$ is the operator with the same symbol and an additional Dirichlet boundary condition at $k^{-1}b$. We shall use also the resolvents $r(k)=(\ell(k)+1)^{-1}$, $r^D(k)=(\ell^D(k)+1)$, denote their kernel by $r(k;x,y)$ and $r^D(k;x,y)$, and introduce the vector $\tau(k)$ defined by

$$(33) \quad \tau(k;x) = \left[r(k;k^{-1}b, k^{-1}b)\right]^{-1/2} r(k;x,k^{-1}b)$$

In this section we want to describe a method relating eigenvalues of $\ell(k)$ and $\ell^D(k)$ which we shall denote by $e(k)$ and $e^D(k)$ respectively. The relation will be given in terms of a convergent power series in what we shall call the tunneling parameter $t(k)$. The results hold for $k>o$ but small enough, $t(k)$ itself can be analyzed for $k\searrow o$ and is bounded by $P(k)\exp{-Ak^{-2}}$ where P is a polynomial in k, A a strictly positive constant. There are two main instruments to carry out this program:

1. Krein's formula for $r(k)$ and $r^D(k)$:

$$r(k;x,y) = r^D(k;x,y) + \tau(k;x)\,\tau(k;y) \tag{34}$$

(M.Krein 1946)

There is a simple proof of formula (34), based on Green's identity. Krein's formula shows that r and r^D differ just by a rank one operator.

2. The formula of Lagrange (Dieudonné.J. (1968)) for the solution of the implicit equation $\lambda=\lambda^D+t\sigma(\lambda)$ for $\lambda(t)$ where $\sigma(\lambda)$ is supposed to be analytic in a neighborhood of λ^D,

$$\lambda = \lambda^D + \sum_1^\infty \frac{t^n}{n!} \left(\frac{d}{dz}\right)^{n-1} (\sigma(z))^n \Big|_{z=\lambda^D} . \tag{35}$$

Application of those two results leads to the following

<u>Theorem 17</u>: Assume that V satisfy all conditions necessary for the applicability of the results in section 1 to $\ell_1(k)$ and $\ell_2(k)$. Let $e^D(k)$ be the n-th eigenvalue of $\ell^D(k)$ assumed to be nondegenerate and $\varphi^D(k)$ the associated eigenfunction. Then for sufficiently small k the n-th eigenvalue $e(k)$ of $\ell(k)$ is given by the formulas (35) and $e(k)= (\lambda(k)+1)^{-1}$ with the definitions $t(k):=|<\varphi^D(k),\tau(k)>|^2$

$$\sigma(k,z):= \left[1 + <\tau(k), \widehat{(r^D(k)-z)^{-1}}\,\tau(k)>\right]^{-1} .$$

<u>Remark 17</u>: The statement can be generalized to the case of degenerate eigenvalues $e^D(k)$.

The proof requires the verification of the conditions for the applicability of Lagrange's formula. They are consequences of exponential fall of eigenfunctions for the operators $\ell_1(k)$ and $\ell_2(k)$ and of the result stated in Lemma 11 on the validity of the harmonic approximation.

4. Application to the Standard Example

In this last section we shall apply the result of the first section to the standard example and prove for this case the validity of equation (1). The nuclear Schrödinger operator $H^{AD}(k)$ for the lowest energy range, i.e. d=1, and restricted to the angular momentum sector ℓ is the differential operator

$$H_\ell^{AD}(k) = \frac{1}{2} k^4 \left(-\frac{d^2}{dx^2} + \frac{\ell(\ell+1)}{x^2} + t^0(x)\right) + V_{eff}(x)$$

with the definition

$$t^0(x) := \langle e_R, (-\Delta_R + \rho p_1^2)\, e_R \rangle$$

The effective potential is drawn in figure 1 and has only one minimum x_1. It satisfies the assumptions of the first section. The microscopic operator of $H^{AD}(k)$ has the form

$$(36)\quad \ell(k) := -\frac{1}{2}\frac{d^2}{dx^2} + \frac{\ell(\ell+1)}{2(x_1+kx)^2} + \frac{1}{2} k^2\, t^0(x_1+kx) + k^{-2}\left[V_{eff}(x_1+kx) - V_{eff}(x_1)\right]$$

It is now easy to compute the harmonic and higher order approximations so that the asymptotic expansion for the eigenvalues can be derived by ordinary perturbation theory according to theorem 11.

In order to show equation (1) we have to make use of the lower bounds described in chapter III, section 3. The operator $H^{AD}(k) - D(E)$ can be considered on the sector with angular momentum ℓ. The corresponding operator is very much the same as $\ell(k)$ defined above (36). It differs from it by two terms, one is a second order differential operator proportional to k^4, the other one a multiplication operator proportional to k^6. So the eigenvalues can again be represented by an asymptotic series in k. The terms in lowest order are however the same for both operators. In particular all terms which go into equation (1) remain unaffected by all contributions coming from D(E).

To show the validity of the Born-Oppenheimer formula (1) and to find explicit expressions for the constants a and b it is therefore enough to compute the Rayleigh-Schrödinger series up to order k^2 of $\ell(k)$. This leads to the following result

$$(37)\quad \begin{aligned} a &= \sqrt{V''(R_0)} \\ b &= \frac{t_0(R_0)}{2} - \left[\frac{V'''(R_0)}{V''(R_0)}\right]^2 \left[\frac{11}{288} + \frac{5n(n+1)}{48}\right] + \frac{V^{(4)}(R_0)}{V''(R_0)}\left(\frac{1}{32} + \frac{n(n+1)}{16}\right) \end{aligned}$$

Footnotes:

1) We will concentrate on the discrete part of the spectrum (bound states) only.

2) The Schrödinger operator for atoms does not contain any small parameter.

3) Exponential fall off of eigenfunctions can be shown using either methods from complex function theory (Combes and Thomas (1973), Deift, Hunziker, Simon and Vock (1979)) or comparison theorems (Protter and Weinberger 1967).

4) The smoothness requirement on V is added here for convenience but is far from necessary.

REFERENCES

Aguilar J. and Combes J.M., 1971; A class of Analytic Perturbations for One-body Schrödinger Hamiltonians, Com.Math.Phys.22,269-279.

Born M. and Oppenheimer R., 1927; Zur Quantentheorie der Molekeln, Annalen der Physik 84,457.

Brattsev V.F.,1965; Dokl.Akad.Nauk. SSSR 160,570.

Brezin E. and Zinn-Justin J., 1979; Expansion of the H_2^+ ground state energy in inverse powers of the distance between the two protons; Le journal de physique, lettres, 40L-511.

Combes J.M. and Seiler R., 1978; Regularity and Asymptotic Properties of the Discrete Spectrum of Electronic Hamiltonians; Int.J. of Quantum Chemistry 14,213.

Combes J.M. and Seiler R., 1980; Spectral Properties of Atomic and Molecular Systems, Ed.G.Wooley, Plenum.

Combes J.M., Duclos P. and Seiler R.,1981; Decoupling and Krein's Formula I, to appear.

Combes J.M. and Thomas L., 1973; Asymptotic Behaviour of Eigenfunctions for Multiparticle Schrödinger Operators; Commun.math.Phys. 34,251.

Deift P., Hunziker W., Simon B., Vock E.,1979; Pointwise Bounds of Eigenfunctions and Wave Packets in N-Body Quantum Systems IV; Com.Math.Phys.64, 1-34.

Dieudonné J., 1968; Calcul Infinitesimal, Chapter VIII.7, Herman, Paris.

Duclos P., Thèse, Dept.de Mathématique Université de Toulon.

Epstein S., 1966; Ground-State Energy of a Molecule in the Adiabatic Approximation, J.Chem.44,863 and 44,4062.

Herring C., 1962; Critique of the Heitler-Landau method of calculating spin coupling at large distances, Rev.Mod.Phys.34, 631-645.

Hoffmann-Osterhoff T., 1980; A Compairison Theorem for Differential Inequalities with Applications in Quantum Mechanics; Journal of Physics AB,417.

Hoffmann-Osterhoff M. and Seiler R., 1981; to appear in Phys. Rev.A.

Kato T., 1966; Perturbation theory for linear Operators, Springer-Verlag, Berlin.

Kato T., 1957; Commun.Pure and Appl.Math.X,151.

Komarov J.N., Ponomarev L.I., Slavianov S.I., 1976; Spherical and Coulomb Spheroidal functions. Nauka, Moscow (in Russian).

Krein M., 1946; Über Resolventen hermitescher Operatoren mit Defektindex (m,m) Doklady Akad.Nauk. SSSR 52 No.8, 657-660.

Lieb E. and Simon B., 1978; Monotonicity of the Electronic Contribution to the Born-Oppenheimer Energy, J.Phys.B. 537.

Morgan J.D. III and Simon B., 1980; Behavior of Molecular Potential Energy Curves for Large Nuclear Seperations; Int.J.Quant. Chemistry, 17,1143.

Pack R.T. and Hirschfelder J.O., 1968; Separation of Rotational Coordinates from the N-electrons Diatomic Schrödinger Equation, Journal of Chemical Physics 49,4009-4020.

Protter M.H. and Weinberger H.F.,1967; Maximum Principles in Differential Equations, Prentice-Hall, Juc.

THOMAS-FERMI AND RELATED THEORIES OF ATOMS AND MOLECULES

Elliott H. Lieb

Departments of Mathematics and Physics
Princeton University, POB 708
Princeton, NJ 08544

I. INTRODUCTION

In recent years some of the properties of the Thomas-Fermi (TF) and related theories for the ground states of non-relativistic atoms and molecules with fixed nuclei have been established in a mathematically rigorous way. The aim of these notes is to summarize that work to date--at least as far as the author's knowledge of the subject goes. In addition, some open problems in the subject will be stated.

TF theory was invented independently by Thomas (1927) and Fermi (1927). The exchange correction was introduced by Dirac (1930), and the gradient correction to the kinetic energy by von Weizsäcker (1935).

No attempt will be made to summarize the voluminous subject of TF theory. That would have to include many varied applications, many formulations of related theories (e.g. relativistic corrections to TF theory, non-zero temperature TF theory) and reams of data and computations. Some reviews exist (March, 1957), (Gombás, 1949), (Torrens, 1972), but they are either not complete or not up to date.

We will concentrate on non-relativistic TF and related theories for the ground state with the following goals in mind:

1) The definition of TF and related theories (i.e. the von Weizsäcker and Dirac corrections). The main question here is whether the theories are well defined and whether the equations to which they give rise have (unique) solutions.

2) Properties of TF and related theories. It turns out that,

 An amended version of this paper appears in Reviews of Modern Physics, Vol. 53, pp. 603-641 (1981).

unlike the correct Schrödinger, quantum (Q) theory, many interesting physical properties of the TF and related theories can be deduced without computation. Some of these properties are physically realistic and some are not, e.g. Teller's no-binding theorem. As will be seen, however, the no-binding result is natural and correct if TF theory is placed in its correct physical context as a large Z (= nuclear charge) theory.

3) The relation of TF theory to Q theory. The main result will be that TF theory is exact in the large Z (nuclear charge) limit. For this reason, TF theory should be taken seriously as one of the cornerstones of atomic physics. The only other regime in which it is possible to make simple, exact statements is the one electron hydrogenic atom. The natural open question is to find the leading correction, in Z, beyond TF theory. This will lead to a discussion of the Scott correction (Scott, 1952) which, while it is very plausible, has not yet been proved. It turns out that Thomas-Fermi-von Weizsäcker (TFW) theory has precisely the properties that Scott predicts for Q theory. Moreover, TFW theory remedies some defects of TF theory: it displays binding, it gives exponential falloff of the density at large distances, and it yields a finite density at the nucleus.

The work reported here originated in (Lieb and Simon, 1973 and 1977, hereafter LS). Subsequently, the ideas were developed by, and in collaboration with, Benguria and Brezis. I am deeply indebted to these coworkers.

Since many unsolved problems remain, these notes are more in the nature of a progress report rather than a textbook. The proofs of many theorems are sketchy, or even absent, but it is hoped that the interested reader can fill in the details with the help of the references. Unless clearly stated otherwise, however, everything presented here is meant to be rigorous.

II. THOMAS-FERMI THEORY

The theories will be stated in this section purely as mathematical problems. Their physical motivation from Q theory will be explained in section V. In order to present the basic ideas as clearly as possible, only TF theory will be treated in this section; the variants will be treated in sections VI, VII and VIII. However, the basic definitions of all the theories will be given in §II.A, and there will be some mention of TFD theory in §II.B and section III.

II.A. The Definitions of Thomas-Fermi and Related Theories

All the theories we will be concerned with start with some energy functional $\mathcal{E}(\rho)$, where ρ is a nonnegative function on 3-space, $\mathbb{R}^3$. ρ is called a density and physically is supposed to be the electron density in an atom or molecule.

The functionals will involve the following function V and constant U.

$$V(x) = \sum_{j=1}^{k} z_j \, |x-R_j|^{-1} \tag{2.1}$$

$$U = \sum_{1\le i<j\le k} z_i z_j \, |R_i-R_j|^{-1} \quad . \tag{2.2}$$

$V(x)$ is the electrostatic potential of k nuclei of charges (in units in which the electron charge $e = -1$) $z_1,\dots,z_k > 0$, and located at $R_1,\dots,R_k \in \mathbb{R}^3$. The R_i are distinct. The positivity of the z_i is important for many of the theorems; while TF theory makes mathematical sense when some $z_i < 0$, it has not been investigated very much in that case. U is the repulsive electrostatic energy of the nuclei.

TF type theories can, of course, be defined for potentials that are not Coulombic, but many of the interesting properties presented here rely on potential theory and hence will not hold for non-Coulombic potentials. This is discussed in section III. There is, however, one generalization of (2.1) and (2.2) that can be made without spoiling the theory, namely the nuclei can be "smeared out". I.e. the following replacements can be made:

$$z_j|x-R_j|^{-1} \to \int dm_j(y)\,|x - R_j + y|^{-1} \tag{2.3}$$

$$z_i z_j|R_i-R_j|^{-1} \to \int dm_i(y)\; dm_j(w)\,|y - w - R_i+R_j|^{-1} \tag{2.4}$$

where m_i is a positive measure (not necessarily spherically symmetric) of mass z_i.

The functional for TF theory is

$$\mathcal{E}(\rho) = \frac{3}{5}\,\gamma \int \rho(x)^{5/3}dx - \int \rho(x)V(x)dx + D(\rho,\rho) + U \tag{2.5}$$

where $D(g,f) = \frac{1}{2}\iint g(x)\; f(y)\,|x-y|^{-1}\; dxdy$. All integrals are three-dimensional.

γ is an arbitrary positive constant but to establish contact with Q theory we must choose

$$\gamma_p = (6\pi^2)^{2/3}\;\hbar^2(2mq^{2/3})^{-1} \tag{2.6}$$

where $\hbar = h/2\pi$, h = Planck's constant, and m is the electron mass. q is the number of spin states (=2 for electrons).

U appears in $\mathcal{E}$ as a constant, ρ independent term. It is unimportant for the problem of minimizing $\mathcal{E}$ with respect to ρ. Nevertheless U will be very important when we consider how the minimum depends on the R_i, e.g. in the no-binding theorem (§IIIC).

For the Thomas-Fermi-Dirac (TFD) theory

$$\mathcal{E}^{TFD}(\rho) = \mathcal{E}(\rho) - \frac{3}{4} C_e \int \rho(x)^{4/3} dx \tag{2.7}$$

with C_e a positive constant. In the original theory (Dirac, 1930), the value $C_e = (6/\pi q)^{1/3}$ was used for reasons which will be explained in section VI. This value is not sacrosanct, however, and it is best to leave C_e as an adjustable constant.

The Thomas-Fermi-von Weizsäcker theory (TFW) is given by (von Weizsäcker, 1935)

$$\mathcal{E}^{TFW} = \mathcal{E}(\rho) + \delta \int [(\nabla \rho^{\frac{1}{2}})(x)]^2 dx \tag{2.8}$$

with $\delta = A \hbar^2/2m$, and A is an adjustable constant. Originally, A was taken to be unity, but in §VII.D it will be seen that A = .186 is optimum from one point of view.

The most complicated, and least analyzed case is the combination of all three (section VIII):

$$\mathcal{E}^{TFDW}(\rho) = \mathcal{E}(\rho) - \frac{3}{4} C_e \int \rho(x)^{4/3} dx + \delta \int [(\nabla\rho^{\frac{1}{2}})(x)]^2 dx. \tag{2.9}$$

The first question to face is the following:

II.B. Domain of Definition of the Energy Functional

Since ρ is supposed to be the electron density we require $\rho(x) \geq 0$ and

$$\int \rho(x)dx = \lambda = \text{electron number} \tag{2.10}$$

is finite. In addition we require $\rho \in L^{5/3}$ in order that the first term in $\mathcal{E}(\rho)$ (called the kinetic energy term) be finite.

Definition: A function f is said to be in L^p if $\int |f(x)|^p dx$ is finite.

If $f \in L^p \cap L^q$ with $p < q$ then $f \in L^t$ for all $p < t < q$.

Proposition 2.1: If $\rho \in L^{5/3} \cap L^1$ then all the terms in $\mathcal{E}$ and $\mathcal{E}^{TFD}$ are finite. If $\int \rho \leq \lambda$ then $\mathcal{E}(\rho)$ and $\mathcal{E}^{TFD}(\rho)$ are bounded below by some constant $C(\lambda)$. Furthermore, for all λ, $\mathcal{E}(\rho) > C > -\infty$ for some fixed C.

Proof: The first part is an easy application of Young's and Hölder's inequalities. The second part requires a slightly more refined estimate of the Coulomb energies. cf. LS. ∎

Remark: Although $\mathcal{E}^{TFW}$ will be seen to be also bounded below by a constant independent of λ, neither $\mathcal{E}^{TFD}$ nor $\mathcal{E}^{TFDW}$ is so bounded. This fact leads to an amusing unphysical consequence of the D theories which will be mentioned later.

A very important fact (which, incidentally is not true for Hartree-Fock theory) is the following.

Proposition 2.2: $\rho \to \mathcal{E}(\rho)$ is strictly convex, i.e. $\mathcal{E}(\lambda\rho_1 + (1-\lambda)\rho_2) < \lambda\mathcal{E}(\rho_1) + (1-\lambda)\mathcal{E}(\rho_2)$ for $0 < \lambda < 1$ and $\rho_1 \neq \rho_2$.

Proof: ρ^p is strictly convex for $p > 1$. $\int V\rho$ is linear in ρ and hence convex. $D(\rho,\rho)$ is strictly convex since the Coulomb kernel $|x-y|^{-1}$ is positive definite. ∎

Remark: $\mathcal{E}^{TFW}$ is also strictly convex but the functionals $\mathcal{E}^{TFD}$ and $\mathcal{E}^{TFDW}$ are not convex because of the $-\int \rho^{4/3}$ term. However, they can be "convexified" in a manner to be described in section VI.

II.C. Minimization of the Energy Functional

The central problem is to compute

$$E(\lambda) = \inf \{ \mathcal{E}(\rho) \mid \rho \in L^{5/3} \cap L^1, \int \rho = \lambda \} \tag{2.11}$$

and
$$e(\lambda) = E(\lambda) - U. \tag{2.12}$$

$E(\lambda)$ is the TF energy for a given electron number, λ. The "inf" in (2.11) is important because, as we shall see, the minimum is not always achieved, although the inf always exists by Proposition 2.1.

Theorem 2.3: $e(\lambda)$ is convex, negative if $\lambda > 0$, non-increasing and bounded below. Furthermore,

$$E(\lambda) = \inf\{\mathcal{E}(\rho) | \rho \in L^{5/3} \cap L^1, \int \rho \leq \lambda\} \tag{2.13}$$

Proof: The first part follows from Prop. 2.2 together with the observation that $V(x) \to 0$ as $|x| \to \infty$. This means that if λ increases we can add some $\delta\rho$ arbitrarily far from the origin so that $\mathcal{E}(\rho + \delta\rho) - \mathcal{E}(\rho) < \varepsilon$ for any $\varepsilon > 0$. (2.13) is a simple consequence of the monotonicity of $e(\lambda)$ and $E(\lambda)$. cf. LS. ∎

(2.13) has an important advantage over (2.11) as Theorem 2.4 shows.

Theorem 2.4: There exists a unique ρ that minimizes $\mathcal{E}(\rho)$ on the set $\int \rho \leq \lambda$.

Note: Uniqueness means, of course, that ρ is determined only almost everywhere (a.e.).

Proof: Cf. LS. Since $\mathcal{E}(\rho)$ is strictly convex, a minimum, if there is one, must be unique. Let $\rho^{(n)}$ be a minimizing sequence for $\mathcal{E}$, namely $\mathcal{E}(\rho^{(n)}) \to E(\lambda)$ and $\int \rho^{(n)} \leq \lambda$. It is easy to see that $\int (\rho^{(n)})^{5/3} \leq c$, where c is some constant; this in fact comes out of the simple estimates used in the proof of Prop. 2.1. We would like to extract a convergent subsequence from the given $\rho^{(n)}$. This cannot be done in the strong topology, but the Banach-Alaogulu theorem tells us that a $L^{5/3}$ weakly convergent subsequence can be found; this will be denoted by $\rho^{(n)}$. We would like to prove

$$\liminf \mathcal{E}(\rho^{(n)}) \geq \mathcal{E}(\rho). \tag{2.14}$$

Since $\rho^{(n)} \rightharpoonup \rho$ weakly in $L^{5/3}$ we have (by the Hahn-Banach theorem for example) that

$$\liminf \int [\rho^{(n)}]^{5/3} \geq \int \rho^{5/3} \tag{2.15}$$

$$\liminf D(\rho^{(n)}, \rho^{(n)}) \geq D(\rho,\rho) . \tag{2.16}$$

The term $-\int V \rho$ requires slightly more delicate treatment. Write $|x-R|^{-1} = f(x) + g(x)$ where $f(x) = |x-R|^{-1}$ for $|x-R| \leq 1$ and $f(x) = 0$ otherwise. $f \in L^{5/2}$ and $\int f \rho^{(n)} \to \int f \rho$ by weak convergence. On the other hand, $g \in L^{3+\varepsilon}$ for all $\varepsilon > 0$. $\rho^{(n)}$ is bounded in $L^{5/3}$ and in L^1 (by λ) so it is bounded in all L^q

with $1 \le q \le 5/3$. Fix $\infty > \varepsilon > 0$ and let q be dual to $3 + \varepsilon$. We can then find a further subsequence if necessary so that $\rho^{(n)} \to \rho$ weakly in L^q as well as in $L^{5/3}$. Then $\int g\, \rho^{(n)} \to \int g\, \rho$. This proves (2.14) which, since $E(\lambda) = \lim\inf \mathcal{E}(\rho^{(n)})$ and $E(\lambda) \le \mathcal{E}(\rho)$, implies that ρ is minimizing provided we can show $\int \rho \le \lambda$. This follows from the fact that if $\int \rho > \lambda$ then there is a bounded set A such that $\int_A \rho > \lambda$. If α is the characteristic function of A then $\alpha \in L^{5/2}$ and $\lambda \ge \int \alpha\, \rho^{(n)} \to \int \alpha\, \rho$ by weak $L^{5/3}$ convergence. ■

Remark: The analogous proof in TFD theory will be harder since $\rho^{5/3} - \rho^{4/3}$ is neither convex nor monotone nor positive; hence we cannot say that $\lim\inf \int [\rho^{(n)}]^{5/3} - [\rho^{(n)}]^{4/3} \ge \int \rho^{5/3} - \rho^{4/3}$. However, in TFW theory a different strategy, using Fatou's lemma, will be employed to deal with these terms. The strategy also works for TFDW theory. Thus, the introduction of the W term (2.8) makes part of the proof easier. It would be desirable to know how to use Fatou's lemma (which does not require convexity) in the TF and TFD proofs.

Since $E(\lambda)$ is non-increasing, bounded, and convex (and hence continuous) we can make the following definition in TF theory.

Definition: λ_c, the critical λ, is the largest λ with the property that for all $\lambda' < \lambda$, $E(\lambda') > E(\lambda)$. Equivalently, if $E(\infty) = \lim_{\lambda \to \infty} E(\lambda)$ then $\lambda_c = \inf\{\lambda \mid E(\lambda) = E(\infty)\}$. In principle λ_c could be $+\infty$, but this will not be the case. λ_c will be shown to be $Z = \sum z_j$ in TF and TFD theory, Theorem 3.18. In TFW theory, $\lambda_c > Z$.

Theorems 2.3, 2.4 and Proposition 2.2 yield the following picture of the minimization problem in TF theory:

Theorem 2.5: For $\lambda \le \lambda_c$ there exists a unique minimizing ρ with $\int \rho = \lambda$. On the set $[0, \lambda_c]$, $E(\lambda)$ is strictly convex and monotone decreasing. For $\lambda > \lambda_c$ there is no minimizing ρ with $\int \rho = \lambda$, and $E(\lambda) = E(\lambda_c)$; the minimizing ρ in Theorem 2.4 is the ρ for λ_c.

Proof: For $\lambda \le \lambda_c$ use the ρ given by Theorem 2.4 and note that if $\lambda' = \int \rho < \lambda$ then $E(\lambda') = \mathcal{E}(\rho) = E(\lambda)$. The strict convexity is trivial: if $\lambda = a\,\lambda_1 + (1-a)\,\lambda_2$ use $a\,\rho_1 + (1-a)\,\rho_2$ as a trial function for λ. On the other hand, for $\lambda > \lambda_c$ the ρ given by Theorem 2.4 will have $\int \rho = \lambda_c$ because if a minimum existed with $\int \rho = \lambda' > \lambda_c$ then $\tilde{\rho} \equiv [\rho + \rho_c]/2$ (with ρ_c being the ρ for λ_c) would satisfy $\lambda_c < \int \tilde{\rho} = \frac{1}{2}(\lambda' + \lambda_c) < \lambda'$ but, by strict convexity

$$\mathcal{E}(\tilde{\rho}) < [\,\mathcal{E}(\rho) + \mathcal{E}(\rho_c)]/2 = E(\lambda_c),$$

which is a contradiction. The strict convexity follows from Prop. 2.2. ■

The general situation is shown in Fig. 1. There Z is shown as less than λ_c; while that is the case for TFW theory, in TF and TFD theory $\lambda_c = Z$. The straight portion to the right of λ_c is horizontal for TF and TFW, but has a negative slope for TFD and TFDW. The slope at the origin is infinite for TF and TFD but finite for TFW theory.

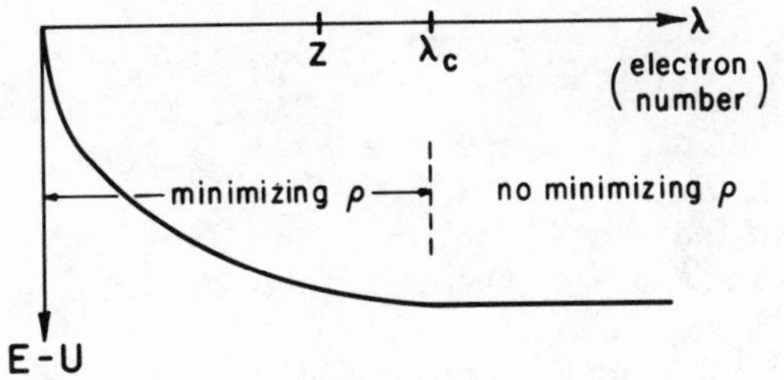

Figure 1

II.D. The Thomas-Fermi-Equation and Properties of the Density

The variational derivative of $\mathcal{E}(\rho)$ is $\delta\mathcal{E}/\delta\rho = \gamma\,\rho^{2/3}(x) - \phi_\rho(x)$ where

$$\phi_\rho(x) = V(x) - \int \rho(y)\ |x-y|^{-1}\ dy. \tag{2.17}$$

A Lagrange multiplier μ should be added to $\delta\mathcal{E}/\delta\rho$ to insure that $\int\rho = \lambda$. It is then expected that $\delta\mathcal{E}/\delta\rho + \mu = 0$ if $\rho(x) > 0$, but $\delta\mathcal{E}/\delta\rho + \mu \geq 0$ if $\rho(x) = 0$ because negative variations of

$\rho(x)$ are not allowed. The two situations can be written as

$$\gamma\rho^{2/3}(x) = \max(\phi_\rho(x) - \mu, 0) \equiv (\phi_\rho(x) - \mu)_+ \tag{2.18}$$

This is the TF equation. [Note that the $(\ )_+$ is very important.] Is all this formal manipulation correct?

Theorem 2.6: If ρ minimizes $\mathcal{E}(\rho)$ with $\int\rho = \lambda \le \lambda_c$ then ρ satisfies (2.18) for some (unique) $\mu(\lambda)$. Conversely if ρ,μ satisfy (2.18) and $\rho \in L^1 \cap L^{5/3}$ then ρ minimizes $\mathcal{E}(\rho)$ for $\lambda = \int\rho$. Hence (2.18) can have at most one solution ρ,μ with $\int\rho = \lambda$. If $\lambda = \lambda_c$ then $\mu = 0$.

Proof: The first part is standard in the calculus of variations. Now let $\rho_i,\mu_i, i=1,2$, satisfy (2.18) with the same λ. Let $F_i(h)=(3\gamma/5)\int h^{5/3} - \int \phi_i h$. It is easy to check that $F_i(h)$ has a unique minimum, F_i, on the set $\int h=\lambda, h\ge 0$; the minimizing h_i is ρ_i. However, $F_1(\rho_2) + F_2(\rho_1) = F_1 + F_2 - D(\rho_1-\rho_2,\rho_1-\rho_2)$. This is a contradiction unless $\rho_1=\rho_2$ (and hence $\mu_1=\mu_2$). The last part (i.e. $\mu=0$) follows by considering the absolute minimum of $\mathcal{E}(\rho)$, in which case no μ is necessary. But this is equivalent to setting $\mu=0$. This minimum occurs for $\lambda\ge\lambda_c$ but as we have shown, only at λ_c is there a minimizing ρ. cf. LS. ■

Remarks: In section III a proof of the uniqueness part of Theorem 2.6 which uses only potential theory will be given. It should be noted that we arrived at the existence of a solution to (2.18) by first considering the minimization problem. A direct attack on (2.18) is rather difficult. Such a direct approach was carried out by (Hille, 1969) in the atomic case, but even in that case he did not prove that the spherically symmetric solution is the only one; our uniqueness result guarantees that.

Theorem 2.7: $E(\lambda)$ is continuously differentiable and $dE/d\lambda = -\mu(\lambda)$ if $\lambda \le \lambda_c$. $dE/d\lambda = 0$ if $\lambda \ge \lambda_c$. Thus, $-\mu(\lambda)$ is the chemical potential.

Proof: The convexity and boundedness of $\mathcal{E}(\rho)$ is used. Cf. LS, Theorem II.10 and Lemma II.27. ■

It will be noted that we have not used the fact that V is Coulombic, only that it vanishes at ∞. Likewise, the only property of the kernel $|x-y|^{-1}$ that was used was its positive definiteness. In section III we will exploit the fact that $|x-y|^{-1}$ is Coulombic and, to a lesser extent, the fact that V is superharmonic. Also, it will be shown that $\lambda_c = Z = \sum_1^k z_j$.

In section III potential theory will shed considerable light on the solution to (2.18). Here we will concentrate on some other aspects of (2.18).

Let us assume that $V(x) = \sum z_j|x-R_j|^{-1}$. ϕ denotes ϕ_ρ for

the solution to (2.18). In section III we show $\phi(x) > 0$. As a distribution,

$$-\Delta\phi(x)/4\pi = \Sigma z_j \delta(x-R_j) - \rho(x)$$

$$= \Sigma z_j \delta(x-R_j) - \gamma^{-3/2} (\phi(x) - \mu)_+^{3/2} . \qquad (2.19)$$

This is the TF differential equation and is equivalent to (2.18). It involves ϕ alone. Since $\rho \in L^{5/3} \cap L^1$, ϕ is continuous away from the R_j (Lemma 3.1) and goes to zero as $|x| \to \infty$.

Theorem 2.8 [LS Theorem IV.5]: (a) Near each R_j

$\rho(x) = (z_j/\gamma)^{3/2} |x-R_j|^{-3/2} + 0(|x-R_j|^{-1/2})$

(b) $\rho(x) \to 0$ as $|x| \to \infty$.

(c) ρ and ϕ are real analytic on $A = \{x | x \neq R_j \text{ all } j, \rho(x) > 0\}$.

(d) In the neutral case $(\mu = 0)$ $\rho(x) > 0$, all x.

(e) In the ionic case $(\lambda < Z, \mu > 0)$ ρ has compact support and ρ and ϕ are C^1 away from the R_j.

Proof: (a) and (b) follow directly from (2.18). ϕ continuous $\Rightarrow \rho$ continuous away from the $R_j \Rightarrow \phi$ is C^1 everywhere. Then ρ is C^1 away from the R_j. [Note: $(\phi - \mu)_+^{3/2}$ is C^1 if ϕ is C^1]. By a bootstrap argument ρ is C^∞ on A. By Theorem 5.8.6 in (Morrey, 1966), ϕ is real analytic away from the R_j and where $\phi > \mu$, namely A. Finally, since $\phi(x) \to 0$ as $|x| \to \infty$, ρ has compact support when $\mu > 0$. The positivity of ϕ is established in section III, so $\rho > 0$ in the neutral case. ∎

In the ionic case $(\lambda < Z)$ the set $\Omega = \{x | \rho(x) > 0\}$ is bounded. What can be said about its boundary, $\partial\Omega$? In the atomic case $\partial\Omega$ is, of course, a sphere. In the general case, the TF equation (2.19) is a "free boundary problem" about which Caffarelli and Friedman (1979) have proved the following result among others.

Theorem 2.9: Consider the generalized TF problem with 5/3 replaced by p and $3/2 < p < 2$. There are at most a finite number of open C^1 curves $\Gamma_1,\ldots,\Gamma_\ell$ such that $\partial\Omega \setminus \{\bigcup_{i=1}^{\ell} \overline{\Gamma}_i\}$ is a $C^{3+\alpha}$ manifold with $\alpha = (2 - p)/(p - 1)$.

The next question to consider is the asymptotics of ρ, in the

neutral case ($\mu = 0$), as $|x| \to \infty$. This involves finding universal bounds on ρ. The function $\psi(x) = \gamma^3(3/\pi)^2|x|^{-4}$ satisfies (2.19) for $|x| > 0$. It is the only "power law" that does so. This was noted by Sommerfeld who concluded that $\psi(x)$ is the asymptotic form of ϕ. Hille (1969), who was possibly the first to make a serious mathematical study of the TF equation, proved this asymptotic law in the atomic case. It is remarkable that ψ, the asymptotic form of ϕ, is <u>independent</u> of z, and it is just as remarkable that the same form holds even in the molecular case.

<u>Theorem 2.10</u> [LS §V.2]: Suppose $\mu = 0$ and $|R_j| < R$, for all j and some R. For $r = |x| > R$ let $\phi_+(r)$ (resp. $\phi_-(r)$) be the max (resp. min) of $\phi(x)$ on $|x| = r$ and $C_\pm(r) = \phi_\pm(r)/\psi(r)$ with $\psi(r) = \gamma^3(3/\pi r^2)^2$. Then $C_\pm(r) \to 1$ as $r \to \infty$. Furthermore, if $R < r$ then

(i) $C_+(R) \geq 1 \Rightarrow C_+(r) \leq C_+(R)$;

(ii) $C_+(R) \leq 1 \Rightarrow C_+(r) \leq 1$;

(iii) $C_-(R) \leq 1 \Rightarrow C_-(r) \geq C_-(R)$;

(iv) $C_-(R) \geq 1 \Rightarrow C_-(r) \geq 1$.

<u>Proof</u>: If f,g are continuous, positive functions on $|x| \geq R$ which go to zero as $|x| \to \infty$, and if $\gamma^{3/2}\Delta f \leq 4\pi f^{3/2}$, $\gamma^{3/2}\Delta g \geq 4\pi g^{3/2}$ for $|x| > R$, and if $f(x) \geq g(x)$ for $|x| = R$, then $f(x) \geq g(x)$ for all $|x| > R$. This is easily proved by a "maximum argument" as in section III. ϕ is of this type with $\gamma^{3/2}\Delta\phi = 4\pi\phi^{3/2}$. If $C_+(R) \geq 1$, compare $\phi(x)$ with $C_+(R)\,\psi(x)$. Then $C_+(r) \leq C_+(R)$, all $r > R$. This proves (i) and similarly (iii). To prove (ii) and (iv) compare ϕ with ψ. It remains to show that $C_\pm(r) \to 1$. C_+ is continuous. We will show that $\limsup C_+(r) \leq 1$; by a similar argument $\liminf C_-(r) \geq 1$. This will complete the proof. If $C_+(R) > 1$, let $R_o = \sup\{r | C_+(r) \geq 1\}$ whence $C_+(r) \geq 1$ for $R \leq r \leq R_o$ and $C_+(r) \leq 1$ for $r \geq R_o$. It is then only necessary to consider $R_o = \infty$. Then, since $C_+(r)$ is decreasing, $C(\infty) = \lim C_+(r)$ exists. Assume $C(\infty) > 1$. Pick $\varepsilon > 0$ and choose R_1 so that $C_+(R_1) < C(\infty) + \varepsilon$. Consider $f(x) = \gamma^3(3/\pi)^2(1 + 2b/3)^2(|x| - bR_1)^{-4}$ for $|x| \geq R_1$ and $b < 1$. $\gamma^{3/2}\Delta f \leq 4\pi f^{3/2}$. Choose $b < 1$ such that $(1 + 2b/3)^2 = C_+(R_1)(1-b)^4$. Then $f \geq \phi$ for $|x| > R_1$ since $f \geq \phi$ when $|x| = R_1$. But this means $C(\infty) \leq (1 + 2b/3)^2 = [C(\infty) + \varepsilon](1-b)^4$. Since $b > B > 0$ satisfying $(1 + 2B/3)^2 = C(\infty)(1-B)^4$, and ε is arbitrary, $C(\infty) \leq 1$. For the C_- problem use $g(x) = \gamma^3(3/\pi)^2(1 - 2b/3)^2(|x| + bR_1)^{-4}$. ■

There are some interesting facts about the possible singularities of TF type differential equations in a ball. These are related to and complement Theorem 2.10.

Theorem 2.11: Let $B = \{x | 0 < |x| < R\}$ and suppose ϕ satisfies $\Delta\phi(x) = G(\phi(x))$ in the sense of distributions on B, and G is C^1. Then

(a) If $\phi \in L^\infty_{loc}(B)$ and G satisfies $G(t) > \varepsilon t^3$ as $t \to \infty$, $G(t) < \varepsilon t^3$ as $t \to -\infty$, $\varepsilon > 0$, then there exists a C^2 function on $0 \leq |x| < R$ which agrees with ϕ a.e. in B. Any singularity is thus removable. In particular $-\Delta\phi + \phi^3 = \delta(x)$ has no solution in $B' = \{x | 0 \leq x < R\}$.

(b) If $\phi \in C^2(B)$, $\phi > 0$, and $G(t) = t^q$, $1 < q < 3$ then either

(i) ϕ has a C^2 extension on B'.

(ii) $\phi(x) \sim C|x|^{-1}$ as $|x| \to 0$, $C > 0$ arbitrary

(iii) $\phi(x) \sim \ell |x|^{-a}$ as $|x| \to 0$ with $a = 2/(q-1)$, $\ell^{q-1} = a(a-1)$. This is called the "strong singularity".

(c) Let $q > 1$ and $B' = \{x | 0 \leq |x| < R\}$. Let $\phi \in L^\infty_{loc}(B')$ satisfy $\Delta\phi = |\phi|^{q-1}\phi$ in B' in the sense of distributions. There is a universal constant $C_q < \infty$ such that $|\phi(0)| \leq C_q R^{-2/(q-1)}$. This implies that if $\phi \in L^\infty_{loc}(B)$ satisfies this equation in B, then $|\phi(x)| \leq C_q |x|^{-2/(q-1)}$ for $2|x| < R$. A stronger bound than this is in (Veron, 1979) and (Brezis and Veron, 1980) for $1 < q < 3$.

Proof: (a) is given in (Brezis and Veron, 1980) and (b) and (c) are given in (Veron, 1979). (c) was given earlier for $q = 3/2$ in (Brezis and Lieb, 1979). ∎

There are other theorems of this type in (Veron, 1979) and (Brezis and Lieb, 1979).

There is another property of ρ which can be derived directly from the variational principle, namely

Theorem 2.12: In the atomic case $\rho(x)$ is a symmetric, decreasing function.

Proof: Assume the nucleus is at the origin and let ρ^* be the symmetric decreasing rearrangement of ρ (for a definition see (Lieb, 1977)). We claim that if $\rho \neq \rho^*$ then $\mathcal{E}(\rho^*) < \mathcal{E}(\rho)$, thereby proving the theorem. $\int(\rho^*)^p = \int\rho^p$, all p. For the Coulomb terms note that when $\int\rho \leq z$ then $f(x) = z|x|^{-1} - |x|^{-1} * (\rho^*)$ is a symmetric decreasing function; hence $\int f\rho \leq \int f\rho^*$. Thus $P(\rho) \equiv D(\rho,\rho) - \int V\rho = D(\rho-\rho^*, \rho-\rho^*) - \int\rho f - D(\rho^*,\rho^*)$ and thus $P(\rho) < P(\rho^*)$ if $\rho \neq \rho^*$. ∎

Remarks: (i) The same theorem (and proof) holds for the TFD, TFW and TFDW theories provided $\lambda = \int\rho \le z$. The only additional fact needed for the W theories is that $\int(\nabla\psi)^2 \ge \int(\nabla\psi^*)^2$, cf. (Lieb, 1977), appendix. In fact, Theorem 2.12 holds for all λ (Theorem 7.25).

(ii) The spherically symmetric (but not the decreasing) property of ρ also follows from the uniqueness of ρ which, in turn, follows from the strict convexity of $\mathcal{E}$. The decreasing property also follows from (2.18) since ϕ is decreasing by Newton's theorem.

II.E. The Virial and Related Theorems

Let us generalize the TF functional $\mathcal{E}$ by multiplying the term $D(\rho,\rho)$ in (2.5) by a parameter $\beta > 0$. $e(\lambda) = E(\lambda) - U$ in (2.12) is then a function of γ, $\{z_j\}$ and β. Define

$$K = \tfrac{3}{5}\gamma \int\rho^{5/3}, \quad R = \beta\, D(\rho,\rho), \; A = \int V\rho \qquad (2.20)$$

with ρ being the minimizing ρ for $\int\rho = \lambda$ with $\lambda \le \lambda_c$. (By scaling, $\lambda_c(\beta) = \lambda_c(\beta{=}1)/\beta$.

Theorem 2.13: $e(\lambda, \gamma, \{z_j\}, \beta)$ is a C^1 function of its k+3 arguments (assuming all are > 0, except for β which is ≥ 0, and $\lambda \le \lambda_c$). e is convex in λ and jointly concave in $(\gamma, \{z_j\}, \beta)$. Moreover, $\partial e/\partial\gamma = K/\gamma$, $\partial e/\partial\beta = R/\beta$, $\partial e/\partial\lambda = -\mu$ $\partial e/\partial z_j = -\int\rho(x)|x-R_j|^{-1}dx$. This implies

$$\partial E/\partial z_j = \lim_{x\to R_j} \{\phi(x) - z_j\,|x-R_j|^{-1}\} . \qquad (2.21)$$

Proof: See LS. The proof uses the convexity of $\rho \to \mathcal{E}(\rho)$. The concavity in the parameters is a trivial consequence of the variational principle and the linearity of $\mathcal{E}$ in the parameters. ■

Now we return to $\beta = 1$.

Theorem 2.14: (a) $5K/3 = A - 2R - \mu\lambda$

(b) $2K = A - R$ for an atom (k = 1).

Proof: (a) Simply multiply the TF equation, (2.18), by ρ and integrate. Alternatively, note that ρ minimizes $G(\rho) = \mathcal{E}(\rho) + \mu\int\rho$ on all of $L^{5/3} \cap L^1$. Therefore $f(t) \equiv G(\rho_t)$, with $\rho_t(x) = t\rho(x)$, has its minimum at t=1. But df/dt = 0 gives (a).

(b) Here, scaling is essential. Consider $\rho_t(x) = t^3\rho(tx)$, so

that $\int \rho_t = \lambda$. Then $f(t) = \mathcal{E}(\rho_t)$ has its minimum at $t=1$ and $df/dt = 0$ gives (b). ∎

Remark: (b) is called the Virial theorem. A priori there is an analogue of (b) for a molecule. Suppose that, with λ fixed, e is stationary with respect to all R_j, i.e. $\nabla_{R_j} e = 0$. Then, by the same scaling argument together with $R_j \to tR_j$, one would conclude that $2K = A - R - U$, equivalently $K + E = 0$. See (Fock, 1932) and (Jensen, 1933). The difficulty with this is that there are no stationary points for $k \geq 2$. The no-binding Theorem 3.23 shows that there are no global minima, and the positivity of the pressure proved in §IV.B shows that there are no local minima (at least for neutral molecules). There it will be shown that for $k \geq 2$, the pressure, P, satisfies

$$3P = K + E > 0, \quad \text{for neutral molecules.} \tag{2.22}$$

For nonneutral molecules, a sharpening of Theorem 4.7 into a strict inequality for the derivative would suffice to show the absence of local minima.

For a neutral atom, (a) and (b) combine to give the following simple ratios:

$$R : K : -e : A = 1 : 3 : 3 : 7 \tag{2.23}$$

Scaling: Suppose the nuclear coordinates R_i are replaced by ℓR_i with $\ell > 0$. If $\underline{z}$, $\underline{R}$ denote the nuclear charges and coordinates, and if $E(\underline{z}, \lambda, \ell\underline{R})$, $-\mu(\underline{z},\lambda, \ell\underline{R})$, $\rho(\underline{z},\lambda, \ell\underline{R};x)$ and $\phi(\underline{z},\lambda, \ell\underline{R};x)$ denote the TF energy, chemical potential, density, and potential with $\int\rho = \lambda$, then

$$\begin{aligned} E(\underline{z},\lambda, \ell\underline{R}) &= \ell^{-7} E(\ell^3 \underline{z}, \ell^3 \lambda, \underline{R}) \\ \mu(\underline{z},\lambda, \ell\underline{R}) &= \ell^{-4} \mu(\ell^3 \underline{z}, \ell^3 \lambda, \underline{R}) \\ \rho(\underline{z},\lambda, \ell\underline{R}; x) &= \ell^{-6} \rho(\ell^3 \underline{z}, \ell^3 \lambda, \underline{R}; \ell^{-1} x) \\ \phi(\underline{z},\lambda, \ell\underline{R}; x) &= \ell^{-4} \phi(\ell^3 \underline{z}, \ell^3 \lambda, \underline{R}; \ell^{-1} x) . \end{aligned} \tag{2.24}$$

This is a trivial consequence of the scaling properties of $\mathcal{E}(\rho)$.

II.F. The Thomas-Fermi Theory of Solids

A solid is viewed as a large molecule with the nuclei arranged periodically. For simplicity, but not necessity, let us suppose that there is one nucleus of charge z per unit cell located on the points of $\mathbb{Z}^3 \subset \mathbb{R}^3$. If Λ is a finite subset of $\mathbb{Z}^3$ we want to know if, as $\Lambda \to \infty$ in a suitable sense, the energy/unit volume, $|\Lambda|^{-1}E_\Lambda$,

has a limit E, and ρ_Λ has a limit, ρ, which is a periodic function. Here, $|\Lambda|$ is the volume of Λ. If so, the equation for ρ and an expression for E in terms of ρ is required. Naturally, it is necessary to consider only neutral systems for otherwise $|\Lambda|^{-1}E_\Lambda \to \infty$. Everything works out as expected except for one mildly surprising thing; a quantity ψ_o appears in the equation for ρ which, while it looks like a chemical potential, and is often assumed to be one, is not a chemical potential. ψ_o is the average electric potential in the solid. All of this is proved in LS, section VI.

Definition: A sequence of domains $\{\Lambda_i\}$ is said to tend to infinity (denoted by $\Lambda \to \infty$) if

(i) $\bigcup_{i=1}^{\infty} \Lambda_i = \mathbb{Z}^3$, (ii) $\Lambda_{i+1} \supset \Lambda_i$, (iii) if $\Lambda_i^h \subset \mathbb{Z}^3$ is the set of points not in Λ_i, but whose distance to Λ_i is less than h, then $|\Lambda_i^h|/|\Lambda_i| \to 0$ for each $h > 0$. $\Gamma = \{x \in \mathbb{R}^3 |\ |x^i| < \tfrac{1}{2}\}$ is the elementary cube centered at the origin.

Theorem 2.15: As $\Lambda \to \infty$ the following limits exist and are independent of the sequence Λ_i:

(i) $\phi(x) = \lim_{\Lambda\to\infty} \phi_\Lambda(x)$; ϕ is periodic, $\gamma\, \rho(x)^{2/3} \equiv \phi(x)$, and the convergence is uniform on compacts in $\mathbb{R}^3$.

(ii) $\phi(x) = \lim_{\Lambda\to\infty} |\Lambda|^{-1} \sum_{y\in\Lambda} \phi_\Lambda(x+y)$

(iii) $\lim_{x\to 0} \phi(x) - z|x|^{-1} = \lim_{\Lambda\to\infty} |\Lambda|^{-1} \sum_{y\in\Lambda} \lim_{x\to y} \phi_\Lambda(x) - z|x-y|^{-1}$

(iv) $\int_\Gamma \rho = \lim_{\Lambda\to\infty} \int_\Gamma \rho_\Lambda = z$

(v) $\int_\Gamma \rho^{5/3} = \lim_{\Lambda\to\infty} |\Lambda|^{-1} \int_{\mathbb{R}^3} \rho_\Lambda^{\ 5/3}$

(vi) $E = \lim_{\Lambda\to\infty} |\Lambda|^{-1} E_\Lambda$ exists.

Definition: $G(x)$ is the periodic Coulomb potential. It is defined up to an unimportant additive constant in Γ by $-\Delta G/4\pi = \delta(x) - 1$. A specific choice is

$$G(x) = \pi^{-1} \sum_{\substack{k\in\mathbb{Z}^3 \\ k\neq 0}} |k|^{-2} \exp[2\pi i\ k\cdot x]$$

Theorem 2.16: ϕ, ρ and E satisfy

(i) $E = (\gamma/10)\int_\Gamma \rho^{5/3} + (z/2) \lim_{x\to 0} \{\phi(x) - z|x|^{-1}\}$ (2.25)

(ii) $\phi(x) = z\,G(x) - \int_\Gamma G(x-y)\,\rho(y) + \psi_o$ (2.26a)
for some ψ_o. Alternatively,

$$- \Delta\phi\,(x)/4\pi = \sum_{y\in\mathbb{Z}^3} \delta(x-y) - \rho(x) \qquad (2.26b)$$

(iii) ϕ and ρ are real analytic on $\mathbb{R}^3\setminus\mathbb{Z}^3$.
(iv) There is a unique pair ρ, ψ_o that satisfies (2.26) with $\gamma\rho^{2/3} = \phi$ and $\int\rho = z$. (cf. Theorem 2.6).

Formula (2.25) may appear strange but it is obtained simply from the TF equation; an analogous formula also holds for a finite molecule.

(2.26) together with $\gamma\rho^{2/3} = \phi$, is the periodic TF equation. ψ_o is not a chemical potential. The chemical potential, $-\mu$, is zero because μ_Λ is zero for every finite system. If (2.26) is integrated over Γ we find, since $\int\rho = z$, that $\psi_o = \int_\Gamma\phi$ = average electric potential. It might be thought that ψ_o could be calculated in the same way that the Madelung potential is calculated: In each cubic cell there is (in the limit) a charge density $z\,\delta(x) - \rho(x)$. Therefore if $\tilde{\phi}(x) \equiv \sum_{y\in\mathbb{Z}^3} g(x-y)$, with $g(x) = z|x|^{-1} - \int_\Gamma\rho(y)|x-y|^{-1}\,dy$, it might be expected that $\tilde{\phi} = \phi$. The correct statement is that $\phi(x) = \tilde{\phi}(x) + d$ and $d \neq 0$ in general. One can show that $\int_\Gamma\tilde{\phi} = 2\pi\int_\Gamma x^2\,\rho(x)\,dx$. cf. LS. The fact that $d \neq 0$ precludes having a simple expression for ψ_o. Why is $d \neq 0$, i.e. why is $\phi \neq \tilde{\phi}$? The reason is that the charge density in the cell centered at $y \in \mathbb{Z}^3$ is $z\,\delta(x-y) - \rho(x-y)$ only in the limit $\Lambda \to \infty$. For any finite Λ there are cells near the surface of Λ that do not yet have this charge distribution. Thus $d \neq 0$ essentially because of a neutral double layer of charge on the surface.

In LS asymptotic formulas as $z \to 0$ and ∞ are given for the various quantities.

Theorems 2.15 and 2.16 will not be proved here. Teller's lemma, which implies that $\phi_\Lambda(x)$ is monotone increasing in Λ, is used repeatedly. Apart from this, the analysis is reasonably straightforward.

II.G. The Thomas-Fermi Theory of Screening

Another interesting solid state problem is to calculate the potential generated by one impurity nucleus, the other nuclei being smeared out into a uniform positive background (jellium model). If Λ is any bounded, measurable set in $\mathbb{R}^3$, and if ρ_B = (const.) > 0 is the charge density of the positive background in Λ, and if the impurity nucleus has charge $z > 0$ and is located at 0, then the potential is

$$V_\Lambda(x) = z|x|^{-1} + \rho_B \int_\Lambda |x-y|^{-1}\,dy \qquad (2.27)$$

The TF energy functional, without the nuclear repulsion, and with $\gamma = 1$ is

$$\mathcal{E}_\Lambda(\rho) = \frac{3}{5}\int \rho^{5/3} - \int V_\Lambda \rho + D(\rho,\rho) \ . \tag{2.28}$$

The integrals are over $\mathbb{R}^3$, not Λ. Let $\rho_\Lambda(x)$ be the neutral minimizing ρ (so that $\int \rho_\Lambda = z + \rho_B \ |\Lambda|$).

<u>Definition</u>: A sequence of domains Λ is said to <u>tend to infinity weakly</u> if every bounded subset of $\mathbb{R}^3$ is eventually contained in Λ.

<u>Remark</u>: This is an extremely weak notion of $\Lambda \to \infty$.

It is intuitively clear that if $\Lambda \to \infty$ weakly and $z = 0$ then $\rho_\Lambda(x) \to \rho_B$. For $z \neq 0$, $\rho_\Lambda(x) - \rho_B$ is expected to approach some function which looks like a Yukawa potential for large $|x|$. This is stated in many textbooks and is correct except for one thing: the coefficient of the Yukawa potential is <u>not</u> z but is some smaller number. In TF theory there is <u>over-screening</u> because of the non-linearities.

<u>Theorem 2.17</u>: Let $\Lambda \to \infty$ weakly and $z = 0$. Then $\phi_\Lambda(x) \to \rho_B{}^{2/3}$ uniformly on compacts in $\mathbb{R}^3$.

The theorem is another example of the effects of "surface charge". Since $\rho_\Lambda \to \rho_B$ and $\phi_\Lambda = \rho_\Lambda{}^{2/3}$, the result is natural. But it means that the average potential is not zero. If, on the other hand, the integrals in (2.28) are restricted to Λ then $\rho_\Lambda(x) = \rho_B$ for all Λ and $x \in \Lambda$, and $\phi_\Lambda(x) = 0$.

<u>Theorem 2.18</u>: Let $\Lambda \to \infty$ weakly and $z > 0$. Let $f(x) = \lim_{\Lambda\to\infty} \phi_\Lambda(x) - \rho_B{}^{2/3}$ and $g(x) = \lim_{\Lambda\to\infty} \rho_\Lambda(x) - \rho_B$.

(i) These limits exist uniformly on compacts;

(ii) $g \in L^1 \cap L^{5/3}$;

(iii) $0 \leq f(x) \leq \phi^{\text{atom}}(x)$;

(iv) f and g are strictly positive and real analytic away from $x = 0$;

(v) $f(x)$ is monotone increasing in z;

(vi) These limits satisfy the TF equation

$$f(x) = z|x|^{-1} - \int |x-y|^{-1} g(y)\, dy \ , \tag{2.29}$$

$$[\rho_B{}^{2/3} + f(x)]^{3/2} - \rho_B = g(x) \ , \tag{2.30}$$

$$\int g = z \ ; \tag{2.31}$$

(vii) Assuming only that $g \in L^1 \cap L^{5/3}$ and $f(x) \geq -\rho_B{}^{2/3}$ there is only one solution to (2.29), (2.30) (without assuming (2.31)).

There is a scaling relation:

$$f(x;z) = \rho_B{}^{2/3}\, F(\rho_B{}^{1/6}|x|;\ \rho_B{}^{-\frac{1}{2}} z)$$

$$g(x;z) = \rho_B\, G(\rho_B{}^{1/6}|x|;\ \rho_B{}^{-\frac{1}{2}} z).$$

Let us write $F(r;z) = q(r;z)\, Y(r)$ where $Y(r) = (1/r)\exp\{-(6\pi)^{\frac{1}{2}} r\}$ is the Yukawa potential.

Theorem 2.19: (i) $q(r;z)$ is monotone decreasing in r and increasing in z;

(ii) $q(0;z) = z$;

(iii) $Q(z) = \lim_{r\to\infty} q(r;z)$ exists. $0 < Q(z) < z$ and Q is monotone increasing. $\limsup_{z\to\infty} Q(z)(bz)^{-2/3} < 1$ with $b = 1.039$.

LS contains graphical plots of $Q(z)$ and $q(r;\ 53.7)$. An asymptotic formula for $Q(z)$ has not been given. In the linearized approximations found in textbooks, $Q(z) = z$, but we see that this is false.

II.H. The Firsov Variational Principle

The problem of minimizing $\mathcal{E}(\rho)$ is a convex minimization problem. It has a dual which we now explore. The advantage of the dual problem is that it gives a lower bound to E. The principle was first given and applied in (Firsov, 1957) in the neutral case ($\mu = 0$) and was first rigorously justified in that case by Benguria (1979). Here we will also state and prove the principle for non-neutral systems; furthermore, in the neutral case our (and Benguria's) principle will contain a slight improvement over Firsov's.

The dual functional to be considered is

$$\mathcal{F}_\mu(f) = -(8\pi)^{-1}\int |\nabla f(x)|^2 dx - \frac{2}{5}\gamma^{-\frac{3}{2}} \int (V(x) - f(x) - \mu)_+^{5/2} dx + U \tag{2.32}$$

where μ is a real parameter. The domain of $\mathcal{F}_\mu$ is $B = \{f \,|\, \nabla f \in L^2,\ |f(x)| < c|x|^{-1}$ for some $c < \infty$ and for $|x| > R$ for some $R\}$. V is assumed to go to zero at ∞ and is such that

the TF problem has a minimum for that V, and the minimizing ρ (with $\int\rho \le \lambda$) satisfies the TF equation (2.18) (for all λ). We define

$$E^F(\mu) = \sup\{\mathcal{F}_\mu(f) \mid f \in B\} \tag{2.33}$$

Remark: When $\mu = 0$, Firsov imposed the additional constraint $V \ge f$. This, as we shall see, is unnecessary provided $(\quad)_+^{5/2}$ is used as in (2.32).

Theorem 2.20: If $\mu < 0$ then $E^F(\mu) = -\infty$. If $\mu \ge 0$ then there is a unique maximizing f for $\mathcal{F}_\mu$. This f is $f_\mu \equiv |x|^{-1} * \rho_\mu$ where ρ_μ is the unique solution to (2.18). If $\lambda = \int\rho_\mu$ then

$$E^F(\mu) = E(\lambda) + \mu\lambda . \tag{2.34}$$

Proof: Suppose $\mu < 0$. Since V and any $f \in B \to 0$ as $|x| \to \infty$, the second term in (2.32) is $-\infty$. Suppose $\mu \ge 0$. Let $\tilde{E}_\mu$ = right side of (2.34). Clearly $\mathcal{F}_\mu(f_\mu) = \tilde{E}_\mu$ by the TF equation (2.18). $f \to \mathcal{F}_\mu(f)$ is strictly concave because $\int(\nabla f)^2$ is strictly convex. Thus, there can be at most one maximizing f, and we therefore must show that if $f \ne f_\mu$ then $\mathcal{F}_\mu(f) \le \tilde{E}_\mu$. By Minkowski's inequality ($|ab| \le 2|a|^{5/2}/5 + 3|b|^{5/3}/5$) we have $(2/5)(V-f-\mu)_+^{5/2} \ge -(3/5)(V-f_\mu-\mu)_+^{5/2} + (V-f-\mu)_+(V-f_\mu-\mu)_+^{3/2}$. But $(V-f-\mu)_+ \ge V-f-\mu$, so $\mathcal{F}_\mu(f) \le \tilde{E}_\mu + h(f)$ where $h(f) = -(8\pi)^{-1}\int(\nabla f)^2 + \int f\rho_\mu - D(\rho_\mu,\rho_\mu)$. By standard methods (e.g. Fourier transforms), $h(f) \le 0$. Furthermore, $h(f) = 0$ only for $f = f_\mu$, which shows once again that the maximizing f is uniquely f_μ. ■

It should be noted that $E^F(\mu)$ is the Legendre transform of $E(\lambda)$. Namely $\lambda \to E(\lambda)$ is convex and

$$E^F(\mu) = \inf_{\lambda\ge 0}\,[E(\lambda) + \lambda\mu],\ \text{all}\ \mu \in \mathbb{R} . \tag{2.35}$$

This shows that $E^F(\mu)$ is concave in μ. On the other hand, Theorem (2.20) displays $E^F(\mu)$ as the supremum (not infimum) of a family of concave functions. Furthermore, since $E(\lambda)$ is convex and bounded it is its own double Legendre transform, viz.

$$E(\lambda) = \sup_\mu\,[E^F(\mu) - \mu\lambda] . \tag{2.36}$$

Theorem 2.21: Fix $\lambda \ge 0$. Then (by (2.36))

$$\sup\{\mathcal{F}_\mu(f) - \mu\lambda \mid f \in B,\ \mu \in \mathbb{R}\} = E(\lambda) . \tag{2.37}$$

Remark: In Theorem 2.20 we refer to the unique ρ_μ satisfying (2.18) for $\mu \geq 0$. This requires some explanation. If $V(x)$ is unbounded (e.g. point nuclei), then as μ goes from ∞ to 0, λ goes from 0 to λ_c and $\rho_{\mu(\lambda)}$ minimizes $\mathcal{E}$ on $\int\rho = \lambda$. If ess sup $V(x) = v < \infty$, then $\rho_\mu \equiv 0$ (and $E^F(\mu) = 0$) for $\infty > \mu \geq v$. In this range $\lambda(\mu) = 0$. Then, as μ goes from v to 0, λ goes from 0 to λ_c and $\rho_{\mu(\lambda)}$ minimizes E on $\int\rho = \lambda$.

III. THE "NO-BINDING" AND RELATED POTENTIAL-THEORETIC THEOREMS

The no-binding theorem was discovered by Teller (1962) and is one of the most important facts about the TF and TFD theories of atoms and molecules. It "explained" the absence of binding found numerically by Sheldon (1955). That this crucial theorem was not proved until 1962--after 35 years of intensive study of TF theory--is remarkable. It can be considered to be a prime example of the fact that pure analysis can sometimes be superior to numerical studies.

While Teller's ideas were correct, his proof was questioned on grounds of rigour. Balàzs (1967) found a different proof for the special case of the symmetric diatomic molecule. A rigorous transcription of Teller's ideas was given in LS . In any case, all proofs of the theorem rely heavily on the fact that the potential is Coulombic.

There are really two kinds of theorems. An example of the first kind is "Teller's lemma" which states that the potential increases when nuclear charge is added. The second, "Teller's Theorem" is the no-binding Theorem 3.23. The second, but not the first, requires the nuclear repulsion U. If U is dropped then the theorem goes the other way. The proof of Teller's Theorem given in LS is complicated in the non-neutral case, but recently Baxter (1980) found a much nicer proof--one which actually produces a variational ρ that lowers the energy for separated molecules. Baxter's Proposition 3.24 will appear again in Lemma 7.22.

In this section we will consider general V and assume that

$$\mathcal{E}(\rho) = \int j(\rho(x))dx - \int V(x)\rho(x)dx + D(\rho,\rho) \tag{3.1}$$

where j is a C^1 convex function with $j(0) = j'(0) = 0$. Note that in this section (only) $\mathcal{E}(\rho)$ does not contain U. This is done partly for convenience, but mainly for the reason that since V is not necessarily Coulombic the definition of U would have no clear meaning.

The Euler-Lagrange equation for (3.1) and $\rho(x) \geq 0$ is (with $\phi_\rho = V - |x|^{-1} * \rho$):

$$\phi_\rho(x) - \mu = j'(\rho(x)) \quad \text{a.e. when } \rho(x) > 0\ ,$$
$$\leq 0 \quad \text{a.e. when } \rho(x) = 0\ . \tag{3.2}$$

Any solution to (3.2) is determined only almost everywhere (a.e.).

We could, in fact, allow more general j's of the form $j(\rho,x)$ (and $\int j(\rho(x),x)dx$ in $\mathcal{E}$) with $j(\cdot,x)$ having the above properties for all x, but we will not do so. An annoying case we must consider, however, is $j'(\rho) = 0$ for $0 < \rho < \rho_o$ and $j'(\rho) > 0$ for $\rho > \rho_o$. This is discussed in some detail in § III.C and is needed for TFD theory (section VI). If $j'(\rho) > 0$, all $\rho > 0$, as it is in TF theory with $j'(\rho) = \gamma\rho^{2/3}$, then (3.2) can be written as

$$(\phi_\rho(x) - \mu)_+ \equiv \max(\phi_\rho(x) - \mu, 0) = j'(\rho(x)), \tag{3.2a}$$

but otherwise (3.2) is stronger than (3.2a).

One aim of this section is to study solutions of (3.2) without considering whether or not (3.2) truly comes from minimizing (3.1) or assuming uniqueness.

Definition: $\mathcal{C} = \{\rho \mid \rho(x) \geq 0,\ \rho \in L^1$ and $\int \rho(y)|x-y|^{-1}dy$ is a bounded, continuous function which goes to zero as $x \to \infty\}$.

We shall be concerned only with solutions to (3.2) in $\mathcal{C}$.

The following lemma [LS, II.25] is useful, in the cases of interest, to guarantee that $\rho \in \mathcal{C}$.

Lemma 3.1 If $f \in L^p$, $g \in L^{p'}$, $1/p + 1/p' = 1$, $p,p' > 1$ then $f*g$ is a bounded continuous function which goes to zero as x goes to infinity. In particular, if $\rho \in L^{3/2+\varepsilon} \cap L^1$ then $\rho \in L^{3/2+\varepsilon} \cap L^{3/2-\delta}$. Since $|x|^{-1} \in L^{3+\varepsilon} + L^{3-\varepsilon}$, $\rho \in \mathcal{C}$.

It will always be assumed that $V(x) \to 0$ as $|x| \to \infty$ (this always means uniformly with respect to direction). Hence, μ cannot be negative in (3.2), for otherwise $\rho \notin L^1$.

III.A. Some Variational Principles and Teller's Lemma

At first it will not be assumed that V is Coulombic.

Theorem 3.2 Fix $\lambda > 0$ and suppose that ρ_λ, μ_λ satisfy (3.2) with $\int \rho_\lambda = \lambda$. Let $\phi_\lambda = \phi_{\rho_\lambda}$ and assume that $\rho_\lambda \in \mathcal{C}$. Then, for all x,

(a) $\phi_\lambda(x) - \mu_\lambda = \sup_\rho\{\phi_\rho(x) - \mu \mid \phi_\rho(y) - \mu \leq j'(\rho(y)) \text{ a.e. } y, \int\rho \leq \lambda, \rho \in \mathcal{C}\}$

(b) $\phi_\lambda(x)-\mu_\lambda = \inf_\rho\{\phi_\rho(x)-\mu \mid \phi_\rho(y)-\mu \geq j'(\rho(y))$ a.e. y when $\rho(y)>0, \int\rho \geq \lambda\ \rho \in \mathcal{C}\}$

(c) $\phi_\lambda(x) = \sup_\rho\{\phi_\rho(x) \mid \phi_\rho(y) - \mu_\lambda \leq j'(\rho(y))$, a.e. y, $\rho \in \mathcal{C}\}$

(d) $\phi_\lambda(x) = \inf_\rho\{\phi_\rho(x) \mid \phi_\rho(y) - \mu_\lambda \geq j'(\rho(y))$. a.e. y when $\rho(y)>0,\ \rho \in \mathcal{C}\}$

Furthermore, in (a) (resp. (b)) there is no ρ satisfying the conditions on the right when $\mu < \mu_\lambda$ (resp. $\mu > \mu_\lambda$). Note that in (a), (b) μ is arbitrary (including $\mu < 0$) and ρ is constrained, while in (c), (d) the opposite is true (except, of course, $\rho(x) \geq 0$).

In the following, a statement such as $\Delta\phi/4\pi = \rho$ is always meant in the distributional sense. We will need [LS II.26].

Lemma 3.3: Let $\rho_1, \rho_2 \in L^1$ with $\rho_i(x) \geq 0$ and $\psi_i = |x|^{-1} * \rho_i$. If $\psi_1(x) \geq \psi_2(x)$, all x, then $\int\rho_1 \geq \int\rho_2$.

Proof of Theorem 3.2: (a) will be proved here; (b), (c) and (d) follow similarly. Since ρ_λ gives equality, $\phi_\lambda - \mu_\lambda \leq \sup\{\quad\}$.

We have to show that if $(\phi_\rho - \mu)_+ \leq j'(\rho)$ a.e. and if $\int\rho \leq \lambda$ then

(i) $\phi_\lambda(x) - \mu_\lambda \geq \phi_\rho(x) - \mu$,

(ii) $\mu \geq \mu_\lambda$. First suppose $\mu \geq \mu_\lambda$ and let $\psi(x) = \phi_\rho(x) - \phi_\lambda(x) + \mu_\lambda - \mu$. Let $B = \{x \mid \psi(x) > 0\}$. B is open since ψ is continuous. As a distribution $-(4\pi)^{-1}\Delta\psi(x) = \rho_\lambda(x) - \rho(x) \leq 0$ a.e. on B since $j'(\rho) \geq \phi_\rho - \mu$, $j'(\rho_\lambda) = \phi_\lambda - \mu_\lambda$ when $\rho_\lambda > 0$ and j' is non-decreasing. Hence ψ is subharmonic on B and takes its maximum on ∂B, the boundary of B, or at ∞. $\psi = 0$ on ∂B. At ∞, $\psi = \mu_\lambda - \mu \leq 0$. Hence B is empty and (i) is proved. Suppose now that $\mu_\lambda - \mu = \delta > 0$. Then $j'(\rho) \geq \phi_\rho - \mu > \phi_\rho - \mu_\lambda$ and, by the previous proof (applied to $\mu = \mu_\lambda$), $\phi_\lambda(x) \geq \phi_\rho(x)$. By Lemma 3.3, $\int\rho \geq \int\rho_\lambda = \lambda$. Hence $\int\rho=\lambda$. At this point there are two possible strategies:

(i) If we assume that $\mathcal{E}(\rho)$ has a minimum that satisfies (3.2) for all $\mu \geq 0$, then we can use the fact (which follows from the strict convexity of $\mathcal{E}(\rho)$) that μ_λ is a continuous decreasing function of λ. Then there exists $\gamma > \lambda$ with $\mu_\lambda > \mu_\gamma > \mu$. Since $j'(\rho) \geq \phi_\rho - \mu_\gamma$, $\int\rho = \gamma$ by what we just proved. But this is a contradiction.

(ii) There is a purely potential theoretic argument without invoking (3.1). There is a (not necessarily unique) f which satisfies $j'(f(x)) = (\phi_\rho(x) - \mu/2 - \mu_\lambda/2)_+$ and $f(x) = 0$ when $(\ )_+ = 0$. Hence $f(x) < \rho(x)$ a.e. when $\rho(x) > 0$, and $f(x) = 0$ when $\rho(x) = 0$. Thus $f \in \mathcal{C}$. Since $\int\rho = \lambda > 0$, $\int f < \lambda$. Let $g = (1-\varepsilon)\rho + \varepsilon f$, $0 < \varepsilon < 1$. Since $|x|^{-1} * \rho$ (and hence $|x|^{-1} * f$) are bounded, $\phi_g(x) \leq \phi_\rho(x) + \varepsilon C$ for some constant, C. Choose $\varepsilon > 0$ so that $\varepsilon C < \delta/2$. Then $j'(g) \geq j'(f) \geq \phi_g - \mu_\lambda$. Since $\int g < \lambda$, g satisfies the condition in (a) with $\mu^g = \mu_\lambda$ but, as we have seen, this implies $\int g = \lambda$. ∎

Teller's lemma (Theorem 3.4) is closely related to Theorem 3.2.

Definition: We say $V \in \mathcal{D}$ if $V \neq 0$, V is superharmonic, vanishing at ∞ (and hence $V > 0$). Moreover, the set $\{x | V(x) = \infty\} = S_V$ (called the singularities of V) is closed, V is continuous on the complement of S_V, and $V(x) \to \infty$ as $x \to S_V$.

Theorem 3.4 Suppose V is replaced by $V' = V + W$ with $W \in \mathcal{D}$. (In the case of interest $W = z|x-R|^{-1}$, which means that we add, or increase, a nuclear charge.) Suppose that for some common μ there are solutions $0 \leq \rho, \rho' \in \mathcal{C}$ to (3.2) with V and with V'. Then $\phi'(x) \geq \phi(x)$ all x and, if j' is strictly monotone or if $\phi - \phi' \in H^2$ (i.e. $\phi - \phi'$ and its first two derivatives are in L^2) away from S_W, then $\rho'(x) \geq \rho(x)$, a.e.

Proof: Let $\psi = \phi' - \phi$ and $B = \{x | \psi(x) < 0\}$. Clearly $B \cap S_W = \emptyset$ so B is open. As a distribution, $\Delta\psi/4\pi \leq \rho' - \rho \leq 0$, so ψ is superharmonic on B. Thus B is empty and $\phi' \geq \phi$. The proof that $\rho' \geq \rho$ is trickier. If j' is strictly monotone it is obvious. Otherwise it can be shown, cf. (Benguria, 1979), that for suitable V,W and j', $\psi \in H^2$ away from S_W. Assuming $\psi \in H^2$, if $\rho'(x) < \rho(x)$ then $x \in C = \{x | \psi(x) = 0\}$. On C, $\Delta\psi = 0$, a.e. (cf. (Benguria, 1979), Thms. 2.19, 3.3). Let $D = C \cap \{x | \rho'(x) < \rho(x)\}$. On D, $0 = \Delta\psi/4\pi \leq \rho' - \rho < 0$ a.e., so D has zero measure. ∎

Remark: If $j'(s)$ is strictly monotone and $W \neq 0$ then $\phi'(x) > \phi(x)$ for all $x \notin S_W$.

A similar proof yields

Theorem 3.5: If $V \in \mathcal{D}$ then $\phi_\lambda(x) \geq 0$. Consequently if $V(x) = \int dM(y)|x-y|^{-1}$, $dM \geq 0$ and $\int dM = Z$, then there is no solution if $\lambda > Z$ because then $\phi_\lambda(x) < 0$ for some large x. Cf. Theorem 6.7.

There are many easy, but important corollaries of Theorem 3.2. We stress that V need not be Coulombic; the important ingredient was that the electron-electron repulsion is Coulombic.

Definition: j' is said to be subadditive if $j'(\rho_1 + \rho_2) \leq j'(\rho_1) + j'(\rho_2)$. j' is subadditive in the TF case.

Corollary 3.6: Suppose $V = V_1 + V_2$ and μ is fixed. Let ϕ, ϕ_1, ϕ_2 be solutions to (3.2) for this μ with V, V_1, V_2 respectively. Suppose $\phi_i \geq 0$ (e.g. $V_i \in \mathcal{D}$) and suppose j' is subadditive. Then $\phi(x) \leq \phi_1(x) + \phi_2(x)$, all x.

Proof: Use Theorem 3.2 (d) with $\rho_1 + \rho_2$ on the right side. ∎

Corollary 3.7: Let $\lambda > 0$. There can be at most one pair ρ, μ satisfying (3.2) with $\rho \in \mathcal{C}$ (in particular for $\rho \in L^{5/3} \cap L^1$) and $\int \rho = \lambda$.

Proof: If ρ_1, ρ_2 are two solutions, use Theorem 3.2(a) twice with ρ_1 and ρ_2 to deduce $\phi_1 - \mu_1 = \phi_2 - \mu_2$. This implies $\mu_1 = \mu_2$ and hence $\phi_1 = \phi_2$. But then $0 = \Delta(\phi_1 - \phi_2) = 4\pi(\rho_2 - \rho_1)$. ∎

This uniqueness result was proved earlier, Theorem 2.6, using the strict convexity of $\mathcal{E}(\rho)$.

Corollary 3.8: If $0 < \lambda' < \lambda$ then

(i) $\phi_{\lambda'} \geq \phi_\lambda$

(ii) $\mu_{\lambda'} \geq \mu_\lambda$

(iii) $\phi_{\lambda'} - \mu_{\lambda'} \leq \phi_\lambda - \mu_\lambda$.

Proof: For (iii) use Theorem 3.2(b) with ρ_λ, μ_λ as trial function for the λ' problem. (iii) $\Rightarrow$ (ii). For (i) use (c) with ρ_λ as variational function for the λ' problem. ∎

Corollary 3.9: Suppose ρ_1, μ and ρ_2, μ (same μ) are two solutions to (3.2) with $\int \rho_1$, $\int \rho_2 > 0$. Then $\phi_1 = \phi_2$ and $\rho_1 = \rho_2$ a.e. Therefore, by Corollary 3.8, whenever $\lambda_2 > \lambda_1$ then $\mu_2 < \mu_1$ (i.e. $\mu_2 = \mu_1$, cannot occur).

Proof: Using Theorem 3.2(d), $\phi_1 = \phi_2$. Then $0 = \Delta(\phi_1 - \phi_2)/4\pi = \rho_1 - \rho_2$ a.e. ∎

Corollary 3.10: Suppose $j_1'(\rho) \leq j_2'(\rho)$, all ρ. Let ρ_λ^1, μ_λ^1 and ρ_λ^2, μ_λ^2 be corresponding solutions to (3.2) with fixed λ, and ρ^1, ρ^2 solutions with fixed μ. $\phi^i_{(\lambda)}(x)$ are the corresponding potentials. Then

(i) $\phi_\lambda^1 - \mu_\lambda^1 \leq \phi_\lambda^2 - \mu_\lambda^2$,

(ii) $\mu_\lambda^{\,1} \geq \mu_\lambda^{\,2}$,

(iii) $\phi^1 \leq \phi^2$.

<u>Proof</u>: For (i) use Theorem 3.2(a) with $\rho_\lambda^{\,1}, \mu_\lambda^{\,1}$ as trial function for the 2 problem. (i) $\Longrightarrow$ (ii). For (iii) use (d) with ρ_2 as trial function for the 1 problem. ■

<u>Lemma 3.11</u>: When $\mu > 0$, ρ has compact support.

<u>Remark</u>: As will be seen in section VI, ρ has compact support in TFD theory even when $\mu = 0$. Cf. Theorem 6.6.

Among the most important consequences of Theorem 3.2 are the <u>variational principles for the chemical potential</u>.

<u>Theorem 3.12</u>: Define the functionals

$$T(\rho) = \operatorname*{ess\,sup}_{x} \; \phi_\rho(x) - j'(\rho(x))$$

$$S(\rho) = \operatorname*{ess\,inf}_{x:\rho(x)>0} \; \phi_\rho(x) - j'(\rho(x)). \tag{3.3}$$

(ess sup means supremum modulo sets of measure zero). Then, whenever there is a solution to (3.2) with $\int\rho = \lambda > 0$,

$$\mu_\lambda = \inf \{T(\rho) \mid \rho \in \mathcal{C}, \int\rho \leq \lambda\} \tag{3.4}$$

$$\mu_\lambda = \sup \{S(\rho) \mid \rho \in \mathcal{C}, \int\rho \geq \lambda\} . \tag{3.5}$$

<u>Corollary 3.13</u>: If $j'(\rho)$ is concave (as in TF theory with $j'=\rho^{2/3}$) then μ_λ and $\mu_\lambda - \phi_\lambda(x)$, for each x, are jointly convex functions of V and λ.

<u>Corollary 3.14</u>: If λ is fixed and $V_1(x) \geq V_2(x)$, all x, then $\mu_\lambda(1) \geq \mu_\lambda(2)$.

By Corollaries 3.9 and 3.14 we know that increasing V increases μ while increasing λ decreases μ. What happens if V and λ are both increased, in particular if we scale up the size of a molecule by $V \to \alpha V$, $\lambda \to \alpha\lambda$? A partial answer is given by the following two corollaries.

<u>Corollary 3.15</u>: Let $V_1, V_2 \in \mathcal{D}$ and $V = V_1 + V_2$. Assume j' is subadditive and suppose (3.2) has solutions to the three problems (V_1,λ_1), (V_2,λ_2), (V,λ) with $\lambda = \lambda_1 + \lambda_2$. Then $\mu \geq \min(\mu_1, \mu_2)$.

<u>Proof</u>: In general, if $W \in \mathcal{D}$ and ρ is a solution to (3.2) with W then $\phi_\rho(x) - j'(\rho(x)) = \mu$ a.e. if $\rho(x) > 0$ and ≥ 0 a.e. if $\rho(x) = 0$ (Theorem 3.5). From this remark it follows that

$S_V(\rho_1 + \rho_2) \geq \min(\mu_1, \mu_2)$. (Here, $S_V(\rho)$ refers to (3.3) with V.) ■

Corollary 3.16: Let $\alpha > 1$ and suppose (3.2) has solutions with (V,λ,μ) and $(\alpha V,\alpha\lambda, \mu(\alpha))$. Assume j' satisfies $j'(\alpha t) \leq \alpha j'(t)$, all t (this holds in TF theory). Then $\mu(\alpha) \geq \alpha\mu$.

Proof: If ρ is the solution to (V,λ,μ) then $S_V(\rho) = \mu$. But $S_{\alpha V}(\alpha\rho) \geq \alpha S_V(\rho)$. ■

Corollary 3.17: Suppose there is a solution to (3.2) for all $\lambda \in (a,b)$ with $a < b$. Then μ_λ is continuous on this interval.

Proof: Let $\lambda_2 = \lambda_1 + \varepsilon$. By Corollary 3.9, $\mu_1 > \mu_2$. Let $\rho = \rho_1 + \varepsilon\chi$ with $\int\chi = 1$, $0 \leq \chi(x) \leq b$ for some b, $\chi(x) = 0$ if $\rho_1(x) = 0$ and $\chi(x) = 0$ if $\rho_1(x) > a$ for some a. Then $\chi \in \mathcal{C}$. Since j' is continuous, $S(\rho) \geq \mu_1 - Q(\varepsilon)$ where $Q(\varepsilon) \downarrow 0$ as $\varepsilon \downarrow 0$. ■

Theorem 3.18: Let $V \in \mathcal{D}$, $V(x) = \int dM(y)|y-x|^{-1}$, $dM \geq 0$, $\int dM = Z > 0$. Suppose that for large t, $j'(t) > c\,\rho^{\frac{1}{2}+\varepsilon}$, with $c,\varepsilon > 0$. By a simple modification of the method of Theorems 2.4, 2.5 and 2.6, $\mathcal{E}(\rho)$ has a unique minimum on the set $\mathcal{C}$ with $\int\rho \leq \lambda$. This ρ satisfies (3.2) and $\int\rho = \lambda$ if $\lambda \leq \lambda_c$, whereas $\int\rho = \lambda_c$ if $\lambda > \lambda_c$. Now assume, in addition, that $j'(t) < d\, t^{\varepsilon+1/3}$, $\varepsilon > 0$, for small t (this is true in all cases of interest). Then $\lambda_c = Z$.

Proof: If $\lambda > Z$ there is no solution by Theorem 3.5. Now suppose $\mu = 0$; we claim $\lambda \geq Z$, and hence that $\lambda = Z$. If so, we are done because $\mathcal{E}(\rho)$ has an absolute minimum. This minimum corresponds to $\mu = 0$ and has $\lambda = \lambda_c$; but $\mu = 0$ implies $\lambda = Z$. Now, to prove that $\lambda \geq Z$, let ϕ be the solution. If $\lambda = Z-3\delta$, let χ be the characteristic function of a ball centered at the origin such that $\int\chi\, dM > Z-\delta$. Then $\phi(x) > \psi(x) = \int[\chi(y)dM(y)-\rho(y)dy]|x-y|^{-1}$. For $|x| >$ some R, $\psi(x) < 2Z|x|^{-1}$. Also $[\psi(x)] =$ (spherical average of ψ) $> 2\delta|x|^{-1}$ for $x > R$. For a given $|x| = r > R$ let $\Omega_+(r)$ be the proportion of the sphere of radius r such that $2Z > r\,\psi(x) > \delta$, and let $\Omega_-(r)$ be the complement. Then $2\delta < r[\psi(x)] < 2Z\,\Omega_+ + \delta\Omega_- = \delta + (2Z-\delta)\,\Omega_+$. Thus, $\Omega_+(r) > \delta/(2Z-\delta)$ for all r. On Ω_+, $\rho^{1/3+\varepsilon} > \delta|x|^{-1}$ for large $|x|$, and therefore $\rho \notin L^1$ if $\delta > 0$. ■

Brezis and Benilan (Brezis, 1978 and 1980) have generalized this. Even if $j(\rho) \sim \rho^{4/3+\varepsilon}$ for large ρ there is a solution to (3.2) if $\lambda \leq Z$ and no solution otherwise. This is noteworthy since if $j(\rho) \sim \rho^a$ for large ρ with $a \leq \frac{3}{2}$ $\mathcal{E}(\rho)$ has no lower bound for point nuclei. There are similar results for other potentials, V, in LS, Theorem II.18.

There is also an "energetic", as distinct from potential theoretic reason that there is no solution if $\lambda > Z$. A solution to (3.2) implies a minimum for the functional $\mathcal{E}(\rho)$, by strict convexity. If $\lambda = \int \rho > Z$ then ϕ_ρ is negative in some set A of positive measure. Then it is easy to see that if ρ is decreased slightly in A to $\tilde{\rho}$, then $\mathcal{E}(\tilde{\rho}) < \mathcal{E}(\rho)$. But $\int \tilde{\rho} < \lambda$ and $E(\lambda)$ is non-increasing.

In the variational principle, Theorem 3.12, ρ_λ gives equality, i.e. $T(\rho_\lambda) = S(\rho_\lambda) = \mu_\lambda$. Is this the only ρ with this property? If $\lambda > Z$ there are many ρ's with $T(\rho) = 0$ and no ρ with $S(\rho) = 0$, (cf. LS). In (Brezis, 1980, §4) it is shown that if j' is concave (as in TF theory) and V has suitable properties (satisfied for $V \in \mathcal{D}$) then when $\lambda < Z$ only ρ_λ satisfies either $T(\rho) = \mu_\lambda$ or $S(\rho) = \mu_\lambda$. If $\lambda = Z$ this uniqueness is lost in general!

Asymptotics of the Chemical Potential

Theorem 3.12 can be used to obtain bounds on μ_λ. In the TF case with point nuclei, the asymptotic formula

$$\mu_\lambda \sim \gamma^{-1}(\pi^2 \Sigma z_j^3/4\lambda)^{2/3} \tag{3.6}$$

holds for λ small (LS, II.31). For λ near Z, (LS, IV.11,12) find upper and lower bounds for μ_λ of the form $\alpha_\pm(Z-\lambda)^{4/3}$ with $Z = \Sigma z_j$. Brezis and Benilan (unpublished) have shown that

$$\alpha = \lim_{\lambda \uparrow Z} \mu_\lambda (Z - \lambda)^{-4/3} \quad \text{exists} \tag{3.7}$$

and is given by solving some differential equation. α is independent of the number of nuclei and their individual coordinates and charges!

(3.7) implies that there is a well defined ionization potential, I, in TF theory (although it probably has nothing to do with the true Schrödinger ionization energy). First observe that if we start with $\Sigma z_j = 1$ and then replace z_j by $Z z_j$, R_j by $Z^{-1/3} R_j$ and λ by $Z\lambda$, then by scaling (2.24),

$$\mu_{Z\lambda} = Z^{4/3} \mu_\lambda . \tag{3.8}$$

Therefore, by (3.7), if we let $\lambda = Z - \varepsilon$ with $\varepsilon > 0$ fixed, and let $Z \to \infty$ then

$$\lim_{Z\to\infty} \mu_{Z-\varepsilon} = \alpha\, \varepsilon^{4/3} . \tag{3.9}$$

The ionization potential is defined to be

$$I = E(\lambda = Z - 1) - E(\lambda = Z). \tag{3.10}$$

By integrating (3.9), and appealing to dominated convergence,

$$I \to 3\alpha/7 \quad \text{as} \quad Z \to \infty . \tag{3.11}$$

Another implication of (3.7) is that an ionized <u>atom</u> has a well defined radius as $Z \to \infty$. This question was raised by Dyson. Suppose $V(x) = Z|x|^{-1}$ and $\lambda = Z - \varepsilon$. The density ρ will have support in a ball of radius $R(Z,\varepsilon)$. At $|x| = R$, $\phi(x) = \mu$. But since ρ is spherically symmetric, $R\phi(x) = Z - \lambda = \varepsilon$ by Newton's theorem. Thus the atomic radius satisfies

$$R = \varepsilon/\mu \quad \text{for all atoms} \tag{3.12}$$

and, by (3.9),

$$\lim_{Z\to\infty} R(Z,\varepsilon) = (\alpha\, \varepsilon^{1/3})^{-1} \tag{3.13}$$

III.B. <u>The Case of Flat j' (TFD)</u>:

In TFD theory, as will be seen in section VI, we have to consider

$$\begin{aligned} j'(\rho) &= 0, \qquad 0 \le \rho \le \rho_o = (5\, C_e/8\gamma)^3 \\ &= \gamma\rho^{2/3} - C_e\, \rho^{1/3} + 15\, C_e{}^2/4^3\gamma, \quad \rho_o \le \rho . \end{aligned} \tag{3.14}$$

j' satisfies all necessary conditions. It is neither concave nor subadditive, however. Let us consider V of the form

$$V(x) = \int dm(y)\, |x-y|^{-1} \tag{3.15}$$

with m being a measure that is not necessarily positive. In the primary case of interest, $dm(x) = \Sigma\, z_j\, \delta(x-R_j)$.

The question we address here (and which will be important in section VI) is this: Does $\rho(x)$ (the solution to (3.2)) take values in $(0,\rho_o)$? It may or may not, depending on m and λ.

<u>Example</u>: Suppose $dm(x) = g(x)\, dx$ with $g(x) \in (0,\rho_o)$ and $\int g = Z < \infty$. Then $\rho(x) = g(x)$ satisfies (3.2) with $\lambda = Z$, and thus $\rho(x) \in (0,\rho_o)$. This ρ also clearly minimizes (3.1).

Nevertheless, in some circumstances $\rho \notin (0,\rho_o)$.

<u>Theorem 3.19</u>: Suppose $j'(\rho) = \alpha$ = constant for $\rho \in F \equiv (\rho_1, \rho_o]$

with $0 \leq \rho_1 \leq \rho_0 < \infty$, and $j'(\rho) > c\,\rho^{\frac{1}{2}+\varepsilon}$ for large ρ. Let V be given by (3.15) and let A be a bounded open set such that as distributions on A either $\rho_0\,dx < dm < (\rho_0 + \text{const.})dx$ or $dm < \rho_1 dx$. Let $\rho \in \mathcal{C}$ satisfy (3.2). Then $\rho(x) \notin F$ a.e. (with respect to Lebesgue measure) on A.

Proof: Cf. (Benguria, 1979, Lemmas 2.19, 3.2). First, it can be shown that $\phi_\rho \in H^2(A)$ (Sobolev space). Let $B = \{x | \rho(x) \in F\} \cap A$. On B, $\phi_\rho - \mu = \alpha$ and since $\phi_\rho \in H^2(A)$, $\Delta\phi_\rho = 0$ a.e. on B. But $\Delta\phi_\rho/4\pi = \rho - dm/dx$. ∎

Remark: Since a solution to (3.2) is determined only a.e., $\rho(x)$ can be chosen $\notin F$ for all $x \in A$.

Corollary 3.20: Consider the TFD problem (3.14) with $V(x) = \Sigma\, z_j\, |x-R_j|^{-1}$. Then any solution to (3.2) can be modified on a set of measure zero so that $\rho(x) \notin (0,\rho_0]$ for all x.

III.C. No-Binding Theorems

Henceforth it will be assumed, as in Theorem 3.18, that j is such that (3.1) has a minimum for $\lambda \leq \lambda_c$ which satisfies (3.2). We will be interested in comparing three (non-zero) potentials, V_1, V_2 and $V_{12} = V_1 + V_2$ with $V_i \in \mathcal{D}$. At first we will consider what happens when the repulsion U is absent. As usual we define $e_a(\lambda) \equiv \inf \mathcal{E}_a(\rho)$ with $\lambda = \int\rho$ and $\mathcal{E}_a$ has V_a. There is no U term in $\mathcal{E}_a$, (3.1). Define

$$\Delta e(\lambda) = e_{12}(\lambda) - \min_{\lambda_1+\lambda_2=\lambda} e_1(\lambda_1) + e_2(\lambda_2) \tag{3.16}$$

Definition: If $\Delta e < 0$ (resp. ≥ 0) we say that in the absence of the repulsion, U, there is binding (resp. no binding).

Theorem 3.21: Suppose j satisfies

$$j(a+b) \leq j(a) + j(b) + a\,j'(b) + b\,j'(a),\ a,b \geq 0 \tag{3.17}$$

[If j' is subadditive then (3.17) is satisfied. $j(t) = t^{5/3}$ satisfies (3.17).] Then $\Delta e < 0$.

Proof: For $i=1,2$ let λ_i minimize in (3.16) and let ρ_i be the minimizing ρ for $\mathcal{E}_i$ with $\int\rho_i \leq \lambda_i$. Recall $e_a(\lambda)$ is monotone non-increasing. Let $\rho \equiv \rho_1 + \rho_2$ be a trial function for e_{12} in $\mathcal{E}_{12}$ and use the variational equations (3.2) for ρ_i and the fact that $\phi_i(x) \geq 0$. ∎

Remark: The condition (3.17) is satisfied in TF theory but not in TFD theory.

Theorem 3.21 says we can (and do if j satisfies (3.17)) have binding if the repulsion U is absent. The no-binding theorem, which we turn to now, relies on the addition of U which, by itself without $\mathcal{E}$, obviously has the no-binding property.

Proposition 3.22: If j is convex and $j(0) = 0$, then j has the superadditivity property: $j(a+b) \geq j(a) + j(b)$. If j' is strictly monotone, then the foregoing inequality is strict when $a,b \neq 0$.

Note: We assumed that j is convex in all cases. Therefore Theorem 3.23 holds in all cases.

Definition: Let $V_i = \frac{1}{|x|} * m_i (m_i$ a measure) be in $\mathcal{D}$. Then $D(m_1,m_2) \equiv \frac{1}{2}\int\int dm_1(x)\, dm_2(y)\, |x-y|^{-1}$.

Theorem 3.23 (no binding): Let m_i, $i=1,2$ be nonnegative measures of finite mass $z_i > 0$ and $V_i \in \mathcal{D}$. Then

$$\Delta E(\lambda) \equiv \Delta e(\lambda) + 2D(m_1,m_2) \geq 0 \;. \tag{3.18}$$

If j is strictly superadditive then >0 holds.

Remarks: Obviously $\Delta E(\lambda)$ is the energy difference when the repulsion U is included. Binding never occurs. In particular, if

$$V_1 = \sum_{j=1}^{n} z_j\,|x-R_j| \quad \text{and} \quad V_2 = \sum_{j=n+1}^{k} z_j\,|x-R_j|^{-1} \quad \text{then}$$

$$m_1 = \sum_{j=1}^{n} z_j\,\delta(x-R_j) \quad \text{and} \quad m_2 = \sum_{j=n+1}^{k} z_j\,\delta(x-R_j).$$ In TFD theory j is not strictly superadditive. As we will see in §IV.C, it is possible to have a neutral diatomic molecule for which equality holds in (3.18).

Proof: We give two proofs. The LS proof in the neutral case $\lambda = z_1 + z_2$ is the following: Clearly $\lambda_1 = z_1$, $\lambda_2 = z_2$, $\mu_1 = \mu_2 = \mu_{12} = 0$. Consider $m_1 \to \alpha m_1$, $\lambda_1 \to \alpha z_1$, $0 \leq \alpha \leq 1$. By Theorem 2.13 we have

$\partial e_1/\partial\alpha = -\int V_1\rho_1$ and $\partial e_{12}/\partial\alpha = -\int V_1\rho_{12}$. Thus

$\partial(e_{12} - e_1 + 2D(\alpha m_1,m_2))/\partial\alpha = \int dm_1(x)[\phi_{12}(x) - \phi_1(x)]$. But

$\phi_{12}(x) \geq \phi_1(x)$, all x (Theorem 3.4 and following remark). When

$\alpha = 0$ $\Delta E = 0$, so this proves the theorem. In the non-neutral case the $\mu_a \neq 0$ and it is necessary to take into account the change of μ_a with α. This is complicated (cf. LS).

The second proof is due to Baxter (1980). For any ρ_{12} with $\int\rho_{12} = \lambda$ we can, by Prop. 3.24, find g, $0 \le g(x) \le \rho_{12}(x)$, and $h(x) \equiv \rho_{12}(x) - g(x)$ such that $\psi_g(x) = \psi_{m_1}(x) = V_1(x)$. a.e. when $h(x) > 0$ and $\psi_g(x) \le V_1(x)$ a.e. when $h(x) = 0$.

Let $a = \int g$, $b = \int h$. Then

$$\begin{aligned}\min\{e_1(\lambda_1) + e_2(\lambda_2) \mid \lambda_1 + \lambda_2 = \lambda\} &\le e_1(a) + e_2(b)\\ \le \mathcal{E}_1(g) + \mathcal{E}_2(h) &\le \mathcal{E}_{12}(\rho_{12}) + 2D(m_1, m_2)\\ + \int h(V_1 - \psi_g)dx - \int (V_1 - \psi_g)dM_2 &\le e_{12}(\lambda) + 2D(m_1\ M_2).\end{aligned} \tag{3.19}$$

The third inequality uses the superadditivity of j. If j' is strictly monotone this superadditivity is strict (and so is the final inequality) provided $g \not\equiv \rho_{12}$. If $g = \rho_{12}$ a.e. then $\psi_{\rho_{12}} \le V_1$ and hence $\lambda \le z_1$ must hold. Choose $\lambda_1 = \lambda$, $\lambda_2 = 0$ and note that $e_1(\lambda) < \mathcal{E}_1(\rho_{12})$ because ρ_{12} does not satisfy (3.2) since $V_2 \not\equiv 0$. (3.19) then gives strict inequality. ■

<u>Proposition 3.24 (Baxter, 1980):</u> Let $V \in \mathcal{D}$ and let $\rho(x) \ge 0$ be a given function with $|x|^{-1} * \rho \equiv \psi_\rho \in \mathcal{D}$. Assume $\rho \in L^p$ for some $p > 3/2$ and $D(\rho,\rho) < \infty$. Then there exists g with $0 \le g(x) \le \rho(x)$ such that $\psi_g = |x|^{-1} * g$ satisfies $\psi_g(x) = V(x)$ a.e. when $\rho(x) - g(x) > 0$ and $\psi_g(x) \le V(x)$ a.e.

<u>Proof</u>: Baxter proves this when ρ and g are measures. We give a simpler proof for functions. Consider $\mathcal{E}(g) = D(g,g) - \int Vg$ and $E = \inf\{\mathcal{E}(g) \mid 0 \le g(x) \le \rho(x)\}$. Let g^n be a minimizing sequence. There exists a subsequence that converges weakly in L^p to some g and, by Mazur's theorem, there exists a sequence h^n of convex combinations of the g^n that converges strongly to g in L^p. Then a subsequence of the h^n converges a.e. to g. Clearly, $0 \le h^n(x) \le \rho(x)$. Since $\mathcal{E}(\cdot)$ is convex (this is crucial) $\mathcal{E}(h^n) \to E$ but, by dominated convergence, $\mathcal{E}(h^n) \to \mathcal{E}(g)$. So g minimizes and satisfies (a.e.): $V(x) = \psi_g(x)$ when $0 < g(x) < \rho(x)$; $\psi_g(x) \le V(x)$ when $g(x) = \rho(x)$ and $\rho(x) > 0$; $\psi_g(x) \ge V(x)$ when $g(x) = 0$ and $\rho(x) > 0$. We have to eliminate the possibility $\psi_g(x) - V(x) \equiv f(x) > 0$ when $g(x) = 0$. We claim $\psi_g \in \mathcal{D}$ and hence f is continuous and goes to zero at ∞. Since $g \le \rho$, $\psi_g \le \psi_\rho$ so $\psi_g \to 0$ at infinity. To examine the continuity at $x = 0$, write $\psi_g = h + (\psi_g - h)$ with $h = |x|^{-1} * (\chi g)$ and χ is the characteristic function of the ball $|x| < 1$. Clearly $\psi_g - h$ is continuous at $x = 0$. Moreover, $\chi g \in L^p \cap L^1$ so $h \in \mathcal{D}$ by Lemma 3.1. (It is here that $p > 3/2$ is used). Now, since $f \in \mathcal{D}$, $B = \{x \mid f(x) > 0\}$ is open and, since $x \in B \Rightarrow g(x) = 0$, f is subharmonic on B. But f vanishes on the boundary of B and at infinity, so B is empty. ■

IV. DEPENDENCE OF THE THOMAS-FERMI ENERGY ON THE NUCLEAR COORDINATES

In the previous sections TF theory was analyzed when the nuclear coordinates $\{R_j\}$ are held fixed. The one exception was Teller's Theorem 3.23 which states that the TF energy is greater than the TF energy for isolated atoms (which is the same as the energy when the R_j are infinitely far apart). Here, more detailed information about the dependence of E on the R_j is reviewed.

Note that in this section (and henceforth) E refers to the total energy, (2.11), including the repulsion U. This is crucial

Although several unsolved problems remain, a fairly complete picture will emerge. The principal open problem is to prove the positivity of the pressure (§IV.B) for subneutral molecules, and to prove it for deformations more general than uniform dilation. The results of this section have been proved only for TF theory and it is not known which ones extend to the variants (cf. the discussion of TFD theory in §IV.C).

IV.A. The Many-Body Potentials

The results here are from (Benguria and Lieb, 1978a). As usual, the two-body atomic energy is defined to be the difference between the energy of a diatomic molecule (with nuclear separation R) and the energy of isolated atoms. Teller's theorem states that this is always positive. We will now investigate the k-body energy which can be defined similarly. The 3-body energy will be shown to be negative, the 4-body positive, etc. In all cases, only neutral systems will be considered; in this case there is a unique way to apportion the electron charge among the isolated atoms, namely make them all neutral. An interesting problem is to treat the k-body energy for subneutral systems.

Definitions: When $c = \{c_1, c_2, \ldots, c_k\}$ is a finite subset of the positive integers with $|c| = k$ elements, $E(c)$ denotes the TF energy for a neutral molecule consisting of nuclear charges $z_{c_i} > 0$ located R_{c_i}. $\phi(c,x)$ denotes the TF potential for this molecule. The z's can all be different.

$$\varepsilon(c) = \sum_{b \subseteq c} (-1)^{|b|+|c|} E(b) \tag{4.1}$$

is the $|c|$ body energy for this molecule. Thus, if $c = \{1,2\}$, $|c| = 2$ and the two-body energy is $\varepsilon(1,2) = E(1,2) - E(1) - E(2)$ as explained above. If $c = \{1,2,3\}$, $|c| = 3$ and the 3-body energy is

$$\varepsilon(1,2,3) = E(1,2,3) - [E(1,2) + E(1,3) + E(2,3] + E(1) + E(2) + E(3).$$

$E(1)$, $E(2)$, $E(3)$ are atomic energies, of course, From (4.1)

$$E(c) = \sum_{b \subseteq c} \varepsilon(b) . \tag{4.2}$$

It is worth remarking that the many-body energies (4.1) are defined in terms of the total energy, E. It is equally possible to use $e = E - U$ on the right side of (4.1). e is the electronic contribution to E, so the corresponding ε's would be the electronic contribution to the many-body potential. However, note that U contains only two-body pieces, $z_i z_j |R_i - R_j|^{-1}$. Therefore the two sets of ε's agree whenever $|c| \geq 3$, i.e. the three and higher body ε's are entirely electronic. As far as the two-body energy is concerned, $\varepsilon(1,2)_{\text{total}} > 0$ (Teller) but

$$\varepsilon(1,2)_{\text{electronic}} = \varepsilon(1,2)_{\text{total}} - U(1,2) < 0 \quad \text{(Theorem 3.23)}.$$

In the following $b \subset c$ means b is a subset of c and $b \neq c$.

<u>Theorem 4.1 (sign of the many-body potential)</u>: If c is not empty

$$(-1)^{|c|}\, \varepsilon(c) > 0 .$$

More generally, if $b \subset c$ and either $|c \setminus b| \geq 2$ or else $|b| = 0$ and $|c| > 0$

$$\tilde{E}(b,c) \equiv \sum_{b \subseteq a \subseteq c} (-1)^{|b|+|a|}\, E(a) > 0 .$$

<u>Theorem 4.2 (Remainder Theorem)</u>: If $2 \leq \gamma \leq |c|$ then the sign of

$$E(c) - \sum_{\substack{b \subset c \\ |b| < \gamma}} \varepsilon(b)$$

is $(-1)^{\gamma}$. In other words if, in (4.2), we sum only over the terms smaller than γ-body, the sign of the error is the sign of the first omitted terms.

<u>Theorem 4.3 (Monotonicity of the many-body potential</u>: Suppose that $b \subset c$ and $|b| \geq 2$. Then

$$(-1)^{|b|}\, \varepsilon(b) > (-1)^{|c|}\, \varepsilon(c) .$$

Theorems 4.1 and 4.3 imply, for example,

$$0 > \varepsilon(1,2,3) > -\min\, [\varepsilon(1,2)\ \varepsilon(1,3),\ \varepsilon(2,3)].$$

<u>Theorem 4.4</u>: If $b \subset c$ and c is not empty

$$\tilde{\phi}(b,c,x) \equiv \sum_{b \subseteq a \subseteq c} (-1)^{|a|+|b|} \phi(a,x) < 0 .$$

Partial Proof: Basically Theorems 4.1, 4.2 and 4.3 are corollaries of Theorem 4.4 through the relation, for $j \in c$, $\partial E(c)/\partial z_j = \lim_{x \to R_j} \{\phi(c,x) - z_j|x-R_j|^{-1}\}$, Theorem 2.13. As an illustration we will prove here that $\varepsilon(1,2,3) < 0$; surprisingly, the proof is much more complicated when $|c| > 3$. The proof for $|c| = 3$ only uses that the function $(j')^{-1}$ is convex (cf. (3.1)). The proof for $|c| > 3$ requires that $j(\rho) = \rho^k$ with $3/2 \le k \le 2$. First note that $\varepsilon(1,2,3) = 0$ when $z_3 = 0$. Thus, it suffices to prove that $\partial\varepsilon(1,2,3)/\partial z_3 = F(R_3) < 0$, where $F(x) \equiv \phi(1,2,3,x) - \phi(1,3,x) - \phi(2,3,x) + \phi(3,x)$. Now $\gamma\Delta F = 4\pi[\rho(1,2,3,x) - \rho(1,3,x) - \rho(2,3,x) + \rho(3,x)]$ and $\rho = \phi^{3/2}$. Let $B = \{x|F(x) > 0\}$. F is continuous, so B is open. We claim F is subharmonic on B which implies B is empty. What is needed is the fact that $a - b - c + d \ge 0 \Rightarrow a^{3/2} - b^{3/2} - c^{3/2} + d^{3/2} \ge 0$ under the conditions that $a \ge b \ge d \ge 0$ and $a \ge c \ge d \ge 0$ (Theorem 3.4). But this is an elementary exercise in convex analysis. Finally, as in the strong form of Theorem 3.4, one can prove that F is strictly negative. ■

It is noteworthy that <u>all</u> the many-body potentials fall off at the same rate, R^{-7}. This will be shown in §IV.C.

IV.B. The Positivity of the Pressure

Teller's Theorem 3.23 suggests that the nuclear repulsion dominates the electronic attraction and therefore a molecule in TF theory should be unstable under local as well as global dilations.

Let us fix the nuclear charges $\underline{z} = \{z_1,\ldots,z_k\}$ and move the R_i keeping λ fixed. Under which deformations does E decrease? We can also ask when $e = E - U$, <u>the electronic contribution to the energy</u>, decreases. A natural <u>conjecture</u> is the following: Suppose $R_i \to R_i'$ with $|R_i' - R_j'| \ge |R_i - R_j|$ for every pair i,j. Then

(i) E decreases and e increases.

(ii) Furthermore if $\lambda_1 < \lambda_2$ then the decrease (increase) in $E(e)$ is smaller (larger) for λ_2 than for λ_1. There is one case in which this conjecture can be proved; it is given in Theorem 4.7 due to Benguria (1981).

One interesting case is that of uniform dilation in which each $R_i \to \ell R_i$. For this case we define the <u>pressure</u> and <u>reciprocal</u>

compressibility to be

$$P(\ell) = -(3\ell^2)^{-1} \, dE(\ell)/d\ell \tag{4.3}$$

$$\kappa^{-1} = -(\ell/3) \, dP(\ell)/d\ell \ , \tag{4.4}$$

where $E(\ell)$ is the energy. This definition comes from thinking of the "volume" as proportional to ℓ^3. If $K(\ell)$ is the kinetic energy (2.20) then

$$3\ell^3 \, P(\ell) = E(\ell) + K(\ell).$$

To see this, define $E(\gamma,\ell)$ to be the energy with the parameter γ thought of as a variable (but with λ fixed). Then, by setting $\rho(x,\ell) = \ell^{-3} \, \tilde{\rho}(x/\ell,\ell)$ one easily sees that $E(\gamma,\ell) = \ell^{-1} \, E(\gamma/\ell,1)$ and $K(\gamma,\ell) = \ell^{-1} \, K(\gamma/\ell,1)$. (4.4) follows from this and Theorem 2.13.

Note that (4.4) is true (for the same reason) in Q theory and also in TFD, TFW and TFDW theories provided K is interpreted there as (2.20) in TFD and as (2.20) + $\delta\int[\nabla\rho^{\frac{1}{2}}]^2$ in TFDW and TFW.

That $e = E - U$ increases under dilation has also been conjectured to hold in Q theory when $\lambda \le Z$. It is known to hold for one electron, but an arbitrary number of nuclei (Lieb and Simon, 1978). There is one simple statement that can be made (in all theories): The (unique) minimum of e occurs when $\ell = 0$ (for any $\lambda > 0$), i.e. all the nuclei are at one point. To prove this, assume $R_1,\ldots,R_k$ are not all identical and let ρ be the minimizing solution. Let $\psi = |x|^{-1} * \rho$. ψ has a maximum at some point R_o. Now place all the nuclei at R_o and use the same ρ as a variational ρ for this problem. Then, trivially $e(R_o,\ldots,R_o) < e(R_1,\ldots,R_k)$, with the strict inequality being implied by the fact that this ρ does not satisfy the variational equation for $R_o,\ldots,R_o$.

It is useful to have a formula for the variation of e with R_i. A natural extension of Theorem 2.13 (a "Feynman-Hellman" type theorem) would be the following: Suppose $V_1,\ldots,V_k \in \mathcal{D}$ with

$$V_i(x) = \int dm_i(y) \, |y-x|^{-1} \tag{4.5}$$

and with m_i a positive measure of mass z_i. Take $V(x) = \sum_{i=1}^{k} V_i(x-R_i)$. Then e is a C^1 function of the R_i and

$$\nabla_{R_i} e = \int \nabla V_i(x-R_i)\rho(x)dx = -\int dm_i(y) \, \nabla\psi(y+R_i) \tag{4.6}$$

with $\psi = |x|^{-1} * \rho$. (4.6) is clearly true, and easy to prove if the m_i are suitably bounded. Benguria (unpublished) proved (4.6) when $V_i(x) = z_i \, |x|^{-1}$ for $|x| \ge a$ and $V_i(x) = z_i \, a^{-1}$ for

$|x| \le a$, with $a > 0$. I.e. $dm_i(y) = z_i(\text{const.})\,\delta(|y| - a)$. In this case, the last equality in (4.6) follows from LS, Lemma IV.4.

For point nuclei, on the other hand, (4.6) has not been proved; indeed, the quantities in (4.6) are not even well defined. We conjecture that the following is true when $V_i(x) = z_i|x|^{-1}$: e is a C^1 function of the R_i on the set where $R_i \neq R_j$, for all $i \neq j$, and

$$\nabla_{R_i} e = -z_i \lim_{a\downarrow 0} \int_{|x-R_i|>a} (x-R_i)|x-R_i|^{-3}\,\rho(x)dx \tag{4.7a}$$

$$= -\lim_{x\to R_i} \nabla_x \{\psi(x) + (z_i/\gamma)^{3/2}\,(16\pi/3)|x-R_i|^{\frac{1}{2}}\}. \tag{4.7b}$$

(4.7a) makes sense because, by Theorem 2.8, $\rho(x) = (z_i/\gamma)^{3/2}|x-R_i|^{-3/2} + O(|x-R_i|^{-\frac{1}{2}})$ near R_i; the angular integration over the first term vanishes. This leading term in ρ implies that near R_i, $\psi(x) \approx (\text{const.}) - (z_i/\gamma)^{3/2}(16\pi/3)|x-R_i|^{\frac{1}{2}}$. The non-differentiable, but spherically symmetric term in ψ is subtracted in (4.7b).

The following theorems have been proved so far. (Theorems 4.5 and 4.6 are in (Benguria and Lieb, 1978b); Theorem 4.7 is in (Benguria, 1981).)

Theorem 4.5 (uniform dilation): Replace each R_i by $\ell\, R_i$ and call the energy $E(\lambda,\ell)$. If $\lambda = Z$ then $E(\lambda,\ell)$ is strictly monotone decreasing and convex in ℓ. In particular, the pressure and compressibility are positive.

Remarks: (i) If $\lambda = 0$ the conclusion is obviously also true. In (Benguria and Lieb, 1978b) it is conjectured that this theorem holds for all λ. That $e = E - U$ is monotone increasing is also conjectured there.

(ii) In (Benguria and Lieb, 1978b) several interesting subadditivity and convexity properties of the energy and potential are also proved.

Theorem 4.6 (molecule with planar symmetry): Suppose the molecule is symmetric with respect to the plane $P = \{(x^1,x^2,x^3)\,|\,x^1 = 0\}$ and suppose no nucleus lies in the plane. Neutrality is not assumed. Let R_i^1 denote the 1 coordinate of nucleus i and, for all i, replace R_i^1 by $R_i^1 \pm \ell$, with $\pm$ if $R_i^1 \gtrless 0$, and $\ell \ge 0$. Then for all fixed $\lambda \le Z$, E is decreasing in ℓ.

Remark: For a homopolar diatomic molecule the dilations in Theorems 4.5, 4.6 are the same. Balàzs (1967) first proved Theorem 4.6 in this case. For a general diatomic molecule, Theorem 4.7 is the strongest theorem.

Theorem 4.7: Suppose there exists a plane P containing $R_1,\ldots,R_m$ and such that all the other R_j (with $j=m+1,\ldots,k$) are on one (open) side of P (call this side P^+). Assume the nuclei at $R_1,\ldots,R_m$ are point nuclei, but the nuclei at $R_{m+1},\ldots,R_k$ are anything in $\mathcal{D}$ and given by (4.5) with the supports of $m_i \in P^+$ (this includes point nuclei). Let $\vec{n}$ be the normal to P pointing away from P^+. Let $\ell_1,\ldots,\ell_m \geq 0$ be given and let $R_i \to R_i + \ell_i\vec{n}$ for $i=1,\ldots,m$. Let $E(\lambda,\underline{\ell})$ denote the energy for fixed $\lambda \leq Z$ and let $\Delta E(\lambda,\underline{\ell}) = E(\lambda,\underline{\ell}) - E(\lambda,0)$ denote the change in energy. Likewise define $\Delta e(\lambda,\underline{\ell}) = \Delta E(\lambda,\underline{\ell}) - \Delta U$. Then

(i) $\Delta e(\lambda,\underline{\ell}) \geq 0$

(ii) $\Delta E(\lambda,\underline{\ell}) \leq 0$

(iii) $\Delta E(\lambda_1,\underline{\ell}) \leq \Delta E(\lambda_2,\underline{\ell})$ if $\lambda_1 \leq \lambda_2$

(iv) $\Delta e(\lambda_1,\underline{\ell}) \leq \Delta e(\lambda_2,\underline{\ell})$ if $\lambda_1 \leq \lambda_2$.

The essential content of Theorem 4.7 is the following. Let $R_1,\ldots,R_k$ be given and let C be their convex hull. C has a surface S which is a (possibly degenerate) polyhedron whose vertices are $R_1,\ldots,R_n$, say. Now consider any displacement of $R_1,\ldots,R_n$ (and not the other R's) which has the property given in the conjecture, namely $|R_i' - R_j'| \geq |R_i - R_j|$ for all i,j. Then the conjectures (i) and (ii) are true for such displacements. In particular, the conjecture holds for molecules such that every atom is a vertex of S.

To prove Theorem 4.7 the following Lemma 4.8, which is of independent interest, is needed.

Lemma 4.8: Assume the plane P, with $R_1,\ldots,R_m$ in P and $R_{m+1},\ldots,R_k$ in P^+ as in Theorem 4.7. However, point nuclei are not assumed. Instead, assume each $V_i \in \mathcal{D}$ and given by (4.5), with m_i required to be spherically symmetric for $i=1,\ldots,m$. This includes point nuclei. Assume also that the support of $m_i \subset P^+$ for $i = m+1,\ldots,k$. If $x \in P^+$ then x^* is defined to be the reflection of x through P. Let ϕ be the potential. For $x \in P^+$, let $\phi_-(x) = \phi(x^*)$ and $f(x) = \phi(x) - \phi_-(x)$. Then

(i) $f(x) > 0$ for $x \in P^+$

(ii) For each $x \in P^+$, $f(x)$ strictly decreases when λ increases.

(iii) $\rho(x) - \rho(x^*) \geq 0$ for $x \in P^+$.

Question: Is it true that $\rho(x) - \rho(x^*)$ is a monotone nonincreasing function of λ ?

Proof: (i) Clearly $f(x) = 0$ on $\partial P^+ = P$ and at ∞. Let $B = \{x \in P^+ | \psi(x) < 0\}$. Since each $V_i(x)$ is symmetric decreasing the singularities of V are not in B. Thus B is open. On B, $-\Delta f(x)/4\pi \geq -\rho(x) + \rho(x^*) > 0$. Thus, f is superharmonic on B so B is empty. By the strong maximum principle $f(x) > 0$, in fact, for $x \in P$.

(ii) Let $\lambda' < \lambda$ with corresponding f' and f. We want to prove $B = \{x \in P^+ | f(x) - f'(x) > 0\}$ is empty. B is open and $f-f'=0$ on P and at ∞. $\Delta(f-f')/4\pi = a_+^{3/2} - b_+^{3/2} - c_+^{3/2} + d_+^{3/2} \equiv h$, where $a = \phi - \mu$, $b = \phi' - \mu'$, $c = \phi_- - \mu$, $d = \phi_-' - \mu'$. By (i) and Corollary 3.8, $a \geq b > d$ and $a > c \geq d$ for all $x \in P^+$. In B, $a + d > b + c$. Thus, $h \geq 0$ in B, whence $f-f'$ is subharmonic on B and hence B is empty. Again, one can prove the stronger result that $f-f' < 0$ for $x \in P$. Trivially, (i) $\Rightarrow$ (iii) through the TF equation. ■

Proof of Theorem 4.7: We may assume all the ℓ_i are equal to some common ℓ, for otherwise if $\ell_1 \leq \ell_2 \leq \ldots \leq \ell_m$ we can first move all the m nuclei by ℓ_1, then move $R_2,\ldots,R_m$ by $\ell_2 - \ell_1$, etc. Next, replace all the point nuclei at $R_1,\ldots,R_m$ by smeared potentials given by (4.5) with $dm_i(x) = z_i\, g^{(n)}(x)dx$ where $g^{(n)}(x) \in C_o^\infty$ and $g^{(n)}$ is symmetric decreasing and with sufficiently small support such that the supports of $dm_i (i=1,\ldots,m)$ are pairwise disjoint and also disjoint from the supports of $dm_i (i=m+1,\ldots,k)$. Under these conditions, e is C^1 in $R_1,\ldots,R_m$ in some neighborhood of the original $R_1,\ldots,R_m$ with derivatives given by (4.6). We will prove

(i)' $\vec{n} \cdot \nabla_{R_i} e \geq 0$,

(ii)' $\vec{n} \cdot \nabla_{R_i} E \leq 0$, and that (iii) and (iv) hold for these

derivatives. Then the theorem is proved because the original point potentials $z_i|x|^{-1}$ can be approximated in $L^{5/2}$ norm by these smeared potentials $z_i|x|^{-1} * g^{(n)}$, and the energies $e^{(n)}$ and $E^{(n)}$ converge to e and E by LS, Theorem II.15. If (i)' holds for $e^{(n)}$, then $(d/d\ell)\, e^{(n)}(\lambda,\ell) \geq 0$ with $R_i \to R_i + \ell\vec{n}$, $i=1,\ldots,m$ and, by integration, (i) holds for $e^{(n)}$. Then, when $n \to \infty$, (i) holds for e. The same applies to (ii) - (iv). Henceforth, the superscript (n) will be suppressed.

Assume $\vec{n} = (1,0,0)$, $P = \{x | x_1 = 0\}$ and thus $(R_i)_1 = 0$ for $i=1,\ldots,m$. Since g is symmetric decreasing

$(\partial g/\partial x_1)(x_1,x_2,x_3) = -x_1\, h(x_1,x_2,x_3)$ with $h(x) \geq 0$ and

$h(x_1,x_2,x_3) = h(-x_1,x_2,x_3)$. Likewise,

$(\partial V_i/\partial x_1)(x_1,x_2,x_3) = -z_i x_1 p(x_1,x_2,x_3)$ and p has the same properties as h. To prove (i)' use (4.6) whence

$\vec{n}\cdot\nabla_{R_i} e/z_i = -\int x_1\, p(x-R_i)\rho(x)dx = -\int_{x_1\le 0} p(x-R_i)[\rho(x)-\rho(x^*)]\, x_1 dx \ge 0$ by Lemma 4.8. To prove (ii)' use the second integral in (4.6), whence $B_i \equiv \vec{n}\cdot\nabla_{e_i} E = \int dm_i(y)\, \vec{n}\cdot\nabla\phi(y+R_i)$, where ϕ is the potential. [Note: $V_i(x-R_i)$ is symmetric in x about R_i so the term $\nabla V_i(x-R_i)$ does not contribute to this integral.] Since V_i is C^∞ it is easy to see that ϕ is also C^∞ near R_i. Now integrate by parts:

$$B_i = -\int \vec{n}\cdot\nabla g(y)\, \phi(y+R_i)\, dy = \int y_1\, h(y)\, \phi(y+R_i)\, dy$$

$$= \int_{y_1\le 0} y_1\, h(y)[\phi(y+R_i) - \phi_-(y+R_i)] \le 0$$

by Lemma 4.8. To prove (iii) note that the last quantity [ ] decreases when λ increases by Lemma 4.8. Clearly (iii) is equivalent to (iv). ■

Proof of Theorem 4.6: Let $\rho(x)$ be the density when $\ell = 0$. For $\ell > 0$ use the variational $\tilde\rho$ given by $\tilde\rho(x^1,x^2,x^3) = \rho(x^1 \mp \ell, x^2, x^3)$ if $x^1 \gtrless \ell$ and $\tilde\rho(x) = 0$ otherwise. Then all terms in the energy $\mathcal{E}(\tilde\rho)$ remain the same except for the Coulomb interaction of the two charge distributions on either side of the plane, P. This term is of the form

$$W(\ell) = \int_{x^1,y^1\ge 0} d^3x d^3y\, f(x)f(y)[(x^1+y^1+2\ell)^2 + (x^2-y^2)^2 + (x^3-y^3)^2]^{-\frac{1}{2}}$$

where $f(x) = -\rho(x) + \Sigma' z_j\, \delta(x-R_j)$ and the Σ' is over those R_i with $R_i^{\,1} > 0$. Since the Coulomb potential is reflection positive (Benguria and Lieb, 1978, Lemma B.2), $W(\ell)$ is a decreasing, log convex function of ℓ. ■

Proof of Theorem 4.5: Let $\underline{z} = (z_1,\ldots,z_k)$ and write $E(\underline{z}), K(\underline{z}), A(\underline{z})$ and $R(\underline{z})$ for the energy and its components (cf. §II.E) of a neutral molecule. These functions are defined on $\mathbb{R}_+^k$. For an atom $3P = E + K = 0$ (Theorem 2.14). By Theorem 3.23, $E \ge \Sigma_1^k E^{atom}(z_j)$ and, by Theorem 4.10, $K \ge \Sigma_1^k K^{atom}(z_j)$. This shows $P \ge 0$. Likewise, by Theorem 4.12, $\kappa^{-1} \ge 0$ and $E(\underline{z},\ell)$ is convex in ℓ (equivalently $\ell^2 P$ is decreasing in ℓ). ■

Definition: Let f be a real valued function on $\mathbb{R}_+^k$ and $\underline{z}_1, \underline{z}_2, \underline{z}_3 \in \mathbb{R}_+^k$. Then f is

(i) weakly superadditive (WSA) $\Leftrightarrow f(\underline{z}_1 + \underline{z}_2) \ge f(\underline{z}_1) + f(\underline{z}_2)$ whenever $(z_1)_i\,(z_2)_i = 0$, all i

(ii) superadditive (SA) $\Leftrightarrow f(\underline{z}_1 + \underline{z}_2) \ge f(\underline{z}_1) + f(\underline{z}_2)$

(iii) strongly superadditive (SSA) $\Leftrightarrow f(\underline{z}_1 + \underline{z}_2 + \underline{z}_3) + f(\underline{z}_1) \ge f(\underline{z}_1 + \underline{z}_2) + f(\underline{z}_1 + \underline{z}_3)$.

Theorems 4.9-4.12 are for neutral molecules.

Theorem 4.9: As a function of $\underline{z} \in \mathbb{R}_+^k$, for each fixed $x \in \mathbb{R}^3$,

(i) $-\phi(\underline{z},x)$ is SSA, convex and decreasing (the latter is Teller's lemma)

(ii) $\phi(\underline{z},x) \in C^1(\mathbb{R}_+^k)$ and $\in C^2(\mathbb{R}_+^k \setminus 0)$

(iii) $\phi_i(\underline{z},x)$ is decreasing in $\underline{z}$ and > 0

(iv) $\phi_{ij}(\underline{z},x) \leq 0$ (all i,j) and is negative semidefinite as a $k \times k$ matrix.

Theorem 4.10: $K(\underline{z}) \in C^1(\mathbb{R}_+^k)$ and $\in C^2(\mathbb{R}_+^k \setminus 0)$

(i) $K_i(\underline{z}) = 3 \lim_{x \to R_i} \{\phi(\underline{z},x) - \sum_{j=1}^{k} z_j \phi_j(\underline{z},x)\}$

(ii) $K_{ij}(\underline{z}) = -3 \sum_{p=1}^{k} z_p \phi_{ij}(\underline{z}, R_p)$

(iii) $K(\underline{z})$, $R(\underline{z})$ and $A(\underline{z})$ are SSA and SA and convex.

(iv) $E(\underline{z})$ is WSA (Teller's Theorem)

Definition $X(\underline{z}) \equiv 3K(\underline{z}) - \sum_{i=1}^{k} z_i K_i(\underline{z})$

Theorem 4.11: $X(\underline{z})$ is SA and SSA and ray convex.
I.e. $X(\lambda \underline{z}_1 + (1-\lambda)\underline{z}_2) \leq \lambda X(\underline{z}_1) + (1-\lambda) X(\underline{z}_2)$, $0 \leq \lambda \leq 1$, when $\underline{z}_1, \underline{z}_2 \in \mathbb{R}_+^k$ and either $\underline{z}_1 - \underline{z}_2$ or $\underline{z}_2 - \underline{z}_1 \in \mathbb{R}_+^k$.

Theorem 4.12: (i) $3\ell^3 P = E + K$

(ii) $9\ell^3 \kappa^{-1} = 6\ell^3 P + 2E + 3X$

(iii) P and κ^{-1} are WSA and nonnegative

(iv) $\ell^2 P$ is decreasing in ℓ. Equivalently, E is convex in ℓ. Equivalently, $2E + 3X \geq 0$
(note: $\partial(\ell^2 P)/\partial \ell = 2\ell P - 3\ell\kappa^{-1} = -(1/3)(2E + 3X)$).

Proof of (iv): $2E + 3X = 0$ for an atom. By Theorem 4.10, $2E + 3X \geq 0$. ∎

The proofs of Theorems 4.9-4.12 are complicated. However if

all necessary derivatives are assumed to exist then an easy heuristic proof can be given (cf. (Benguria and Lieb, 1978b)). We illustrate this for K being SSA, which will then prove $P \geq 0$. Since $K(0) = 0$ this is equivalent to $K_{ij} \geq 0$, all i,j. First we show $\phi_{ij} \leq 0$ and then Theorem 4.10 (ii).

Differentiate the TF differential equation $(\Delta\phi/4\pi = -\Sigma z_j \delta(x-R_j) + (\phi/\gamma)^{3/2}$, which holds for any neutral system) with respect to z_i and then z_j:

$$\mathcal{L}\phi_i = \delta(x-R_i) \tag{4.8}$$

$$\mathcal{L}\phi_{ij} = -(3/4\,\gamma^{3/2})\,\phi^{-1/2}\,\phi_i\,\phi_j \tag{4.9}$$

with $\mathcal{L} = -\Delta/4\pi + (3\,\gamma^{-3/2}/2)\,\phi(x)^{1/2}$. The kernel for $\mathcal{L}^{-1}$ is a positive function, so $\phi_i \geq 0$. Likewise $\phi_{ij} \leq 0$ and ϕ_{ij} is a negative semi-definite matrix.

Next, $K = (3\gamma^{-3/2}/5)\int\phi^{5/2}$, so $K_{ij} = (3\gamma^{-3/2}/2)\{\int\phi^{3/2}\,\phi_{ij} + (3/2)\int\phi^{1/2}\,\phi_i\,\phi_j\}$. Using (4.9) and integrating by parts

$$K_{ij} = 3\int\phi_{ij}[\Delta\phi/4\pi - (\phi/\gamma)^{3/2}]$$

$$= -3\sum_{p=1}^{k} z_p\,\phi_{ij}(R_p) \geq 0.$$

IV.C. The Long Range Interaction of Atoms

In §IV.B it was shown that the energy of a molecule decreases monotonically under dilation (at least for neutral molecules). If the $R_i \to \ell R_i$ then, for small ℓ, E is dominated by U so $E \approx \ell^{-1}$. To complete the picture it is necessary to know what happens for large ℓ. We define

$$\Delta E = E^{\text{molecule}} - \sum_{j=1}^{k} E^{\text{atom}}. \tag{4.10}$$

For large ℓ it is reasonable to consider only neutral molecules, for otherwise $\Delta E \approx \ell^{-1}$ because of the unscreened Coulomb interaction. In the neutral case $\Delta E \approx \ell^{-7}$, as proved by Brezis and Lieb (1979). This result (ℓ^{-7}) is not easy to ascertain numerically (Lee, Longmire and Rosenbluth, 1974), so once again the importance of pure analysis in the field is demonstrated. Some heuristic remarks about the result are given at the end of this section.

A surprising result is that all the many-body potentials are

$\approx \ell^{-7}$. Thus, in TF theory it is not true that the interaction of atoms may be approximated purely by pair potentials at large distances.

An interesting open problem is to find the long range interaction of polyatomic molecules of fixed shape. Presumably this is also $\approx \ell^{-7}$.

Theorem 4.13: For a neutral molecule, let the nuclear coordinates be ℓR_i with $\{R_i, z_i\} = (\underline{R},\underline{z})$ fixed and $z_i > 0$. Then

$$\Delta E(\ell,\underline{z},\underline{R}) \equiv \ell^{-7}\, C(\ell,\underline{z},\underline{R})$$

where C is increasing in ℓ and has a finite limit, $\Gamma(\underline{R}) > 0$ as $\ell \to \infty$. Γ is independent of $\underline{z}$. Furthermore, if A denotes a subset of the nuclei (with coordinates $\underline{R}_A$) and $\varepsilon(A)$ is the many body potential of (4.1) then, by (4.1), for $|A| \geq 2$

$$\ell^7\, \varepsilon(A) \to \sum_{B \subseteq A} (-1)^{|A|-|B|}\, \Gamma(\underline{R}_B) \tag{4.11}$$

and the right side of (4.11) is strictly positive (negative) if $|A|$ is even (odd).

Proof of first part: By scaling (2.24), $\Delta E(\ell,\underline{z},\underline{R}) = \ell^{-7}\{E(\ell^3\,\underline{z},\underline{R}) - \Sigma_j\, E^{atom}\,(\ell^3\, z_j)\}$. Therefore, C increasing is equivalent to $f = E^{mol} - \Sigma\, E^{atom}$ increasing in $\underline{z}$. But $\partial f/\partial z_j = \lim_{x\to R_j} \phi^{mol}\,(x) - \phi^{atom}\,(x)$, and this is positive by Teller's lemma. All that has to be checked is that C is bounded above. This is done by means of a variational ρ for E^{mol}. Let B_i be a ball of radius ℓr_i centered at ℓR_i; the r_i are chosen so that the B_i are disjoint. Let $\rho_i(x) = \rho^{at}\,(x - \ell R_i)$ be the TF atomic densities for z_i, and let $\rho(x) = \rho_i(x)$ in B_i and $\rho(x) = 0$ otherwise. Of course $\int\rho < \Sigma z_j$ but this is immaterial for a variational calculation since the minimum molecular energy occurs when $\int\rho = \Sigma\, z_j$. It is easy to check that $f <$ (const.) ℓ^{-7}. Finally, since f is monotone in each z_i, $\lim_{\ell\to\infty} f$ must be independent of the z_j. ■

Remarks: (i) The variational calculation shows clearly why Γ is independent of the z_j. The long range interaction comes, in some sense, from the tails of the atomic ρ's, but these tails are independent of z, namely $\rho(x) \approx (3\gamma/\pi)^3\, |x|^{-6}$.

(ii) At first sight it might appear counter-intuitive that the interaction is $+\ell^{-7}$ and not $-\ell^{-6}$, as would be obtained from a dipole-dipole interaction. The following heuristic remark might be useful in this respect. Consider two neutral atoms separated by a

large distance R. In the quantum theory, as in all the theories discussed in this paper, there is no static polarization of the atoms; i.e. there is no polarization of the single particle density, ρ. TF theory is therefore correct as far as the density is concerned. The reason there is no polarization is that the formation of a dipole moment, d, increases the atomic energy by $+\alpha d^2$, with $\alpha > 0$. The dipole-dipole energy gain is $-(\text{const.})d^2 R^{-3}$. Hence, if R is large enough the formation of dipoles does not decrease the energy. In quantum theory there is, in fact, a $-R^{-6}$ dipolar energy but this effect is a correlation, and not a static effect. There are two ways to view it. In second order perturbation theory there are virtual transitions to excited, polarized states. Alternatively, the electrons in each atom are correlated so that they go around their respective atoms in phase, but spherically symmetrically. This correlated motion increases the internal atomic energy only by αd^4, not d^2. In short, the $-R^{-6}$ interaction arises from the fact that the density, ρ, is not that of a structureless "fluid" but is the average density of many separate particles which can be correlated. This fact poses a serious problem for any "density functional approach". It is necessary to predict a $-R^{-6}$ dipolar interaction, yet predict zero static polarization.

An explicit formula for $\Gamma(\underline{R})$ does not seem to be easy to obtain. Two not very explicit formulas are given in (Brezis and Lieb, 1979). One is simply to integrate the formula for $\partial f/\partial \ell = 3\ell^2 \Sigma z_j \, \partial f/\partial z_j$ given in the above proof. Another is obtained by noting that Γ is related to ϕ in the limit $z \to \infty$. This limiting ϕ can be defined and satisfies the TF differential equation, but with a strong singularity at R_i instead of the usual $z|x-R_i|^{-1}$ singularity. As we saw in Theorem 2.11, the only other singularity allowed for the TF equation is $\phi(x) \approx \gamma^3(3/\pi)^2|x-R_i|^{-4}$. Therefore that peculiar solution to the TF equation does have physical interest; it is related to the asymptotic behavior of the interatomic interaction.

TFD theory: Here the interaction for large ℓ is precisely zero and not ℓ^{-7}. To be precise, $\Delta E = 0$ when the spacing between each pair $|R_i - R_j|$ exceeds a critical length, $L(z_i) + L(z_j)$. The same is a-fortiori true for the many-body potentials, ε.

The reason is the following. In TFD theory an atomic ρ has compact support, namely a ball of radius $L(z)$. Cf. Theorem 6.6. When $|R_i-R_j| > L(z_i) + L(z_j)$, then $\rho(x) = \sum_j \rho(x-R_j; z_j)$ where $\rho(\cdot;\cdot)$ is the TFD atomic ρ. Since each atom is neutral, there is then no residual interaction, by Newton's theorem. One may question whether the ρ just defined is correct. It is trivial to check that it satisfies the TFD equation and, since the solution is unique, this must be the correct ρ.

V. THOMAS-FERMI THEORY AS THE $Z \to \infty$ LIMIT OF QUANTUM THEORY

Our goal in this section is to show that TF theory is the $Z \to \infty$ limit of Q theory and that it correctly describes the cores of heavy atoms. This is the perspective from which to view TF theory, and in this light it is seen to be a cornerstone of many body theory, just as the theory of the hydrogen atom is an opposite cornerstone useful for thinking about light atoms. We shall not review the stability of matter question here (cf. (Lieb, 1976)).

In units in which $\hbar^2/2m = 1$ and $|e| = 1$ the Hamiltonian for N electrons is

$$H_N = \sum_{i=1}^{N} \{-\Delta_i + V(x_i)\} + \sum_{1 \le i < j \le N} |x_i - x_j|^{-1} + U \tag{5.1}$$

E_N, $\rho_N(x)$ and μ will denote the TF energy, ρ and μ corresponding to this problem with $\lambda = N$ electrons if $N \le Z = \sum_{j=1}^{k} z_j$. Of course, γ is taken to be γ_p (cf. (2.6)). If $N > Z$ then these quantities are defined to be the corresponding TF quantities for $N = Z$. E_N^Q denotes the ground state energy of H_N (defined to be $\inf \operatorname{spec} H_N$) on the physical Hilbert space $\mathcal{H}_N = \bigwedge_1^N L^2(\mathbb{R}^3; \mathbb{C}^q)$ (antisymmetric tensor product). q is the number of spin states (= 2 for electrons), but it is convenient to have it arbitrary, but fixed. The TF quantities also depend on q. through γ_p.

V.A. The $Z \to \infty$ Limit for the Energy and Density

Let us first concentrate on the energy; later on we will investigate the meaning of $\rho(x)$. For simplicity the number of nuclei is fixed to be k; it is possible to derive theorems similar to the following if $k \to \infty$ in a suitable way (e.g. a solid with periodically arranged nuclei) but we shall not do so here. In TF theory the relevant scale length is $Z^{-1/3}$ and therefore we shall consider the following limit:

Fix $\{\underline{z}^\circ, \underline{R}^\circ\} = \{z_j^\circ, R_j^\circ\}_{j=1}^{k}$ and $\lambda > 0$. For each $N = 1,2,\ldots$ define a_N by $\lambda a_N = N$ and in H_N replace z_j by $a_N z_j^\circ$ and R_j by $a_N^{-1/3} R_j^\circ$. Thus $\lambda = N/\Sigma z_j$, and a_N is the scale parameter. The TF quantities scale as (2.24)

$$E_{\lambda a}(a\underline{z}^\circ, a^{-1/3}\underline{R}^\circ) = a^{7/3} E_\lambda(\underline{z}^\circ, \underline{R}^\circ)$$
$$\rho_{\lambda a}(a^{-1/3}x, a\underline{z}^\circ, a^{-1/3}\underline{R}^\circ) = a^2 \rho_\lambda(x, \underline{z}^\circ, \underline{R}^\circ) \tag{5.2}$$

In this limit the nuclear spacing decreases as $a_N^{-1/3} \sim N^{-1/3} \sim Z^{-1/3}$. This should be viewed as a refinement rather than as a necessity. If instead the R_j are fixed $= R_j^\circ$, then in the limit one has isolated atoms. The only thing that really matters are the limits $N^{1/3}|R_i - R_j|$.

Theorem 5.1: With $N = \lambda a_N$ as above

$$\lim_{N\to\infty} a_N^{-7/3} E_N^Q(a_N \underline{z}^\circ, a_N^{-1/3} \underline{R}^\circ) = E_\lambda(\underline{z}^\circ, \underline{R}^\circ)$$

The proof is via upper and lower bounds for E_N^Q. The upper bound is greater than the Hartree-Fock energy, which therefore proves that Hartree-Fock theory is correct to the order we are considering, namely $N^{7/3}$.

Upper Bound for E_N^Q: The original LS proof used a variational calculation with a determinantal wave function; this is cumbersome. Baumgartner (1976) gave a simpler proof (both upper and lower bounds) which intrinsically relied on the same Dirichlet-Newmann bracketing ideas as in LS. Here, we will give a new upper bound (Lieb, 1981a) that uses coherent states; these will also be very useful for obtaining a lower bound.

Let $y = (x,\sigma)$ denote a single space-spin pair and $\int dy \equiv \sum_{\sigma=1}^{q} \int dx$. Let $K(y,y')$ be any admissible single particle density matrix for N fermions, namely $0 \le K \le I$ (as an operator on $L^2(\mathbb{R}^3; \mathbb{C}^q)$) and $\mathrm{Tr}\, K = N$. Let h be the single particle operator $-\Delta + V(x)$. Then (Lieb, 1981a)

$$E_N^Q \le E_N^{HF} \le \tilde{E}(K) \quad \text{with} \tag{5.3}$$

$$\tilde{E}(K) = \mathrm{Tr}\, Kh + \tfrac{1}{2}\int\int dy\, dy' |x-x'|^{-1}\{K(y,y)K(y',y') - |K(y,y')|^2\}. \tag{5.4}$$

In (5.3), E_N^{HF} is the Hartree-Fock energy. Since $|x-x'|^{-1}$ is positive we can drop the "exchange term", $-|K|^2$, in (5.4) for the purposes of an upper bound.

First, suppose $N \le Z$. To construct K, let $g(x)$ be any function on $\mathbb{R}^3$ such that $\int |g|^2 = 1$ and let $M(p,r)$ be any function on $\mathbb{R}^3 \times \mathbb{R}^3$ such that $0 \le M(p,r) \le 1$ and $(2\pi)^{-3} \int M\, dp dr = N/q$. Then the coherent states in $L^2(\mathbb{R}^3)$ which we will use are

$$f_{pr}(x) = g(x-r) \exp[ip \cdot r] \tag{5.5}$$

and

$$K(y,y') = I_\sigma (2\pi)^{-3} \int dpdr\ g(x-r)g(x'-r)M(p,r)\exp[ip\cdot(r-r')]. \tag{5.6}$$

I_σ is the identity operator in spin space. It is easy to check that $\text{Tr}\, K = N$ and that for any normalized ϕ in L^2, $(\phi, K\phi) \le 1$ by using Parseval's theorem and the properties of g and M. Thus, K is admissible.

We choose (with $\rho = \rho_{\min(N,Z)}$ in (5.7) - (5.26))

$$M(p,r) = \theta(\gamma_p\ \rho(r)^{2/3} - p^2) \tag{5.7}$$

where $\theta(t) = 1$ if $t \ge 0$ and $\theta(t) = 0$ otherwise. γ_p is given in (2.6). One easily computes

$$K(y,y) = q^{-1}\ I_\sigma\ \rho_g(x) \tag{5.8}$$

$$\text{Tr}(-\Delta)K = (3\gamma_p/5) \int \rho(x)^{5/3} dx + N \int |\nabla g(x)|^2\ dx \tag{5.9}$$

$$\text{Tr}\ VK = \int V_g(x)\ \rho(x)\ dx \tag{5.10}$$

where $\rho_g = |g|^2 * \rho$ and $V_g = V * |g|^2$.

For $g(x)$ we choose

$$g(x) = (2\pi R)^{-\frac{1}{2}} |x|^{-1} \sin(\pi |x|/R) \tag{5.11}$$

for $|x| \le R$, and $g = 0$ otherwise, and with $R = N^{2/5}\ Z^{-1}$. Then $\int |\nabla g|^2 = \pi^2/R^2 = \pi^2\ Z^2\ N^{-4/5}$.

The electron-electron interaction term in (5.4) is less than $D(\rho,\rho)$ because, as an operator, $[|g|^2 * |x|^{-1} * |g|^2](x - x') < |x - x'|^{-1}$. To see this, use Fourier transforms. Thus

$$E_N^Q \le \tilde{E}(K) \le E_N + \pi^2\ N^{1/5}\ Z^2 + \int [V(x) - V_g(x)]\ \rho(x) dx. \tag{5.12}$$

To bound the last term in (5.12) note that, by Newton's theorem, $|x|^{-1} - |g|^2 * |x|^{-1} = 0$ for $x \ge R$. Furthermore, with the scaling we have employed, $|R_i - R_j| > 2R$ for all $i \ne j$ and N large enough. Since $\gamma_p \rho^{2/3}(x) < V(x)$, for sufficiently large N and for $|x - R_i| \le R$ we have $\gamma_p\ \rho(x)^{2/3} < 2z_i\ |x-R_i|^{-1}$. Thus, the last integral in (5.12) is bounded above for large N by

$$2\gamma_p^{-3/2} \sum_{j=1}^{k} z_j^{5/2}\ A \quad \text{with}$$

$$A = \int_{|x|\leq R} |x|^{-5/2}\, dx = 8\pi R^{\frac{1}{2}} = 8\pi N^{1/5} Z^{-\frac{1}{2}} .$$

Thus, if $N \leq Z$, we have established an adequate upper bound, namely

$$E_N^Q - E_N \leq (\text{const.})\, N^{1/5} Z^2 \qquad (5.13)$$

Since $Z \approx N$, this error is $\approx N^{11/5}$, and this is small compared to E which is $\approx N^{7/3}$.

If $N > Z$ we use $K = K^1 + K^\infty$ where K^1 is given above (with $N = Z$) and K_∞ is a density matrix (really, a sequence of density matrices) whose trace is $N - Z$ and whose support is a distance, d, arbitrarily far away from the origin. K_∞ does not contribute to $E(K)$ in the limit $d \to \infty$.

<u>Lower Bound for E_N^Q</u>: In LS a lower bound was constructed by decomposing $\mathbb{R}^3$ into boxes and using Neumann boundary conditions on these boxes. However, control of the singularities of V caused unpleasant problems. Here we will use coherent states again (cf. (Thirring, 1981)).

Let $\psi(x_1,\ldots,x_N;\, \sigma_1,\ldots,\sigma_N)$ be any normalized function in $\mathcal{H}_N$ and let

$$\rho_\psi(x) = N \sum_{\sigma=1}^{q} \int |\psi(x,x_2,\ldots,x_N;\, \sigma_1,\ldots,\sigma_N)|^2\, dx_2 \cdots dx_N \qquad (5.14)$$

$$E_\psi = (\psi,\, H_N\, \psi) \qquad (5.15)$$

$$T_\psi = (\psi,\, -\Sigma\, \Delta_i\, \psi) \qquad (5.16)$$

It is known that (Lieb, 1979, Lieb and Oxford, 1981)

$$I_\psi = (\psi,\, \sum_{i<j} |x_i - x_j|^{-1}\, \psi) \geq D(\rho_\psi,\rho_\psi) - (1.68) \int \rho_\psi(x)^{4/3} dx\ . \qquad (5.17)$$

Choose any $\tilde{\rho}(x) \geq 0$ and $\tilde{\phi} = V - |x|^{-1} * \tilde{\rho}$. Since $D(\rho_\psi - \tilde{\rho},\, \rho_\psi - \tilde{\rho}) \geq 0$, we have for any $0 \leq \varepsilon < 1$

$$E_\psi \geq (\psi,\, \sum_{i=1}^{N} h_i\, \psi) + U - D(\tilde{\rho},\tilde{\rho}) - (1.68) \int \rho_\psi^{\,4/3} + \varepsilon T_\psi \qquad (5.18)$$

with $h = -(1-\varepsilon)\Delta - \tilde{\phi}(x)$ (5.19)

being a single particle operator.

We will choose $\tilde{\rho}$ to be the TF density for the problem with γ_p replaced by $(1-\varepsilon)\,\gamma_p$, and with the same $\lambda = \min(N,Z)$. $-\tilde{\mu}$

and $\tilde{E}$ are the corresponding chemical potential and energy.

Let f_{pr} be the coherent states in $L^2(\mathbb{R}^3)$ given by (5.5) and π_{pr} = (projection onto f_{pr}) $\otimes I_\sigma$. For any function $m(y) = m(x,\sigma)$ in $L^2(\mathbb{R}^3;\mathbb{C}^q)$ we easily compute:

$$(m,m) = (2\pi)^{-3}\int dpdr\ (m,\pi_{pr}m)$$

$$\int |\nabla m|^2\, dz = (2\pi)^{-3}\int dpdr\ p^2(m,\pi_{pr}m) - (m,m)\int |\nabla g(x)|^2\, dx$$

$$\int |m|^2\,\tilde{\phi}_g(x)dz = (2\pi)^{-3}\int dpdr\ \tilde{\phi}(r)\ (m,\pi_{pr}m) \tag{5.20}$$

with $\tilde{\phi}_g = |g|^2 * \tilde{\phi}$.

Write $\tilde{\phi} = \tilde{\phi}_g + (\tilde{\phi} - \tilde{\phi}_g)$ and $h^g = -(1-\varepsilon)\Delta - \tilde{\phi}_g(x)$. Let us first concentrate on $e_1 = \inf e_1(\psi)$ where $e_1(\psi) = (\psi, \sum_{i=1}^N h_i^g\,\psi)$. Since Σh^g is a sum of single particle operators we need only consider ψ's which are determinants of N orthonormal single particle functions. If $m_1,\ldots,m_N$ are such then $M(p,r) = \sum_{i=1}^N (m_i, \pi_{pr}m_i)$ has the property that $0 \le M(p,r) \le q$ and $(2\pi)^{-3}\int dpdr\ M(p,r) = N$. Therefore

$$e_1(\psi) = (2\pi)^{-3}\iint dpdr\{(1-\varepsilon)p^2 - \tilde{\phi}(r)\}M(p,r) - N\int |\nabla g|^2. \tag{5.21}$$

The minimum of the right side of (5.21) over all M with the stated properties is given as follows: $M(p,r) = q\theta(\tilde{\phi}(r) - (1-\varepsilon)p^2-\mu)$ for some $\mu \ge 0$. μ is the smallest μ such that $(2\pi)^{-3}\int M(p,r) \le N$. Since $\tilde{\phi}$ is the TF potential (for $(1-\varepsilon)\gamma_p$) we see that $\mu = \tilde{\mu}$ and

$$e_1 - D(\tilde{\rho},\tilde{\rho}) + U \ge \tilde{E} - N\int |\nabla g|^2 . \tag{5.22}$$

Next, let us consider the missing piece $e_2 = -\int (\phi - \phi_g)\,\rho_\psi$. The second piece of $\tilde{\phi}$, namely $-\tilde{\psi} = -|x|^{-1} * \tilde{\rho}$, has the property that $\tilde{\psi} - |g|^2 * \tilde{\psi} \ge 0$ since $\tilde{\psi}$ is superharmonic and $|g|^2$ is spherically symmetric. Therefore we can ignore this piece in e_2. $V-V_g$ is bounded above, as before, by $\Sigma z_j|x-R_j|^{-1}\,\theta(R-|x-R_j|)$. For large N, $|R_i-R_j| > 2R$ and, using Hölder's inequality,

$$e_2 \ge -\,||\rho_\psi||_{5/3}\,[8\pi R^{\frac{1}{2}}\,\Sigma\, z_j^{5/2}]^{2/5} \ge -\,||\rho_\psi||_{5/3}\{(8\pi)^{2/5}R^{1/5}Z\}. \tag{5.23}$$

The negative term e_2 is controlled by the εT_ψ term through an inequality of Lieb and Thirring (1975 and 1976), cf. (Lieb, 1976).

$$T_\psi \geq L \int \rho_\psi(x)^{5/3}\, dx \tag{5.24}$$

with $L = (3/5)(3\pi/2q)^{2/3}$. Furthermore, by the Schwarz inequality, $\int \rho_\psi^{4/3} \leq \{N \int \rho_\psi^{5/3}\}^{\frac{1}{2}}$. If we write $\int \rho_\psi^{5/3} = X$ then $e_2 \geq -X^{3/5} D$, with $D = \{\ \}$ in (5.23), and

$$e_2 + \varepsilon T_\psi - (1.68) \int \rho_\psi^{4/3} \geq \min_{X>0} -DX^{3/5} + \varepsilon LX - (1.68) N^{\frac{1}{2}} X^{\frac{1}{2}} \equiv Y. \tag{5.25}$$

(5.22) contains $\tilde{E}$ instead of E; we must bound the difference. Using $\tilde{\rho}$ as a trial function for E, $E \leq \tilde{E} + [\varepsilon/(1-\varepsilon)]\, \tilde{K}$, where $\tilde{K} = (3(1-\varepsilon)/5)\gamma_p \int \tilde{\rho}^{5/3}$. Choose $\varepsilon = Z^{-1/30}$ (this is not optimum). For large Z, $\varepsilon < \frac{1}{2}$ and it is easy to see that $\tilde{K} < (\text{const.})\, Z^{7/3}$ for all N, Z. Thus

$$0 > \tilde{E} - E > -(\text{const.})\, Z^{7/3}\, Z^{-1/30} \tag{5.26}$$

Choose $R = Z^{-\frac{1}{2}}$, which is a different choice from the upper bound calculation. Then $D \approx Z^{9/10}$ and

$$Y \geq -(\text{const.})\, Z^{7/3}\, Z^{-1/30} \tag{5.27}$$

(It is easy to see that the term $-(1.68)\, N^{\frac{1}{2}} X^{\frac{1}{2}}$ is negligible as long as N/Z is fixed.)

Finally $-N \int |\nabla g|^2 \approx -N R^{-2} \approx Z^2$. Combining all these bounds, we find

$$E^Q - E > -(\text{const.})\, Z^{7/3-1/30}$$

which is the desired result. ■

Clearly there is room for a great deal of improvement for it is believed that $E^Q - E > 0$ as explained in § V.B. But first let us turn to the correlation functions.

In analogy with (5.14) we define

$$\rho_\psi^j(x_1,\ldots,x_j) = j!\, \binom{N}{j} \sum_\sigma \int |\psi(x_1,\ldots,x_N;\sigma_1,\ldots,\sigma_N)|^2\, dx_{j+1} \cdots dx_N \tag{5.28}$$

We wish to obtain a limit theorem for ρ_ψ^j when ψ is a ground state of H_N. But there may be no ground state (inf spec H_N may

not be an eigenvalue) or there may be several. In any case it is intuitively clear that the limit of ρ_ψ^j should not depend upon ψ being exactly a ground state, but only upon ψ being "nearly" a ground state.

Definition: Let $\psi_1, \psi_2, \ldots$ be a sequence of functions with $\psi_N \in \mathcal{H}_N$ for N particles. This sequence is called an approximate ground state if $|(\psi_N, H_N\psi_N) - E_N^Q| a_N^{-7/3} \to 0$ as $N \to \infty$. H_N always has k nuclei.

Theorem 5.2: Let $\{\psi_N\}$ be an approximate ground state with the scaling given before (5.2), and let $\rho_N^j(x)$ be given by (5.28) with ψ_N, and $\hat{\rho}_N^j(x_1,\ldots,x_N) \equiv a_N^{-2j}\, \rho_N^j(a_N^{-1/3} x_1,\ldots,a_N^{-1/3} x_N)$.

Let $\rho^j(x_1,\ldots,x_N) = \rho(x_1) \cdots \rho(x_N)$ with ρ being the solution to the TF problem for λ and $\{z_j^\circ, R_j^\circ\}$. (Note that $\lambda = N/Z$ is now fixed). Then

$$(\psi_N, -\Sigma\Delta_i\, \psi_N)\, a_N^{-7/3} \to \gamma_p \int \rho(x)^{5/3}\, dx$$

$$(\psi_N, \Sigma\, V(x_i)\, \psi_N)\, a_N^{-7/3} \to \int \rho V$$

$$(\psi_N, \Sigma\, |x_i - x_j|^{-1}\, \psi_N)\, a_N^{-7/3} \to D(\rho^{TF}, \rho^{TF})$$

Moreover, $\hat{\rho}_N^j(x) \to \rho^j(x)$ in the sense that if Ω is any bounded set in $\mathbb{R}^{3j}$ then

$$\int_\Omega \hat{\rho}_N^j(x)\, d^{3j}x \to \int_\Omega \rho^j(x)\, d^{3j}\, x.$$

If $\lambda \leq Z = \Sigma z_j$, the restriction that Ω be bounded can be dropped and $\hat{\rho}_N^j \to \rho^j$ in the weak L^1 sense.

Proof: The reader is referred to LS, Theorem III.5 for details. The basic idea is to consider a function $U(x_1,\ldots,x_y) \in C_o^\infty(\mathbb{R}^{3j})$ and add $\alpha \int \rho^j U\, d^{3j}x$ to the TF functional, $\mathcal{E}(\rho)$. On the other hand, the potential $\alpha\, a_N^{4j/3} \sum_{\substack{i_1 \cdots i_j \\ \text{unequal}}} U(a_N^{1/3} x_{i_1},\ldots,a_N^{1/3} x_{i_j})$

is added to H_N. By the aforementioned methods the energies are shown to converge on the scale of $a_N^{7/3}$. But

$\partial E/\partial\alpha\big|_{\alpha=0} = \int \rho^j U$. By concavity of $E(\alpha)$ the derivatives and the limits $a_N \to \infty$ can be interchanged. Thus, for all such U, $\int \hat{\rho}_N^j U \to \int \rho^j U$. ■

One of the assertions of Theorem 5.2 is that, as $N \to \infty$, correlations among any finite number of electrons disappear. A-posteriori thus is the justification for replacing the electron-electron repulsion $\Sigma |x_i - x_j|^{-1}$ by $D(\rho,\rho)$ in TF theory.

V.B. The Scott Conjecture for the Leading Correction

We have seen that $E^{TF} = -CZ^{7/3}$ under the assumption that the nuclear coordinates R_j and charges z_j scale as $Z^{-1/3} R_j°$ and $Z z_j°$, $\Sigma z_j° = 1$, and $\lambda = N/Z > 0$ is fixed. C depends on λ, $\underline{z}°$, $\underline{R}°$. What is the next correction to the energy? While this question takes us to some extent outside TF theory, we will briefly mention the interesting conjecture of Scott (1952) and a generalization of that conjecture. None of these conjectures have been proved.

The basic idea of Scott is that in the Bohr atom (no electron repulsion) the electrons close to the nuclei each have an energy $\sim -Z^2$. This should also be true in some sense even with electron repulsion. Since TF theory cannot yield exactly the right energy near the singularities of V, the leading correction should be $O(Z^2)$.

The leading correction should have three properties:

(i) It is the same with or without electron repulsion because the repulsive part of $\phi(x)$, namely $|x|^{-1} * \rho$, is $O(Z^{4/3})$ for all x.

(ii) It is independent of N/Z, provided $N/Z > 0$ and fixed. This is so because the correction comes from the core electrons whose distance from the nucleus is $O(Z^{-1})$. The number of electrons thus involved is small compared to Z.

(iii) It should be additive over a molecule. If the correction is Dz^2 for an atom then the total leading correction should be

$$\Delta E = D \sum_{j=1}^{k} z_j^2 \tag{5.29}$$

and

$$E^Q = E^{TF} + \Delta E + o(Z^2) \,. \tag{5.30}$$

Of course E^{TF} depends on whether electron repulsion is present or not, but ΔE supposedly does not change. To calculate D let

us first calculate E^{TF} for an atom without repulsion. The general theory goes through as before, but now the TF equation is $\gamma\rho^{2/3} = (V-\mu)_+$, $V(x) = z/|x|$, $\int\rho = N$ and $\mu > 0$, even when $N = z$. It is found (Lieb, 1976, p. 560) that $\mu = z/R$, $R = 3\gamma(4N/\pi^2)^{2/3}/5z$ and $E^{TF}_{Bohr} = -3z^2 N^{1/3}(\pi^2/4)^{2/3}/\gamma$. Using γ_p, $E^{TF}_{Bohr} = -z^{7/3}(3N/z)^{1/3}(2mq^{2/3}/\hbar^2)/4$. The quantum energy is computed by adding up the Bohr levels. For each principal quantum number, n, the energy is $e_n = m/2\hbar^2 n^2$ and it is qn^2-fold degenerate. The result (Lieb, 1976) is

$$E^Q_{Bohr} = E^{TF}_{Bohr} + qz^2/8 + 0(z^{5/3}) \tag{5.31}$$

thus

$$D = qz^2/8 \tag{5.32}$$

in the Scott conjecture. Scott's (1952) derivation was slightly different from the above, but his basic idea was the same.

The Scott conjecture about the energy can be supplemented by the following about the density. Let $f_{n\ell m}(z,x)$ be the normalized <u>bound state</u> eigenfunctions for the hydrogenic atom with nuclear charge z, and define

$$\rho^H(z,x) = q \sum_{n\ell m} |f_{n\ell m}(z,x)|^2. \tag{5.33}$$

This sum converges and represents the quantum density for a Bohr atom with infinitely many electrons. It is being tabulated and studied by Heilmann and Lieb. A graphical plot of ρ^H shows that it is monotone decreasing and has almost no discernable shell structure. Clearly, $\rho^H(z,x) = z^3\rho^H(1,zx)$ and is spherically symmetric. By our previous analysis of the $z \to \infty$ limit (which strictly speaking is not applicable when $N = \infty$, but which can be suitably modified)

$$z^{-2}\rho^H(z,z^{-1/3}x) \to z^{-2}\rho^{TF}_{Bohr}(z,z^{-1/3}x) \tag{5.34}$$

as $z \to \infty$. But $\rho^{TF}_{Bohr}(z,x) = (z/\gamma_p|x|)^{3/2}$ when $\mu = 0$, as we have just seen. Thus

$$\rho^H(1,y) \to (\gamma_p\,|y|)^{-3/2} \tag{5.35}$$

as $y \to \infty$. (5.35) is not obvious, but it can be proved from (5.33).

Thus, $\rho^H(z,x)$, whose scale length is z^{-1}, agrees nicely with $\rho^{TF}(z,x)$, whose scale length is $z^{-1/3}$, in the overlap region $z^{-1} << |x| << z^{-1/3}$. This is true even when electron repulsion is included in ρ^{TF} because of Theorem 2.8(a). The common value is $\rho(z,x) = (z/\gamma_p |x|)^{3/2}$. Because of this we are led to the following.

Conjecture: Suppose the sequence $\{\psi_N\} \in \mathcal{H}_N$ is an approximate ground state for a molecule (with repulsion) in the strong sense that

$$|(\psi_N, H_N\psi_N) - E_N^Q| \, a_N^{-2} \to 0 \quad \text{as} \quad N \to \infty .$$

Let $\rho_N^Q(x)$ be given by (5.14). Recall that $R_j = a_N^{-1/3} R_j°$. Fix $\lambda = N/Z > 0$ and $x \neq R_j°$, all j. Then, as $N \to \infty$,

$$a_N^{-2} \rho_N^Q(a_N^{-1/3} x) \to \rho^{TF}(x) \tag{5.36}$$

where ρ^{TF} is the TF density for λ, $z_j°$, $R_j°$. On the other hand, for all fixed y,

$$a_N^{-3} \rho_N^Q(a_N^{-1/3} R_j° + a_N^{-1} y) \to (z_j°)^3 \rho^H(1, z_j° y). \tag{5.37}$$

(5.36) has already been proved in §V.A.

TFW Theory: It is a remarkable fact that the TFW correction, which has no strong a-priori justification, has, as its chief effect, precisely the kind of correction (i), (ii), (iii) above predicted by Scott. If δ is chosen correctly in (2.8), even the constant D in (5.32) can be duplicated. This will be elucidated in section VII. TFW theory also (accidentally ?) improves TF theory in two other ways: negative ions can be supported and binding occurs.

V.C. A Picture of a Heavy Atom

With the real and imagined information at our disposal we can view the energy and density profile of a heavy, neutral, non-relativistic atom as being composed of seven regions.

(1) The inner core: Distances are $O(z^{-1})$ and ρ is $O(z^3)$. For large σ, the number of electrons out to $R = \sigma/z$ is $\sim \sigma^{3/2}$, while the energy $\sim z^2 \sigma^{1/2}$. If $1 << zr << z^{2/3}$, $\rho(r)$ is well approximated by $(z/\gamma_p r)^{3/2}$. $\rho(r)$ is infinity on a scale of z^2 which is the appropriate scale for the next, or TF region. The leading corrections, beyond TF theory, comes from this region. None of this has been proved.

(2) The core: Distances are $O(z^{-1/3})$ and ρ is $O(z^2)$. TF theory is exact to leading order. The energy is $E^{TF} \sim -z^{7/3}$ and

almost all the electrons are in this region. This is proved.

(3) The core mantle: Distances are order $\sigma z^{-1/3}$ with $\sigma >> 1$. $\rho(r) = (3\gamma_p/\pi)^3 r^{-6}$, the Sommerfeld asymptotic formula. ρ is still $O(z^2)$. This is proved.

(4) A transition region to the outer shell: This region may or may not exist.

(5) The outer shell: In the Bohr theory, $z^{1/3}$ shells are filled. The outer shell, if it can be defined, would presumably contain $O(z^{2/3})$ electrons and each electron in the shell would "see" an effective nuclear charge of order $z^{2/3}$. This picture would give a radius unity for the last shell and an average density $\sim z^{2/3}$ in the shell. On the same basis the average electron energy would be $O(z^{2/3})$ and thus the energy in the shell would be $O(z^{4/3})$. All this is conjectural, for reliable estimates are difficult to obtain.

(6) The surface: Here the potential is presumably $O(1)$, and so is the energy of each electron. Chemistry takes place here.

TF theory, which is unreliable in this region, nevertheless predicts a surface radius of $O(1)$. We thank J. Morgan for this remark. His idea is that if the surface radius R_s is defined to be such that outside R there is one unit of electron charge then $R_s = O(1)$ because the TF density is $\rho(r) = (3\gamma_p/\pi)^3 r^{-6}$, independent of z, for large r. Likewise, if R_o is defined such that between R_o and R_s there are $z^{2/3}$ electrons, then the average TF density in this "outer shell" is $z^{2/3}$ in conformity with (5). Finally, the energy needed to remove one electron is $O(1)$ as (3.11) shows. The radius of this ionized atom is also $O(1)$ as (3.13) shows.

In no sense is it being claimed that TF theory is reliable at the surface, or even that the existence of the surface, as described, is proved. We are only citing an amusing coincidence. It is quite likely that the surface radius of a large atom has a weak dependence on z.

(7) The region of exponential falloff:

$\rho(r) \sim K \exp[-2(2me/\hbar^2)^{\frac{1}{2}}(r-R)]$, where e is the ionization potential, K is the density at the surface and R is the surface radius. An upper bound for ρ of this kind has been proved by many people, of which the first was O'Connor (1973). See also (Deift, Hunziker, Simon and Vock, 1978, and M. Hoffmann-Ostenhof, T. Hoffmann-Ostenhof, R. Ahlrichs and J. Morgan, 1980) for recent developments and bibliographies of earlier work.

The density profile of a heavy atom, as described above, is shown schematically in Fig. 2

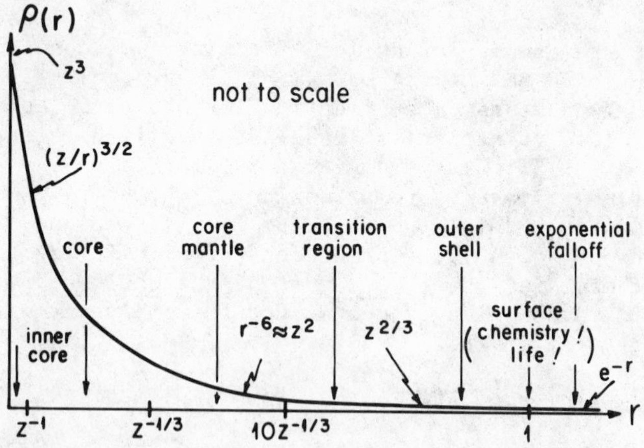
ρ(r)
z³
not to scale
(z/r)^3/2
core
core mantle
transition region
outer shell
exponential falloff
surface (chemistry! life!)
inner core
r⁻⁶≈z²
z^2/3
e⁻ʳ
r
z⁻¹
z^-1/3
10z^-1/3
1

Figure 2

VI. THOMAS-FERMI-DIRAC THEORY

The previous sections contain most of the mathematical tools for the analysis of this model; the main new mathematical idea to be introduced here will be the j model and its relation to TFD theory.

The TFD functional is

$$\mathcal{E}^{TFD}(\rho) = \int J(\rho(x))dx - \int V\rho + D(\rho,\rho) + U , \tag{6.1}$$

and

$$J(\rho) = \frac{3}{5}\gamma\rho^{5/3} - \frac{3}{4}C_e\rho^{4/3} . \tag{6.2}$$

The term $-D = -(3C_e/4)\int\rho^{4/3}$ was suggested by Dirac (1930) to account for the "exchange" energy. The true electron repulsion, I, in (5.17) is expected to be less than $D(\rho,\rho)$ because the electrons are correlated. For an ideal Fermi gas at constant density I is computed to be $D(\rho,\rho) - D$ with $C_e = (6/\pi q)^{1/3}$. There is, however, no fundamental justification for the Dirac approximation; it can even lead to unphysical results, as will be seen shortly. In particular, I is always positive but $D(\rho,\rho) - D$ can be arbitrarily negative. As remarked in (5.17), there is a lower bound of this form $D(\rho,\rho) - D$ (Lieb, 1979) (Lieb and Oxford, 1981) with $3C_e/4 = 1.68$ (independent of q). In any event, it should be remembered that D is part of the Coulomb energy even though it is mathematically convenient to combine it with the kinetic energy as in (6.2).

For simplicity we assume

$$V(x) = \sum_{i=1}^{k} V_i(x-R_i) \tag{6.3}$$

with $V_i \in \mathcal{D}$: $V_i = |x|^{-1} * m_i$ (with m_i a nonnegative measure) and $|m_i| = z_i$.

Henceforth, the superscript TFD will be omitted. All quantities in this section refer to TFD, and not TF, theory, unless otherwise stated.

VI.A. The TFD Minimization Problem

The function space is the same as for TF theory, namely

$$\mathcal{J} = \{\rho | \rho \in L^1 \cap L^{5/3}, \rho(x) \geq 0\} . \tag{6.4}$$

The energy is

$$E(\lambda) = \inf\{\mathcal{E}(\rho) | \int\rho = \lambda, \rho \in \mathcal{J}\} \tag{6.5}$$

Theorem 6.1: $E(\lambda)$ is finite, non-increasing in λ, and

$$E(\lambda) = \inf\{\mathcal{E}(\rho) \mid \int \rho \leq \lambda,\ \rho \in \mathcal{J}\}. \tag{6.6}$$

Moreover, $e(\lambda) \equiv E(\lambda) - U < 0$ when $\lambda > 0$.

Proof: Same as Proposition 2.1 and Theorem 2.3. The crucial fact to note is that $J(0) = J'(0) = 0$, which permits us to place "surplus charge density" at infinity. ■

It is not immediately obvious that $E(\lambda)$ is convex because J is not convex. The proof of convexity is complicated and will be given later (Theorem 6.9).

A second difficulty is that $E(\lambda)$ is not bounded below for all λ. This is so because J is not positive. This latter difficulty can be dealt with in the following way. Introduce

$$\mathcal{E}_\alpha(\rho) = \mathcal{E}(\rho) + \alpha \int \rho \tag{6.7}$$

with

$$\alpha = 15\, C_e^{\,2} (64\gamma)^{-1}. \tag{6.8}$$

This amounts to replacing J by

$$J_\alpha(\rho) = J(\rho) + \alpha\rho. \tag{6.9}$$

Note that

$$J_\alpha(\rho) \geq 0 \quad \text{and} \quad J_\alpha(\rho_0) = 0 = J'_\alpha(\rho_0) \quad \text{for } \rho_0 = (5C_e/8\gamma)^3. \tag{6.10}$$

Correspondingly, introduce

$$E_\alpha(\lambda) = \inf\{\mathcal{E}_\alpha(\rho) \mid \int \rho = \lambda,\ \rho \in \mathcal{J}\}. \tag{6.11}$$

Theorem 6.2: $E_\alpha(\lambda)$ is non-increasing in λ and has a lower bound, independent of λ. Moreover

$$E_\alpha(\lambda) = E(\lambda) + \alpha\lambda \tag{6.12}$$

and

$$E_\alpha(\lambda) = \inf\{\mathcal{E}_\alpha(\rho) \mid \int \rho \leq \lambda,\ \rho \in \mathcal{J}\}. \tag{6.13}$$

$e_\alpha(\lambda) \equiv E_\alpha(\lambda) - U = e(\lambda) + \alpha\lambda < 0$ when $\lambda > 0$ and $e_\alpha(\lambda) \to \inf_\lambda e_\alpha(\lambda) \equiv e_\alpha(\infty)$ as $\lambda \to \infty$.

Proof: Again the proof is the same as for Proposition 2.1 and Theorem 2.3. Here, however, $J'_\alpha(0) > 0$; the fact that

$J_\alpha(\rho_0) = J'_\alpha(\rho_0) = 0$ is used instead. The fact that $J_\alpha \geq 0$ is responsible for the lower bound. ■

Remark: One consequence of Theorem 6.2 is that $dE(\lambda)/d\lambda \leq -\alpha$ (if the derivative exists). Another is that when λ is large enough so that $e_\alpha(\lambda) = e_\alpha(\infty)$ then $e(\lambda) = e_\alpha(\infty) - \alpha\lambda$. As will be seen, this happens when $\lambda \geq \lambda_c = Z = \Sigma z_j$. Thus, the graph of $e_\alpha(\lambda)$ is similar to that for $e^{TF}(\lambda)$ in Fig. 1. $e(\lambda)$ then has a negative slope, $-\alpha$, at λ_c and afterwards $e(\lambda)$ has the same constant negative slope. This is a highly unphysical feature of TFD theory which arises from the fact that one can have spatially small "clumps" of density in which $\rho = \rho_0$, arbitrarily far apart. These "clumps" have an energy approximately $-\alpha\,\rho_0\cdot$ (volume) and are physically nonsensical because the $-\rho^{4/3}$ term, which causes this effect, is a gross underestimate of the positive electron repulsion which it is meant to represent. There is no minimizing ρ for these "clumps" because for no ρ is the energy exactly $-\alpha\,\rho_0\cdot$(volume).

VI.B. The j Model

Now we must deal with the fact that J_α is not convex. To this end we follow Benguria (1979) who introduced the "convexified" j model. With its aid, Benguria was the first to place the TFD theory on a rigorous basis for a certain class of amenable potentials in (6.3), which is defined in §VI.C. This class includes the point nuclei. It will turn out that the j model also permits us to analyze TFD theory for all potentials, not just the amenable class. However, for non-amenable potentials, the analysis is complicated and the final result has an unexpected feature, namely that a minimizing ρ for E may not exist, even if $\lambda < \lambda_c$. The j model is explored in detail here because as will be seen in §VI.C, its energy is the same as $E_\alpha(\lambda)$ for the TFD model. Moreover, for amenable potentials the density, ρ, of the two models is also the same.

Definition:

$$\begin{aligned} j(\rho) &= J_\alpha(\rho)\,, \quad \rho \geq \rho_0 = (5\,C_e/8\gamma)^3 \\ &= 0 \qquad\;\;, \quad 0 \leq \rho \leq \rho_0\,. \end{aligned} \tag{6.14}$$

The derivative of this convex function is given in (3.14). $\mathcal{E}_j(\rho)$ is given by (6.1) with J replaced by j. $E_j(\lambda)$ is defined by (6.5) with $\mathcal{E}$ replaced by $\mathcal{E}_j$.

By the methods of Sections II and III the j model has many of the same properties as TF theory.

Theorem 6.3: If V is given by (6.3) and if $\mathcal{E}$ is replaced by

$\mathcal{E}_j$, E by E_j and e by $e_j = E_j - U$, then the following results of TF theory hold for the j model(they also hold for TF theory, of course, with this V). [Note: Ignore any mention of TFD and TFDW theory in the cited theorems.]:

Propositions 2.1 and 2.2; Theorems 2.3 and 2.4; the definition of λ_c; Theorem 2.5; Theorem 2.6 (with (2.18) replaced by (3.2)); Theorem 2.7; Theorem 2.12 (for a point nucleus); Theorem 2.13 without the γ dependence (for point nuclei). The last two equations in this theorem have an obvious generalization for non-point nuclei; Theorem 2.14 (for point nuclei) is changed to (a) $5K/3 = A-2R-\mu\lambda-\alpha\lambda+4D/3$, (b) $2K = A-R+D$ for an atom, with $D = (3C_e/4) \int \rho^{4/3}$ (note that Theorem 3.19 must be used in the proof); Equation (2.22); Theorem 3.2; Theorems 3.4,3.5 (Benguria (1979) has shown that if W is the potential of point nuclei then $\phi'-\phi \in H^2$ away from S_W) ; Corollaries 3.7, 3.8, 3.9, and 3.10 (note, in particular, that $\phi^{TF} > \phi^{j\ \text{model}}$ for fixed μ); Lemma 3.11; Theorem 3.12, Corollaries 3.14 and 3.17; Theorem 3.18 (i.e., $\lambda_c = Z$); Equation (3.6); §III.B; Theorem 3.23 (but note that equality can occur. Cf. remark at the end of §IV.C).

Remarks: (i) Theorem 2.8 (a) holds in the sense that $\rho(x) \approx (z_j/\gamma)^{3/2}|x-R_j|^{-3/2}$ near R_j .

(ii) There is no simple scaling for the j model, as in (2.24) for TF theory.

(iii) We emphasize that a minimizing ρ exists if and only if $\lambda \leq Z$. This ρ is unique and satisfies the Thomas-Fermi-Dirac equation (3.2).

(iv) Question: Under what conditions do the conclusions of Corollaries 3.13, 3.15, and 3.16 and Theorem 3.21 hold for the j model? Question: To what extent do the results of section IV carry over to the j model?

(v) To prove the analogue of (2.15), Mazur's theorem can be used, as in the proof of Proposition 3.24.

There are some useful additional facts about the j model not mentioned in Theorem 6.3.

Theorem 6.4: If C_e increases then (i) $\phi(x) - \mu(\lambda)$ decreases and $\mu(\lambda)$ increases for fixed λ ; (ii) $\phi(x)$ decreases for fixed μ.

Proof: By Corollary 3.10, since $j'(\rho)$ decreases with C_e for fixed ρ. ■

Theorem 6.5: For all λ, $E^{TF}(\lambda) > E_j(\lambda) > \mathbf{E}(\lambda)$ since $J(\rho) < j(\rho) < 3\gamma\rho^{5/3}/5$. On the other hand, suppose V is the potential of k point nuclei as in (2.1). Then for $\lambda \leq \lambda_c = Z$,

$$E^{TF}(\lambda) \le E_j(\lambda) - \alpha\lambda + (3C_e/4)\lambda^{\frac{1}{2}}\{(5\varepsilon_1/2\gamma)Z^{7/3}\}^{\frac{1}{2}} + 27\, C_e^{\,2}\lambda/(10\gamma)\ , \tag{6.15}$$

where $-\varepsilon_1$ is the TF energy for a neutral atom with $z = 1$.

Remarks: (i) When $\lambda > Z$ then $(E^{TF}-E_j)(\lambda) = (E^{TF}-E_j)(Z)$.
(ii) Clearly (6.15) can be improved. But it does show that the effect of the Dirac term is to decrease the energy by $0(Z^{5/3})$ for large Z. (Note: by Theorem 6.8, $E_j - \alpha\lambda = E$.)

Proof: Let ρ be the minimizing density for E_j, and ρ^{TF} that for E^{TF}. Use ρ as a trial function for E^{TF}. Noting that $\rho(x) \notin (0,\rho_0]$ a.e. (Theorem 3.19), we have $E^{TF} \le \mathcal{E}^{TF}(\rho) = E_j - \alpha\lambda + (3C_e/4)\int\rho^{4/3}$. By Theorem 3.19, $\gamma\rho^{2/3} - C_e\rho^{1/3} + \alpha = \phi - \mu$ when $\rho > 0$. But by Corollary 3.10 $\phi - \mu \le \phi^{TF} - \mu^{TF} \le \gamma(\rho^{TF})^{2/3}$. Thus $\rho^{2/3} \le (\rho^{TF})^{2/3} + (C_e/\gamma)\rho^{1/3}$. Squaring this and using $\int(\rho^{TF})^{2/3}\rho^{1/3} \le \lambda$(Hölder), and $\int\rho^{2/3} \le \lambda(\rho_0)^{-1/3}$, and $\int(\rho^{TF})^{4/3} \le [\lambda\int(\rho^{TF})^{5/3}]^{\frac{1}{2}}$, we obtain (6.15), but with $5K^{TF}/3\gamma$ in place of $\{\ \}$. By Theorem 2.14(a), $2K^{TF}/3 < -e^{TF}$ and by the remark preceding (4.5), $e^{TF} > e^{TF}$ (all nuclei at one point). ∎

The next theorem states that ρ always has compact support. even when $\lambda = \lambda_c$. When $\lambda < \lambda_c$ this is also true in TF theory (Lemma 3.11). The proof we give seems unnecessarily complicated; a simpler one must be possible.

Theorem 6.6: Suppose $V = |x|^{-1} * m \in \mathcal{D}$, with m a nonnegative measure of compact support and $\int dm = Z$. Let ρ be the minimizing TFD density for $\lambda \le \lambda_c = Z$. Then ρ has compact support Moreover, suppose $\text{supp}(m) \subset B_R$, the ball of radius R centered at 0. Then $\text{supp}(\rho) \subset B_r$ for some r depending on R and Z, but independent of λ. $r \le 2R + tZ(\rho_0 R^2)^{-1}$ for some universal constant, t, independent of all parameters.

Proof: The strategy is to construct a function, f, such that $\text{supp}(\rho) \subset \text{supp}(f)$. Let $S_R = \partial B_R$ be the sphere of radius R. There exists a function (surface charge distribution) σ on S_{2R} such that, V_σ, the potential of σ outside B_{2R} is V, i.e. $V(x) = V_\sigma(x) \equiv 4\pi(2R)^2\int d\Omega\sigma(\Omega)|x-2R\Omega|^{-1}$ for $|x| > 2R$, where Ω denotes a point on S_1 and $\int d\Omega = 1$. It is easy to see that σ is a bounded, continuous function since $\text{supp}(m) \subset B_R$, and $|\sigma(\Omega)| \le s\,Z\,R^{-2}$ for some universal constant s. Let $\Sigma(\Omega) = -\sigma(\Omega) + s\,Z\,R^{-2} \ge 0$, and let $dM(x) = dm(x) + \Sigma(x/2R)\delta(|x|-2R)\,dx$. If $V_M = |x|^{-1} * M$, we see that $V_M(x) \ge V(x)$, all x, and

$V_M(x) = Q|x|^{-1}$ for $|x| > 2R$, with $0 \leq Q \leq (1+16\pi s)Z$. $V_M - V$ is a bounded function in $\mathcal{D}$. Now let $f(x) = \rho_0$ for $2R < |x| \leq r$ and $f(x) = 0$ otherwise, with $\int f = Q$. Let $\psi = |x|^{-1} * \rho$, $g = |x|^{-1} * f$, $\phi = V - \psi$, $h = V_M - g$ where ρ satisfies (3.2). We claim $u(x) \equiv h(x) - \phi(x) + \mu \geq 0$. If not, let $B = \{x | u(x) < 0\}$. B is open since $h - \phi$ is continuous. $-\Delta u/4\pi \geq -f + \rho$. But $f \leq \rho$ on B since, for $|x| \leq 2R$, $f(x) = 0$; for $|x| > 2R$, $\rho(x)$ is either 0 or $> \rho_0$ a.e. by Theorem 3.19. If $\rho(x) = 0$ then $\phi - \mu \leq 0$ by (3.2) and $h(x)$ is clearly ≥ 0, so $x \notin B$; if $\rho(x) \geq \rho_0$, $\rho - f \geq 0$. Thus, u is superharmonic on B and, since $u = 0$ on ∂B and $u = \mu \geq 0$ at infinity, B is empty. Now consider $A = \{x | r < |x|\}$. In A, $-\Delta u = 4\pi\rho \geq 0$ and $u \geq 0$ on ∂A and at infinity. Therefore, either (i) $\rho = 0$ a.e. in A or (ii) $u > 0$ everywhere in A. In case (ii), $\phi < \mu$ in A because $h = 0$ in A. But then, by (3.2), $\rho = 0$ a.e. in A. The bound for r is obtained by $4\pi (r-2R)(2R)^2 < (4\pi/3)(r^3-(2R)^3) = Q/\rho_0$. ∎

Remark: For an atom with nucleus located at the origin, R can be chosen to be any positive number. If the inequality for r is minimized then we find $\rho(x) = 0$ for $|x| > 3(tZ/\rho_o)^{1/3}$.

Theorem 6.7: Suppose $V \in \mathcal{D}$ and $\rho \in \mathcal{C}$ are such that the second line of (3.2) holds with $\mu = 0$, in the sense that $\phi_\rho(x) \leq 0$ a.e. when $\rho(x) = 0$ and $\rho(x) = 0$ a.e. when $\phi(x) < 0$. Let A be the complement of the support of ρ. Then $\phi \equiv 0$ on $\bar{A}$, the closure of A.

Remark: Theorem 6.7 does not mention j. However, the theorem is meaningful only if $\text{supp}(\rho)$ is not all of $\mathbb{R}^3$. This does not happen in TF theory when $\mu = 0$, but it does happen for the j model if the hypothesis of Theorem 6.6 holds. The significance of Theorem 6.7 is that there is total shielding in TFD theory. This is in contrast to TF theory where there is under-screening in the neutral case in the sense that the potential falls off only with a power law. One consequence of Theorem 6.7 is that two or more molecules, each of fixed shape, do not interact when their supports are disjoint. Cf. the remark at the end of section IV.

Proof: Let $B = \{x | \phi(x) < 0\}$. Clearly the singularities of V are not in B, so B is open. On B, $\Delta\phi \leq 0$ since $\rho = 0$ a.e. in B. But $\phi = 0$ on ∂B and at infinity so B is empty. Therefore $\phi \geq 0$ everywhere. Let $D = \{x | \rho(x) = 0\}$. $\phi \leq 0$ a.e. on D. Since $A \subset D$ is open, and ϕ is continuous on $\{x | \phi(x) < 1\}$, $\phi \equiv 0$ on A, and hence on $\bar{A}$. ∎

VI.C. The Relation of the j Model to TFD Theory

We will show that the energy of the j model is exactly $E_\alpha(\lambda) = E(\lambda) + \alpha\lambda$ for the TFD problem. Thus, all the facts about

the energy in Theorems 6.3 and 6.5 hold for TFD theory. However, the densities may be different!

Let us start with the simplest case studied in (Benguria, 1979).

Definition: A (nonnegative)measure m is said to be amenable if $dm(x) = \sum_{i=1}^{k} z_i\delta(x-R_i)dx + g(x)dx$ with $z_i > 0$ and g satisfies: (i) $g \geq 0$, (ii) $g \in L^{\infty}_{loc}$, (iii) If $A = \{x|g(x)=0\}$ and $\sim A$ is its complement then $\mathbb{R}^3 \setminus [\text{Interior } (A) \cup \text{Interior } (\sim A)]$ has zero Lebesgue measure, (iv) $g(x) \geq \rho_0$ for $x \notin A$. (v) $V = |x|^{-1}*m \in \mathcal{D}$. m is strongly amenable if $g(x) > \rho_0$ for $x \notin A$.

Remark: This amenable class is more restrictive than necessary for Theorem 6.8. Technicalities aside, (iv) is the crucial point.

Theorem 6.8: Suppose that in (6.1) $V = |x|^{-1}*m$ and m is amenable. Then $E_\alpha(\lambda) = E(\lambda) + \alpha\lambda$ for the TFD problem equals $E_j(\lambda)$ for the j model. Moreover, there is a minimizing ρ for the TFD problem if and only if $\lambda \leq \lambda_c = Z = \int dm$. This ρ is unique and is the same as the ρ for the j model. It satisfies (3.2).

Proof: Clearly $E_\alpha \geq E_j$ since $J_\alpha(\rho) \geq j(\rho)$. First suppose $\lambda \leq \lambda_c = Z$ and let ρ be the unique minimum for the j problem. By Theorem 3.19, $\rho(x) \notin (0,\rho_0)$ so $E_\alpha(\lambda) \leq \mathcal{E}_\alpha(\rho) = \mathcal{E}_j(\rho) = E_j(\lambda)$. Thus, $E_\alpha(\lambda) = E_j(\lambda)$. Let ρ satisfy $\int\rho = \lambda$ and $\mathcal{E}_\alpha(\rho) = E_\alpha(\lambda)$. Then since $\mathcal{E}_\alpha(\rho) \geq \mathcal{E}_j(\rho) \geq E_j(\lambda)$ we conclude that ρ minimizes $\mathcal{E}_j(\rho)$. But there is only one such ρ. Next, suppose $\lambda > \lambda_c$. Then $E_j(\lambda) = E_j(\lambda_c) = E_\alpha(\lambda_c)$. But $E_\alpha(\lambda) \leq E_\alpha(\lambda_c)$ by Theorem 6.2, and $E_\alpha(\lambda) \geq E_j(\lambda)$. Hence $E_\alpha(\lambda) = E_j(\lambda)$. By the above argument, any minimizing ρ for $\mathcal{E}_\alpha$ would have to minimize $\mathcal{E}_j$, but no such ρ exists. ■

Remark: By Theorem (3.19), $\rho(x) \notin (0,\rho_0)$ a.e. if m is amenable, and $\rho(x) \neq \rho_0$ a.e. if m is strongly amenable. If ρ is merely amenable, $\rho(x)$ can be ρ_0 with positive measure. An example is $dm(x) = \rho_0 B_R(x)dx$ with B_R being the characteristic function of a ball of radius R centered at 0. Then $\rho_\lambda(x) = \rho_0 B_r(x)$ with $4\pi\rho_0 r^3/3 = \lambda$ for $\lambda \leq \lambda_c = 4\pi\rho_0 R^3/3$. This ρ_λ is easily seen to satisfy (3.2).

If m is not amenable the situation is much more complicated, but more amusing mathematically. First let us consider the energy.

Theorem 6.9: If $V = |x|^{-1}*m \in \mathcal{D}$, then $E_\alpha(\lambda) = E_j(\lambda)$ for all λ. In particular, $\lambda_c = Z = \int dm$ and E_α is convex in λ. If there is a minimizing ρ for $E(\lambda)$, it is unique and it is the ρ for the j model.

A number of lemmas are needed for the proof.

Lemma 6.10: Let $A \subset \mathbb{R}^3$ be a measurable set and let ρ be a function in L^1 with $0 \leq \rho(x) \leq 1$ for $x \in A$, and $\rho(x) = 0$ for $x \notin A$. (This implies $\rho \in L^p$, all p.) Then there exists a sequence of functions $f^n \in L^1$ such that (i) $f^n \rightharpoonup \rho$ weakly in every L^p with $1 < p < \infty$; (ii) f^n is the characteristic function of some measurable set $F^n \subset A$; (iii) $\int f^n = \int \rho$.

Proof: For $\delta > 0$ and $y \in \mathbb{Z}^3$ let $B(\delta,y) = \{x \in \mathbb{R}^3 \mid -\delta/2 < x^i - \delta y^i \leq \delta/2\}$ be the elementary cubes of side δ. Let $C(\delta,y) = A \cap B(\delta,y)$. Partition $C(\delta,y)$ into two disjoint measurable sets, C^+ and C^-, such that $|C^+(\delta,y)| = \int_{B(\delta,y)} \rho$. Let $F^\delta = \bigcup_{y \in \mathbb{Z}^3} C^+(\delta,y)$, and let f^δ be the characteristic function of F^δ. Clearly $f^\delta \in L^p$ with norm $(\int \rho)^{1/p}$, and f^δ satisfies (ii) and (iii). Let $1/p + 1/q = 1$. Since C_0, the continuous functions of compact support, are dense in L^q, and $\|f^\delta\|_p =$ constant, it suffices to prove that $I(\delta,g) = \int g(f^\delta - \rho) \to 0$ as $\delta \to 0$ for every $g \in C_0$. But g is uniformly continuous, so for any $\varepsilon > 0$ $|g(x) - g(\delta y)| < \varepsilon$ (uniformly) for $x \in B(\delta,y)$ when δ is small enough. Since

$$\int_{B(\delta,y)} (f^\delta - \rho) = 0, \quad |I(\delta,g)| < 2\varepsilon \int \rho . \; \blacksquare$$

Lemma 6.11: Suppose $\rho \in \mathcal{J}$ and

$$A(\rho) = \{x \mid 0 < \rho(x) < \rho_0\} \tag{6.16}$$

has positive measure. Then there exists $\tilde{\rho} \in \mathcal{J}$ satisfying (i) $\mathcal{E}(\tilde{\rho}) < \mathcal{E}(\rho)$; (ii) $A(\tilde{\rho})$ is empty; (iii) $\tilde{\rho}(x) = \rho(x)$ if $x \notin A(\rho)$; (iv) $\int \tilde{\rho} = \int \rho$.

Proof: Apply Lemma 6.10 to the function $h(x) = \rho(x)/\rho_0$ if $x \in A(\rho)$, $h(x) = 0$ otherwise. Let $\tilde{\rho}^n(x) = \rho_0 f^n(x)$ if $x \in A(\rho)$, $\tilde{\rho}^n(x) = \rho(x)$ otherwise. Then $\int \tilde{\rho}^n = \int \rho$ and $\int J_\alpha(\tilde{\rho}^n) - \int J_\alpha(\rho) = -K$ with $K > 0$ and independent of n (since $J_\alpha(t) > 0$ for $0 < t < \rho_0$, and $0 = J_\alpha(0) = J_\alpha(\rho_0)$). Now, as in the proof of Theorem 2.4, $\int V\tilde{\rho}^n \to \int V\rho$. $\lim D(\tilde{\rho}^n, \tilde{\rho}^n) = D(\rho,\rho)$ (easy proof). Hence, for any $\varepsilon > 0$ there is some n such that, $\int V\tilde{\rho}^n > \int V\rho - \varepsilon$ and $D(\tilde{\rho}^n,\tilde{\rho}^n) < D(\rho,\rho) + \varepsilon$. ■

The following is a corollary of Lemma 6.11.

Theorem 6.12: (a) If $\rho \in \mathcal{J}$ minimizes $\mathcal{E}(\rho)$ on $\int \rho = \lambda$ then measure $\{A(\rho)\} = 0$. (b) Even if there is no minimizing ρ, a minimizing sequence for $E(\lambda)$ with $\int \rho = \lambda$ can be chosen such that $A(\rho^n)$ is empty for all n.

Theorem 6.13: Let ρ minimize $\mathcal{E}_j$ for $\int \rho = \lambda$. $A(\rho)$ may not be empty, but for any $\varepsilon > 0$ there exists a $\tilde{\rho}$ such that $\int \tilde{\rho} = \lambda$, $A(\tilde{\rho})$ is empty and $\mathcal{E}_j(\tilde{\rho}) < E_j(\lambda) + \varepsilon$.

Proof: Again, use Lemma 6.10 and mimic the proof of Theorem 6.11. ■

Proof of Theorem 6.9: That $E_\alpha(\lambda) = E_j(\lambda)$ follows from Theorem 6.13 and an imitation of the argument in Theorem 6.8. By Theorem 6.12 any minimizing ρ has $A(\rho) = 0$ and thus minimizes $\mathcal{E}_j(\rho)$. There can be only one such ρ since the minimizing ρ for $\mathcal{E}_j$ is unique. ■

In summary: $E(\lambda) = E_j(\lambda) - \alpha\lambda$ always, but a minimizing ρ may or may not exist for the TFD problem. It exists if and only if the minimizing ρ for the j model (which always exists when $\lambda \le \lambda_c$) satisfies $\rho(x) \notin (0,\rho_0)$ a.e. A sufficient condition on V for this to occur is that V be amenable; a necessary condition seems to be difficult to find. The example of §III.B illustrates the non-existence phenomenon: If $dm(x) = g(x)dx$ with $g(x) \in (0,\rho_0)$ and $g \in L^1$, then $\rho_j(x) = g(x)$ in the neutral case, $\lambda = \lambda_c = \int g$. But $g(x)$ does not minimize $\mathcal{E}(\rho)$. A sequence of minimizing ρ's for $\mathcal{E}(\rho)$ are functions which on the average locally imitate g but which oscillate rapidly between the two values 0 and ρ_0.

VII. THOMAS-FERMI-VON WEIZSÄCKER THEORY

TFW theory was originally suggested by von Weizsäcker (1935) with $\delta = \delta_1 = \hbar^2/2m$ in (7.1). There is no fundamental justification for TFW theory in the sense that there is for TF theory, i.e. there is no theorem that the correction to the energy or density caused by the Weizsäcker term with $\delta = \delta_1$ agrees with the real quantum problem. We will see, however, that if $\delta = (.186)\,\delta_1$ then the energy correction proposed by Scott (§V.B), but not the density correction conjectured in (5.37) is realized in TFW theory. If Scott is correct then a-posteriori TFW theory has some fundamental meaning for atoms and molecules.

We were able to make a great deal of progress with TF and TFD theories essentially because of the pointwise relation between $\phi(x)$ and $\rho(x)$. In TFW theory this relation is lost and therefore TFW theory is much more difficult mathematically. However, the physical consequences of TFW theory are much richer and qualitatively more nearly parallel the physics of real atoms and molecules. In addition to the above mentioned Z^2 energy correction, TFW theory remedies three defects of TF (and TFD) theory:

(i) ρ will be finite at the nuclei.

(ii) binding of atoms occurs and negative ions are stable (i.e. $\lambda_c > Z$). These two facts are closely related.

(iii) ρ has exponential falloff if $\lambda < \lambda_c$, e.g. for neutral atoms and molecules.

The theory presented here was begun by Benguria (1979) and then further developed by Benguria, Brezis and Lieb (1981), (BBL), to which we will refer for technical details. Some newer results will also be given, especially that $\lambda_c > Z$ for molecules, the Z^2 correction to the energy (§VII.D), and the binding of equal atoms. Many interesting problems are still open, however.

The TFW energy functional is

$$\mathcal{E}(\rho) = A\int(\nabla\rho^{\frac{1}{2}}(x))^2dx + (\gamma/p)\int\rho(x)^p dx - \int V(x)\rho(x)dx + D(\rho,\rho) + U. \tag{7.1}$$

This agrees with (2.8) in units in which $\hbar^2/2m = 1$. A closely related functional, obtained by writing $\psi^2 = \rho$, is

$$\mathcal{E}'(\psi) = A\int(\nabla\psi)^2 + (\gamma/p)\int\psi^{2p} - \int V\psi^2 + D(\psi^2, \psi^2) + U \tag{7.2}$$

Notes: (i) In this section all quantities refer to TFW theory unless otherwise stated.

(ii) (7.1) is defined for $\rho(x) \geq 0$ while in (7.2) $\psi(x)$ only has to be real.

(iii) As in section VI, we will assume for simplicity that V is given by (6.3) <u>et</u>. <u>seq</u>. Later on a slightly stronger hypothesis (7.12) will be used.

(iv) $p > 1$ is a parameter; it will not be indicated explicitly unless necessary. Recall that E^{TF} was finite for point nuclei only if $p > 4/3$. E is finite in TFW theory for <u>all</u> $p > 0$. We need $p > 1$ for Theorems 7.1 and 7.2, among other reasons. Even though we are interested in $p = 5/3$, we allow p to be arbitrary because the dependence on p is interesting. It will turn out that $p = 5/3$, the case of physical interest, is special--at least it is so for the proof that $\lambda_c > Z$.

VII.A. <u>The TFW Minimization Problem</u>

The function space is

$$G'_p = \{\psi \mid \nabla\psi \in L^2,\ \psi \in L^6 \cap L^{2p},\ D(\psi^2, \psi^2) < \infty\} \tag{7.3}$$

We say $\rho \in G_p$ if $\rho(x) \geq 0$ and $\rho^{\frac{1}{2}} \in G'_p$. G'_p contains

$$F'_p = G'_p \cap L^2 = \{\psi \mid \nabla\psi \in L^2, \psi \in L^6 \cap L^{2p} \cap L^2\} . \tag{7.4}$$

Even though we are interested in $\rho \in L^1$ (or $\psi \in L^2$), the larger space G'_p is used for technical reasons in order to prove that $\lambda_c < \infty$; in other words we will eventually find that all ρ's of interest are in F_p (defined analogously to G_p).

Remark: By Sobolev's inequality, $||\nabla\psi||_2 \geq L\,||\psi||_6$ with $L = 3^{\frac{1}{2}}(\pi/2)^{2/3}$ (cf. (Lieb, 1976)) when $\nabla\psi$ and $\psi \in L^2$. Thus, when $\psi \in F'_p$ the restrictions that $\psi \in L^6$ and $\psi \in L^{2p}$ (if $p \leq 3$) are unnecessary. In short, $F'_p = H^1 \cap L^{2p}$ where $H^1 = \{\psi | \nabla\psi$ and $\psi \in L^2\}$.

As usual,

$$E(\lambda) = \inf\{\mathcal{E}(\rho) | \rho \in F_p, \int \rho = \lambda\}$$

$$E'(\lambda) = \inf\{\mathcal{E}'(\psi) | \psi \in F'_p, \int \psi^2 = \lambda\}\ . \qquad (7.5)$$

Theorem 7.1: $\mathcal{E}(\rho)$ is strictly convex in ρ.

Proof: The only term that has to be checked is $\int(\nabla\rho^{\frac{1}{2}})^2$. If $\psi_i = \rho_i^{\frac{1}{2}}$, $i = 1,2$ and $\psi = (\Sigma\, \alpha_i^2 \rho_i)^{\frac{1}{2}}$, $\Sigma\, \alpha_i^2 = 1$, then $\psi\nabla\psi = \Sigma(\alpha_i \psi_i)(\alpha_i \nabla\psi_i)$ and $(\psi\nabla\psi)^2 \leq (\Sigma\alpha_i^2 \psi_i^2)(\Sigma\, \alpha_i^2(\nabla\psi_i)^2)$. Assuming $\psi(x) > 0$ everywhere, we are done. Otherwise, the result follows by approximation. ■

Remark: $\mathcal{E}'(\psi)$ is not convex in ψ because of the $-\int V\psi^2$ term.

Theorem 7.2: For all finite λ

(i) $E(\lambda) = E'(\lambda)$

(ii) $E(\lambda) = \inf\{\mathcal{E}(\rho) | \rho \in F_p, \int\rho \leq \lambda\}$
and similarly for $E'(\lambda)$

(iii) $E(\lambda)$ is convex and monotone nonincreasing in λ

Proof: (i) Given ρ, we can always construct $\psi = \rho^{\frac{1}{2}}$, so $E'(\lambda) \leq E(\lambda)$. Conversely, given ψ let $f = |\psi|$. But $\nabla f = (\nabla\psi)(\mathrm{sgn}\,\psi)$ so $\int(\nabla f)^2 = \int(\nabla\psi)^2$. Thus $\mathcal{E}'(f) = \mathcal{E}'(\psi)$. Choosing $\rho = f^2$, $E(\lambda) \leq E'(\lambda)$. (ii) As before, "excess charge" can be put at "infinity". Here, $p > 1$ is essential. (iii) $\mathcal{E}(\rho)$ is convex so $E(\lambda)$ is convex. Monotonicity is implied by (ii). ■

Remark: (i) relates the two problems defined by (7.1) and (7.2).

To obtain the convexity (iii), $\mathcal{E}$ and Theorem (7.1) were used. We will use $\mathcal{E}'$ to obtain the existence of a minimum, and then the TFW equation for this minimum.

Lemma 7.3: Let $V = |x|^{-1} * m$, with m a measure and $|m| = Z < \infty$. Let $\rho(x) \geq 0$. Then there exists a constant, C, independent of m and ρ such that for every $\varepsilon > 0$

$$\int V\rho \leq \varepsilon Z||\rho||_3 + Z\,\varepsilon^{-\frac{1}{2}}\,C\,D(\rho,\rho)^{\frac{1}{2}} .$$

Proof: By regarding $\mathbb{R}^3$ as the union of balls of unit radius centered on the points of $(\frac{1}{2})\,\mathbb{Z}^3$ it suffices to assume $\mathrm{supp}(m) \subset B_1$, where $B_R = \{x \,|\, |x| \leq R\}$ and χ_R is the characteristic function of B_R. In the following, irrelevant constants will be suppressed. Write $V = V_- + V_+$, where $V_- = V\chi_2$ and similarly for ρ. First consider $I_- = \int V_-\rho_-$. By Young's inequality (and writing $|x-y|^{-1} = |x-y|^{-1}\chi_4$ for $x,y \in B_2$) $||V_-||_2 \leq Z$. Thus,

$$I_- \leq Z||\rho_-||_2 \leq Z||\rho||_3{}^{3/4}||\rho_-||_1{}^{1/4} \leq Z\{||\rho||_3 + ||\rho_-||_1\}.$$

But $D(\rho,\rho) \geq D(\rho_-,\rho_-) \geq ||\rho_-||_1^2$ since $|x-y|^{-1} > 1/8$ in B_2. Now, outside B_2, $V(x) < 2Z/|x| \equiv W(x)$. Let $Q = q\chi_2$ be the constant charge distribution such that $|x|^{-1} * Q = W(x)$ outside B_2. Then $I_+ \leq 2D(Q,\rho_+) \leq 2D(Q,Q)^{\frac{1}{2}}\,D(\rho_+,\rho_+)^{\frac{1}{2}}$. But $D(Q,Q) = Z^2$. Therefore, (on the whole of $\mathbb{R}^3$) $I \leq Z||\rho||_3 + ZC\,D(\rho,\rho)^{\frac{1}{2}}$. Now replace $\rho(x)$ by $\varepsilon^3\,\rho(\varepsilon x)$ and $dm(x)$ by $\varepsilon^3 dm(\varepsilon x)$. Then, $\varepsilon I = I_\varepsilon \leq Z\,\varepsilon^2||\rho||_3 + ZC\,\varepsilon^{\frac{1}{2}}\,D(\rho,\rho)$. ■

From the Sobolev inequality, the following is obtained

Corollary 7.4: There are constants a and $b > 0$ such that for every $\psi \in G_p'$

$$\mathcal{E}'(\psi) \geq a[||\nabla\psi||_2 + ||\rho||_p^p + ||\rho||_3 + D(\rho,\rho)] + U - b. \quad (7.6)$$

with $\rho = \psi^2$. In particular, $\mathcal{E}'$ is bounded below on G_p' and $E(\lambda)$ is bounded below.

It is obvious that $E(\lambda) \geq E^{TF}(\lambda)$, with the same p. If $p > 4/3$, $E^{TF}(\lambda)$ is finite for point nuclei. The following illustrates the sort of lower bound for $E(\lambda)$ that can be obtained with the Sobolev inequality.

Theorem 7.5: Let $p = 5/3$. Let $E^{TFW}(A,\gamma,\lambda)$ denote the TFW energy and $E^{TF}(\gamma,\lambda)$ denote the TF energy. Let $\tilde{L} = 9.578$. Then

$$E^{TFW}(A,\gamma,\lambda) \geq E^{TF}(\gamma + A\tilde{L}\lambda^{-2/3},\,\lambda) \quad (7.7)$$

In particular for an atom with a point nucleus, $E^{TF}(\gamma,\lambda) \sim \gamma^{-1}$ whence, for an atom,

$$E^{TFW}(A,\gamma,\lambda) \geq \gamma(\gamma + A\tilde{L}\lambda^{-2/3})^{-1} E^{TF}(\gamma,\lambda) \tag{7.8}$$

Proof: $\int|\nabla\psi|^2 \geq \tilde{L}(\int|\psi|^{10/3})(\int|\psi|^2)^{-2/3}$. cf. (Lieb, 1976). ■

Remark: The right side of (7.8) has two properties: (i) Its slope at $\lambda = 0$ is finite. (ii) It is strictly monotone decreasing for all λ. To some extent, E^{TFW} will be seen to mimic this: E^{TFW} has a finite slope at $\lambda = 0$ and is strictly decreasing up to $\lambda_c > Z$.

Theorem 7.6: (i) $\mathcal{E}'(\psi)$ has a minimum on the set $\psi \in F_p'$ and $\int\psi^2 \leq \lambda$.

(ii) $\mathcal{E}'(\psi)$ has a minimum on G_p'.

(iii) The same is true for $\mathcal{E}(\rho)$ on $F_p(\int\rho \leq \lambda)$ and G_p. Furthermore ρ and ψ are related by $\rho(x) = \psi(x)^2$. The minimizing ρ is unique.

Proof: The proof we will give is different from the proof of Theorem 2.4 because Fatou's lemma will be used, as stated in the remark after Theorem 2.4. Let ψ^n be a minimizing sequence. By Corollary 7.4 all quantities in (7.6) are bounded. By passing to a subsequence we can demand, by the Banach-Alaogulu theorem, that $\nabla\psi^n \to f$ weakly in L^2 and $\rho^n \to \rho$ weakly in L^3 and in L^p (where $\rho^n = (\psi^n)^2$). Furthermore, for any bounded ball, B, $\psi \in L^2(B)$ since $\psi \in L^6(B)$. Moreover, $H^1(B)$ is relatively (norm) compact in $L^2(B)$. Thus, by passing to a further subsequence, we may assume $\psi^n \to \psi$ strongly in $L^2_2(B)$ for every B and pointwise a.e. Then it is clear that $\rho = \psi^2$ and $f = \nabla\psi$. As before, $\liminf ||\nabla\psi^n||_2 \geq ||\nabla\psi||_2$. From the pointwise convergence and Fatou's lemma, $\liminf D(\rho^n,\rho^n) \geq D(\rho,\rho)$ and $\liminf ||\rho^n||_p \geq ||\rho||_p$. For the V term we write $m = m_1 + m_2$ with $m_1 = m\chi_R$ and choose R large enough so that $|m_2| < \delta$ (since $|m| = Z < \infty$). If $V_2 = m_2 * |x|^{-1}$ then $\int V_2|\rho-\rho^n| < \delta(\text{const.})$ by Lemma 7.3 (with $\varepsilon = 1$). Next, write $V_1 = V_- + V_+$ with $V_- = V_1\chi_r$. If $r > 2R$, $V_+(x) < 2Z/|x|$. Let Q_r be the uniform charge distribution inside B_r so that $Q_r * |x|^{-1} = 2Z/|x|$ outside B_r. Then $\int V_+|\rho - \rho^n| \leq D(Q_r,Q_r)^{\frac{1}{2}} D(|\rho - \rho^n|, |\rho - \rho^n|)^{\frac{1}{2}}$. Choose r large enough so that $D(Q_r,Q_r) < \delta^2$. $V_- \in L^{3/2}$, so $\int V_-(\rho - \rho^n) \to 0$. Since δ was arbitrary, $\int V(\rho - \rho^n) \to 0$. Combining all this, $\liminf \mathcal{E}'(\psi^n) \geq \mathcal{E}'(\psi)$. Finally, if $\int\rho^n \leq \lambda$ then $\int\rho \leq \lambda$ as in the proof of Theorem 2.4 (but using L^3). As remarked in the proof of Theorem 7.2, we can choose $\psi^n(x) \geq 0$

everywhere; hence $\psi(x) \geq 0$ and $\rho(x) = \psi(x)^2$ minimizes $\mathcal{E}(\rho)$. The uniqueness of ρ follows from the strict convexity of $\mathcal{E}(\rho)$. ■

Definition: λ_c can be defined as in section II, namely $\lambda_c = \sup\{\lambda | E(\lambda) = \lim_{\lambda\to\infty} E(\lambda)\}$. A simple variational calculation, which exploits the fact that $V(x) \approx -Z/|x|$ for large $|x|$, shows that $\lambda_c > 0$.

Theorem 7.7: There is a minimizing ρ on F_p' with $\int\rho = \lambda$ if and only if $\lambda \leq \lambda_c$. The minimizing ρ in Theorem 7.6 when $\lambda > \lambda_c$ is the ρ for λ_c. $E(\lambda)$ is strictly convex on $[0,\lambda]$.

Proof: Same as for Theorem 2.5. ■

Theorem 7.8: (i) Any minimizing $\psi \in F_p'$ for $\mathcal{E}'(\psi)$ on the set $\int\psi^2 \leq \lambda$ satisfies the TFW equation (in the sense of distributions):

$$[-A\Delta + W_\psi(x)]\,\psi(x) = -\mu\,\psi(x) \tag{7.9}$$

where $$W_\psi(x) = \gamma\,\psi(x)^{2p-2} - \phi_\rho(x) \tag{7.10}$$

and $\phi_\rho = V - |x|^{-1} * \rho$ with $\rho = \psi^2$.

(ii) If ψ minimizes $\mathcal{E}'(\psi)$ on G_p' then ψ satisfies (7.9) with $\mu = 0$.

(iii) $E(\lambda)$ is continuously differentiable and $-\mu = dE/d\lambda$ for $\lambda \leq \lambda_c$ while $0 = dE/d\lambda$ for $\lambda \geq \lambda_c$. In particular $\mu \geq 0$.

(iv) If $\rho \in G_p$ satisfies (7.9) and $\int\rho = \lambda$ (possibly ∞) then ρ minimizes $\mathcal{E}(\cdot)$ on the set $\int\rho \leq \lambda$.

(v) Fix λ. There can be at most one pair ρ,μ (with $\rho(x) \geq 0$) that satisfies (7.7) with $\int\rho = \lambda$.

Proof: (i) and (ii) are standard. Just consider $\psi + \varepsilon f$ with $f \in C_o^\infty$ and $(f,\psi) = 0$ and set $d\mathcal{E}/d\varepsilon = 0$. For the absolute minimum we do not require $(f,\psi) = 0$. The proof of (iii) is as in Theorem 2.7 (cf. LS Theorem II.10 and Lemma II.27). The proof of (iv) and (v) imitates that of Theorem 2.6. ■

We will eventually prove that the minimizing ψ is unique (we already know that ρ, and hence ψ^2, is unique), but Theorem 7.9 is needed first. We also want to prove that $\lambda_c < \infty$, i.e. the ψ that satisfies (7.9) with $\mu = 0$ satisfies $\int\psi^2 < \infty$. This will be done in § VII.B.

Theorem 7.9: If $\psi \in G_p'$ satisfies (7.9) (as a distribution) and $\psi(x) \geq 0$ for all x then (i) ψ is continuous. More precisely $\psi \in C^{0,\alpha}$ for every $\alpha < 1$ (i.e. for every bounded ball, B, $|\psi(x) - \psi(y)| < M|x-y|^{\alpha}$ for some M and all $x,y \in B$). (ii) If V is C^{∞} on some open set, Ω, then ψ is C^{∞} on Ω. For point nuclei, $\Omega = \mathbb{R}^3\setminus\{R_i\}$. (iii) Either $\psi \equiv 0$ or $\psi(x) > 0$ everywhere. (iv) $W_{\psi} \in L^{3-\varepsilon}_{loc}$ for every $\varepsilon > 0$.

Proof: Clearly $V \in L^{3-\varepsilon}_{loc}$ (all $\varepsilon > 0$) and, since $\psi \in L^6$, $-A\Delta\psi \leq f$ with $f = V\psi \in L^{2-\varepsilon}_{loc}$ (all $\varepsilon > 0$). Choosing $\varepsilon < \frac{1}{2}$, we can apply a result of Stampacchia (1965, Theorem 5.2) to conclude $\psi \in L^{\infty}_{loc}$ and hence $\psi^{2p-1} \in L^{3-\varepsilon}_{loc}$ (all $\varepsilon > 0$). Now, $g = |x|^{-1} * \rho \in L^6$ (since $\Delta g = -4\pi\rho \Rightarrow K||g||_6^2 \leq \int(\nabla g)^2 = 8\pi\, D(\rho,\rho)$). Therefore $-\Delta\psi \in L^{3-\varepsilon}_{loc}$ (all $\varepsilon > 0$). Then (Adams, 1975, p. 98) $\psi \in C^{0,\alpha}$. (ii) follows by a bootstrap argument as in Theorem 2.8. For (iii) we note that $-\Delta\psi = b\psi$ and $b \in L^q_{loc}$, $q > 3/2$. The conclusion follows from Harnack's inequality (Gilbarg and Trudinger, 1977). ∎

We know that $\rho^{\frac{1}{2}} \geq 0$ satisfies (7.9) so $\rho^{\frac{1}{2}}$ enjoys the above properties. Since ρ is unique we will henceforth denote (7.10) simply by W. We will also use the notation

$$H = -A\Delta + W \tag{7.11}$$

Theorem 7.10: The minimizing ψ is unique up to a sign which is fixed by $\psi(x) = \rho(x)^{\frac{1}{2}} > 0$ everywhere. ψ is also the unique ground state eigenfunction of $H = -A\Delta + W(x)$ and μ is its ground state eigenvalue.

Proof: If ψ is minimizing then $\psi^2 = \rho$ and H are uniquely determined. $\psi = \rho^{\frac{1}{2}}$ satisfies $H\psi = -\mu\psi$. Since ψ is nonnegative, it is the ground state of H, and the ground state of H is unique up to sign (cf. Reed and Simon, § XIII.12). ∎

Remarks: (i) It is not claimed that the TFW equation (7.9) and (7.10) has no solution other than the positive one. Infinitely many other solutions probably exist. They have been found for certain non-linear equations which have some resemblance to the TFW equation (Berestycki and Lions, 1980), but the TFW equation itself has not been analyzed in this regard. These other solutions correspond, in some vague sense, to "excited states".

(ii) The interplay between $\mathcal{E}'(\psi)$ and $\mathcal{E}(\rho)$ should be noted. Apart from the somewhat pedantic question of the uniqueness of ψ,

$\mathcal{E}$ was used to get the uniqueness of $\rho = \psi^2$ and the convexity of $E(\lambda)$. $\mathcal{E}'$ was used to get the TFW equation in which it is not necessary to distinguish between $\rho(x) > 0$ and $\rho(x) = 0$ as in the TF equation (2.18). The ψ of interest automatically turns out to be positive. For purposes of comparison, the TF equation is $(W + \mu)\,\psi = 0$ if $\psi > 0$, and $(W + \mu) \geq 0$ if $\psi = 0$. The TFW equation is $(W + \mu)\,\psi = A\Delta\psi$ everywhere.

(iii) Note that there is a solution even for $\mu = 0$. For this ρ, $H = -A\Delta + W$ has zero as its ground state eigenvalue with an L^2 eigenfunction, ψ (Theorem 7.12). This is unusual. Zero is also the bottom of the essential spectrum of H.

To complete the picture of $E(\lambda)$ we have to know how $E(\lambda)$ behaves for small λ. Since μ is a decreasing function of λ, μ has its maximum at $\lambda = 0$.

Theorem 7.11: $\mu(\lambda = 0) = -e_o$ where $e_o < 0$ is the ground state energy of the Hamiltonian $H_o = -A\Delta - V(x)$. In particular, for a point nucleus $\mu(\lambda = 0) = z^2/(4A)$.

Proof: $\mu(0) = \lim_{\lambda\to o} E(0) - E(\lambda)$. $E(0) = U$. Let f be the normalized ground state of H_o : $H_o f = e_o f$. Let $\rho = \lambda f^2$. Clearly, $\mathcal{E}(\rho) = \lambda\, e_o + U + o(\lambda)$, since $p > 1$. On the other hand, for any ρ with $\int\rho = \lambda$, $\mathcal{E}(\rho) \geq \lambda\, e_o + U$. ■

In §VII.B we will see that $Z < \lambda_c < \infty$. Therefore the behavior of $E(\lambda)$ can be summarized as follows:

$e(\lambda) = E(\lambda) - U$ in TFW theory looks like Fig. 1 with two important changes: (i) $\lambda_c > Z$ (at least for $p \geq 5/3$). $e(\lambda)$ is strictly convex for $0 \leq \lambda \leq \lambda_c$. (ii) The slope at $\lambda = 0$ is finite. [In TF theory $e(\lambda) \approx \lambda^{1/3}$.]

VII.B. Properties of the Density and λ_c

Our main concern here will be to estimate λ_c. For energetic reasons, it is intuitively clear that $\lambda_c \geq Z$ for large enough p because otherwise the energy could always be lowered by adding some additional charge far out. Benguria (1979) proved this for $p \geq 4/3$. We will also see that $\lambda_c > Z$ for $p \geq 5/3$.

What is far from obvious, however, is that λ_c is finite. There is no energetic reason why $E(\lambda)$ could not steadily decrease (and be bounded, of course). It is easy to construct a $\rho(x)$ with $\int\rho = \infty$ so that all the terms in the energy and also $\phi(x)$, except near the nuclei, are finite. $\rho(x) = (1 + x^2)^{-3/2}$ is an example. That $\lambda_c < \infty$ is a subtle fact. The same question arises

in quantum theory and it has not been proved there that λ_c is finite.

In the following ψ always means $+\rho^{1/2}$. For simplicity we will henceforth assume the following condition in addition to $V \in \mathcal{D}$:

$$V(x) \leq C/|x| \tag{7.12}$$

for some $C < \infty$ and for all $|x| >$ some R. The fact that $V = |x|^{-1} * m$ and $|m| = Z$ does not guarantee (7.12). If, however, m has compact support, then (7.12) holds.

Theorem 7.12: $\lambda_c < \infty$ for all $p > 1$.

Proof: Let ρ give the absolute minimum of $\mathcal{E}(\rho)$ on G_p. ψ satisfies (7.9) with $\mu = 0$. We will prove that this ρ has $\lambda = \int\rho < \infty$, thereby proving that $E(\lambda)$ has an absolute minimum at λ and hence that $\lambda_c = \lambda$. Assume $\lambda = \infty$. Then for $|x| \geq$ some R (which is bigger than the R in (7.12)), $|x|^{-1} * \rho > 2C/|x|$. Thus, for $|x| > R$, $-A\Delta\psi \leq -C\psi/|x|$. Now we use a comparison argument. Let $f(x) = M\exp\{-2[C|x|/A]^{1/2}\}$ with $M > 0$. f satisfies $-A\Delta f \geq -C\, f/|x|$, for $|x| \neq 0$, so $-A\Delta(\psi - f) \leq -C(\psi - f)/|x|$. Fix M by $f(x) \geq \psi(x)$ for $|x| = R$. If we knew that $\psi(x) \to 0$ as $|x| \to \infty$ we could conclude, from the maximum principle, that $\psi < f$ for $|x| \geq R$. This implies that $\psi \in L^2$. Unfortunately, we only know that $\psi(x) \to 0$ in a weak sense (namely, L^6). This, it turns out, is good enough. Cf. BBL for details. ∎

Now that we know $\int\rho < \infty$, even for the absolute minimum $(\mu = 0)$, we can prove

Theorem 7.13: ψ is bounded on $\mathbb{R}^3$ and $\psi(x) \to 0$ as $|x| \to \infty$. Also, $\psi \in H^2$ (i.e. $\psi, \nabla\psi$ and $\Delta\psi \in L^2$).

Proof: $-A\Delta\psi < V\psi$ so $(-A\Delta + 1)\psi \leq (V + 1)\psi$. Since $(V + 1)\psi \in L^2$, $\psi \leq (-A\Delta + 1)^{-1}[(V + 1)\psi]$ and this is bounded and goes to zero as $|x| \to \infty$ (Lemma 3.1). Finally, $\psi^{2p-1} \leq d\psi$ for some d, and $g = |x|^{-1} * \rho_2 \in L^6$ together with $\psi \in L^6$ imply $g\psi \in L^2$. Hence $\Delta\psi \in L^2$. ∎

Theorem 7.14: If $p \geq 3/2$ then, for all x,

$$\gamma\rho(x)^{p-1} \leq V(x). \tag{7.13}$$

In particular, if $p = 5/3$, $\rho(x) < [V(x)/\gamma]^{3/2}$.

Proof: The essential point is that since V is superharmonic,

so is V^t for $t \le 1$. Let $f = \psi - (V/\gamma)^t$ with $t = 1/(2p-2)$. Let $B = \{x | f(x) > 0\}$. Since ψ and V are continuous on B, B is open. On B, $W > 0$ so $-\Delta f < 0$. $f = 0$ at ∞ and on ∂B, so B is empty. ∎

Remark: The bound in (7.13) also holds trivially in TF theory from (2.18).

Theorem 7.15: If $p \ge 4/3$ then $\lambda_c \ge Z$.

Proof: Suppose $\lambda_c = Z - \varepsilon$. Since $H = -A\Delta + W$ has zero as its ground state energy, $(f, Hf) \ge 0$ for any $f \in C_0^\infty$. Let $f_1(x) \not\equiv 0$ be spherically symmetric with support in $1 \le |x| \le 2$, $f_1(x) \le 1$, and $f_n(x) = f(x/n)$. Then $\int f_n{}^2\phi = \int f_n{}^2[\phi]$, where $[\phi]$ is the spherical average of ϕ. It is easy to see that for $|x| \ge$ some R, $[\phi] \ge \varepsilon/2|x|$ since $\int\rho = Z - \varepsilon$. Therefore $\int f_n{}^2\phi \ge (\text{const.})\, n^2$ for large n. $\int(\nabla f_n)^2 = (\text{const.})n$. The crucial quantity is $D_n = \int f_n{}^2 \rho^{p-1}$. If $p \ge 2$, $D_n \le (\text{const.})\int\rho$. If $p < 2$ use Hölder's inequality: $D_n \le X_n{}^{p-1} Y_n{}^{2-p}$ where $X_n = \int f_n{}^2\rho$ and $Y_n = \int f_n{}^2$. Clearly $X_n \to 0$ as $n \to \infty$ since $\rho \in L^1$. $Y_n = (\text{const.})\, n^3$. Now let $n \to \infty$, whence $(f_n, Hf_n) \to -\infty$. ∎

Remarks: (i) The basic reason that $p \ge 4/3$ is needed in the proof of Theorem 7.15 is that we want to be able to ignore the ρ^{p-1} term in W and thereby obtain a negative energy bound state for H when $\lambda < Z$. However, if $\rho \in L^1$ then (essentially) $\rho(x) \sim |x|^{-3} f(x)$, where $f(x)$ can be slowly decreasing. Hence we can be certain that ρ^{p-1} is small compared to $|x|^{-1}$ only if $3(p-1) \ge 1$.

(ii) In Theorems 7.16 and 7.19 we prove that $\lambda_c > Z$. The underlying idea is that to have a zero energy L^2 bound state, $W(x)$ has to be positive for large $|x|$. Essentially, $W(x)$ has to be bigger than $|x|^{-2}$; this requirement is clear if we assume $\psi(x) \sim |x|^{-a}$ for large $|x|$. If $\lambda_c = Z$, then ϕ is (essentially) positive for large $|x|$, so the repulsion has to come from ρ^{p-1}. But if $p - 1 \ge 2/3$ then ρ^{p-1} cannot be sufficiently big since $\rho \in L^1$. The theorem that $\lambda_c > Z$ when $p \ge 5/3$ was proved for an atom in BBL. We give that proof first in Theorem 7.16 in order to clarify the ideas. Then, after Lemma 7.18, we give a proof (which is not in BBL) of the general case in Theorem 7.19. Some condition on p really is needed to have $\lambda_c > Z$. In BBL it is proved that if $p = 3/2$, $\gamma = 1$, V is given by (2.1) and $A \le 1/16\pi$, then $\lambda_c = Z$.

Theorem 7.16: Suppose $p \ge 5/3$ and suppose $V = |x|^{-1} * m$ where m is a nonnegative measure that satisfies (i) m is spherically symmetric; (ii) the support of m is contained in $B_R = \{x | |x| < R\}$.

Then $\lambda_c > Z$.

Proof: Assume that $\lambda_c \le Z$. By Newton's theorem, $\phi(x) \ge 0$ for $|x| > R$. Then when $\lambda = \lambda_c$, $-\Delta\psi \ge -\gamma\,\psi^{2p-1}$ for $|x| > R$. ψ is spherically symmetric and $\psi(R) > 0$. Let $f(x) = C|x|^{-3/2}$ which satisfies $-A\Delta f \le -\gamma\, f^{2p-1}$ for $|x| \ge R$ provided $0 < C \le D$ with $D^{2p-2} = (3A/4\gamma)R^{3p-5}$. Let $C = \min(D, \psi(R))$. Then $\psi(x) \ge f(x)$ for all $|x| > R$, because $-A\Delta(\psi - f) \ge -\gamma(\psi^{2p-1} - f^{2p-1})$ which would imply that $\psi - f$ is superharmonic on the set where $\psi - f < 0$. Since ψ and f go to zero at infinity, and $\psi - f \ge 0$ at $|x| = R$, this is impossible. Hence $\psi \notin L^2$ which contradicts $||\psi||_2 \le Z$. ■

In the foregoing we used a comparison argument which, in turn, relied on the fact that the positive part of W, namely ρ^{p-1}, was simply related to ψ. In the proof of Theorem 7.19 we will not have that luxury and so the more powerful Lemma 7.18 is needed.

Lemma 7.17: Let S_R denote the sphere $\{x \,|\, |x| = R\}$ and let $d\Omega$ be the normalized invariant spherical measure on S_1. For any function, h, let $[h](r) = \int h(r,\Omega)d\Omega$ be the spherical average of h. Now suppose $\psi(x) > 0$ is C^2 in a neighborhood of S_R. Let $f(r) = \exp\{[\ln\psi](r)\}$. Then, for r in some neighborhood of R, $[\Delta\psi/\psi](r) \ge (\Delta f/f)(r) = \{d^2f/dr^2 + (2/r)\, df/dr\}/f(r)$.

Proof: Let $g(x) = \ln\psi(x)$. Then $\Delta\psi/\psi = \Delta g + (\nabla g)^2$. Clearly $[\Delta g] = \Delta[g]$. Moreover, $(\nabla g)^2 \ge \{\partial g(r,\Omega)/\partial r\}^2$, and $[(\partial g/\partial r)^2] \ge (d[g]/dr)^2$ by the Schwarz inequality. Thus, $[\Delta\psi/\psi] \ge \Delta[g] + (\nabla[g])^2 = \Delta f/f$. ■

Lemma 7.18: Suppose $\psi(x) > 0$ is a C^2 function in a neighborhood of the domain $D = \{x \,|\, |x| > R\}$ and ψ satisfies $\{-\Delta + W(x)\}\,\psi(x) \ge 0$ on D. Let $[W]$ be the spherical average of W and write $[W] = V_+ - V_-$ with $V_+(x) = \max([W](x), 0)$. Suppose $V_+ \in L^{3/2}(D)$. Then $\psi \notin L^2(D)$. (Note: no hypothesis is made about V_-.)

Remarks: Simon (1981, Appendix 3) proves a similar theorem for $D = \mathbb{R}^3$, except that $[W] = V_+ - V_-$ is replaced by $W = V_+ - V_-$ with $V_+ = \max(W,0)$. Simon does not requires the technical restrictions that $\psi(x) \ge 0$ and ψ is C^2. Simon's theorem will be used in our proof. Lemma 7.18 improves Simon's result in two ways:
(i) It is sufficient to consider D and not all of $\mathbb{R}^3$.
(ii) It is only necessary that $[W]_+$, and not W_+, be in $L^{3/2}$; the latter distinction is important. As an example, suppose that for large $|x|$ the potential, W, is that of a dipole, i.e. $W(x_1,x_2,x_3) = x_1|x|^{-3}$. $W_+ \notin L^{3/2}$ but, since $[W]_+ = 0$, Lemma 7.18 says that this W cannot have a zero energy L^2 bound state.

Proof: Let $f = \exp\{[\ln \psi]\}$ as in Lemma 7.17. Then $-\Delta f/f + [W] \geq [-\Delta\psi/\psi + W] \geq 0$. By Jensen's inequality $\int f^2 \leq \int \psi^2$, so if $f \notin L^2$ then $\psi \notin L^2$. Therefore, it suffices to consider $\{-A\Delta + [W](x)\}f \geq 0$ and to prove $f \notin L^2$ under the stated condition on $[W]$. First, suppose $D = \mathbb{R}^3$. Then this is just Simon's (1981) theorem. [However, since we are now dealing with spherically symmetric [W] and f, it is likely that a direct, ordinary differential equation proof can be found to replace Simon's proof.] Next, suppose $R > 0$. Let $g(x) > 0$ be any C^2 function defined in $\mathbb{R}^3$ such that $g(x) = f(x)$ for $|x| \geq R$. Then $\{-A\Delta + U(x)\}g \geq 0$ on $\mathbb{R}^3$ where $U = [W]$ for $|x| \geq R$ and U is bounded for $|x| \leq R$. Clearly $[W]_+ \in L^{3/2}(D)$ if and only if $U_+ \in L^{3/2}(\mathbb{R}^3)$. Apply Simon's theorem to U. ∎

Theorem 7.19: Let the hypothesis be the same as in Theorem 7.16 except that (i) is omitted. (In other words, a molecule is now being considered.) Then $\lambda_c > Z$.

Proof: For $|x| > R$, $V(x)$ is C^∞ so $\psi(x) > 0$ and $\psi \in C^2$ by Theorem 7.9. Assume $\lambda_c \leq Z$. The hypotheses of Lemma 7.18 is satisfied with $[-A\Delta + \bar{W}(x)]\psi = 0$. To obtain a contradiction we have to show $[W]_+ \in L^{3/2}$. Consider ϕ. Even if ϕ is negative somewhere, $[\phi](r) > 0$ in D by Newton's theorem. Therefore it suffices to show $[\rho^{p-1}] \in L^{3/2}$. If $p \geq 5/3$ then $p-1 \geq 2/3$ and $[\rho^{p-1}](r) < C[\rho^{2/3}](r)$, since ρ is bounded. But $[\rho^{2/3}](r) \leq \{[\rho](r)\}^{2/3}$, by Hölder, and $\int [\rho]^{(2/3)(3/2)} = \int \rho < \infty$. ∎

We know that $\lambda_c > Z$ in the physically interesting case $p = 5/3$. How large is $\lambda_c - Z$? In other words, how negative can ions be? This seems to be a very difficult question, even for an atom. To obtain qualitative agreement with quantum theory, it would be desirable if $\lambda_c - Z = O(1)$, at least for Z up to 100, say. The only available bound, at present, is Theorem 7.23. First Lemmas 7.20, 7.21 and 7.22 are needed. The lemmas were inspired by the work of R. Benguria (private communication) who proved the lemmas and Theorem 7.23 in the spherically symmetric case (which corresponds to the atom in TFW theory).

Lemma 7.20: Let ψ and f be two real valued functions on $\mathbb{R}^3$ which satisfy $-\Delta\psi = f$ in the sense of distributions. Let r denote the function $|x|$. Suppose $\psi \in L^2$, $f \in L^2$ and $r\psi f \in L^1$. Then, for any constant $d \geq 0$,

$$\int (|x|^2 + d)^{\frac{1}{2}} \psi(x) f(x) dx \geq 0.$$

Proof: Using dominated convergence, it is sufficient to consider only $d > 0$. Let $R = (r^2 + d)^{\frac{1}{2}} \in C^\infty$. We have $\Delta R = 2R^{-1} + dR^{-3}$ and $|(R/r) \nabla R|^2 = 1$. Suppose $\phi \in C_0^\infty$ (infinitely differentiable functions of compact support). We claim $I = -\int R\phi\Delta\phi \geq 0$. To see

this, integrate by parts: $I = A+B$ with $A = \int(\nabla\phi)^2 R$ and $B = \int\phi\nabla\phi\cdot\nabla R = \int(\nabla\phi\cdot\nabla R)(R/r)(r/R)\phi$. By Schwarz, and $\{(\nabla\phi\cdot\nabla R)(R/r)\}^2 \le (\nabla\phi)^2$, we have $B^2 \le AC$ with $2C = 2\int\phi^2 r^2 R^{-3} \le \int\phi^2\Delta R$. However, $2B = \int\nabla\phi^2\cdot\nabla R = -\int\phi^2\Delta R$, and hence $|B| \le A$, which proves the lemma. Now, suppose ψ and $f \notin C_o^\infty$ have compact support. Given $\varepsilon > 0$ there exists $g \in C_o^\infty$ such that $||\psi-g||_2 < \varepsilon$, $||\nabla\psi-\nabla g||_2 < \varepsilon$ and $||\Delta\psi-\Delta g||_2 < \varepsilon$. (Note: since ψ and $\Delta\psi \in L^2$, so is $\nabla\psi$.) Then $gR \in C_o^\infty$ and $\int gRf = -\int gR\Delta\psi \equiv -\int\psi\,\Delta(Rg) = -\int g\Delta(Rg) - M = -\int Rg\Delta g - M$ with $M = \int(\psi-g)\,\Delta(Rg)$. It suffices to show that $M \to 0$ as $\varepsilon \to 0$ because $\int gRf \to \int\psi Rf$. But $M = \int(\psi-g)\{g\Delta R + 2\nabla g\cdot\nabla R + R\Delta g\}$. We can assume $\mathrm{supp}(g)$ is in some fixed ball, independent of ε. Since $g, \nabla g$ and Δg are uniformly L^2 bounded, $M \to 0$. For the general case, let $h \in C_o^\infty$ satisfy $1 \ge h \ge 0$, $(\nabla h)(0) = 0$, $h(0) = 1$, $h(x) = 0$ for $|x| > 1$. Let $h_n(x) = h(x/n)$ and $\psi_n = h_n\psi$. Then, as a distribution, $-\Delta\psi_n = h_n f - 2\nabla h_n\cdot\nabla\psi - \psi\Delta h_n \equiv K_n$. By the previous result, $T_n \equiv \int R\psi_n K_n \ge 0$. But $\int h_n^2\,\psi fR \to \int\psi fR$ by dominated convergence. $Rh_n\,\Delta h_n = n^{-1}\,P_n(x/n)$ with $P_n(x) = (|x|^2 + dn^{-2})^{\frac{1}{2}}(h\Delta h)(x) < a$, so $\int R\psi_n\,\psi\Delta h_n \to 0$ since $\psi \in L^2$. Similarly, $R\nabla h_n^{\,2} \equiv L_n$ is uniformly bounded and converges pointwise to zero. Since ψ and $\nabla\psi \in L^2$, $\int\psi\nabla\psi L_n \to 0$ by dominated convergence. ■

<u>Remarks</u>: (i) Lemma 7.20 is useful for L^2 solutions to the Schrödinger equation $(-\Delta + W(x))\psi = -\mu\psi$. Then $\int\psi^2(W+\mu)r \le 0$ under some mild conditions on W (e.g. $W(x) < c/r$ for large r, $W \in L^2_{loc}$ and $r\psi^2 \in L^1$ if $\mu \ne 0$.)

(ii) The essential properties of R that were used were $R > 0$ and $R\Delta R \ge 2(\nabla R)^2$. Therefore Lemma 7.20 will hold for functions other than $(r^2 + d)^{\frac{1}{2}}$ having these properties. Formally, this means that if $R(x) = 1/V(x)$ then we require $V > 0$ and $\Delta V < 0$. We state this as Lemma 7.21 whose proof imitates the proof of 7.20.

<u>Lemma 7.21</u>: Suppose $V = |x|^{-1} * m$ with m a nonnegative measure, $|m| = Z < \infty$ and V satisfies (7.12). Then if ψ and f satisfy the hypotheses of Lemma 7.20, $\int dx\,\psi(x)f(x)/V(x) \ge 0$.

<u>Lemma 7.22</u>: Let $\rho(x) \ge 0$, $\int\rho = \lambda < \infty$ and $V = |x|^{-1} * m$ where m is any nonnegative measure with $|m| = Z < \infty$. Then $I \equiv \int\rho(x)\,\rho(y)\,|x-y|^{-1}\,V(x)^{-1}\,dxdy \ge \lambda^2/2Z$.

<u>Proof</u>: Take $Z=1$ and let $0 < \varepsilon < 1$. By Proposition 3.24, $\rho = \rho_1 + \rho_2$ with $\rho_1, \rho_2 \ge 0$, and $H_i = |x|^{-1} * \rho_i$ satisfies $H_1 \le \varepsilon\lambda V$ and $H_1 = \varepsilon\lambda V$ when $\rho_2 > 0$. Clearly, $\int\rho_1 \le \varepsilon\lambda$ by Lemma 3.3. Then $\int\rho_2 \ge (1-\varepsilon)\lambda$ and $I \ge \int\rho_2(H_1+H_2)/V \ge \varepsilon(1-\varepsilon)\lambda^2 + \int\rho_2\,H_2/V$. Repeat the argument with ρ_2 (using $\int\rho_2 \ge (1-\varepsilon)\lambda$), and so on ad infinitum. Then

$I \geq \lambda^2 \varepsilon(1-\varepsilon)\Sigma_{j=0}^{\infty} (1-\varepsilon)^{2j} = \lambda^2(1-\varepsilon)/(2-\varepsilon)$. Now let $\varepsilon \to 0$. ∎

Remarks: (i) Benguria proved Lemma 7.22 when $V = 1/r$. In this case one can simply use the fact that $(|x| + |y|)|x-y|^{-1} \geq 1$.

Theorem 7.23: Assume V satisfies 7.12. Then $\lambda_c < 2Z$, for all $p > 1$.

Proof: We know $\lambda_c < \infty$. Let ψ be the minimizing solution of (7.9) with $\mu = 0$. Then, by Theorem 7.13, ψ and $f = (\gamma\rho^{p-1} - \phi)\psi$ satisfy the hypotheses of Lemma 7.21. Thus, $0 < \int \rho\phi/V = \lambda_c - I$ with $I = \int H\rho/V$ and $H = |x|^{-1} * \rho$. But $I \geq \lambda_c{}^2/2Z$. ∎

Remark: This bound does not involve the value of A in (7.1). It also does not utilize the $\gamma\rho^{p-1}$ term in W. There is considerable room for improvement.

The next two theorems are about the asymptotics of ψ.

Theorem 7.24: Let ψ be the positive solution to (7.9), for any p.

(i) Let $\mu > 0$. Then for every $t < \mu$ there exists a constant, M, such that $\psi(x) \leq M \exp[-(t/A)^{\frac{1}{2}}|x|]$.

(ii) Let $\mu = 0$ (i.e. $\lambda = \lambda_c$), and assume $\lambda_c > Z$, as is certainly the case when $p \geq 5/3$. Assume also, that m has compact support. Then, for every $t < \lambda_c - Z$ there is a constant M such that $\psi(x) \leq M \exp[-2(t\,|x|/A)^{\frac{1}{2}}]$.

Proof: (i) is standard. Since ψ and $V \to 0$ as $|x| \to \infty$ we have $\psi = -(-A\Delta + t)^{-1}(W + \mu-t)\psi$. For $|x| >$ some R, $W + \mu-t > 0$. Therefore, since $\psi > 0$, $\psi(x) \leq \int Y(x-y)\, \chi_R(y)(W(y) + \mu-t)\psi(y)dy$ where $Y(x) = (4\pi A|x|)^{-1} \exp[-(t/A)^{\frac{1}{2}}|x|]$. The proof of (ii) is the same as the proof of Theorem 7.12. It is only necessary to note that since m has compact support (in B_R, say) then $V(x) \leq Z/(|x|-R)$ for $|x| > R$ and this is $\leq (Z+\varepsilon)/|x|$ for $|x|$ large enough. ∎

The next theorem is the well-known cusp condition (Kato, 1957).

Theorem 7.25: Let $V(x) = \Sigma z_j|x-R_j|^{-1}$ be the potential of point nuclei. Then at each R_j

$$z_j\psi(R_j) = -2A \lim_{r\downarrow 0} \int_{|x-R_j|=r} r^{-1}(x-R_j) \cdot \nabla\psi(x)\, d\Omega$$

where $d\Omega$ is the normalized uniform measure on the sphere. This holds for any λ. In particular, for an atom with nuclear charge z located at the origin, ψ is spherically symmetric and

$$z\,\psi(0) = -2A \lim_{r\downarrow 0} (d\psi/dr)(r).$$

Proof: Recall that, by Theorem 7.9, ψ is C^∞ away from the R_j and ψ is Hölder continuous everywhere. The theorem is proved by integrating (7.9) in a small ball B_r and then integrating by parts. The spherical symmetry in the atomic case is implied by uniqueness. ■

Theorem 7.26: Let $V(x) = z/|x|$ be the potential of an atom with a point nucleus. Then, for any λ, $\psi(r)$ is a strictly decreasing function of r.

Proof: In Theorem 2.12 and the following remark we proved this for $\lambda \le z$ by using rearrangement inequalities. Here we give a different proof for $\lambda \le z$ which extends to $\lambda > z$. Recall that ψ is continuous and positive and that ψ is C^∞ for $r > 0$. Also, $\mu \ge 0$. Let $Q(r) = \int \chi_r \rho$ be the electronic charge inside the ball B_r. By Newton's theorem, the potential, ϕ, satisfies:

(i) $\phi(r) \le [z - Q(r)]/r$. (ii) $\dot\phi = [Q(r) - z]/r^2$ (dots denote d/dr). (iii) If $\lambda \le z$, $\phi(r) \ge 0$ and $\phi(r)$ is decreasing. (iv) If $\lambda > z$ there is a unique $R > 0$ such that $\phi(r) \ge 0$ and decreasing for $r \le R$, and $\phi(r) < 0$ for $r > R$. $Q(R) < z$.
$\lambda \le z$: By Theorem 7.24 $\dot\psi(r) < 0$ near $r = 0$. If ψ is not monotone then since $\psi(r) \to 0$ as $r \to \infty$, there are two points $0 < r_1 < r_2$ such that $\psi(r_1) \le \psi(r_2)$, $\ddot\psi(r_1) \ge 0$, $\ddot\psi(r_2) \le 0$ and $\dot\psi(r_1) = \dot\psi(r_2) = 0$. Since ψ does not have compact support, $Q(r) < z$, all r. Hence $W(r_2) > W(r_1)$. Since $\ddot\psi(r_1) = [W(r_1) + \mu]\,\psi(r_1) \ge 0$, we have $W(r_1) + \mu \ge 0$. But then $0 \ge \ddot\psi(r_2) = [W(r_2) + \mu]\,\psi(r_2)$ is impossible.

$\lambda > z$: There is an $\varepsilon > 0$ such that $W(r) > 0$ for $r > R-\varepsilon$. Let $D_r = \{x \in \mathbb{R}^3 \,|\, |x| > r\}$. Take $r > R-\varepsilon$. On $D_r, -\Delta\psi < 0$. Since $\psi > 0$ is subharmonic on D_r, ψ has its unique maximum on ∂D_r, namely $|x| = r$. This proves the theorem on the domain $\{r \,|\, r > R-\varepsilon\}$. To prove the theorem on the domain $\{r \,|\, 0 \le r < R\}$ the argument in the $\lambda \le z$ case can be used since $Q(r) < z$ in this domain. ■

Conjecture: In the point nucleus, atomic case ψ is convex, possibly even log convex.

VII.C. Binding in TFW Theory

In TF theory binding never occurs when the repulsion, U, is included. In TFW theory binding is a common phenomenon. We conjecture that every neutral system (molecule or atom) binds to every other neutral system. In the following, the occurence of binding will be proved in enough cases so as to render the conjecture plausible. It will also be seen that binding in TFW theory is intimately connected with the existence of negative ions, i.e. $\lambda_c > Z$. We will assume here that $\lambda_c > Z$ for all the systems under consideration. $p \geq 5/3$ guarantees this, but no requirement on p other than $\lambda_c > Z$ will be made. $V = |x|^{-1} * m$, with m a nonnegative measure of compact support (so that (7.12) is satisfied) with $|m| = Z$, and with m being spherically symmetric in the atomic case.

First, let us define what binding means. Suppose we have two systems (not necessarily atoms) with potentials V_1 and V_2 and a combined system with $V(x) = V_1(x) + V_2(x-R)$ for some vector R. The combined system is neutral, i.e. $\lambda = Z \equiv Z_1 + Z_2$. Let $E(R)$ denote the energy of the combined system and $E_i(\lambda)$ denote the energies of the subsystems with arbitrary electron charge λ and $E_i = E_i(\lambda = Z_i)$ (note the difference in notation). Then

$$E(\infty) \equiv \min_{0\leq\lambda\leq Z} E_1(\lambda) + E_2(Z - \lambda). \tag{7.14}$$

Let μ_i be the chemical potentials of the subsystems when they are neutral, i.e. $\lambda_i = Z_i$. We know that $\mu_i > 0$. If $\mu_1 = \mu_2$ then $E(\infty) = E_1 + E_2$. Otherwise, $E(\infty) < E_1 + E_2$. In general, λ in (7.14) is determined by $\mu_1(\lambda) = \mu_2\ (Z - \lambda)$ if this equation has a solution for $0 \leq \lambda \leq Z - \lambda$; otherwise $\lambda = 0$ if $\mu_1(0) \leq \mu_2(Z)$ and $\lambda = Z$ if $\mu_2(0) \leq \mu_1(Z)$.

If $\mu_1 \neq \mu_2$ then the subsystems spontaneously ionize when they are infinitely far apart. This is not considered to be binding. For real atoms the phenomenon of spontaneous ionization apparently never occurs, because it seems to be the case that the lowest ionization potential among all atoms is less than the largest electron affinity. (I thank J. Morgan III for pointing this out to me). In real atoms, λ and $Z - \lambda$ in (7.14) are restricted to be integral, but no such restriction occurs in TFW theory. In TF theory the phenomenon never occurs because μ_i is always zero.

In TFW theory it is possible for λ to be zero in (7.14), i.e. one subsystem is completely stripped of electrons. Let $V_i = z_i/r$ with $z_2 >> z_1$. If $\mu_1(\lambda{=}0) \leq \mu_2(\lambda{=}z_1 + z_2)$, then $\lambda = 0$ in (7.14). By Theorem 7.11, $\mu_1(\lambda) \leq \mu_1(0) = z_1^{\,2}/4A$.

Since $\lambda_c(2) > z_2$ and $\mu_2 > 0$, the above inequality will hold for any fixed z_2 if z_1 is chosen small enough. This case was cited in BBL as an example where binding occurs (cf. Theorem 7.27).

Definition: Binding is said to occur if $E(R) < E(\infty)$ for some R.

Theorem 7.27: Suppose the chemical potentials of the neutral subsystems are unequal, i.e. $\mu_1 \neq \mu_2$. Then binding occurs. (This holds for all $p > 1$, even if $\lambda_c = Z$ for one or more of the three systems).

Proof: Suppose $\lambda < Z_1$ in (7.14). Then when $R = \infty$ subsystem 1 is positively charged (with charge $Q = Z_1 - \lambda$) and subsystem 2 has charge $-Q$. Let ρ_i be the TFW densities (with λ and $Z-\lambda$, respectively). By Theorem 7.24, ρ_i has exponential falloff. For the combined system (at R) consider the variational ρ defined by $\rho(x) = \rho_1(x) + \rho_2(x-R)$. The first term in (7.1) is subadditive (by convexity). For large R the total Coulomb energy decreases essentially by $-Q^2/R$ because of the exponential falloff of each ρ_i. The $\int\rho^p$ term is superadditive, but it increases only by a term of order $\exp[-(\text{const.})R]$ for large R. We omit the easy proof of these last two assertions. Thus, for large enough, but finite, R, $E(R) < E(\infty)$. ■

The difficult case is $\mu_1 = \mu_2$. Henceforth we confine our attention to atoms.

Conjecture: If $z_1 < z_2$ for two atoms then

$\mu_1(\lambda = bz_1) < \mu_2(\lambda = bz_2)$ for all b such that $bz_2 < \lambda_c(2)$. In particular, $\mu_1 < \mu_2$ and $\lambda_c(1) < \lambda_c(2)$.

If this conjecture is correct then only the homopolar case has to be considered. This we do in Theorem 7.28. However, we have already showed that binding occurs if $z_1 \ll z_2$, so it is likely that binding always occurs, even if the conjecture is wrong.

Theorem 7.28: Binding occurs for two equal atoms for any nuclear charge, z, and for any $p > 1$ provided $\lambda_c > z$ for the atom.

Proof: We will construct a variational $\tilde{\rho}$ for the combined system, with $\int\tilde{\rho} = 2z$, such that $\mathfrak{E}(\tilde{\rho})$ for the combined system at some R is less than $E(\infty) = 2E_1$. First, consider the atom with the nucleus at the origin and with $\lambda = z+\varepsilon$, where $z < \lambda < \lambda_c$. Let ρ be the TFW density. Denote E_1 by E and μ_1 by μ. Center the nucleus at the point $(-R,0,0)$, where $R > 0$, depending on ε, is such that $\int\chi_-\rho = z$, with χ_- being the characteristic functions of the half space, $H=\{(x_1,x_2,x_3)|x_1 \leq 0\}$. Assume, for the moment, that the nuclear m has support in $B_{R/2}$, i.e. the displaced m

has support in $\{x_1 \le -R/2\}$. Center the second atom at $(R,0,0)$. Its corresponding density is ρ^*, where $*$ means reflection through the plane $x_1 = 0$. Choose the variational $\tilde{\rho} = \rho_- + \rho_-^*$ with $\rho_- = \chi_- \rho$. Clearly $\tilde{\rho}$ is continuous across the plane $x_1 = 0$, and $\int\tilde{\rho} = 2z$, so it is a valid variational function. In the following bookkeeping of $\mathcal{E}(\tilde{\rho})$ we will use the terminology "energy gain" (resp. "loss") to mean that the contribution to $\mathcal{E}(\tilde{\rho})$ is negative (resp. positive) relative to $2E$. Before the χ_- cutoff, we start with $2E_1(\lambda) \le 2E - (2\varepsilon)(\mu/2)$ if ε is small enough, so we have gained $\varepsilon\mu$. This linear term in ε is the crucial point; it exists because $\lambda_c > z$. After the cutoff we gain the kinetic energy (first two terms in (7.1)) contributions from the missing pieces of ρ and ρ^*. Next, we lose on the $-\int V\rho$ term (for each atom separately) because of the missing pieces. Each missing charge is ε and its distance to its atomic origin is R. Since the atomic $V(r) \le z/r$, the energy loss is at most $2(\varepsilon z/R)$. Clearly we gain on the missing atomic repulsion, $D(\rho,\rho)$ term. Finally, if $dM(x) = dm(x+R) - \rho_-(x)dx$ is the total charge density in H, we lose the atom-atom interaction $\Delta = 2D(M,M^*)$. By reflection positivity, $\Delta \ge 0$. On balance, the net energy gain is at least $\varepsilon(\mu - 2z/R) - \Delta$.

Now we claim two things: (i) As $\varepsilon \to 0$, $R \to \infty$. ii) $\Delta < Cz\varepsilon/R$ for some constant, C. (Actually, it is possible to prove $\Delta < o(\varepsilon)\ z/R$.) Using (i) and (ii) we are done, because for sufficiently small ε the gain is positive and the assumption on $\mathrm{supp}(m)$ is justified.

<u>Proof of (i)</u>: Let ρ_n be the atomic density for $\lambda = z+\varepsilon_n$, with $\varepsilon_n \to 0$. As in the proof of Theorem 7.6, we can find a weakly convergent subsequence, $\rho_n \to \bar{\rho}$, so that $E = \lim \mathcal{E}(\rho_n) \ge \mathcal{E}(\bar{\rho})$. But $\int\bar{\rho} \le z$, so $\bar{\rho}$ must be ρ_z, the atomic density with $\lambda = z$. If R_n does not tend to ∞ then, for large enough n, $\int\chi_-\rho_n < z$ by the weak convergence, which is a contradiction.

<u>Proof of (ii)</u>: This is messy. Let B_r be the ball of radius r centered at $(-R,0,0)$ and let χ_r be its characteristic function. Write $\rho = \rho^a + \rho^b$ where $\rho^a = \chi_{3R/2}\,\rho$. By elementary geometry, $d\int\rho^b \ge \int\rho^b_-$ with $d < 1$. Since $\int\rho^b(1 - \chi_-) < \varepsilon$, $\int\rho^b < \varepsilon/(1-d)$. The contribution of ρ^b to Δ is $4D(\rho^b_-, M^*) - 2D(\rho^b_-,\rho^{b*}_-) \le 4d(\rho^b_-,M^*) \le 4D(\rho^b_-,\rho^*_-)$

$\le 4D(\rho^b,\rho^*_-) \le 2z(\int\rho^b)/R$, since the potential of ρ^b is everywhere less than $(3R/2)^{-1}\int\rho^b$. Henceforth, we can assume $\rho_- = \rho^a_-$ and $z > \int\rho^a_- > z - t\varepsilon$. This assumption changes M to M^a. Let $d\tilde{M}(x) = dm(x+R) - \rho^a(x)dx$. (Note: $\mathrm{supp}(\tilde{M})$ extends outside H, but is inside $\{x_1 \le R/2\}$. $\phi \equiv |x|^{-1} * M^{a*}$ is subharmonic on $\mathrm{supp}(\tilde{M})$ and harmonic on $\mathrm{supp}(m)$ so $D(\tilde{M},M^{a*}) \le (\int d\tilde{M})\, D(\delta,M^{a*})$, where δ is a delta function at

$(-R,0,0)$. This is $\leq |\int d\tilde{M}|\ z/R \leq (t+1)\varepsilon z/R$, since the distance of $\mathrm{supp}(M^{a*})$ to $(-R,0,0)$ is R. Finally, $D(M^a - \tilde{M}, M^{a*}) = D(\rho^a - \rho^a_-, M^{a*}) \leq D(\rho^a - \rho^a_-, m^*) = D(\rho^a - \rho^a_-, z\delta^*) \leq 2\varepsilon z/R$ since $\int \rho^a - \rho^a_- \leq \varepsilon$ and the distance of $\mathrm{supp}(\rho^a)$ to $(R,0,0)$ is $R/2$. ■

I thank J. Morgan III for valuable discussions about Theorem 7.28.

VII.D. The Z^2 Correction and the Behavior near the Nuclei

Here we consider point nuclei with potential given by (2.1). The question we address is what is the principal correction to the TF energy and density caused by the first term in (7.1)? This term, $A\int(\nabla\rho^{\frac{1}{2}})^2$ will henceforth be denoted by T. For simplicity we confine our attention to $p = 5/3$, the physical value of p.

$E^{TF} \sim Z^{7/3}$. In particular, for a neutral atom,

$$E^{TF} = -\ 3.67874\ z^{7/3}/\gamma \tag{7.15}$$

(I thank D. Liberman for this numerical value). At first sight, it might be thought that the leading energy correction is $O(z^{5/3})$. If $\rho^{TF}(z,r) = z^2\rho^{TF}(1,z^{1/3}r)$ is inserted into T then, by scaling, $T(z) = z^{5/3}\,T(z=1)$. But $T(z=1) = \infty$ since $\rho^{TF} \sim r^{-3/2}$ for small r. Thus, for point nuclei, T cannot be regarded as a small perturbation.

The actual correction is $+O(z^2)$ and bounds of this form can easily be found. The following bounds are for an atom, and can obviously be generalized for molecules. Upper bound: Use a variational $\tilde{\rho}$ for TFW of the form $\tilde{\rho}(r) = \rho^{TF}(r)$ for $r \geq 1/z$ and $\tilde{\rho}(r) = \rho^{TF}(1/z)$ for $r \leq 1/z$. Lower bound: Let $b > 0$ and write $V(r) = \tilde{V}(r) + H(r)$ where $H(r) = z/r - z^2/b$ for $zr < b$ and $H(r) = 0$ otherwise. For small enough b, $-A\Delta + H > 0$, since $||H||_{3/2} \sim b$. Now $\tilde{V} = |x|^{-1} * m$ with $m \geq 0$ and $|m| = z$. Let $\tilde{\rho}$ minimize $\mathcal{E}^{TF}(\tilde{V},\rho)$ with energy $E^{TF}(\tilde{V})$. Then $E^{TFW} \geq E^{TF}(\tilde{V})$. But $E^{TF}(V) \leq \mathcal{E}^{TF}(V,\tilde{\rho}) = E^{TF}(\tilde{V}) - \int\tilde{\rho}H$. It is not hard to prove, from the TF equation with $\tilde{V}$, that this last integral is $O(z^2)$.

The foregoing calculations show that the main correction in TFW theory comes from distances of order z^{-1} near the nuclei. The calculations, if carried out for arbitrary λ, also show that the correction is essentially independent of λ. We now show how this correction can be exactly computed to leading order in z, namely $O(z^2)$.

Let us begin by considering the atom without electron-electron repulsion. The TF theory of such an atom was presented in §V.B following (5.30). The analogous TFW equation (with $\delta = A\hbar^2/2m$ and $\hbar^2/2m = 1$) is

$$(-A\Delta + W(x))\ \psi = -\mu\ \psi \tag{7.16}$$

with $W(x) = \gamma\rho(x)^{2/3} - z/|x|$, and $\rho = \psi^2$. The absolute minimum, which corresponds to $\lambda = \infty$, has $\mu = 0$, namely

$$(-A\Delta + \gamma|\psi|^{4/3} - z|x|^{-1})\psi = 0 \tag{7.17}$$

The first task is to analyze (7.17). By simple scaling, any solution scales with A, γ and z as

$$\psi(z,\gamma,A;x) = (z^2/A\gamma)^{3/4}\ \psi(1,1,1;zx/A). \tag{7.18}$$

Henceforth we take $z=\gamma=A=1$. Consider the functional

$$\mathcal{F}'(\psi) = T(\psi) + P(\psi) \tag{7.19}$$

$$T(\psi) = \int(\nabla\psi)^2,\ P(\psi) = \int k(\psi(x),x)dx \tag{7.20}$$

$$k(\psi,x) = 3|\psi|^{10/3}/5 + 2|x|^{-5/2}/5 - |\psi|^2|x|^{-1}. \tag{7.21}$$

Note that $k \geq 0$ and, for each x, k has a minimum at $\psi = |x|^{-3/4}$. The function space for $\mathcal{F}'$ is

$$G' = \{\psi|\nabla\psi \in L^2,\ P(\psi) < \infty\}. \tag{7.22}$$

G' is not convex since $0 \notin G'$. Clearly, (7.17) is the variational equation for $\mathcal{F}'$. We can also define $G = \{\rho|\rho \geq 0,\ \rho^{\frac{1}{2}} \in G'\}$ and $\mathcal{F}(\rho) = \mathcal{F}'(\rho^{\frac{1}{2}})$. G is convex and $\rho \to \mathcal{F}(\rho)$ is convex.

Theorem 7.29: $\mathcal{F}'(\psi)$ has a minimum on G'. This minimizing ψ is unique, except for sign, and satisfies: (i) $\psi > 0$ (ii) ψ is spherically symmetric. (iii) ψ satisfies (7.17). (iv) ψ is the only nonnegative solution to (7.17) in G'. (v) ψ is C^∞ for $|x| > 0$. (vi) ψ satisfies the cusp condition $2(d\psi/dr)(0) = -\psi(0)$. (vii) for large $r = |x|$, ψ has the asymptotic expansion (which can be formally deduced from (7.17)),

$$\psi(r) = r^{-3/4} - (9/64)r^{-7/4} - (3/2)(21/64)^2 r^{-11/4} + 0(r^{-15/4}).$$

$$\rho(r) = r^{-3/2} - (9/32)r^{-5/2} - (621/2^{11})r^{-7/2} + 0(r^{-9/2}). \tag{7.24}$$

(viii) Any solution, f, to (7.17) in G' satisfies $|f(x)| < |x|^{-3/4}$. (ix) By (viii), ψ is superharmonic, and thus $\psi(r)$ is decreasing.

The proof of Theorem 7.29 follows the methods of §VII.A and B,

and is given in (Lieb, 1981b). The following numerical values, together with a tabulation of ψ, are in (Liberman and Lieb, 1981). $\rho = \psi^2$.

$$\psi(0) = 0.9701330$$

$$I_1 = \int(\nabla\psi)^2 = 8.5838197$$

$$I_2 = \int\{r^{-5/2} - \rho^{5/3}\} = 42.92$$

$$I_3 = \int[r^{-3/2} - \rho]/r = 34.34. \tag{7.25}$$

From (7.17) one has $I_1 + I_3 = I_2$. By dilating $\psi(r) \to t^{3/4}\,\psi(tr)$ in (7.19), a "virial theorem" is obtained: $5I_1 + 3I_2 = 5I_3$. Thus,

$$I_1 : I_2 : I_3 = 1 : 5 : 4 \tag{7.26}$$

$$\Delta E \equiv \mathfrak{F}(\psi) = I_1 - 3I_2/5 + I_3 = 2I_1. \tag{7.27}$$

If the parameters are reintroduced

$$I_1(z,\gamma,A) = A\int(\nabla\psi)^2 = z^2\, A^{\frac{1}{2}}\, \gamma^{-3/2}\, I_1 . \tag{7.28}$$

Let us denote the ρ we have just obtained in Theorem 7.29 (with the parameters reintroduced according to (7.18)) by ρ_∞. The scale length of ρ_∞ is z^{-1} and, for <u>large</u> r, $\rho_\infty(r)$ agrees to leading order with $\rho^{TF}(r)$ for <u>small</u> r (on a scale of $z^{-1/3}$), namely $(z/\gamma r)^{3/2}$. We claim that ρ_∞ can be spliced together with ρ^{TF} in the overlap region, $r = 0(z^{-2/3})$, and the result is ρ^{TFW} to leading order in z. The splicing is independent of λ provided $\lambda/z > (\text{const.}) > 0$. The change in energy for an atom is then, to leading order, ΔE of (7.27), and is independent of λ. An analogous situation holds for a molecule; near each nucleus ρ^{TF} is spliced together with ρ_∞ for the appropriate z_j. This is formalized in the following theorem.

<u>Theorem 7.30</u>: Let $V(x) = \Sigma z_j\,|x-R_j|^{-1}$. Consider the $Z \to \infty$ limit with the scaling given before (5.2), except that the electron charge, N, is not restricted to be integral. $\lambda = N/Z > 0$ is fixed. $z_j = az_j{}^\circ$, $R_j = a^{-1/3}\,R_j{}^\circ$ with $a\lambda = N$. Then, as $N \to \infty$,

(i) $$E^{TFW}(N) = E^{TF}(N) + D\sum_{j=1}^{k} z_j{}^2 + o(a^2), \tag{7.29}$$

with $$D = 2\,A^{\frac{1}{2}}\,\gamma^{-3/2}\,I_1 . \tag{7.30}$$

(ii) $a^{-4/3}\,\mu^{TFW}(N) \to \mu^{TF}(\lambda,\underline{z}^\circ,\underline{R}^\circ)$.

(iii) Fix x with $x \neq R_j$, all j. Then

$$a^{-2}\,\rho^{\mathrm{TFW}}(N,\underline{z},\underline{R};\,a^{-1/3}x) \to \rho^{\mathrm{TF}}(\lambda,\underline{z}°,\underline{R}°;x), \tag{7.31}$$

with convergence in the sense of weakly in L^1 if $\lambda \le Z$ and weakly in L^1_{loc} if $\lambda > Z$.

(iv) Fix y. For each j

$$z_j^{-3}\,\rho^{\mathrm{TFW}}(N,\underline{z},\underline{R};\,R_j + z_j^{-1}y) \to (A\gamma)^{-3/2}\,\psi^2(y/A), \tag{7.32}$$

where ψ is the solution to (7.17) with $A = z = \gamma = 1$ given by Theorem 7.29. The convergence is pointwise and in L^1_{loc}. A refinement of (7.32) is given in Theorems 7.32 - 7.35.

Before proving Theorem 7.30 let us comment on its significance.

(i) (7.29) states that the energy correction in TFW theory is exactly of the form of the quantum correction conjectured by Scott (5.29). In particular, since $\gamma^{3/2} \sim q^{-1}$, the q dependence is the same. In order to obtain the conjectured coefficient 1/8 of (5.32), with $\gamma = \gamma_p$, we must choose

$$A = q^2\,\gamma_p^{\,3}\,[16\,I_1]^{-2} = 0.18590919. \tag{7.33}$$

This number was mentioned after (2.8).

(ii) The density, on a length scale $Z^{-1/3}$ agrees with quantum (and TF) theory, Theorem 5.2.

(iii) On a length scale z^{-1} near each nucleus, (7.32) states that ρ^{TFW} converges to a universal function. This phenomenon is the same as we conjectured in (5.37) for quantum theory. The universal functions are not exactly the same, however, but they are very close. For large values of the argument they agree, namely $(\gamma_p y)^{-3/2}$, independent of A. Since the convergence in (7.32) is <u>pointwise</u>, it makes sense to ask what happens at $y = 0$. Using γ_p and A given by (7.33), the right side of (7.32) is obtained from (7.18) and (7.25) as

$$q^{-1}\,z_j^{-3}\,\rho^{\mathrm{TFW}}(x=R_j) \to 0.19827149 \tag{7.34}$$

On the other hand, ρ^H in (5.33) can be evaluated at $x = 0$ since only S-waves contribute. At $x = 0$, $f_{n\ell m}(0)^2 = (2\pi n^3)^{-1}$. Thus (5.37), if correct, would state that

$$q^{-1}\,z_j^{-3}\,\rho^{Q}(x=R_j) \to \zeta(3)/2\pi = 0.19131330. \tag{7.35}$$

Remarkably, the difference between (7.34) and (7.35) is only 4%.

To prove Theorem 7.30, Theorems 7.32 - 7.35, which are independently interesting, are needed. To prove them we need the following comparison theorem which was proved by Morgan (1978) in the spherically symmetric case and by T. Hoffmann-Ostenhof (1980) in the general case.

Lemma 7.31: Let $B \subset \mathbb{R}^3$ be open, and let f and g be continuous functions on the closure of B that satisfy Δf and $\Delta g \in L^1(B)$ and $f(x)$ and $g(x) \to 0$ as $|x| \to \infty$ if B is unbounded. Assume $\Delta f \leq Ff$ and $\Delta g \geq Gg$ as distributions on B, where F,G are functions satisfying $F(x) < G(x)$ a.e. in B. Assume $f(x) > 0$ in B and $f(x) \geq g(x)$ for all $x \in \partial B$. Then $f(x) \geq g(x)$ for all $x \in B$.

Theorem 7.32: Let $V = z/|x|$, and let ψ_∞ be the positive solution to (7.17) given in Theorem (7.29). Let ψ be the positive solution to the TFW equation, (7.9), for some $\mu \geq 0$ and $p = 5/3$. Then, for all x, $\psi(x) \leq \psi_\infty(x)$.

Proof: Let $B = \{x|\psi(x) - \psi_\infty(x) > 0\}$. Take $f = \psi_\infty$ and $g = \psi$ in Lemma 7.31. Since f and g are continuous and, by Theorem 7.24, $g(x) < f(x)$ for large $|x|$, B is open and bounded. Hence Δf and $\Delta g \in L^1(B)$. On B, $A\Delta f = Ff$, $A\Delta g > Gg$ with $F = -V + f^{4/3}\gamma$ and $G = -V + g^{4/3}\gamma$. Since $F < G$ in B and $f = g$ on ∂B, $f \geq g$ in B. Therefore, B is empty. ■

For a molecule, an upper bound to ψ, which is not as nice as Theorem 7.32 but which is sufficient for Theorem 7.30, can also be obtained. We always assume $p = 5/3$.

Theorem 7.33: Let V be as in Theorem 7.30 with the scaling given there. Let ψ be the solution to the TFW equation and let B be the ball $\{x|Z^{-2/3} > |x-R_1|\}$. Then, for sufficiently large a,

$$\psi(x) \leq \psi_\infty(x-R_1) \quad \text{for } x \in B \tag{7.36}$$

where ψ_∞ is the positive solution to (7.17) with

$z = z_1 + d\,Z^{2/3}$ and $d = 1 + 2(Z^\circ)^{-1/3}/\min\{|R_j^\circ - R_1^\circ|\,|\,j=2,\ldots,k\}$.

Proof: By Theorem 7.14 and (7.24), we can choose a large enough so that $\psi_\infty(x) > \psi(x)$ when $x \in \partial B$ and so that $R_2,\ldots,R_k \notin B$. The proof is then the same as for Theorem 7.32, with $f = \psi_\infty$ and $g = \psi$, provided we can verify that $M(x) \equiv z|x-R_1|^{-1} - V(x) > 0$ when $x \in B$. But M, being superharmonic in B, has its minimum on ∂B. This minimum is positive for large enough a. ■

To obtain a lower bound to ψ, the following is needed.

Theorem 7.34: Assume the hypothesis of Theorem 7.30 with $\lambda > 0$ and let ψ be the positive solution to the TFW equation. Then there is a constant d, independent of λ, such that

(i) $h(x) \equiv |x|^{-1} * \rho < d\, a^{4/3}$.

(ii) $\mu < d\, a^{4/3}\, \lambda^{-2/3}$.

Proof: For (i) we use Theorem 7.14 together with the fact that $\int\rho = a\lambda$. For any x, let B be the ball of radius $a^{-1/3}$ centered at x. The contribution to h from $\chi_B\rho$ is bounded by $(\text{const.})(Z°)^{3/2}\, a^{4/3}$. The contribution from $(1-\chi_B)\rho$ is bounded by $a^{1/3}\int\rho$. For (ii), since μ is decreasing in λ, $\mu \le -e(N)/N$. However, $e(N) > e^{TF}(N)$. But $-e^{TF}(N)$ scales as $a^{7/3}\, f(\lambda)$ and $f(\lambda) \le (\text{const.})\, \lambda^{1/3}$ by (3.6). ■

Theorem 7.35: Assume the same hypothesis as in Theorem 7.33. Then, for sufficiently large a,

$$\psi(x) \ge \psi_\infty(x-R_1)\, \sigma(x-R_1) \quad \text{for all } x \tag{7.37}$$

where ψ_∞ is the positive solution to (7.17) with $z = z_1 - 4t\, a^{2/3} A$ and

$$\sigma(x) = [1 - a^{2/3} t|x|]\, \exp(-a^{2/3} t|x|).$$

Here, $At^2 = d(1+\lambda^{-2/3})$ with d given in Theorem 7.34.

Proof: Let $f = \psi$ and g = right side of (7.37). We have to verify (7.37) only in $B = \{x \mid a^{2/3}\, t|x-R_1| < 1\}$ because $g \le 0$ otherwise. Since, by Theorem 7.29, both ψ_∞ and σ are symmetric decreasing, $\Delta g > \sigma\Delta\psi_\infty + \psi_\infty\Delta\sigma$. But $(\Delta\sigma)(x) = (a^{4/3}\, t^2 - 4a^{2/3}\, t/|x|)\sigma$, and $\psi_\infty{}^{4/3} \ge g^{4/3}$ since $\sigma \le 1$. Therefore, to imitate the proof of Theorem 7.32 it is only necessary to verify that $a^{4/3}At^2 > h(x) + \mu$, but this is clearly true. ■

Proof of Theorem 7.30: (iv) is a trivial consequence of Theorems 7.33 and 7.35. (iii) is proved in the same way as Theorem 5.2 if we note that the energy can be controlled to $o(z^{7/3})$ by the variational upper bound given in the paragraph after (7.15). (ii) is proved by noting that, by the proof of (iii) just given, $a^{-7/3}\, E^{TFW}(a, N = a\lambda) \to E^{TF}(a = 1, \lambda)$. The limit of the derivative of a sequence of convex (in λ) functions is the derivative of the limit function.

The proof of (i) is complicated. Upper and lower bounds to E of the desired accuracy, $O(Z^2)$, are needed. First, let us make a remark. Consider E as a function of A. By standard arguments used earlier, E(A) is monotone increasing, concave, and hence differentiable almost everywhere for $A > 0$. $dE/dA = T/A$, a.e. and $E^{TFW} - E^{TF} = \int_0^A (T/A)\, dA$. If we can find a lower (or upper) bound to T/A of the form $T/A = A^{-1/2} \gamma^{-3/2} I_1 \Sigma z_j^2 +$ (lower order) then (7.29) will be proved. We can, indeed, find a lower bound of this form, and hence a lower bound to E. We cannot find an upper bound of this form and therefore must resort to a direct variational calculation to obtain an upper bound to E.

Upper bound: By the monotonicity of E in N, it is only necessary that $\int\rho \le N$. There are several ways to construct a variational ρ, which we call f. The details of the calculation of $\mathcal{E}(f)$ are left to the reader. One construction is to define $B = \{x | \rho^{TF}(x) > Z^{5/2}\}$. For large a, B is the union of k connected components which are approximately spheres centered at R_j. Call these B_j. Let $\psi_{\infty j}$ be the solution to (7.17) centered at R_j and with $z = z_j - t\, a^{2/3}$. Let $C_j = \{x | \psi_{\infty j} > Z^{5/4}\}$. For large enough, but fixed t, $C_j \subset B_j$ for large a. The variational f is defined by $f(x) = \rho^{TF}(x)$ for $x \notin B$, $f(x) = Z^{5/2}$ for $x \in B_j \setminus C_j$ and $f(x) = \psi_{\infty j}(x)^2$ for $x \in C_j$.

Lower bound: We construct a lower bound to T/A. Suppose $P_1, \ldots, P_k$ are orthogonal, vector valued functions. Then $T/A \ge \Sigma_j L_j^2 / \int P_j^2$, where $L_j = \int \nabla\psi \cdot P_j$. We take $P_j(x) = \nabla\psi_{\infty j}(x)\, \chi_j(x)$ where χ_j is the characteristic function of $D_j = \{x \,|\, |x-R_j| < t\, z_j^{-2/3}\}$, and t is some fixed constant. For large a, the D_j are disjoint so the P_j are orthogonal. Clearly, $\int P_j^2 = \int \nabla\psi_{\infty j}^2 + o(Z^2)$. Now multiply (7.17) for $\psi_{\infty j}$ by ψ and integrate over D_j. Then $L_j = -\int W_{\infty j}\psi_{\infty j}\psi\chi_j + A\int \psi\nabla\psi_{\infty j} \cdot n\, ds$. By the bound (7.37), the first integral is $(T_{\infty j}/A) + o(Z^2)$. It is not difficult to show that the second integral is $o(Z^2)$. This can be done by using (7.24) whence, for some $t \in [\frac{1}{2}, 1]$, $d\psi_{\infty j}/dr > -10\, z_j^{3/4}\, r^{-7/4}$ at $r = t\, z_j^{-2/3}$. ∎

VIII. THOMAS-FERMI-DIRAC-VON WEIZSÄCKER THEORY

This theory has not been as extensively studied as the other theories. The results presented here are from unpublished work by Benguria, Brezis and Lieb done in connection with their 1981 paper.

The energy functional is

$$\mathcal{E}'(\psi) = A\int(\nabla\psi)^2 + \int J(\psi^2) - \int V\psi^2 + D(\psi^2,\psi^2) + U \qquad (8.1)$$

in units in which $\hbar^2/2m = 1$.

$$J(\rho) = (\gamma/p)\,\rho^p - (3C_e/4)\,\rho^{4/3}. \qquad (8.2)$$

For convenience we assume $p > 4/3$ (not $p > 1$). $\mathcal{E}(\rho) \equiv \mathcal{E}'(\rho^{\frac{1}{2}})$. The function space for ψ is the same as for TFW theory, namely G_p' of (7.3).

Note that $\mathcal{E}(\rho)$ is not convex because of the $-\int\rho^{4/3}$ term.

As in TFD theory (6.7) - (6.10), we introduce

$$J_\alpha(\rho) = J(\rho) + \alpha\rho \qquad (8.3)$$

and α is chosen so that $J_\alpha(\rho) \geq 0$ and $J_\alpha(\rho_o) = 0 = J'_\alpha(\rho_o)$ for some ρ_o, namely

$$\rho_o^{\,p-4/3} = C_e\, p\, [4\gamma(p-1)]^{-1}$$

$$\alpha = (3p-4)\,[4(p-1)]^{-1}\,\rho_o^{\,1/3}\,C_e. \qquad (8.4)$$

The necessity of $p > 4/3$ for this construction is obvious. $\mathcal{E}_\alpha'$ and $\mathcal{E}_\alpha$ are defined by using J_α in (8.1).

The energy for $\lambda \geq 0$ is

$$E(\lambda) = \inf\,\{\mathcal{E}(\rho)\,|\,\rho \in G_p,\ \int\rho = \lambda\}. \qquad (8.5)$$

and similarly for $E_\alpha(\lambda)$ and $E'(\lambda)$, $E_\alpha'(\lambda)$ using $\mathcal{E}'$. If the condition $\int\rho = \lambda$ is omitted in (8.5) we obtain E, E_α, E', E'_α.

<u>Theorem 8.1</u>: (i) The four functions $E(\lambda)$, $E_\alpha(\lambda)$, $E'(\lambda)$ and $E_\alpha'(\lambda)$ are finite, continuous, and satisfy

$$E(\lambda) = E'(\lambda) = E_\alpha(\lambda) - \alpha\lambda = E_\alpha'(\lambda) - \alpha\lambda. \qquad (8.6)$$

(ii) E_α is finite.

(iii) ρ minimizes $\mathcal{E}(\rho)$ on $\int\rho = \lambda$ if and only if $\psi = \rho^{\frac{1}{2}}$

minimizes $\mathcal{E}'(\psi)$ on $\int\psi^2 = \lambda$. This ρ and ψ also obviously minimize $\mathcal{E}_\alpha$ and $\mathcal{E}_\alpha'$.

Proof: The same as for Theorems 2.1, 6.2 and 7.2. Note that $[\int\rho^{4/3}]^{3p-3} \le [\int\rho]^{3p-4} \int\rho^p$. ■

Theorem 8.2: Let ψ minimize $\mathcal{E}'(\psi)$ on the set $\int\psi^2 = \lambda$. Then ψ satisfies the TFDW equation:

$$(-A\Delta + W(x))\psi = -\mu\ \psi \tag{8.7}$$

with $W = \gamma\rho^{p-1} - C_e\ \rho^{1/3} - \phi + \alpha$, (8.8)

$\phi = V - |x|^{-1} * \rho$ and $\rho = \psi^2$. Apart from a sign, $\psi(x) > 0$ for all x, and ψ satisfies the conclusions of Theorem 7.9. ψ is the unique ground state of $H = -A\Delta + W(x)$ and μ is its ground state eigenvalue. E is differentiable at λ and $\mu = -dE_\alpha/d\lambda = -dE/d\lambda - \alpha \le 0$. $\mu = 0$ if and only if E_α has an absolute minimum at this λ.

Proof: The proof is basically the same as for Theorems 7.8 - 7.10. Although it is not known that $\rho = \psi^2$ is unique, this is not really necessary. By considering the variation of $\mathcal{E}'(\psi)$, ψ satisfies (8.7), (8.8). If ψ is minimizing, then so is $|\psi|$ (cf. Theorem 7.2). Hence $|\psi|$ satisfies (8.9) with the same W. But, as in Theorem (7.10), the ground state of $H = -A\Delta + W$ is unique and nonnegative and therefore ψ may be taken to be ≥ 0 for all x. The rest follows by the methods of Theorem 7.9 (Note: $\rho^{1/3}\psi \in L^3$ since $\psi \in L^6 \cap L^2$.) ■

Remark: As in section VII, the role of $\mathcal{E}'$, as distinct from $\mathcal{E}$, is solely to prove (8.7) in which no explicit reference to $\rho \ge 0$ is made.

Now we turn to a difficult and serious problem. We do not know that $E_\alpha(\lambda)$ is monotone non-increasing. Therefore, if we define

$$\hat{E}_\alpha(\lambda) = \inf\{\mathcal{E}_\alpha(\rho)\,|\,\rho \in G_p,\ \int\rho \le \lambda\} \tag{8.10}$$

we do not know that $E_\alpha(\lambda) = \hat{E}_\alpha(\lambda)$. By definition, $\hat{E}_\alpha(\lambda)$ is monotone non-increasing. The source of the difficulty is this: Although $J_\alpha(\rho_o) = J_\alpha'(\rho_o) = 0$ (as in TFD theory), we cannot simply add small clumps of charge, of amplitude ρ_o, at ∞. This is so because such a clump would then have $\int(\nabla\psi)^2 = \infty$. Neverless, we can add clumps with $\mathcal{E}_\alpha$ energy strictly less than $\alpha\int\rho$, as the following theorem shows

Theorem 8.3: Set $V = 0$ in $\mathcal{E}'$. There are C^∞ functions of compact support such that $\mathcal{E}'(\psi) < 0$.

Proof: Let f be any function in C_o^∞ and let $\psi(x) = b^2 f(bx)$. For some sufficiently small, but positive b, $\mathcal{E}(\psi) < 0$. To see this, note that $\int(\nabla\psi)^2$ scales as b^3, $\int\rho^{5/3}$ as $b^{11/3}$, $D(\rho,\rho)$ as b^3, while $\int\rho^{4/3}$ scales as $b^{7/3}$. ■

As a corollary we have

Theorem 8.4: $E(\lambda)$ is strictly monotone decreasing in λ. Hence

$$E(\lambda) = \inf\{\mathcal{E}(\rho) \,|\, \rho \in G_p, \int\rho \le \lambda\}. \tag{8.9}$$

$$E = \inf_\lambda E(\lambda) = -\infty .$$

We conjecture that $E(\lambda)$ is convex. Unfortunately, the "convexification" trick of section VI, in which J_α is replaced by j, is not helpful. Because of the gradient term any minimizing ψ will be continuous and therefore ψ cannot omit the values $(0, \rho_o)$, even for point nuclei. While the energy for j is, indeed, convex, it is strictly smaller than E_α for all λ.

Theorem 8.5 states that E_α and $\hat{E}_\alpha$ have absolute minima at some common, finite λ. For all we know, there may be several such λ, but all these λ are bounded. Furthermore, for every λ there is a minimizing ρ for $\hat{E}_\alpha(\lambda)$. Unfortunately, for no λ are we able to infer that $\int\rho = \lambda$.

Theorem 8.5: (i) There exists a minimizing ρ for $\mathcal{E}_\alpha(\rho)$ on G_p, and $\psi = \rho^{\frac{1}{2}}$ minimizes $\mathcal{E}_\alpha'(\psi)$. Every such $\rho \in L^1$, and $\int\rho \le$ some constant which is independent of ρ.

(ii) There exists a minimizing ρ for $\mathcal{E}_\alpha(\rho)$ on the set $\int\rho \le \lambda$.

Remark: It is not claimed (but it is conjectured) that the minimizing ρ is unique.

Proof: The proofs of (i) and (ii) are the same, so we concentrate on (i). The proof merely imitates the proof of Theorem 7.6. The only new point is that $\rho \in L^1$. Each term in $\mathcal{E}(\rho)$ is finite and, in particular, $I = \int J_\alpha(\rho) < \infty$. But $J_\alpha(\rho) \ge k\rho$ when $0 \le \rho \le \beta$ for some $k, \beta > 0$. If χ is the characteristic function of $\{x \,|\, \rho(x) \le \beta\}$, $k\int\chi\rho < \infty$. On the other hand, $\beta^2(1-\chi)\rho \le \rho^3$, so $\beta^2\int(1-\chi)\rho \le \int\rho^3 < \infty$ since $\rho \in L^3$. It is easy to see from (7.6) that the bound on $\int\rho$ is independent of ρ. ■

Remark: It is surprising that the fact that $\lambda < \infty$ for any absolute minimum is obtained so easily. Recall that in TFW theory the proof of this fact (Theorem 7.12) required an analysis of the TFW equation.

An important question is whether λ, for an absolute minimum, always satisfies $\lambda \geq Z$.

A few things can be said about the properties of any minimizing ρ on $\int\rho = \lambda$.

Theorem 8.6: In the atomic case, $V(r) = z/r$, any minimizing ψ is symmetric decreasing when $\lambda \leq z$. (Conjecture: this also holds for all λ.)

Proof: The rearrangement inequality proof of Theorem 2.12 is applicable. ■

Theorem 8.7: The conclusions of Theorem 7.13 hold for any minimizing ψ. Moreover, for every $t < \mu_1 + \alpha$ there exists a constant, M, such that $\psi(x) \leq M \exp[-(t/A)^{\frac{1}{2}} |x|]$.

Proof: Same as for Theorems 7.13 and 7.24. ■

Theorem 8.6: Every minimizing ψ satisfies Theorem 7.25.

Plainly, TFDW theory is not in a satisfactory state from the mathematical point of view. In TFD theory we were able to deal with the lack of convexity by means of the J_α trick. In TFW theory, the presence of the gradient term does not spoil the general theory because $\mathcal{E}$ is convex. When taken together, however, the two difficulties present an unsolved mathematical problem.

Acknowledgement. I am grateful to the U.S. National Science Foundation (grant PHY-7825390-A02) for supporting this work.

References

Adams, R.A., 1975, Sobolev Spaces (Academic Press, New York).

Balàzs, N., 1967, "Formation of stable molecules within the statistical theory of atoms," Phys. Rev. 156, 42-47.

Baumgartner, B., 1976, "The Thomas-Fermi theory as result of a strong-coupling-limit," Commun. Math. Phys. 47, 215-219.

Baxter, J.R., 1980, "Inequalities for potentials of particle systems," Ill. J. Math. 24, 645-652.

Benguria, R., 1979, "The von Weizsäcker and exchange corrections in Thomas-Fermi theory," (Ph.D. thesis, Princeton University).

Benguria, R., 1981, "Dependence of the Thomas-Fermi Energy on the Nuclear Coordinates," Commun. Math. Phys., to appear.

Benguria, R., H. Brezis and E. H. Lieb, 1981, "The Thomas-Fermi-von Weizsäcker theory of atoms and molecules," Commun. Math. Phys. 79, 167-180 (1981).

Benguria, R., and E. H. Lieb, 1978, "Many-body potentials in Thomas-Fermi theory," Ann. of Phys. (NY) 110, 34-45.

Benguria, R., and E. H. Lieb, 1978, "The positivity of the pressure in Thomas-Fermi theory," Commun. Math. Phys. 63, 193-218. Errata 71, 94 (1980).

Berestycki, H. and P. L. Lions, 1980, "Existence of stationary states in nonlinear scalar field equations," in Bifurcation Phenomena in Mathematical Physics and Related Topics, C. Bardos and D. Bessis (eds.) (D. Reidel, Dordrecht), 269-292. See also "Nonlinear Scalar field equations, Parts I and II," Arch. Rat. Mech. Anal., 1981, in press.

Brezis, H., 1978, "Nonlinear problems related to the Thomas-Fermi equation," in Contemporary Developments in Continuum Mechanics and Partial Differential Equations, G. M. de la Penha, L. A. Medeiros (eds.), (North Holland, Amsterdam), 81-89.

Brezis, H., 1980, "Some variational problems of the Thomas-Fermi type," in Variational Inequalities and Complementarity Problems: Theory and Applications, R. W. Cottle, F. Giannessi and J-L. Lions (eds.), (Wiley, New York), 53-73.

Brezis, H., and E. H. Lieb, 1979, "Long range atomic potentials in Thomas-Fermi theory," Commun. Math. Phys. 65, 231-246.

Brezis, H., and L. Veron, 1980, "Removable singularities of nonlinear elliptic equations," Arch. Rat. Mech. Anal. 75, 1-6.

Caffarelli, L.A. and A. Friedman, 1979, "The free boundary in the Thomas-Fermi atomic model," J. Diff. Equ. 32, 335-356.

Deift, P., W. Hunziker, B. Simon and E. Vock, 1978, "Pointwise Bounds on eigenfunctions and wave packets in N-body quantum systems IV," Commun. Math. Phys. 64, 1-34.

Dirac, P.A.M., 1930, "Note on exchange phenomena in the Thomas-Fermi atom," Proc. Camb. Phil. Soc. 26, 376-385.

Fermi, E., 1927, "Un metodo statistico per la determinazione di alcune priorieta dell'atome," Rend. Accad. Naz. Lincei 6, 602-607.

Firsov, O. B., 1957, "Calculation of the interaction potential of atoms for small nuclear separations," Zh. Eksper. i Teor. Fiz. 32, 1464. [English transl. Sov. Phys. JETP 5, 1192-1196 (1957]. See also Sov. Phys. JETP 6, 534-537 (1958) and 7, 308-311 (1958).

Fock, V., 1932, "Über die Gültigkeit des Virialsatzes in der Fermi-Thomas'schen Theorie," Phys. Z. Sowjetunion 1, 747-755.

Gilbarg, D. and N. Trudinger, 1977, Elliptic Partial Differential Equations of Second Order, (Springer Verlag, Heidelberg).

Gombas, P., 1949, Die statistischen Theorie des Atomes und ihre Anwendungen (Springer Verlag, Berlin).

Hille, E., 1969, "On the Thomas-Fermi equation," Proc. Nat. Acad. Sci. (USA) 62, 7-10.

Hoffmann-Ostenhof, M., T. Hoffmann-Ostenhof, R. Ahlrichs and J. Morgan, 1980, "On the exponential falloff of wave functions and electron densities," Springer Lecture Notes in Physics, 116, 62-67.

Hoffmann-Ostenhof, T., 1980, "A comparison theorem for differential inequalities with applications in quantum mechanics," J. Phys. A13, 417-424.

Jensen, H., 1933, "Über die Gültigkeit des Virialsatzes in der Thomas-Fermischen Theorie," Z. Phys. 81, 611-624.

Kato, T., 1957, "On the eigenfunctions of many-particle systems in quantum mechanics," Commun. Pure Appl. Math. 10, 151-171.

Lee, C. E., C. L. Longmire and M. N. Rosenbluth, 1974, "Thomas-Fermi calculation of potential between atoms," Los Alamos Scientific Laboratory Report LA-5694-MS.

Liberman, D. A. and E. H. Lieb, 1981, "Numerical calculation of the Thomas-Fermi-von Weizsäcker function for an infinite atom without electron repulsion," Los Alamos National Laboratory Report -- in preparation.

Lieb, E. H., 1974, "Thomas-Fermi and Hartree-Fock theory," Proc. Int. Cong. Math., Vancouver, vol. 2, pp. 383-386.

Lieb, E. H., 1976, "The stability of matter," Rev. Mod. Phys. 48, 553-569.

Lieb, E. H., 1977, "Existence and uniqueness of the minimizing solution of Choquard's nonlinear equation," Stud. in Appl. Math. 57, 93-105.

Lieb, E. H., 1979, "A lower bound for Coulomb energies," Phys. Lett. 70A, 444-446.

Lieb, E. H., 1981a, "A variational principle for many-fermion systems," Phys. Rev. Lett. 46, 457-459.

Lieb, E. H., 1981b, "Analysis of the Thomas-Fermi-von Weizsäcker equation for an atom without electron repulsion," in preparation.

Lieb, E. H. and S. Oxford, 1981, "An improved lower bound on the indirect Coulomb energy," Int. J. Quant. Chem. 19, 427-439.

Lieb, E. H. and B. Simon, 1977, "The Thomas-Fermi theory of atoms, molecules and solids," Adv. in Math. 23, 22-116. These results were first announced in "Thomas-Fermi theory revisited," Phys. Rev. Lett. 31, 681-683 (1973). An outline of the proofs was given in (Lieb, 1974).

Lieb, E. H. and B. Simon, 1978, "Monotonicity of the electronic contribution to the Born-Oppenheimer energy," J. Phys. B11, L537-542.

Lieb, E. H. and W. Thirring, 1975, "A bound for the moments of the eigenvalues of the Schroedinger Hamiltonian and their relation to Sobolev inequalities," in Studies in Mathematical Physics: Essays in Honor of Valentine Bargmann, edited by E. H. Lieb, B. Simon and A. S. Wightman (Princeton University Press, Princeton), 269-303.

March, N. H., 1957, "The Thomas-Fermi approximation in quantum mechanics," Adv. in Phys. 6, 1-98.

Morgan III, J., 1978, "The asymptotic behavior of bound eigenfunctions of Hamiltonians of single variable systems," J. Math. Phys. 19, 1658-1661.

Morrey, C. B., Jr., 1966, Multiple Integrals in the calculus of variations (Springer, New York).

O'Connor, A. J., 1973, "Exponential decay of bound state wave functions," Commun. Math. Phys. 32, 319-340.

Reed, M. and B. Simon, 1978, Methods of Modern Mathematical Physics, Vol. 4 (Academic Press, New York).

Scott, J.M.C., 1952, "The binding energy of the Thomas-Fermi atom," Phil. Mag. 43, 859-867.

Sheldon, J. W., 1955, "Use of the statistical field assumption in molecular physics," Phys. Rev. 99, 1291-1301.

Simon, B., 1981, "Large time behavior of the L^p norm of Schroedinger semigroups," J. Func. Anal. 40, 66-83.

Stampacchia, G., 1965, "Equations elliptiques du second ordre a coefficients discontinus" (Presses de l'Univ., Montreal).

Teller, E., 1962, "On the stability of molecules in the Thomas-Fermi theory," Rev. Mod. Phys. 34, 627-631.

Thirring, W., 1981, "A lower bound with the best possible constant for Coulomb Hamiltonians," Commun. Math. Phys. 79, 1-7 (1981).

Thomas, L. H., 1927, "The calculation of atomic fields," Proc. Camb. Phil. Soc. 23, 542-548.

Torrens, I. M., 1972, Interatomic Potentials (Academic Press, New York).

Veron, L., 1979, "Solutions singulières d'equations elliptiques semilinéaire," C. R. Acad. Sci. Paris 288, 867-869. This is an announcement; details will appear in "Singular solutions of nonlinear elliptic equations," J. Non-Lin. Anal.

von Weizsäcker, C. F., 1935, "Zur Theorie der Kernmassen," Z. Phys. 96, 431-458.

THE STABILITY OF MATTER

Walter Thirring

Institut für Theoretische Physik
Universität Wien
Boltzmanng. 5, A-1090 Wien, Austria

1. INTRODUCTION

Real matter is composed of electrons and nuclei and non-relativistic quantum mechanics is supposed to give a correct description of its coarse structure. In this theory such a system is governed by the Hamiltonian

$$H_N = \sum_{i=1}^{N} \frac{p_i^2}{2m_i} + \sum_{i>j} (e_i e_j - \kappa m_i m_j)|x_i - x_j|^{-1} \tag{1.1}$$

(Notation: $(x_i, p_i, m_i; e_i)$ are position, momentum, mass and charge of the i-th particle, N is their total number, κ = gravitational constant.)

One of the fundamental properties of matter is the extensive nature of the energy. It is believed to be a consequence of the saturation properties of the chemical forces and ought to be deducible from (1.1).

Thus one wants energy $A > 0$ independent of N which gives the lower bound

$$H_N \overset{?}{\geq} - AN . \tag{1.2}$$

Operator inequalities mean that the inequality holds for all expectation values and thus (1.2) is equivalent to the statement that the spectrum of H_N is above $- AN$. A result of the type (1.2) has

first been derived by Dyson and Lenard [1]. A better understanding of the problem resulted from a serious study of Thomas-Fermi (TF) theory. I shall start my lecture with some heuristic considerations. Since the more recent history has been mentioned by Prof. Fröhlich I shall give a sketch of the older history long forgotten by most people. Next I will show a simple proof of (1.2) due to E. Lieb and myself [2]. Finally I shall give the details of some recent work [3] which sharpens the inequality to the point that for large nuclear charges the TF-energy appears as lower limit.

2. HEURISTICS

1. Since H_N contains a double sum with $\sim N^2$ terms there must be a remarkable cancellation if we want something $\sim N$. In particular if the system has a total charge

$$\sum_{i=1}^{N} e_i = Q \neq 0$$

   and is in a volume with radius R there will be an electrostatic energy $\sim Q^2/R$ which may not be $\sim N$. (Since this is > 0, (1.2) would still be satisfied.)

2. Since gravity does not neutralize there is no hope of proving (1.2) if $\kappa > 0$. In fact it can be shown that in this case (if $Q = 0$) $H_N \sim - N^3$ for bosons and $\sim - N^{7/3}$ for fermions. This behaviour can be understood as follows. The ground state energy is the best possible compromise between the quantum mechanical zero point energy and the potential energy. For sufficiently high N gravity compresses the system to a high density plasma so that the electric potential is neutralized and the gravitational energy becomes $\sim - N^2/R$. (We shall consider the equal mass case and use units $m_i = \hbar = 1$.) The zero point energy is $N/(\Delta x)^2$ where Δx is the linear dimension of the space available for a particle to move around. For bosons we may identify Δx with R, the radius of the system. Then the energy becomes

$$\sim \frac{N}{R^2} - \kappa \frac{N^2}{R}$$

   which assumes its minimum $\sim - N^3$ for $R \sim N^{-1}$. For fermions the exclusion principle requires $(\Delta x)^3 = R^3/N$. In that case the energy

$$\sim \frac{N^{5/3}}{R^2} - \kappa \frac{N^2}{R}$$

   has its minimum $\sim - N^{7/3}$ at $R \sim N^{-1/3}$.

3. Even if $Q = \kappa = 0$ H_N is not $\sim N$ if positive and negative bosons

are present. This can be made plausible by a similar argument. For a neutral system the potential energy per particle will be of the order $-1/r$ if now $e_i = 1$ and r is the distance to the next neighbour: $r = R/N^{1/3}$. Thus for fermions the energy will be

$$\frac{N^{5/3}}{R^2} - \frac{N^{4/3}}{R} .$$

This has its minimum $\sim -N$ at $R \sim N^{1/3}$ which is fine. For bosons we have

$$H_N \sim \frac{N}{R^2} - \frac{N^{4/3}}{R}$$

which goes like $-N^{5/3}$ for $R \sim N^{-1/3}$. Of course, these heuristic considerations are not conclusive. In particular, up to date it is undecided whether for bosons the ground state energy goes $\sim -N^{5/3}$. To get an upper bound for it by using trial functions one has found only $-N^{7/5}$. This comes about because the required correlations between the particles cost extra kinetic energy.

4. One should be able to prove stability by using Fermi statistics for the electrons only: Helium or deuterium appear to be perfectly stable. The relevant energy being Ry with the electron mass, (1.2) should remain valid in the limit nuclear mass $\to \infty$.
5. If one uses the relativistic kinetic energy the situation becomes even more catastrophic. In this case for $Q = 0$, $\kappa > 0$ and N sufficiently big H_N is no longer bounded from below. The reason is, that then the kinetic energy is $\sim |\vec{p}| \sim 1/R$ and thus for $\kappa N^2 \geq N$ (or $N^{4/3}$ for fermions) the infimum $-\infty$ is approached for $R \to 0$. Since relativistic quantum electrodynamics has not yet been formulated in a mathematically sound way (in 4 dimensions) it is everybody's guess how the stability situation is in reality. The empirical argument that matter appears to be stable is not convincing because we may be separated by a sufficiently thick potential barrier from the abyss.
6. The danger for stability does not come from the long range of the Coulomb potential, but from its short distance behaviour. Smoothing it at $r = 0$ by $1/r \to (1 - \exp -\mu r)/r = v(r)$ renders it stable (even for bosons). To see this one has to realize that $v(r)$ is a function of positive type (i.e. its Fourier transform $\tilde{v}(k) > 0$). Now

$$\sum_{i>j} e_i e_j v(x_i - x_j) = -\frac{N}{2} v(0) + \int \frac{d^3k}{(2\pi)^3} \tilde{v}(k) \left|\sum_j e_j e^{i\vec{k}\vec{x}_j}\right|^2$$

and the last term is > 0. Thus, if $\kappa = 0$, the potential energy and hence H_N is bounded from below by $-\frac{N}{2} v(0)$. In case the

energy is not extensive the system must collapse to take advantage of the $1/r$ singularity. As a consequence the stability problem for a Yukawa potential is as serious as for a Coulomb potential.

7. Whereas potentials v of positive type finite at the origin give Bose-stability, if $|v(r_o)| > |v(0)|$ for some $r_o > 0$ the system is instable even for fermions. To prove this, construct a trial function for

$$H_N = \frac{1}{2} \sum_{i=1}^{N} p_i^2 + \sum_{i>j} e_i e_j \, v(x_i - x_j)$$

as follows. Put the N/2 positive and the N/2 negative particles in balls with radius εr_o, $\varepsilon << 1$, and separate the balls by r_o. The attraction $- e^2 v(r_o) N^2/4$ wins between the balls over the repulsion $e^2 v(0) N(N-2)/4$ inside the balls and also over the kinetic energy $\sim N^{5/3}$. Thus the total energy goes $\sim - N^2$.

These remarks suffice to show that the stability problem is rather subtle and cannot be settled by cheap arguments.

3. THE EARLY HISTORY OF STABILITY CONSIDERATIONS

As mentioned relativistically H_N is not bounded from below if N is sufficiently large. The recognition that the fixed stars are not as stable as they seem to be but many of them are doomed to disappear in a giant catastrophe, is one of the major scientific events of the past decades. It is not only of obvious philosophical interest but also provides a meeting ground of the major branches of physics like statistical physics, quantum theory and relativity theory. Already in 1926, the year of the discovery of Schrödinger's equation, Fowler [5] realized that it was the zero point pressure of the electrons which keeps a star from collapsing, that "a star was like a giant molecule in its ground state". However, Fowler considered only non-relativistic kinematics and the gravitational pressure can overcome the zero point pressure only once the electrons become relativistic. The first mention of this fact was published in a paper by I. Frenkel [6] in 1928. He uses the correct physical argument but not a complete mathematical theory, only order of magnitude estimates. Apparently independently W. Anderson [7] published in 1929 in the same journal a similar observation but used an incorrect expression for the relativistic Fermi pressure. This was corrected by E. Stoner [8] in 1930 and he derived in a reasonably complete manner the critical mass. A year later Chandrasekhar [9,10] gave a mathematically even more consistent deduction of the same quantity which is now named after him. Shortly after (January 1932) L. Landau [11] published a discussion of the problem which not only emphasizes the physics behind the collapse of a star of supercritical mass but for the first time also the incredible astronomical importance of this phenomenon. In fact, he did not

trust the underlying laws but resorted to the pet-idea of Bohrs of energy-nonconservation to save the stars.

It is interesting to note that none of the authors so far mentioned quotes any of the previous work on that problem. Landau also does not mention that the correct answer, why stars with supercritical mass still exist, had been given three years earlier by Atkinson and Houtermans [12]. These authors point out that in the interior of stars conditions are such that nuclear processes can be ignited and the energy lost by radiation can be supplied by nuclear fuel. Their paper involves the correct physical principles although at that time not enough nuclear physics was known to identify the dominant reactions. This had been done only about ten years later by C. Weizsäcker [13] and H. Bethe [14]. Therefore by the beginning of the thirties it was clear that a large star could release tremendous energies by shrinking to a small object. In fact it was conjectured by W. Baade and F. Zwicky [15] that the end product would be a neutron star. What was still missing was the effective mechanism for carrying the energy off since electromagnetic radiation is far too slow a process to make the event dramatic. There the correct possibility was pointed out by G. Gamow and M. Schoenberg in two very clear papers [16]. They realize that once $p + e \to n + \nu$ becomes energetically possible the neutrinos can carry off the energy instantaneously and a star would collapse in a matter of seconds. Thus by that time all principles involved in gravitational collapse had been correctly understood. Yet it took more than twenty years until this phenomenon was taken seriously.

Regarding stability without gravitational interactions the saturation of chemical (and of nuclear) forces was for a long time part of the folklore without any serious attempt of a proof. To my knowledge L. Onsager was the first to derive an inequality which shows stability of extended charges. His result is a special case of the theorem about positive type potentials. For the later development of this problem see Fröhlich's lectures.

4. STABILITY FOR $\kappa = 0$

Let us first rewrite the Hamiltonian (1.1) for N electrons (coordinates x_i, momenta p_i, charges $e = -1$) and M nuclei (coordinates X_α, momenta P_α, charges Z_α):

$$H_N = \sum_{j=1}^{N} p_j^2 - \sum_{\alpha,j} \frac{Z_\alpha}{|x_j - X_\alpha|} + \sum_{i>j} |x_i - x_j|^{-1} + \sum_{\alpha>\beta} Z_\alpha Z_\beta |X_\alpha - X_\beta|^{-1} + \\ + \sum_{\alpha=1}^{M} \frac{P_\alpha^2}{2m_\alpha} \quad . \tag{4.1}$$

The strategy for proving $H_N \geq - AN$ is to show that the TF-energy-functional

$$E_N(\sigma) = \frac{3}{5}(3\pi^2)^{2/3} \int d^3x\, \sigma^{5/3}(x) - \sum_\alpha Z_\alpha \int \frac{d^3x\, \sigma(x)}{|x-X_\alpha|} + $$

$$+ \frac{1}{2}\int d^3x\, d^3y \frac{\sigma(x)\sigma(y)}{|x-y|} + \sum_{\alpha>\beta} Z_\alpha Z_\alpha |X_\alpha - X_\beta|^{-1} \tag{4.2}$$

with modified constants provides a lower bound to (4.1).

Trying to relate the Hamiltonian (4.1) to the TF-functional (4.2) it appears to be obvious that σ should be

$$\sigma(x_1) = \int d^3x_2 \ldots d^3x_N\, |\psi(x_1, x_2, \ldots, x_N)|^2$$

ψ being the ground state wave function. If we want to show something like $\langle\psi|H_N\psi\rangle \geq E(\sigma)$ two problems arise:

1. What has $\int\sigma^{5/3}$ got to do with $\int d^3x_1 \ldots d^3x_N \sum_i |\nabla_i\psi|^2$.

2. What is the error in replacing $\langle\psi| \sum_{i>j} |x_i - x_j|^{-1}\psi\rangle$ by

$\frac{1}{2}\int \frac{d^3x\, d^3y}{|x-y|}\sigma(x)\,\sigma(y)$, that is neglecting correlations.

Curiously enough we may use the no-binding theorem (see Lieb's lectures)

$$\frac{3}{5\gamma}\int d^3x\, \sigma^{5/3} - \sum_{k=1}^{M} Z_k \int \frac{d^3x\, \sigma(x)}{|x-X_k|} + \frac{1}{2}\int \frac{d^3x\, d^3y}{|x-y|}\sigma(x)\,\sigma(y) +$$

$$+ \sum_{n>m} \frac{Z_n Z_m}{|X_n - X_m|} \geq - 3.68 \sum_{k=1}^{M} Z_k^{7/3}\gamma \quad \forall\, \gamma \geq 0, \quad \sigma \in L^1 \cap L^{5/3}, \tag{4.3}$$

to solve the second problem. Its special case $Z_i = 1$, $M = N$, can be written

$$\sum_{i>j} |x_i - x_j|^{-1} \geq \sum_i \int \frac{d^3x\, \sigma(x)}{|x_i - x|} - \frac{1}{2}\int \frac{d^3x\, d^3y}{|x-y|}\sigma(x)\,\sigma(y) - 3.68\gamma N -$$

$$- \frac{3}{5\gamma}\int d^3x\, \sigma^{5/3}(x) \quad . \tag{4.4}$$

This is an electrostatic inequality and tells you that the Coulomb

repulsion between equal charges is always more than their energy in a potential produced by a charge distribution σ minus three correction terms, two depending on σ. It allows us to minorize (4.1) by a σ-dependent Hamiltonian H_σ where the electrons do no longer interact but all move in a common potential V_σ.

$$H_N \geq H_\sigma := \sum_{i=1}^{N} h_i - \frac{1}{2}\int \frac{d^3x\, d^3y}{|x-y|}\,\sigma(x)\,\sigma(y) - \frac{3}{5\gamma}\int d^3x\, \sigma^{5/3} - 3.68\gamma N + \\ + \sum_{n>m} \frac{Z_n Z_m}{|X_n - X_m|} , \tag{4.5}$$

$$h_i = p_i^2 - \sum_{k=1}^{M} \frac{Z_k}{|x_i - X_k|} + \int \frac{d^3x\, \sigma(x)}{|x-x_i|} .$$

Thus we have reduced the problem to solving the 3-dimensional Schrödinger equation with V_σ and then choose σ judiciously. At this point the Fermi-statistics of the electrons enter because it allows us to use the sum of the lowest N eigenvalues of $h(\sigma)$ to get a bound for H_N.

A rather crude bound for this sum is obtained by using a result of Birman and Schwinger for the number of bound states.

Theorem:
Let e_j be the j-th negative eigenvalue of the Hamiltonian $p^2 + V(x)$. Then

$$\sum_j |e_j| \leq \frac{4}{15\pi}\int d^3x\, |V|_-^{5/2} , \tag{4.6}$$

where
$$|V|_- = \begin{cases} |V(x)| & \text{if} \quad V(x) < 0 \\ 0 & \text{if} \quad V(x) \geq 0 \end{cases} .$$

Proof:
First we notice that all eigenvalues of $p^2 + V$ are lowered if V is replaced by its negative part $-|V|_-$. Furthermore the number of negative eigenvalues is not decreased if V is multiplied by $\lambda \geq 1$ because

$$p^2 + \lambda V = \lambda(p^2 + V) + (1-\lambda)p^2 < \lambda(p^2 + V) . \tag{4.7}$$

Hence the number of negative eigenvalues of $p^2 + V$ is majorized by the one of $p^2 - \lambda|V|_-$, $\lambda \geq 1$. Rewriting the corresponding Schrödinger equation as

$$(p^2 - E)^{-1}\, |V|_-\, \psi = \frac{1}{\lambda}\, \psi$$

we realize that the number N(E) of eigenvalues of $p^2 + V$ below E is $\leq$ the number of eigenvalues ≥ 1 of $(p^2 - E)^{-1} \, |V|_-$.

We could also use the operator $|V|_-^{1/2} \frac{1}{p^2 - E} |V|_-^{1/2}$, which is > 0 for $E \leq 0$, or even better

$$|V - \tfrac{E}{2}|_-^{1/2} \frac{1}{p^2 - \frac{E}{2}} |V - \tfrac{E}{2}|_-^{1/2} \quad .$$

In particular N(E) is less than the trace of the square of this latter operator. In an x-representation we have

$$N(-\alpha) \leq \operatorname{Tr} \frac{1}{p^2 + \frac{\alpha}{2}} |V + \tfrac{\alpha}{2}|_- \frac{1}{p^2 + \frac{\alpha}{2}} |V + \tfrac{\alpha}{2}|_- =$$

$$= (4\pi)^{-2} \int d^3x \, d^3y \, |V(x) + \tfrac{\alpha}{2}|_- |x-y|^{-2} e^{-\sqrt{2\alpha}|x-y|} |V(y) + \tfrac{\alpha}{2}|_- \leq$$

$$\leq \frac{1}{4\pi\sqrt{2\alpha}} \int d^3x \, |V(x) + \tfrac{\alpha}{2}|_-^2 \quad . \tag{4.8}$$

In the last step Young's inequality (* = convolution)

$$||f * g||_r \leq ||f||_p ||g||_q \quad , \quad 1 + \frac{1}{r} = \frac{1}{p} + \frac{1}{q} \quad ,$$

has been used. To complete the proof one only has to realize that

$$\sum_i |e_i| = \int_{-\infty}^{0} dE \; N(E)$$

and carry out the integration in (4.8).

REMARKS

1. The constant in (4.6) is not the best possible one and has in fact been improved by almost a factor 2 by E. Lieb
2. The classical expression

$$\int_{h<0} d^3x \, d^3p \, |h(x,p)| = \frac{1}{15\pi^2} \int d^3x \, |V|_-^{5/3} \tag{4.9}$$

is smaller than (4.6) by 4π. Experience gained from computer studies leads us [4] to conjecture that (4.9) is actually a bound.
3. Using the present strategy we will not have to answer question (1). However it is not hard to see that (4.6) is equivalent to

$$\int \sum_i |\nabla_i \psi|^2 d^3x_1 \ldots d^3x_N \geq \left(\frac{1}{4\pi}\right)^{2/3} \frac{3}{5} (3\pi^2)^{2/3} \int d^3x \, [\rho(x)]^{5/3} \quad . \tag{4.10}$$

With (4.6) we get the following numerical bound for H_N (with a factor 2 for spin)

$$H_\sigma \geq -\frac{4\cdot 2}{15\pi}\int d^3x \left|\sum_n \frac{-Z_n}{|x-X_n|} + \int \frac{d^3y\ \sigma(y)}{|x-y|}\right|_-^{5/2} - \frac{1}{2}\int \frac{d^3x\, d^3y}{|x-y|}\sigma(x)\sigma(y) -$$

$$- 3.68\gamma N - \frac{3}{5\gamma}\int d^3x\ \sigma^{5/3}(x) + \frac{1}{2}\sum_{n>m}\frac{Z_n Z_m}{|X_n - X_m|} \quad . \tag{4.11}$$

If we now choose $\sigma(x)$ to obey the Thomas-Fermi equation we may express $\int V^{5/2}$ as $\int \sigma^{5/3}$ or $\int \sigma V$ and rewrite (4.11) in a form similar to (4.2)

$$H_N \geq \frac{3}{5}\left[\left(\frac{3\pi}{4}\right)^{2/3} - \frac{1}{\gamma}\right]\int d^3x\ \sigma^{5/3}(x) - 3.68\gamma N - \int d^3x \sum_n \frac{Z_n\sigma(x)}{|x-X_n|} +$$

$$+ \frac{1}{2}\int \frac{d^3x\ d^3y}{|x-y|}\sigma(x)\ \sigma(y) + \frac{1}{2}\sum_{n>m}\frac{Z_n Z_m}{|X_n-X_m|} \quad . \tag{4.12}$$

Using again the no-binding theorem (4.3) and optimizing with respect to γ finally gives

$$H_N \geq - 3.68\ \left[\gamma N + \left(\left(\frac{3\pi}{4}\right)^{2/3} - \frac{1}{\gamma}\right)^{-1} \sum_n Z_n^{7/3}\right] \rightarrow$$

$$H_N \geq - 2.08\ N\left[1 + \left(\frac{1}{N}\sum_{n=1}^{M} Z_n^{7/3}\right)^{1/2}\right]^2 \quad . \tag{4.13}$$

REMARKS

1. If we have a neutral system with equal nuclear charges $Z_i = N/M$ we get, for large Z, a bound $\sim MZ^{7/3}$. This corresponds to the Z-dependence of the energies of atoms. For $N \ll ZM$ our bound is not optimal, one should expect something $\sim NZ^2$.
2. If we have q species of fermions the same argument leads, in the case $Z_i = 1$, $M = N$, to a bound $\sim q^{2/3}\ N$. If $q = N$ this means that without symmetry requirement on ψ we have a bound $\sim N^{5/3}$. A fortiori this proves the lower bound for bosons of our intuitive argument. The same conclusions can also be reached from the trivial fact $|e_i| \leq \sum_j |e_j|$.
3. If (4.9) were a bound the constant in front of (4.13) would be the one for single atoms in Thomas-Fermi theory. This would be the best possible constant in (4.13) since for $Z \rightarrow \infty$ this value is actually approached.

5. THE TF-ENERGY AS A BOUND

The asymptotic exactness of TF-theory states

$$\lim_{Z\to\infty} E(M,Z)/Z^{7/3} = -M\,\varepsilon_{TF}$$

where $\varepsilon_{TF} = 0.77\ me^2/\hbar^2$ is the TF-energy. We shall finally prove the uniform statement

$$E(M,Z) \geq -M\,\varepsilon_{TF}\,Z^{7/3}(1 + O(Z^{-2/33}))$$

which for $Z \to \infty$ sharpens our previous inequality by $(4\pi)^{2/3}$ which we lost using the Schwinger-Birman bound.

Our general strategy is to split the Coulomb potential into a regularized long-range part v_r and a short range singular part v_s

$$\frac{1}{r} = v_r(r) + v_s(r), \quad v_r = g^2 * \frac{1}{r} * g^2, \quad g^2(x) = e^{-\mu r}\frac{\mu^3}{8\pi}\ . \quad (5.1)$$

For v_r the classical bound with ε_{TF} will hold for $Z \to \infty$ whereas the corrections due to v_s go with a lower power of Z. We keep the previous notation (x_i, p_i, σ_i = coordinate and momentum and spin of the i-th electron, R_k, Z_k = coordinate and charge of the k-th nucleus) but study the family of Hamiltonians

$$H(\alpha) = \sum_i p_i^2 - \sum_{i,k} Z_k(|x_i - R_k|^{-1} + (\alpha-1)v_s(x_i - R_k)) + \sum_{i>j}(|x_i - x_j|^{-1} +$$

$$+ (\alpha-1)v_s(x_i - x_j)) + \sum_{k>m} Z_k Z_m(|R_k - R_m|^{-1} + (\alpha-1)v_s(R_k - R_m))\ . \quad (5.2)$$

We are actually interested in H(1) and shall use the concavity of the ground state energy [17] $E(\alpha)$ of $H(\alpha)$ in the form

$$E(1) \geq E(0) + \alpha^{-1}(E(\alpha) - E(0)) \quad \forall\, \alpha \geq 1\ . \quad (5.3)$$

The No-Binding Theorem

For our estimates we shall need the no-binding theorem of TF-theory with Yukawa potentials $\exp(-\mu r)/r$. The proof for $\mu > 0$ proceeds exactly as for $\mu = 0$, the subharmonicity argument being also applicable [18] since $\Delta \exp(-\mu r)/r + 4\pi\delta(x) = \mu^2 \exp(-\mu r)/r > 0$. The theorem states that the TF-energy is always above the sum of the energies of the isolated atoms and reads in formulae

$$\frac{3\gamma}{5}\int d^3x\, n^{5/3}(x) - \sum_{k=1}^{M} Z_k \int \frac{d^3x\, e^{-\mu|x-R_k|}}{|x-R_k|} n(x) + \frac{1}{2}\int \frac{d^3x\, d^3y}{|x-y|} e^{-\mu|x-y|}$$

$$\cdot\, n(x)\, n(y) + \sum_{k>m} Z_k Z_m \frac{e^{-\mu|R_k-R_m|}}{|R_k-R_m|} \geq - \sum_{k=1}^{M} \varepsilon(\mu,\gamma,Z_k) \qquad (5.4)$$

where

$$-\,\varepsilon(\mu,\gamma,Z) = \inf_n \{\gamma \frac{3}{5}\int d^3x\, n^{5/3}(x) - Z\int \frac{d^3x}{|x|} e^{-\mu|x|} n(x) +$$

$$+ \frac{1}{2}\int \frac{d^3x\, d^3y}{|x-y|} e^{-\mu|x-y|} n(x)\, n(y)\} \quad .$$

By scaling one sees that ε involves only an unknown function of one variable, $\varepsilon(\mu,\gamma,Z) = \gamma^6\mu^7 f(\frac{Z}{\gamma^3\mu^3})$ and TF-theory tells us

$$\varepsilon(0,\gamma,1) = \frac{1}{\gamma}\,\varepsilon_{TF}(3\pi^2)^{2/3} = 3.68/\gamma \quad . \qquad (5.5)$$

For large μ the repulsion becomes unimportant. Then the bound

$$-f(Z) > \frac{3}{5}||n||_{5/3}^{5/3} - Z\,||\frac{e^{-x}}{x}||_{5/2}\,||n||_{5/3} \geq -Z^{5/2}\, 4(\frac{2\pi}{5})^{3/2} \equiv -c\, Z^{5/2} \qquad (5.6)$$

and thus $\varepsilon < c\,\gamma^{-3/2}\,\mu^{-1/2}\,Z^{5/2}$ will be sufficient.

<u>Bound for E(Q)</u>

Taking (5.4) for $\mu = 0$, $Z_k = 1$, $M = N$, and integrating over $\int d^3R_1 \ldots d^3R_N\, g^2(x_1-R_1)\ldots g^2(x_N-R_N)$ we learn

$$\sum_{i>j} v_r(x_i-x_j) \geq \sum_i \int d^3x\, n(x)\, v_g(x-x_i) - \frac{1}{2}\int \frac{d^3x\, d^3x'}{|x-x'|} n(x)\, n(x') -$$

$$- N\,\frac{3.68}{\gamma} - \frac{3\gamma}{5}\int d^3x\, n^{5/3}(x), \qquad (5.7)$$

$$v_g = \frac{1}{r} * g^2$$

or, upon substitution in (5.2)

$$H(0) \geq \sum_i \{p_i^2 - \sum_k Z_k\, v_r(x_i-R_k) + \int d^3x\, n(x)\, v_g(x-x_i)\} - \frac{3.68}{\gamma} N -$$

$$- \frac{3\gamma}{5} \int d^3x \, n^{5/3}(x) + \sum_{k>m} Z_k Z_m v_r(R_k - R_m) - \frac{1}{2} \int \frac{d^3x \, d^3x'}{|x-x'|} n(x)n(x') \equiv$$

$$\equiv \sum_i \{h_i\} + C \,.$$

Here h_i is the Hamiltonian of the i-th particle in an external field. Using the coherent states $|q,k\rangle$ with wave function $e^{ikx} g(x-q)$ it can be written as

$$h = \int \frac{d^3q \, d^3k}{(2\pi)^3} \left(k^2 - \sum_k Z_k v_g(q-R_k) + \int d^3x \frac{n(x)}{|q-x|}\right) |q,k\rangle \langle q,k| - \frac{\mu^2}{4} \,. \tag{5.8}$$

For spin 1/2-electrons we obviously have

$$\sum h_i \geq - 2 \operatorname{tr} |h|_- \tag{5.9}$$

and general trace inequalities [18] tell us

$$\operatorname{tr} \left| \int \frac{d^3q \, d^3k}{(2\pi)^3} h(q,k) |q,k\rangle \langle k,q| \right|_- \leq \int \frac{d^3k \, d^3q}{(2\pi)^3} |h(q,k)|_- \,. \tag{5.10}$$

In our case the k-integral is easily performed and we find

$$H(0) \geq - \frac{2}{15\pi^2} \int d^3q \left| -\sum_j Z_j v_g(q-R_j) + \int \frac{d^3x \, n(x)}{|x-q|} \right|_-^{5/2} - \left(\frac{3.68}{\gamma} + \frac{\mu^2}{4}\right) N -$$
$$- \frac{1}{2} \int \frac{d^3x \, d^3x'}{|x-x'|} n(x)n(x') + \sum_{j>m} Z_j Z_m v_r(R_j - R_m) - \frac{3\gamma}{5} \int d^3x \, n^{5/3}(x) \,. \tag{5.11}$$

So far the inequality holds for any $n \in L^{5/3}$ and we shall now optimize

$$- \frac{2}{15\pi^2} \int d^3q \left| -\sum_j Z_j v_g(q-R_j) + \int \frac{d^3x \, n(x)}{|x-q|} \right|_-^{5/2} - \frac{1}{2} \int d^3x \, d^3y \frac{n(x)n(y)}{|x-y|} .$$

The formal variational derivation vanishes when

$$n(q) = \frac{1}{3\pi^2} \left| -\sum_j Z_j v_g(q-R_j) + \int \frac{d^3x \, n(x)}{|x-q|} \right|_-^{3/2} \tag{5.12}$$

and TF-theory guarantees that $\sup_n$ is actually attained. From (5.12)

we learn

$$\int d^3q \; |-\sum_j Z_j v_g(q-R_j) + \int \frac{d^3x\, n(x)}{|x-q|}|_-^{5/2} = \int d^3q \; |-\sum_j Z_j v_g(q-R_j) +$$

$$+ \int \frac{d^3x\, n(x)}{|x-q|}|_- \; 3\pi^2 \, n(q) = (3\pi^2)^{5/3} \int d^3q \; n^{5/3}(q) \quad .$$

Substituting a linear combination of the two versions back into (5.11) and using the generalization of (5.7) for $Z_k \geq 1$, $N \geq M$, we find the desired bound:

$$H(0) \geq \frac{3}{5}[(3\pi^2)^{2/3} - \gamma] \int d^3x \; n^{5/3}(x) - \sum_j Z_j \int d^3x \; n(x) \; v_g(x-R_j) -$$

$$- (\frac{3.68}{\gamma} + \frac{\mu^2}{4})N + \frac{1}{2} \int \frac{d^3x \; d^3y}{|x-y|} n(x)n(y) + \sum_{k>m} Z_k Z_m v_r(R_k - R_m) \geq$$

$$\geq - \varepsilon_{TF}(\frac{1}{\gamma_1} N + \sum_k Z_k^{7/3}/(1-\gamma_1)) - \mu^2 N/4 \; , \qquad 0 < \gamma_1 < 1 \; . \qquad (5.13)$$

<u>Correction due to v_s</u>

Since g^2 has the Fourier-transform $\tilde{g}^2(k) = \frac{\mu^4}{(k^2+\mu^2)^2}$, we have

$$\tilde{v}_r(k) = \frac{\mu^8}{(k^2+\mu^2)^4} \frac{4\pi}{k^2}$$

and therefore

$$\tilde{v}_s(k) = \frac{4\pi}{k^2+\mu^2} \left[1 + \frac{\mu^2}{k^2+\mu^2} + \frac{\mu^4}{(k^2+\mu^2)^2} + \frac{\mu^6}{(k^2+\mu^2)^3}\right] \; .$$

Thus

$$v_s = \sum_{\rho=1}^{4} f_\rho * \frac{e^{-\mu r}}{r} * f_\rho$$

with

$$f_1 = \delta(x), \qquad f_2 = \mu \int \frac{d^3k \; e^{ikx}}{(k^2+\mu^2)^{1/2}} = \frac{\mu}{2\pi^2 r^2} \int_{\mu r}^{\infty} d\zeta \; K_o(\zeta) \; \zeta \; ,$$

$$f_3 = \frac{\mu^2 e^{-\mu r}}{4\pi r} \; , \qquad f_4 = \frac{\mu^3}{2\pi^2} K_o(\mu r) \quad .$$

Since the f_ρ are positive we may again multiply (5.4) with

$$\prod_{k=1}^{M} f_\rho(x_k - R_k)$$

and integrate: Substituting $n = m * f_\rho$ and using

$$||m * f||_{5/3} \leq ||m||_{5/3}||f||_1 = ||m||_{5/3}$$

we get a "no-binding" theorem for the

$$f_\rho * \frac{e^{-\mu r}}{r} * f_\rho\text{-potential,}$$

however with the binding energy of the Yukawa potential. Adding these four inequalities to the one with $\mu = 0$ gives a no-binding theorem for the $v_r + \alpha v_s$-potential:

$$\frac{3\gamma}{5}\int d^3x\, m^{5/3}(x) - \sum_{k=1}^{M} Z_k \int d^3x\, m(x)(|x-x_k|^{-1} + (\alpha-1)v_s(x-x_k)) +$$

$$+ \frac{1}{2}\int d^3x\, d^3y\, m(x)\, m(y)(|x-y|^{-1} + (\alpha-1)v_s(x-y)) + \sum_{k>m} Z_k Z_m(|x_k-x_m|^{-1} +$$

$$+ (\alpha-1)v_s(x_k-x_m) \geq - \sum_{k=1}^{M}\{\frac{3.68}{\gamma-\gamma'} Z_k^{7/3} + 4(\alpha-1)\ \varepsilon(\mu,\frac{\gamma'}{4(\alpha-1)},Z_k)\} \tag{5.14}$$

$$\forall\, m \in L^{5/3}, \quad x_k \in \mathbb{R}^3, \quad 0 < \gamma' < \gamma\ .$$

Upon inserting this for $Z_k = 1$, $M = N$, into (5.2) we generalize (5.7) to

$$H(\alpha) \geq \sum_i p_i^2 - \sum_{i,k} Z_k(|x_i-R_k|^{-1} + (\alpha-1)v_s(x_i-R_k)) + \sum_i \int d^3x m(x)(|x-x_i|^{-1}$$

$$+ (\alpha-1)v_s(x-x_i)) + \sum_{k>m} Z_k Z_m(|R_k-R_m|^{-1} + (\alpha-1)v_s(R_k-R_m)) -$$

$$- \frac{3\gamma_2}{5}\int d^3x\, m^{5/3}(x) - \frac{1}{2}\int d^3x\, d^3y\, m(x)\, m(y)(|x-y|^{-1} + (\alpha-1)v_s(x-y)) -$$

$$- N(\frac{3.68}{\gamma_2-\gamma_3} + 4(\alpha-1)\ \varepsilon(\mu,\frac{\gamma_3}{4(\alpha-1)},1))\ . \tag{5.15}$$

Taking the expectation value of $H(\alpha)$ with any $\Psi(x_1,\dots x_N;\sigma_1,\dots\sigma_N)$ the potential will be integrated with the one-particle density

$$\rho(x) = N \sum_{\sigma_1 \ldots \sigma_N} \int d^3x_2 \ldots d^3x_N \, |\Psi(x, x_2, \ldots x_N, \sigma_1, \ldots \sigma_N)|^2 . \quad (5.16)$$

For the kinetic energy we employ the sharpened version of (4.10)

$$\langle\Psi| \sum_{i=1}^{N} p_i^2 |\Psi\rangle \geq \frac{3}{5f}(3\pi^2)^{2/3} \int d^3x \, \rho^{5/3}, \qquad f = (4\pi)^{2/3}/1.5 . \quad (5.17)$$

Thus, setting $m(x) = \rho(x)$, using (5.14) again and the bound (5.6) for ε we conclude

$$E(\alpha) = \inf_{||\Psi||=1} \langle\Psi|H(\alpha)|\Psi\rangle \geq \frac{3}{5}\left(\frac{(3\pi^2)^{2/3}}{f} - \gamma_2\right) \int d^3x \, \rho^{5/3}(x) -$$

$$- \sum_k Z_k \int d^3x \, \rho(x)(|x-R_k|^{-1} + (\alpha-1)v_s(x-R_k)) + \frac{1}{2} \int d^3x \, d^3y \, \rho(x) \, \rho(y) \cdot$$

$$\cdot(|x-y|^{-1} + (\alpha-1)v_s(x-y)) + \sum_{k>m} Z_k Z_m (|R_k-R_m|^{-1} + (\alpha-1)v_s(R_k-R_m)) -$$

$$- N\left(\frac{3.68}{\gamma_2-\gamma_3} + 4(\alpha-1)\,\varepsilon\left(\mu, \frac{\gamma_3}{4(\alpha-1)}, 1\right)\right) \geq - \varepsilon_{TF} \sum_{k=1}^{M} \left\{\frac{f \, Z_k^{7/3}}{1 - f(\gamma_2+\gamma_4)} +\right.$$

$$\left.+ \frac{c(4(\alpha-1))^{5/2}}{3\pi^2 \, \varepsilon_{TF}\mu^{1/2}} \frac{Z_k^{5/2}}{\gamma_4^{3/2}}\right\} - N \, \varepsilon_{TF}\left(\frac{1}{\gamma_2-\gamma_3} + \frac{c(4(\alpha-1))^{5/2}}{3\pi^2 \, \varepsilon_{TF}\mu^{1/2}\gamma_3^{3/2}}\right) . \quad (5.18)$$

At the end we have rescaled the γ's to introduce ε_{TF} according to (5.5). Inserting (5.18) and (5.13) into (5.3) gives the

<u>Lower bound for the Coulomb Hamiltonian</u>

$$H(1) \geq - \varepsilon_{TF} \left\{ \sum_{k=1}^{M} Z_k^{7/3}\left(\frac{\alpha-1}{\alpha} \frac{1}{1-\gamma_1} + \frac{1}{\alpha} \frac{f}{1 - f(\gamma_2+\gamma_4)}\right) + \right.$$

$$+ \sum_{k=1}^{M} Z_k^{5/2} \frac{c(4(\alpha-1))^{5/2} \alpha^{-1}}{3\pi^2 \, \varepsilon_{TF}\mu^{1/2}\gamma_4^{3/2}} + N\left(\frac{\alpha-1}{\gamma_1\alpha} + \frac{1}{\gamma_2-\gamma_3} + \frac{(\alpha-1)\mu^2}{4\alpha \, \varepsilon_{TF}} + \right.$$

$$\left.\left.+ \frac{c(4(\alpha-1))^{5/2} \alpha^{-1}}{3\pi^2 \, \varepsilon_{TF}\mu^{1/2}\gamma_3^{3/2}}\right)\right\} \quad (5.19)$$

$$\forall \, \alpha > 1, \; 0 < \gamma_1 < 1, \; 0 < \gamma_3 < \gamma_2 < \frac{1}{f} - \gamma_4, \; \gamma_4 > 0, \; \mu > 0,$$

$$f = (4\pi)^{2/3}/1.5 = 3.6, \qquad c = 4(2\pi/5)^{3/2} = 5.6 .$$

REMARKS

1. For $\alpha = 1$, $\gamma_3, \gamma_4 \to 0$, $\gamma_2^{-1} = f(1+(\sum Z_k^{7/3}/N)^{1/2})$ we recover the result (4.13)

$$E(1) \geq - f\, \varepsilon_{TF}\, N(1 + (\sum_k Z_k^{7/3}/N)^{1/2})^2 \quad .$$

2. Optimizing (5.19) over all parameters is an extensive numerical job. However, the allegation made at the beginning is obtained easily by putting $\alpha = Z^{2/33}$, $\mu = Z^{7/11}$, and, say, $\gamma_1 = Z^{-1/2}$, $\gamma_2 = \gamma_4 = 2\gamma_3 = 1/3f$. Then the first term is

$$- \varepsilon_{TF} \sum_{k=1}^{M} Z_k^{7/3}(1 + O(Z^{-2/33}))$$

and all the others are $O(Z^{-2/33})$.
3. The stability proof can be extended from potentials of the form

$$\sum_{i>j} e_i e_j\, v(x_i - v_j)$$

to potentials with spin and isospin, f.i.

$$\sum_{i>j} (\vec{\tau}_i \vec{\tau}_j)(\vec{\sigma}_i \vec{\sigma}_j)\, v(x_i - x_j) \ .$$

Then (5.7) would prove stability of nuclear matter with Yukawa potentials but without hard core.
4. If one could prove the no-binding theorem for v_s in the form that ε is the atomic energy for v_s and not for the Yukawa potential then the numbers in the terms $O(Z^{-2/33})$ could be improved.
5. Stability and no-binding require potentials of positive type because $\tilde{v} \geq 0$ implies $v(0) \geq v(r)\ \forall\, r > 0$.
6. The physical reason why the contribution of v_s does not increase as fast as $Z^{7/3}$ is that $e^{-\mu r}/r$ does not bind as many particles. Even neglecting the electron repulsion the atomic energy is sum of all binding energies $\geq - c||v_s||_{5/2}^{5/2} \sim Z^{5/2}\, \mu^{-1/2}$. Thus if μ increases faster than $Z^{1/3}$ then v_s does not contribute to the leading order in Z.

REFERENCES

1. F.J. Dyson, A. Lenard, J. Math. Phys. 8, 423 (1967)
2. E.H. Lieb, W.E. Thirring, Phys. Rev. Lett. 35, 687 (1975), see ibid. 1116 for errata
3. W. Thirring, A Lower Bound with the Best Possible Constant for Coulomb Hamiltonians, Vienna preprint UWThPh-80-7, to appear in Comm. Math. Phys.
4. E.H. Lieb, W.E. Thirring, Inequalities for the Moments of the Eigenvalues of the Schrödinger Hamiltonian and Their Relation to Sobolev Inequalitites, in the volume dedicated to V. Bargmann, Princeton University Press 1976
5. R. Fowler, Monthly Notices 87, 114 (1926)
6. I. Frenkel, Z. f. Physik 50, 234 (1928)
7. W. Anderson, Z. f. Physik 56, 851 (1929)
8. E. Stoner, Phil. Mag. 9, 944 (1930)
9. S. Chandrasekhar, Phil. Mag. 11, 592 (1931)
10. S. Chandrasekhar, Astrophys. J. 74, 81 (1931)
11. L. Landau, Phys. Z. d. Sowjetunion 1, 285 (1932)
12. R. Atkinson, F. Houtermans, Z. f. Physik 54, 656 (1929)
13. C. Weizsäcker, Z. f. Physik 39, 633 (1938)
14. H. Bethe, Phys. Rev. 55, 434 (1939)
15. W. Baade, F. Zwicky, Proc. Nat. Acad. 20, 259 (1934)
16. G. Gamow, M. Schoenberg, Phys. Rev. 58, 1117 (1940) and ibid. 59, 539 (1941)
17. W. Thirring, Quantenmechanik von Atomen und Molekülen, Springer, Wien 1979, Equ. (3.5,23)
18. W. Thirring, Quantenmechanik großer Systeme, Springer, Wien 1980, Equ. (2.2,11) and (4.1,46;3)

PHASE DIAGRAMS AND CRITICAL PROPERTIES OF (CLASSICAL) COULOMB SYSTEMS†

Jürg Fröhlich[1] and Thomas Spencer[2*]

[1] Institut des Hautes Etudes Scientifiques
35, Route de Chartres
F-91440 Bures-sur-Yvette

[2] Courant Institute of Mathematical Sciences
251 Mercer Street
New York, N.Y. 10012

I. INTRODUCTION

It seems good to start these lecture notes with the cautioning remark that, in a sense, their contents do not really fit into the topic "Rigorous Atomic and Molecular Physics", although the results we shall discuss are certainly rigorous and do concern atoms, ions and molecules to some extent. Traditionally, the underlying theoretical framework for atomic and molecular physics is thought to be quantum mechanics, and the number of degrees of freedom of the atomic and molecular systems which are considered is finite, at least if the radiation field is neglected which is what workers in that field do almost always.

---

† Lectures presented at the International School of Mathematical Physics "Ettore Majorana", Erice, Sicily, June 1980. (Lectures delivered by J.F.)

* Work partially supported by NSF DMR 79-04355 and A. Sloan Foundation.

In these lecture notes, quantum mechanics is not heard of, except that some models which we will mention briefly do provide idealized descriptions of certain quantum mechanical systems. In those instances, however, we will employ the imaginary-time, Feynman-Kac formulation of quantum mechanics which makes it look like classical, statistical mechanics.

The basic, theoretical framework underlying our lectures is the classical statistical mechanics of systems with infinitely many degrees of freedom. We think that we do describe methods and results which are relevant for the physics of systems composed of very many atoms and molecules, namely for condensed matter physics, but we leave it to the reader to judge.

Due to various circumstances it was not possible for us to produce somewhat detailed lecture notes which would include careful statements of results, proofs and discussion. Thus we can only hope the present notes will motivate the reader to consult the literature that is quoted in the text.

1.1. What are Coulomb systems ?

The underlying theory for the description of matter composed of nuclei and electrons forming ions, atoms and molecules is Quantum Electrodynamics (QED). Since in atomic, molecular and condensed matter physics the energies and velocities of the constituent particles are usually very moderate, non-relativistic QED [1] ought to provide a sufficient description. This point of view bears, however, some problems. Non-relativistic QED cannot correctly and consistently account for spin-dependent interactions. In particular, magnetic dipole interactions among constituent particles and the interactions of their magnetic moments with the quantized magnetic field have to be ignored or cutoff in some phenomenological way in order to prevent the theory from becoming mathematically meaningless. Some of these problems are briefly discussed e.g. in [2].

Thus, non-relativistic QED can only be expected to be an accurate description if

- nuclear charges are moderate;
- temperatures and densities are moderate;
- the interactions between charged particles and the radiation field do not excite high frequency modes of the electromagnetic field;
- spin-dependent forces are weak.

For the major part of these notes, spin and the radiation field will be neglected altogether, leaving us with charged particles interacting via two-body Coulomb forces. We shall call systems for which these approximations are valid Coulomb systems. Some of the models we shall discuss would normally not be called Coulomb systems. However, in those examples it turns out that one can isolate certain particle-like excitations with long-range Coulomb interactions. Among those models we shall mention ones which do feature an ultraviolet cutoff electromagnetic field : Simplified Landau-Ginsburg type models of superconductors.

Most of the models appearing in these lecture notes are quite naive caricatures of more realistic theories. Many of them are classical lattice models. However, in many instances, neither the assumption that the models be classical, nor the replacement of space by a lattice are really important, but are made because they are reasonable, or in order not to obscure the simplicity of an argument. The only serious requirement is that in order to render a classical, three-dimensional Coulomb system in thermal equilibrium mathematically meaningful, the Coulomb potential must be regularized at short distances. (The lattice is often a pedagogically and physically attractive regularization).

In spite of all these crude approximations we believe that the models, methods and results we discuss in these notes have quite a lot to do with physics and are chosen so as to exhibit certain interesting physical phenomena in a pure and simple form. (It is a good tradition in theoretical physics to replace a theory if it turns out to evade our comprehension by an approximate model which can be analysed in satisfactory depth).

Our most interesting methods, results and speculations appear in Sections III - V. Among them is a rigorous version of real-space renormalization group techniques powerful enough to establish the existence of Kosterlitz-Thouless (plasma → dipolar phase) transitions in a large variety of situations. Furthermore, we comment on "liquid crystal" phases in hard core Coulomb gases and the possible transitions in the three-dimensional Coulomb gas. See Sections II.2 and III.

The reader familiar with the basic definitions and notions concerning Coulomb systems may skip the remainder of the introduction and proceed to Section II.

I.2. Stability

The ν-dimensional Coulomb potential is defined to be the Green's function, $V(x,y)$, of a ν-dimensional Laplacean, $-\Delta$. If particles in a Coulomb system are confined to a bounded region, Λ , some boundary conditions (b.c.) need to be specified at the boundary, $\partial\Lambda$, of the box. (Different b.c. in the Coulomb potential can yield different thermodynamic limits of the corresponding systems). We shall consider two types of b.c. :

(BC1) Insulating, or free b.c.

$V(x,y) \equiv V(x-y)$ is the Green's function of the infinite volume Laplacean. The arguments, x and y , are constrained to be inside Λ . Physically, this corresponds to putting up walls at $\partial\Lambda$ which are perfect insulators.

(BC2) Conducting, or Dirichlet b.c.

$V(x,y)$ is the Green's function of the Laplacean with 0-Dirichlet data at $\partial\Lambda$. The physical interpretation of these b.c. is that the walls of Λ are perfect conductors.

(In two dimensions, (BC1) and (BC2) can result in different thermodynamic limits of finite temperature, finite density Coulomb gases, [3]. We shall discuss further b.c. with yet more drastic effects in the thermodynamic limit, "roughening", in Section III).

For $\Lambda = \mathbb{R}^\nu$ we have

$$V(x) = \begin{cases} 1/2|x| \quad , & \nu = 1 \\ 1/2\pi \, \ln(1/|x|) \quad , & \nu = 2 \\ 1/\sigma_\nu |x|^{-(\nu-2)} \quad , & \nu \geqq 3 \, , \end{cases} \tag{I.1}$$

where σ_ν is the surface of the $(\nu-1)$ dimensional unit sphere. If configuration space is replaced by a lattice, $\mathbb{Z}^\nu$, we shall define the Coulomb potential to be the Green's function of the

finite difference Laplacean (with free-, resp. Dirichlet b.c.). Instead, one could also define it to be the restriction of the continuum Coulomb potential to the lattice, with $V(x = 0) \overset{\text{e.g.}}{=} 0$. The long range behaviour of the lattice Coulomb potential is still given by (I.1).

We also introduce a dipole potential : The potential between a dipole pointing in the $+\alpha$-direction at position x and one pointing in the $+\beta$-direction at y, both of unit strength, is given by

$$\mathring{W}_{\alpha\beta}(x,y) = -(\partial_\alpha \partial_\beta V)(x,y) \tag{I.2}$$

where $\partial_\alpha = \partial/\partial x^\alpha$, $\alpha = 1,\ldots,\nu$. (On the lattice ∂_α is a finite difference derivative). In the continuum, $\mathring{W}$ must be regularized at short distances : When $\Lambda = \mathbb{R}^\nu$

$$\mathring{W}_{\alpha\beta}(x,y) = (2\pi)^{-\nu/2} \int e^{ik(x-y)}(k_\alpha k_\beta/k^2) d^\nu k \ .$$

We replace $\mathring{W}$ by

$$W_{\alpha\beta}(x,y) = (2\pi)^{-\nu/2} \int e^{ik(x-y)}(k_\alpha k_\beta/k^2) f(k) d^\nu k \ , \tag{I.3}$$

where f is a non-negative function of rapid fall off, so that $||W_{\alpha\beta}||_\infty$ is finite. Of course, regularization is unnecessary on the lattice.

The Hamilton function (resp.- operator) of a system consisting of N point particles with masses $m_1,\ldots,m_N$, charges $q_1,\ldots,q_N$ and dipole moments $\mu_1,\ldots,\mu_N$, (where $\mu_j \propto S_j$, S_j is the spin operator of the jth particle), is given by

$$\left.\begin{aligned} H^{(N)} &= T^{(N)} + U^{(N)} \ , \\ T^{(N)} &= \sum_{j=1}^{N} (1/2m_j) p_j^2 \ , \\ U^{(N)} &= \sum_{1 \leq i<j \leq N} \{q_i q_j V(x_i - x_j) + \sum_{\alpha,\beta=1}^{\nu} \mu_i^\alpha \mu_j^\beta W_{\alpha\beta}(x_i - x_j)\} \end{aligned}\right\} \tag{I.4}$$

If interactions with the radiation field, described by a vector potential, A, in the Coulomb gauge (i.e. $\nabla \cdot A = 0$), are to be taken into account, $p_j^2/2m_j$ is replaced by $(p_j - q_j A(x_j))^2/2m_j$, and a term $\sum_{j=1}^{N} \mu_j \cdot B(x_j)$, $B(x) = (\nabla \wedge A)(x)$, is added.

The first basic problem to be studied is the stability problem. One assumes that m_j, q_j and $\|\mu_j\|$ are bounded uniformly in $j = 1,\ldots,N$ and $N = 2,3,\ldots,$ and asks whether

$$H^{(N)} \geqq -\text{const. } N , \tag{I.5}$$

for some finite, N-independent constant. A system satisfying (I.5) is said to be H-stable. Classical, continuum Coulomb systems of point particles are never H-stable, unless $\nu = 1$, or $q_j \geqq 0$, for all j. If all charges are positive the system does however not behave thermodynamically because of the long range of the Coulomb potential. In fact, overall neutrality is important.

For three-dimensional, quantum mechanical systems with $\mu_j = 0$, for all j, H-stability has been established, provided all negatively charged particles are Fermions, and is known to fail if all particles are Bosons. These matters are discussed in W. Thirrings contribution and in [2].

We emphasize that H-stability depends only on the short range singularity of the the two-body potential, i.e. H-stability is the ultraviolet (not the infrared) problem of statistical mechanics; see e.g. [2,4]. If the Coulomb potential is cutoff at short distances - as we have done with the dipole potential - H-stability holds, and the proof is very simple, [5]. Thus, on the lattice, (I.5) is always true.

Three-dimensional, non-relativistic, quantum-mechanical matter, with negatively charged particles assumed to be Fermions, coupled to an ultraviolet cutoff, quantized electromagnetic field is stable if the spin of all particles is zero, but unstable if spin is in-

cluded. It is unlikely that stability is restored when the ultraviolet cutoff is removed. (We thank Erhard Seiler for a discussion which helped to clarify this point).

Another notion of stability, equally basic for statistical mechanics, is Ξ-stability : Consider a system of m different species of particles, the total number of particles being arbitrary. The ℓth species is supposed to consist of particles with mass m_ℓ, charge $q_\ell, \ldots$ and activity (= fugacity) $z_\ell = e^{-\beta\mu_\ell}$, where $\beta = 1/kT$ is the inverse temperature and μ_ℓ the chemical potential. Let $\Xi_\Lambda(\beta, z_1, \ldots, z_m)$ denote the grand canonical partition function of this system, the Hamiltonian being given by (I.4). See [4,5,6] for the definition of Ξ_Λ. The system is said to be Ξ-stable if

$$(1 \leqq) \Xi_\Lambda(\beta, z_1, \ldots, z_m) \leqq e^{\text{const.}|\Lambda|}, \qquad \text{(I.6)}$$

for some finite constant; $|\Lambda|$ is the volume of the box Λ containing the system.

The notions of H-stability and Ξ-stability are not equivalent : The two-dimensional, classical neutral Coulomb gas with two species of particles of charge $\pm q$ is never H-stable, but is Ξ-stable if $\beta q^2 < 4\pi$. This is the result of [4] . (For some extensions see [6]).

However, if the Coulomb potential is regularized at short distances, a Coulomb system of finitely many species of Bosons, with positive and negative charges, is H-stable, but fails to be Ξ-stable when some of the activities are large enough; (in fact Ξ_Λ is infinite when some of the activities exceed critical values). See [7] .

Non-relativistic, quantum-mechanical matter in three dimensions, with negatively charged particles = Fermions, is Ξ-stable [8] ; (see also [7]). Classical, H-stable systems are always Ξ-stable, [5] ; in particular classical lattice Coulomb gases are Ξ-stable.

(Quantum mechanically, the implication tends to go the other way around).

I.3. Thermodynamic Functions

The basic results concerning the existence of the thermodynamic functions of Coulomb systems are due to Lieb and Lebowitz [8] . For various extensions of their methods see [2] and refs. given there. The problem of the thermodynamic limit is the "infrared problem" of statistical mechanics, and it is equally hard classically and quantum mechanically. Under certain restrictive conditions, the proof of existence of the thermodynamic limit for e.g. the pressure of Coulomb systems is simple, (much simpler than the proofs in [8], although the results are not quite as strong) :

A system composed of $2m$ species of particles is said to be charge conjugation invariant iff

$$m_{2j} = m_{2j+1} \ , \ q_{2j} = -q_{2j+1} \ , \ z_{2j} = z_{2j+1} \ ,$$

and if the system is quantum mechanical the statistics of the particles in the $2j^{th}$ and $(2j+1)^{st}$ species are the same; for all $j = 1,\ldots,m$. (If, in addition, the particles have dipole moments, μ , it is required that μ_{2j} and $-\mu_{2j+1}$ have identical distributions, $j = 1,\ldots,m$) .

For charge conjugation invariant systems a simple proof of existence for the thermodynamic limit of the pressure has been given in [7] , extending an idea of Griffiths [9] . For such systems, the screening properties of the Coulomb potential emphasized in [8,2] are actually unimportant for the existence of the thermodynamic limit of the pressure, although sensitive dependence on shape and boundary conditions must be expected for potentials like the dipole potential which cannot be screened. As an example, we mention that in three dimensions the thermodynamic limit of the pressure of a charge conjugation invariant system with two-body potential

$V(x) \underset{|x|\to\infty}{\approx} |x|^{-\varepsilon}, \varepsilon > 0$, of positive type exists, although for $\varepsilon \neq 1$ there is no screening. See [7] . For additional methods involving correlation inequalities see [6] .

I.4. Equilibrium States

A third basic problem concerning Coulomb systems is the question of existence and properties of the thermodynamic limit of equilibrium states, in particular of the correlation functions of classical systems, resp. the reduced density matrices or imaginary-time Green's functions of quantum mechanical systems.

For a rather large class of classical and quantum Coulomb systems locally normal equilibrium states in the thermodynamic limit can be constructed by means of a weak compactness argument, provided suitable boundary conditions (periodic b.c.) are imposed. (A proof of this can be based on constructive field theory methods of Glimm and Jaffe). In many physically interesting situations not even such a weak result is known to hold ! Moreover, the problem of constructing the time evolution for infinite Coulomb systems "near equilibrium" is essentially entirely open, except in very special, physically unrealistic cases.

After these rather depressing remarks we now recall some positive results among which the most impressive ones are due to Brydges [10] and Brydges and Federbush [11] : For a large class of classical, dilute Coulomb systems in two or more dimensions they have constructed the thermodynamic limit of the correlation functions (with Dirichlet, i.e. conducting b.c.), and they have established Debye screening in the form of exponential cluster properties. This remarkable development is reviewed in detail in the lectures given by D. Brydges.

Another construction of the thermodynamic limit of correlation functions, resp. reduced density matrices or imaginary-time Green's

functions valid for all values of the thermodynamic parameters for which the system behaves thermodynamically is given in [6,7] . That method is based on correlation inequalities first used in a related context in [12] . The hypotheses under which those inequalities are known to hold are unfortunately rather restrictive :

- Exact charge conjugation invariance.
- The two-body potential is of positive type; (n-body potentials vanish for $n > 2$).
- The system is classical or quantum mechanical with Boltzmann - or Bose-Einstein statistics.

The first and the third hypothesis are physically awkward. However, the inequalities hold for arbitrary values of β and z and a large class of potentials including the Coulomb potential and ones with slower decrease than the Coulomb potential. Moreover, they are strong enough to provide some general information about the properties of the thermodynamic limit, [6,7]. They also permit to include the radiation field and supply some general information of interest in superconductivity and Bose-Einstein condensation, [7]. In spite of the many encouraging results alluded to above and discussed in more detail in the lectures by Aizenman, Brydges, Lebowitz, Lieb and Thirring it should be clear that the mathematical foundations of the theory of Coulomb systems and non-relativistic matter - starting from first principles - are still quite incomplete. Several topics, such as non-relativistic QED, may have been undeservedly neglected.

In the remainder of these notes we shall study highly idealized systems of excitations with Coulomb interactions about which detailed statements can be made. We shall concentrate on the discussion of the Kosterlitz-Thouless transition and other aspects of the phase diagram of two - (and higher) dimensional Coulomb systems.

II. Generalities about Classical Coulomb Gases

Throughout the remaining sections we study classical lattice Coulomb gases, but many of our results extend to continuum gases, provided the Coulomb potential is regularized at short distances, some also to quantum mechanical gases. We concentrate our attention on monopole gases but at various places mention results on dipole gases.

We first recall the sine-Gordon (or Siegert) transformation [3,4,6,13]. The end of the section contains an outlook on what is discussed in subsequent sections, in particular a phase diagram of a hard core Coulomb lattice gas in two, resp. three dimensions which we shall establish in part.

II.1. The sine-Gordon transformation

We consider Coulomb gases on the lattice $\mathbb{Z}^\nu$. The Coulomb potential, V, is the Green's function of the finite difference Laplacean, Δ. Unless stated otherwise, free, i.e. insulating, b.c. are imposed. (Other b.c. are treated in the references quoted in the text).

We start by considering systems in a finite region $\Lambda \subset \mathbb{Z}^\nu$. A configuration of such a system is a function

$$q_\Lambda : \Lambda \to \mathbb{Z} \ , \ \Lambda \ni j \to q(j) \in \mathbb{Z} \ ,$$

where $q(j)$ is interpreted as the total electric charge concentrated at site j. The a priori distribution of $q(j)$ is given by a measure $d\lambda$ on $\mathbb{Z}$. We shall be interested in the following choices of $d\lambda$:

A) Hard core gas

$$d\lambda(q) = \{\delta(q)+z/2[\delta(q-1)+\delta(q+1)]\}dq \ , \tag{2.1}$$

where δ is the Dirac function and z the (bare) activity.

B) Standard lattice gas without hard cores

$$d\lambda(q) = \{ \sum_{n \in \mathbb{Z}} I_n(z)\delta(q-n)\}dq \quad , \tag{2.2}$$

where $I_n(z)$ is the nth modified Bessel function, i.e. the nth Fourier coefficient of $\exp(z\cos\phi)$, and z is a (bare) activity.

C) Villain gas :

$$d\lambda(q) = \{ \sum_{n \in \mathbb{Z}} \delta(q-n)\}dq \tag{2.3}$$

We note that this measure is the limit of $I_o(z)^{-1}d\lambda(q)$, as $z \to +\infty$, with $d\lambda$ given by (2.2).

Clearly there are other interesting choices for $d\lambda$, but here $d\lambda$ will usually be given by A).

$$\begin{aligned} E(q_\Lambda) &= \tfrac{1}{2} \sum_{i,j} q(i)\, q(j)\, V(i-j) \\ &\equiv \tfrac{1}{2} \left(q_\Lambda, (-\Delta)^{-1} q_\Lambda\right). \end{aligned} \tag{2.4}$$

The functional $E(q_\Lambda)$ is the electrostatic self-energy of the configuration q_Λ , self-energies of charges included.

The equilibrium distribution for the configuration q_Λ is given by

$$\left.\begin{aligned} &Z_\Lambda^{-1} \exp[-\beta E(q_\Lambda)] \prod_{j\in\Lambda} d\lambda(q(j)) \\ &Z_\Lambda = \int \exp[-\beta E(q_\Lambda)] \prod_{j\in\Lambda} d\lambda(q(j)). \end{aligned}\right\} \tag{2.5}$$

Note, by a finite redefinition of $d\lambda$, self-energies of charges can be excluded in the definition of $E(q_\Lambda)$.

Next we consider the Fourier transform of the equilibrium measure introduced in (2.5). Let $\phi : \mathbb{Z}^\nu \to \mathbb{R}$ be a Gaussian random field on $\mathbb{Z}^\nu$ with distribution

$$d\mu_{\beta V}(\phi) = N^{-1} \exp[1/2\beta\, (\phi, \Delta\phi)] \prod_j d\phi(j), \tag{2.6}$$

where

$$(\phi, \Delta\phi) = -\sum_{|i-j|=1} (\phi(i) - \phi(j))^2,$$

and

$$N = \det(-\Delta/2\pi\beta)^{-1/2}$$

is a normalization factor. Mathematically, $d\mu_{\beta V}$ is defined to be the Gaussian measure with mean 0 and covariance βV. In one and two dimensions, $d\mu_{\beta V}$ is only defined, a priori, when integrated against bounded functions of

$$\{\phi(f) : \text{supp} f \text{ bounded}, \sum_j f(j) = 0\}, \tag{2.7}$$

with $\phi(f) \equiv \sum_j \phi(j) f(j)$. This is because

$$\hat{V}(k) = [2(\nu - \sum_{\alpha=1}^{\nu} \cos k^\alpha)]^{-1}, \tag{2.8}$$

(the Fourier transform of $V(j)$) is not integrable at $k = 0$ when $\nu = 1$ or 2. See e.g. [4] for details. Thus

$$\int d\mu_{\beta V}(\phi) e^{i\phi(f)} = \begin{cases} \exp[-\beta/2 (f, Vf)], & \sum_j f(j) = 0, \\ 0, & \sum_j f(j) \neq 0. \end{cases} \tag{2.9}$$

In $\nu \geqq 3$ dimensions, V is positive definite and no constraints arise. By (2.4) and (2.9)

$$\exp[-\beta E(q_\Lambda)] = \int d\mu_{\beta V}(\phi) e^{i\phi(q_\Lambda)}, \tag{2.10}$$

provided $Q(q_\Lambda) \equiv \Sigma q(j) = 0$ when $\nu = 1,2$. (If $Q(q_\Lambda) \neq 0$, $\nu = 1$ or 2, we set $E(q_\Lambda) \equiv +\infty$, and (2.10) remains true). Note that the variable $\phi(j)$ is conjugate to the charge variable $q(j)$. Thus

$$Z_\Lambda = \int \exp[-\beta E(q_\Lambda)] \prod_{j\in\Lambda} d\lambda(q(j)) = \int d\mu_{\beta V}(\phi) \prod_{j\in\Lambda} \hat{\lambda}(\phi(j)), \tag{2.11}$$

where

$$\hat{\lambda}(\phi) = \int d\lambda(q)\, e^{iq\phi}. \tag{2.12}$$

In the hard core gas (2.1) ,

$$\hat{\lambda}(\phi) = 1 + z\cos\phi \tag{2.1'}$$

In the standard gas (2.2),

$$\hat{\lambda}(\phi) = \exp[z\cos\phi], \tag{2.2'}$$

and in the Villain gas (2.3)

$$\hat{\lambda}(\phi) = \sum_{n\in\mathbb{Z}} \delta(\phi - 2\pi n). \tag{2.3'}$$

We denote by $\langle - \rangle_\Lambda(\beta,\lambda)$ both, expectations in the equilibrium measure (2.5), and expectations in the (generally non-positive) measure

$$Z_\Lambda^{-1} \prod_{j\in\Lambda} \hat{\lambda}(\phi(j))\, d\mu_{\beta V}(\phi). \tag{2.13}$$

The interpretation of correlations $\langle F(q_\Lambda)\rangle_\Lambda(\beta,\lambda)$, where F is a bounded function of q_Λ , is obvious. (It is an expectation of a sort we are familiar with from lattice spin systems).

In order to interpret expectations such as

$$\langle \exp i\phi(f)\rangle(\beta,\lambda),$$

note that, by (2.9) - (2.11),

$$\langle \exp i\phi(f)\rangle_\Lambda(\beta,\lambda) = Z_\Lambda^{-1} \int \exp[-\beta E(q_\Lambda + f)] \prod_{j\in\Lambda} d\lambda(q(j)). \tag{2.14}$$

Thus $\langle \exp i\phi(f)\rangle_\Lambda(\beta,\lambda)$ measures the correlations between external charges, $f(j)$, located at different sites $j \in \Lambda$, which are put into the system from the outside. More precisely, $-1/\beta \log\langle e^{i\phi(f)}\rangle_\Lambda$ is the average amount of free energy needed to pump the charges $\{f(j)\}_{j\in\Lambda}$ into a system of charges in thermal equilibrium at inverse temperature β. Of particular interest in the following is the <u>fractional charge correlation</u>

$$G_\Lambda(x) = \langle \exp i\gamma(\phi(0)-\phi(x))\rangle_\Lambda(\beta,\lambda) \tag{2.15}$$

which measures the correlation between two fractional charges, one charge γ located at $0 \in \Lambda$ and one, $-\gamma$, located at $x \in \Lambda$, $0 < \gamma < 1$, put into the system. The behaviour of the fractional charge correlation $G_\Lambda(x)$ reflects the screening properties of the Coulomb gas; see [3] and Section III.

Next, we briefly sketch how to extend the formalism developed here to the simplest example of a dipole gas. See [3,6] for more details. As our dipole potential we choose e.g.

$$W_{\alpha\beta}(x,y) = -(\partial_\alpha \partial_\beta V)(x-y), \tag{2.16}$$

where ∂_α is the finite difference derivative in direction α. Let $\Lambda_d \subseteq \Lambda$ be some sublattice of Λ, e.g. $\Lambda_d = \ell\mathbb{Z}^\nu \cap \Lambda$, $\ell = 1,2,3,\ldots$. A <u>configuration</u> of a lattice dipole gas is described by a function $\mu_\Lambda : \Lambda_d \to \mathbb{R}^\nu$, $\Lambda_d \ni j \to \mu(j) \in \mathbb{R}^\nu$, where $\mu(j)$ is the total dipole moment at site $j \in \Lambda_d$. The dipoles are non-overlapping if $\ell \geqq 2$. The energy of μ_Λ is given by

$$E_d(\mu_\Lambda) = \tfrac{1}{2} \sum_{i,j} \sum_{\alpha,\beta} \mu^\alpha(i)\mu^\beta(j) W_{\alpha\beta}(i,j), \tag{2.17}$$

the equilibrium distribution by

$$Z_\Lambda^{-1} \exp[-\beta E_d(\mu_\Lambda)] \prod_{j\in\Lambda_d} d\lambda(\mu(j)),$$

where $d\lambda$ is a finite measure on $\mathbb{R}^\nu$, e.g.

$$d\lambda(\mu) = \{\delta(\mu) + z\,\delta(\mu^2 - 1)\}\,d^\nu\mu \tag{2.18}$$

Combining (2.9) with (2.16) and (2.17) one gets

$$\exp[-\beta E_d(\mu_\Lambda)] = \int d\mu_{\beta V}(\phi) \prod_{j\in\Lambda_d} \exp[i\mu(j)\cdot(\nabla^*\phi)(j)], \tag{2.19}$$

where $\nabla^* \equiv (\partial_1^*, \ldots, \partial_\nu^*)$, and ∂_α^* is the adjoint of ∂_α . For $j \in \Lambda_d$, let $\{\phi\}_j = \{\phi(i) : |i-j| \leqq 1\}$. We set

$$\hat{\lambda}(\{\phi\}_j) = \int d\lambda(\mu) \exp[i\mu\cdot(\nabla^*\phi)(j)].$$

The dipole measure in the ϕ-variables is then given by

$$Z_\Lambda^{-1} \prod_{j\in\Lambda_d} \hat{\lambda}(\{\phi\}_j)\, d\mu_{\beta V}(\phi) . \tag{2.20}$$

The definition and interpretation of the expectations $\langle - \rangle_\Lambda(\beta,\lambda)$ and correlations $\langle F(\mu_\Lambda)\rangle_\Lambda(\beta,\lambda)$, $\langle \exp i\phi(\nabla\cdot h)\rangle_\Lambda(\beta,\lambda)$, h an $\mathbb{R}^\nu$-valued function on Λ , is analogous as in the previous case of monopole gases. Note that, by (2.6) and (2.20), dipole gases have the continuous symmetry $\phi(\cdot) \mapsto \phi(\cdot)$ + const. which is always broken by the boundary conditions. As a consequence one can show [3] that there exist Goldstone excitations and that <u>correlations in dipole gases do not decay exponentially</u>.

We conclude this subsection by recalling the standard integration by parts formula, (e.g. [3] and refs. given there).

We do this for the monopole gases; for dipoles see e.g. [3]. First, we recall the well known identity

$$\int \phi(j)\, G(\phi)\, d\mu_{\beta V}(\phi) = \beta \sum_i V(j-i) \int \frac{\partial G}{\partial \phi(i)}\, d\mu_{\beta V}(\phi) \tag{2.21}$$

which a physicist calls Wick's theorem.

Clearly

$$\frac{\partial \hat{\lambda}}{\partial \phi}(\phi) = i\int d\lambda(q)\, q\, e^{iq\phi}$$

Thus, by (2.13) and (2.21)

$$\langle \phi(j)\, G(\phi) \rangle_\Lambda (\beta,\lambda) =$$

$$\beta \sum_i V(j-i) \left\{ \left\langle \frac{\partial G}{\partial \phi(i)}(\phi) \right\rangle_\Lambda (\beta,\lambda) + i \langle q(i)\, G(\phi) \rangle_\Lambda (\beta,\lambda) \right\} \tag{2.22}$$

This equation shows that $(1/i\beta)\phi(j)$ is the effective potential felt by an infinitesimal test charge at site j. By setting $G(\phi) = \phi(\ell)$ and repeating (2.22) one gets

$$\langle \phi(j)\phi(\ell) \rangle_\Lambda (\beta,\lambda) = \beta V(j-\ell) - \beta^2 \sum_{i,m} V(j-i)\, V(\ell-m) \langle q(i) q(m) \rangle_\Lambda (\beta,\lambda) \tag{2.23}$$

If on the r.s. of (2.23) integrations over q and ϕ are interchanged one obtains

$$\begin{aligned} \langle \phi(j)\phi(\ell) \rangle_\Lambda (\beta,\lambda) &= \beta V(j-\ell) \\ &- \beta^2 \sum_i V(j-i)\, V(\ell-i) \langle \gamma(\phi(i)) \rangle_\Lambda (\beta,\lambda) \\ &+ \beta^2 \sum_{i,\ell} V(j-i)\, V(\ell-m) \langle \sigma(\phi(i))\, \sigma(\phi(m)) \rangle_\Lambda (\beta,\lambda) \end{aligned} \tag{2.24}$$

for some functions γ and σ on the real line which are determined by $\hat{\lambda}$ and are real-valued if $\hat{\lambda}$ is real, [3] . If $\hat{\lambda}$ is non-negative (resp. a "renormalized" version of $\hat{\lambda}$ is non-negative, see Section II.2) formulas (2.23) and (2.24) provide a surprisingly powerful tool.

Finally, we wish to add a remark on the existence of the thermodynamic limit : For a large variety of Coulomb monopole and dipole

gases the correlation inequalities in [6] can be used to construct the thermodynamic limit of the states $\langle - \rangle_\Lambda(\beta,\lambda)$, as $\Lambda \nearrow \mathbb{Z}^\nu$.

In particular, assume that

$$\hat{\lambda}(\phi) = \exp G(\phi) \tag{2.25}$$

where $G(\phi)$ is real-valued and of positive type. Impose free (insulating) or Dirichlet (conducting) b.c. on the Coulomb potential V . Then

$$\lim_{\Lambda \nearrow \mathbb{Z}^\nu} \langle - \rangle_\Lambda (\beta;\lambda) \equiv \langle - \rangle (\beta,\lambda) \tag{2.26}$$

exists. The limiting state, $\langle - \rangle(\beta,\lambda)$, is translation invariant and, in $\nu \geq 3$ dimensions, clustering, [6] . (In two dimensions it clusters on observables which are functions of

$$\{ \phi(f) : \text{supp} f \text{ bounded }, \sum_j f(j) = 0 \}).$$

Necessary conditions for (2.25) to hold are exact charge conjugation invariance of the system, i.e. $d\lambda(q) = d\lambda(-q)$, and positivity of $\hat{\lambda}(\phi)$. It is easy to see that (2.25) holds for the standard and the Villain gas for which $d\lambda$ is given by (2.2), (2.3), respectively. It fails for the hard core gas, although that gas is charge conjugation invariant, and $\hat{\lambda}(\phi) > 0$, for $z < 1$.

For charge conjugation invariant, strictly neutral systems with periodic b.c. at $\partial\Lambda$, a translation-invariant, thermodynamic limit, $\langle - \rangle(\beta,\lambda)$, can be constructed by passing to subsequences. Every limiting state obtained in this way has strong regularity properties - as a state on bounded functions of the charge variables $\{q(j)\}_{j \in \mathbb{Z}^\nu}$ - reminiscent of superstability estimates. This can be proven by means of chessboard estimates [3,14] . For any translation-invariant limiting state, $\langle - \rangle(\beta,\lambda)$, we obtain from (2.23) by Fourier transformation

$$\langle|\hat{\phi}(k)|^2\rangle(\beta,\lambda) = \beta\hat{V}(k) - \beta^2\hat{V}(k)^2\langle|\hat{q}(k)|^2\rangle(\beta,\lambda)$$

with

$$\hat{V}(k) = \left[2\left(\nu - \sum_{\alpha=1}^{\nu} \cos k^{\alpha}\right)\right]^{-1} \tag{2.27}$$

Existence of thermodynamic limits will not be discussed any further (see [6-11]), but there certainly are still many interesting open problems.

II.2. The phase diagram of lattice Coulomb gases

We consider the hard core lattice Coulomb gas with $d\lambda$ given by (2.1). The equilibrium state of this system is henceforth denoted by $\langle - \rangle(\beta,z)$. Our purpose here is to describe what is known about the phase diagram of this interesting system in two and three dimensions. The relevant thermodynamic parameters are the inverse temperature β and the activity z. Here are portraits of the phase diagrams in $\nu = 2$ and 3 dimensions.

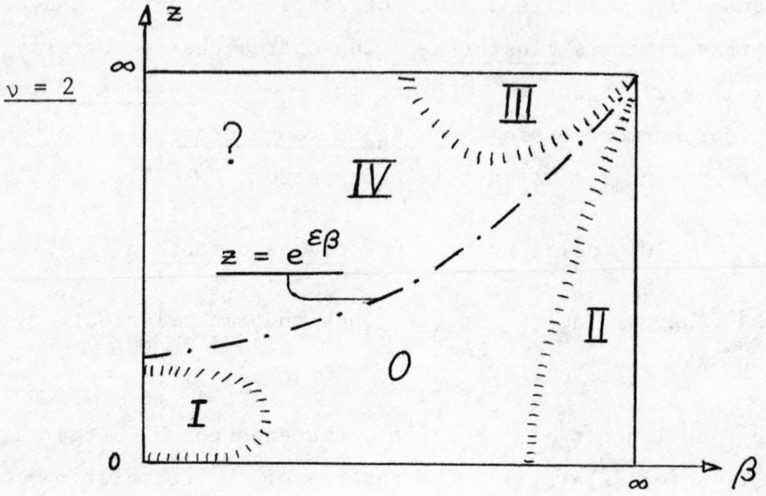

Fig. 1

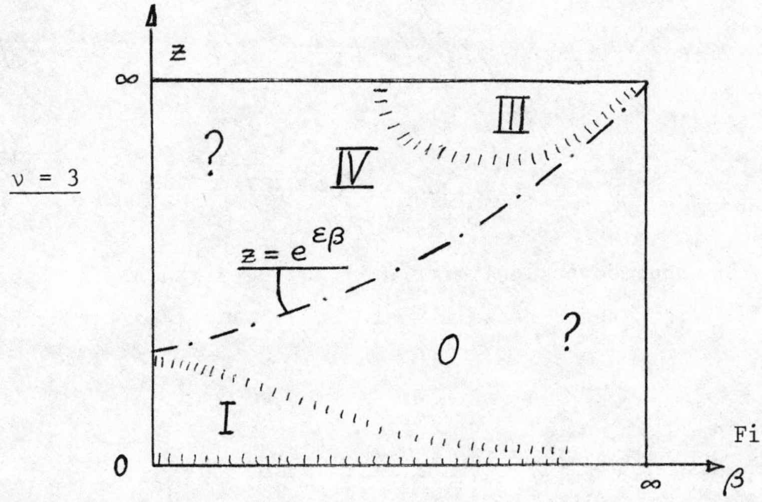

Fig. 2

We first discuss common features of the two- and three dimensional diagrams and then discuss striking differences known to arise within domain 0 .

Domain 0 is an open region bounded by the lines $\beta = 0$, $z = 0$ $\beta = \infty$ and $z = e^{\varepsilon\beta}$, with $\varepsilon \geqq 1/8\nu$ (= 1/16 , for $\nu = 2$, = 1/32, for $\nu = 3$) . It is characterized by the existence of a translation invariant state which is a limit of states with periodic b.c. (For $z < 1$ that limit is clustering). The charge-charge correlation $\langle q(0)q(x)\rangle(\beta,z)$ tends to 0 , as $|x| \to \infty$, (absence of long range order). Furthermore there is no short range order, in the sense that for $x = ne$, e a unit lattice vector, $n = 1,2,3,...$,

$$\langle q(0)q(ne)\rangle(\beta,z) \text{ is negative and concave.} \tag{2.28}$$

The local charge, $q_\Lambda = \sum_{j\in\Lambda} q(j)$, has abnormal fluctuation, i.e. $\langle q_\Lambda^2\rangle \leqq O(\partial\Lambda)$.

On each line $z = z_o$ = const., the exponential decay rate, $m(\beta)$, of $\langle q(0)q(x)\rangle(\beta,z)$, the inverse of the correlation length $\xi(\beta)$, is known to satisfy the inequality

$$m(\beta) \leqq const.\, e^{-\delta\beta}, \tag{2.29}$$

for some $\delta > 0$ which depends on z_o and ν. These results are proven in [3].

The basic tools used in the proofs are the existence of self-adjoint transfer matrices, i.e. reflection positivity in the ϕ- and q-representations, [3,14], and the fact that for $z < e^{\varepsilon\beta}$

$$\langle |\hat{\phi}(k)|^2 \rangle (\beta, z) \geq 0 . \tag{2.30}$$

Thus by (2.27)

$$\langle |\hat{q}(k)|^2 \rangle (\beta, z) \leq [\beta \hat{V}(k)]^{-1} \leq \beta^{-1} k^2 . \tag{2.31}$$

Absence of long range order and abnormality of the fluctuations of q_Λ follow from (2.31) by Fourier transformation. For $z < 1$, the state $\langle - \rangle(\beta, z)$ is given by a positive measure in the ϕ-variables, so that (2.30) holds. When $z > 1$, $\langle - \rangle(\beta, z)$ does not correspond to a positive measure in the ϕ's, and (2.30) is not obvious. In order to deal with large bare activities z, $1 < z < e^{\varepsilon\beta}$, we must apply a simple renormalization transformation : Between two nearest neighbor sites, i and j, we introduce an additional site, ij, and replace the factor $\exp -1/2\beta(\phi(i)-\phi(j))^2$ in the Gaussian measure $d\mu_{\beta V}(\phi)$ by

$$\text{const.} \exp -{}^1\!/_\beta (\phi(i) - \psi(ij))^2 \exp -{}^1\!/_\beta (\phi(j) - \psi(ij))^2 \, d\psi(ij)$$

We then interchange the integrations over ψ- and ϕ-variables, first integrating out all $\phi(j)$, $j \in \mathbb{Z}^\nu$. The ϕ-integrations can be done explicitly by using the identities

$$\int e^{iq\phi} \prod_{\alpha=1}^{2\nu} e^{-(1/\beta)(\phi - \psi_\alpha)^2} d\phi = e^{-\beta q^2/8\nu} \, e^{iq\bar{\psi}} \tag{2.32}$$

where $I_\nu(\psi) = \int \prod_{\alpha=1}^{2\nu} \exp[-(1/\beta)(\phi - \psi_\alpha)^2] \, d\phi$,

$\bar{\Psi} = (1/2\nu) \sum_{\alpha=1}^{2\nu} \psi_\alpha$, and each ψ_α stands for a variable $\psi(ij)$

associated with the new site in the middle of the link ij . By differentiating in q and adding the resulting identities for ±q we obtain

$$\int \phi \cos(q\phi) \prod_{\alpha=1}^{2\nu} e^{-1/\beta(\phi-\psi_\alpha)^2} d\phi$$
$$= e^{-\beta q^2/8\nu}[\bar{\psi}\cos(q\bar{\psi}) - (\beta q/4\nu)\sin(q\bar{\psi})]\, I_\nu(\psi). \tag{2.33}$$

Thus, under this renormalization transformation, the activity, $\lambda(\{q\})$, of $e^{iq\phi}$ is multiplied by $\exp-(\beta q^2/8\nu)$, in particular, in the hard core gas,

$$1 + z\cos\phi(i) \longmapsto 1 + z e^{-\beta/8\nu} \cos\bar{\psi}(i)$$
$$\phi(i)\,(1 + z\cos\phi(i)) \longmapsto \tag{2.34}$$
$$\bar{\psi}(i)\,(1 + z e^{-\beta/8\nu}\cos\bar{\psi}(i)) - (\beta q/4\nu) e^{-\beta/8\nu} \sin\bar{\psi}(i)$$
$$\bar{\psi}(i) = \sum_{j:|j-i|=1/2} \psi(ij), \qquad i \in \mathbb{Z}^\nu.$$

In the ψ-variables, the state $\langle - \rangle(\beta,z)$ is given by a positive measure, provided $ze^{-\beta/8\nu} < 1$, i.e. $z < e^{\varepsilon\beta}$, with $\varepsilon = (8\nu)^{-1}$. Then (2.34) clearly implies (2.30).

A sequence of renormalization transformations of this type, driving down bare activities, is a crucial tool in [15]. See also Section IV and § 5 of [3] .

Next we discuss domain I which is contained in domain 0 . Its main characteristics is the existence of a state with exponential Debye screening : The infinite volume limit, $\langle - \rangle(\beta,z)$, of the family of states $\{\langle - \rangle_\Lambda(\beta,z)\}$ with Dirichlet b.c. exists, and correlations in $\langle - \rangle(\beta,z)$ cluster exponentially. For small β , $m(\beta,z) \approx \sqrt{\beta z}$, [10,11], to be compared with the large β behaviour (2.29). What we have described here corresponds to a plasma phase of the Coulomb gas. It is discussed in detail in the lectures of D. Brydges.

We now describe the differences in the behaviour of the two- and three-dimensional Coulomb gas, including domain II for the two-dimensional gas. The two-dimensional Coulomb potential between a positive and a negative charge separated by a distance ℓ grows like $(1/2\pi)\ln(\ell+1)$. This is a confining potential, and in the absence of other charges the two charges form a tightly bound, neutral dipole. At finite temperature and density, this dipole may break up, due to interactions with other charges in the system. The probability of this event can be estimated heuristically as follows : The Boltzmann factor of the two charges is

$$\exp\left[-\beta/2\pi \ln(\ell+1)\right] \tag{2.35}$$

The entropy S of the configuration is

$$S \propto \ln(\ell \cdot v) \tag{2.36}$$

where ℓ estimates the order of magnitude of the number of possible positions of the negative charge, for a fixed position of the positive charge, and v is the mean area over which the position of the positive charge may vary. It is shown in [15] that $v \propto \ell^p$, for some $p > 0$. At densities low enough that the lattice structure is not felt on large scales, dimensional analysis gives $v \propto \ell^2$. Now observe that

$$\ell^{p+1} \exp\left[-(\beta/2\pi) \ln(\ell+1)\right] \tag{2.37}$$

is summable in ℓ , for $\beta > 2\pi(p+2)$, (i.e. for $\beta > 8\pi$ if $v \propto \ell^2$ which is exact in the continuum limit). This means that the probability of separating the negative charge from the positive one by a distance ℓ tends to 0 , as $\ell \to \infty$, "integrably fast", provided β is large enough, i.e. stable, neutral "molecules" are expected to form among which neutral dipoles may be expected to be the dominant configurations of the gas if the density is low enough. We have shown in [3] that dipole gases have correlations with power law decay, i.e. Debye screening breaks down in this low temperature-

low density dipolar phase. A refined version of this somewhat rough picture is justified rigorously in [15] by means of an inductive renormalization group scheme. Thus, in two dimensions, domain II corresponds to the Kosterlitz-Thouless dipolar phase characterized by power fall-off of correlations and scaling. It is clear that - and why - the mechanism described here fails in $\nu \geqq 3$ dimensions : The Boltzmann factor for a neutral multipole of point charges, e.g. a dipole of lenght ℓ , $\propto \exp(+\beta/4\pi\ell)$, does not tend to 0 , as the diameter $d\ (=\ell)$ tends to ∞ . For this reason, all neutral multipoles are unstable, no matter how small the temperature and density are, and the gas is a mixture of free charges forming a plasma and "unstable molecules". Therefore one expects that exponential Debye screening persists throughout a domain in the (β,z) plane essentially as large as domain 0 . This is not quite what Brydges and Federbush [10,11] are able to show. Their methods only establish screening for small enough densities, depending on the value of β . Instead of the Berezinski-Kosterlitz-Thouless transition which the two-dimensional gas undergoes when (β,z) is moved from domain I to domain II one expects that the three-dimensional gas exhibits what one might call a <u>roughening transition</u> when β and z are increased. We shall briefly comment on this kind of transition in Section III.

We now continue our discussion of the phase diagrams with <u>domain III,</u> corresponding to low temperatures and high densities. It is characterized by the existence of at least two <u>ordered states</u>, $\langle - \rangle_{\pm}(\beta,z)$, with

$$\langle q(x)\rangle_{\pm}(\beta,z) = \pm(-1)^{|x|}, \quad |x| \equiv \sum_{\alpha=1}^{\nu} x^{\alpha}, \tag{2.38}$$

i.e. the charge density is staggered and the charges are arranged in a crystal of the $NaC\ell$ type. On the boundary of region III <u>at least three states</u> coexists, two ordered ones and a state describing a low density phase. This has been proven in [14] by means of a Peierls argument inspired by the one in [16] . An analogous (more

difficult) result for hard core lattice dipole gases has been proven in [3] .

It is reasonable to conjecture that domain IV contains a region of parameters (β,z) corresponding to equilibrium states that describe a high density liquid phase with short range order, ($\langle q(0)q(x)\rangle(\beta,z)$ is staggered in x) , but without long range order in the charge-charge correlation $\langle q(0)q(x)\rangle(\beta,z)$. Presently, we know of no analytical method that would permit to investigate domain IV rigorously, except that some of the ideas in [15] may be useful. This is proposed to the reader as a challenging open problem.

Finally, if - instead of the hard core Coulomb gas - the standard or the Villain gas are considered, domain 0 extends over the whole quadrant $\{(\beta,z) : \beta > 0 , z > 0\}$, because in these models $\hat{\lambda}(\phi) \geqq 0$, so that reflection positivity in the ϕ- and the q-representation and the inequality $\langle|\hat{\phi}(k)|^2\rangle(\beta,z) \geqq 0$ hold. The boundary line $z = \infty$ of the standard model is the Villain model, (e.g. [3]) . Domains I and II extend up to that line. For general results on dipole gases, see [3,6].

III. Screening and Roughening

In this section we sketch some of the features of the equilibrium states when (β,z) is in the Debye-Hückel domain I, resp. in 0 . Our remarks are intended to be somewhat complementary to the lectures of D. Brydges.

First, we give several different characterizations of Debye screening and then we speculate on a "roughening transition" in the three-dimensional Coulomb gas. We consider the gases introduced in (2.1) - (2.3).

i) Strong screening [10,11]

Let $A(\phi),B(\phi)$ be bounded functions depending only on finitely

many of the random variables $\{\phi(j)\}_{j\in\mathbb{Z}^\nu}$. Let $B(\phi)_x$ denote the translation of $B(\phi)$ by a vector $x\in\mathbb{Z}^\nu$. Strong screening is the statement that

$$\langle A(\phi)B(\phi)_x\rangle(\beta,\lambda) \xrightarrow[|x|\to\infty]{} \langle A(\phi)\rangle(\beta,\lambda)\langle B(\phi)\rangle(\beta,\lambda) \tag{3.1}$$

exponentially fast.

By interchanging the order of the q- and ϕ-integrations, one derives from this exponential clustering of q-correlations. In one-dimensional Coulomb gases this strong form of screening always fails for suitably chosen A and B. (See the lectures by Aizenman and refs. given there). For $\nu \geqq 2$ dimensional gases, (3.1) is established in [10,11] in the (β,z)-domain which we have denoted by I, and for Dirichlet b.c.. We now interpret this result for the fractional charge correlation

$$G(x)= \langle \exp i\gamma(\phi(0)-\phi(x))\rangle(\beta,\lambda), \tag{3.2}$$

$0<\gamma<1$, introduced in (2.15). Let $\langle A;B\rangle$ be a short hand for $\langle AB\rangle - \langle A\rangle\langle B\rangle$. By (3.1)

$$G^c(x)\equiv \langle e^{i\gamma\phi(0)}; e^{-i\gamma\phi(x)}\rangle(\beta,\lambda) \leqq \text{const}\, e^{-m|x|}, \tag{3.3}$$

for some $m>0$, provided $(\beta,z)\in I$.

Next, we derive a lower bound on $G_\Lambda(x)$. Let

$$\delta\rho_{ox} \equiv \gamma(\delta_{jo}-\delta_{jx}) \quad \text{and suppose} \quad d\lambda(q) = d\lambda(-q) \tag{3.4}$$

By equations (2.4) and (2.14)

$$G_\Lambda(x) = Z_\Lambda^{-1} \int \prod_{j\in\Lambda} d\lambda(q(j)) \exp[-\beta E(q_\Lambda)] \cdot$$
$$\cdot \exp[-\beta(q_\Lambda, (-\Delta)^{-1}\delta\rho_{ox})] \cdot$$
$$\cdot \exp[-\beta/2 (\delta\rho_{ox}, (-\Delta)^{-1}\delta\rho_{ox})]$$

$$\geq \exp[-\beta \langle (q_\Lambda, (-\Delta)^{-1}\delta\rho_{ox}) \rangle_\Lambda(\beta,\lambda)] \cdot$$
$$\cdot \exp[-\beta/2 (\delta\rho_{ox}, (-\Delta)^{-1}\delta\rho_{ox})] \tag{3.5}$$

$$= \exp[-\beta\gamma^2 (V(0) - V(x))] , \tag{3.6}$$

where (3.5) follows from Jensen's inequality, and (3.6) from the fact that $\langle(q_\Lambda, (-\Delta)^{-1}\delta\rho_{ox})\rangle_\Lambda(\beta,\lambda) = 0$, by (3.4). Thus, for $\nu \geqq 2$, $G(x)$ does <u>not</u> approach 0 exponentially fast, as $|x| \to \infty$. This and (3.3) imply that in the thermodynamic limit

$$G(x) \xrightarrow[|x|\to\infty]{} |\langle e^{i\gamma\phi(0)}\rangle(\beta,\lambda)|^2 > 0 \tag{3.7}$$

The interpretation of inequalities (3.3) and (3.7)is that the Coulomb potential of a pair of <u>fractional</u> charges brought into the Coulomb gas from the outside is screened exponentially fast by the particles in the gas, <u>although their charge is integer</u>. In particular, the mean free energy needed to bring the pair of fractional charges corresponding to $\delta\rho_{ox}$ into the system does not diverge, as $|x| \to \infty$, as one would at first expect in two dimensions, because of the logarithmic growth of the Coulomb potential and the fact that fractional charges are not screened easily by integer charges. Thus the pair of external charges can break up in two essentially free charges. Note that by (3.6),

$$G(x) \xrightarrow[|x|\to\infty]{} \text{const.} \geqq \exp[-\beta\gamma^2 V(0)] > 0 \tag{3.8}$$

in three dimensions, for arbitrary β and z, i.e. a fractional-charge dipole can always break up. In contrast, in one dimension

$$\exp[-\beta\gamma^2|x|] \leqq G(x) \leqq \exp[-c|x|], \tag{3.9}$$

for some constant c which is positive for all β and z. Thus, in one dimension the electrostatic potential of fractional charges is never screened.

The behaviour of the two-dimensional gas interpolates between the one of the one- and the one of the three-dimensional gas, as (β,z) is varied. We have shown in [15] that in domain II, the low temperature, low density Kosterlitz-Thouless domain,

$$a(1+|x|)^{-\beta\gamma^2/2\pi} \leqq G(x) \leqq b(1+|x|)^{-c}, \tag{3.10}$$

for some constants a,b and $c > 0$. Together with (3.3) and (3.7) this proves the existence of a Kosterlitz-Thouless transition which is further discussed in Section IV.

ii) Screening of integral charges

This is (3.1) for observables, A and B, which are periodic in $\phi(j)$ with period 2π, for all j. I.e. the Fourier transforms of A and B only contain integral charges in their support. The one-dimensional Coulomb gas generally does screen integral charges, for arbitrary β and z. This is discussed in Aizenman's lectures and refs. given there.

iii) Weak screening of external charges

This notion of screening involves studying the expectation of the charge density, $q(j)$, in the presence of external charges, described by a charge density, $\rho(j)$, of bounded support, not as-

sumed to be integer-valued. Let

$$Z(\rho) \equiv \langle e^{i\phi(\rho)} \rangle (\beta,\lambda).$$

We consider

$$I(j) \equiv Z(\rho)^{-1} (-i) \langle \phi(j) e^{i\phi(\rho)} \rangle (\beta,\lambda) \tag{3.11}$$

We apply integration by parts, namely identity (2.22). This yields

$$I(j) = \beta \sum_{\ell} V(j-\ell) \{ \rho(\ell) + Z(\rho)^{-1} \langle q(\ell) e^{i\phi(\rho)} \rangle (\beta,\lambda) \} \tag{3.12}$$

Fourier transformation yields

$$\hat{I}(k) = \beta \hat{V}(k) \{ \overline{\hat{\rho}(k)} + Z(\rho)^{-1} \langle \overline{\hat{q}(k)} e^{i\phi(\rho)} \rangle (\beta,\lambda) \}$$

We now assume that

$$I(j) \xrightarrow[|x| \to \infty]{} -i \langle \phi(0) \rangle (\beta,\lambda),$$

by a power $\varepsilon > 0$ faster than $V(x)$ decays. This is clearly true in domain I, where strong screening (3.1) holds. Then

$\hat{I}(k)\, \hat{V}^{-1}(k) \longrightarrow 0, \quad$ as $\; |k| \longrightarrow 0$, so that

$$\beta^{-1} [\hat{I}\hat{V}^{-1}](0) = 0 = Q(\rho) + Z(\rho)^{-1} \langle \hat{q}(0) e^{i\phi(\rho)} \rangle (\beta,\lambda)$$

i.e.

$$\sum_{j} Z(\rho)^{-1} \langle q(j) e^{i\phi(\rho)} \rangle (\beta,\lambda) = -Q(\rho), \tag{3.13}$$

where $Q(\rho) = \Sigma \rho(\ell) = \hat{\rho}(0)$ is the total charge of ρ. This says that the external charges, $\{\rho(\ell)\}_{\ell \in \mathbb{Z}^{\nu}}$, are screened completely, asymptotically, by particles in the system (even if ρ is not integer-valued). Stronger statements, e.g. on the decay of the effective electric field of ρ, are obtained if the decay assumptions for $I(j) + i\langle\phi(0)\rangle(\beta,\lambda)$, as $|j| \to \infty$, are refined. For a related, "axiomatic" discussion, see [17].

By using the sine-Gordon (ϕ-) representation, it is easy to

see that an evaluation of $I(j)$ in mean field approximation leads precisely to the Debye-Hückel equation. The methods of Brydges and Federbush [10,11] can be used, in principle, to estimate systematic corrections.

iv) Dipole layers

An interesting variant of the discussion in iii) is the following : Let

$$D_\Lambda(\gamma\phi) \equiv \prod_{j\in\Lambda\subset\mathbb{Z}^{\nu-1}} e^{i\gamma(\phi(j,0)-\phi(j,-1))},$$

where Λ is a square array in the $j^\nu = 0$ plane. The state

$$\langle - \rangle_\gamma(\beta,\lambda) \equiv \lim_{\Lambda\nearrow\mathbb{Z}^{\nu-1}} \frac{\langle - D_\Lambda(\gamma\phi)\rangle(\beta,\lambda)}{\langle D_\Lambda(\gamma\phi)\rangle(\beta,\lambda)},$$

$\langle - \rangle(\beta,\lambda)$ an infinite volume state with Dirichlet b.c., describes a ν-dimensional Coulomb gas in the presence of an infinite, planar dipole layer located on the $j^\nu = 0$ plane. We wish to estimate the effective potential

$$\psi_\gamma(j) \equiv -i/_\beta \langle\phi(j)\rangle_\gamma(\beta,\lambda). \tag{3.14}$$

In vacuo, the graph of ψ_γ as a function of $m \equiv j^\nu$ is as shown in Fig.3.

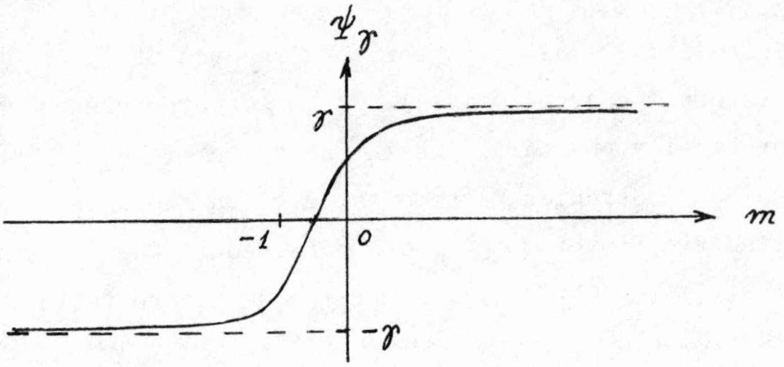

Fig. 3

If the particles in the Coulomb gas form a perfect plasma the graph of ψ_γ has the shape displayed in Fig. 4,

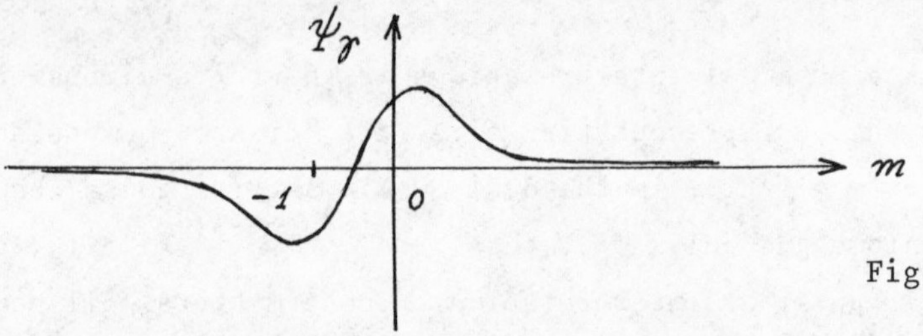

Fig.4

$$\text{and} \quad \psi_\gamma(m) \xrightarrow[|m|\to\infty]{} 0 \,, \quad \text{exponentially fast.} \tag{3.15}$$

One way of analyzing transitions in the three-dimensional Coulomb gas in domain 0 of Fig.2 is to investigate the behaviour of $\psi_\gamma(m)$ for different values of (β,z). It seems likely, that one can prove (3.15) in the domain of convergence of the expansion of [10,11]. (For the Villain gas (2.3) a proof is actually simple). Outside domain I, $\psi_\gamma(m)$ may approach 0 only like some inverse power of $|m|$, (or have the shape shown in Fig.3). We have no idea of how to prove this. It is more rewarding to replace γ by $-i\gamma$ and $-i\phi(j)$ by $\phi(j)$. One then considers the function

$$\psi_{-i\gamma}(m) = 1/\beta \, \langle \phi(\cdot, m) \rangle_{-i\gamma} \, (\beta,\lambda) \tag{3.16}$$

which is real-valued. For the Dirichlet b.c. state of the Villain gas in the thermodynamic limit, the graph of $\psi_{-i\gamma}$ is just as shown in Fig.4, with $\psi_{-i\gamma}(m) \xrightarrow[|m|\to\infty]{} 0$, exponentially fast, provided β is small enough.

However, there are indications that

$$\psi_{-i\gamma}(m) \longrightarrow \pm\gamma\,, \text{ as } m \longrightarrow \pm\infty\,, \tag{3.17}$$

if β is large. This phenomenon would be the analogue of the roughening transition in the three-dimensional Ising model [18]. In two dimensions, (3.17) is the expected behaviour for all β, but the surface tension, positive for small β, is expected to

vanish for large β. The functions $\psi_{-i\gamma}(m)$ and

$$\langle e^{i\alpha(\phi(m)-\phi(-m))}\rangle_{-i\gamma}(\beta,\lambda)$$

have nice physical interpretations in the three-dimensional Villain gas in the ϕ-representation, (i.e. the "discrete Gaussian model"), and, for $\gamma = 1/\beta$, in the dual U(1) lattice gauge theory. (In the Coulomb gas only ψ_γ, resp. $\langle e^{i\alpha(\phi(m)-\phi(-m))}\rangle_\gamma(\beta,\lambda)$ have a natural, physical interpretation). These matters will have to be discussed in more detail, elsewhere.

IV. The Kosterlitz-Thouless Transition in the Two-Dimensional Coulomb Gas

In this section we briefly sketch a rigorous argument [15] establishing the Berezinski-Kosterlitz-Thouless transition [19] in a class of models, including the two-dimensional Coulomb gas, the plane rotator and higher dimensional, abelian lattice gauge theories.

The main idea is to use the sine-Gordon (ϕ-) representation to rewrite correlations in the Coulomb gas as convex combinations of correlations in dilute gases of neutral multipoles of variable size, at random positions. Such gases are known not to exhibit Debye screening [3,15]. Here we study the behaviour of the fractional charge correlation $G(x)$ defined in (2.15) which we discussed already in Section III.1.

Our aim is to sketch the proof of (3.10), i.e.

$$a\,(1+|x|)^{-\beta\gamma^2/2\pi} \leqq G(x) \leqq b\,(1+|x|)^{-c}, \tag{4.1}$$

for some $c > 0$, provided $z < e^{\varepsilon'\beta}$ and $\beta > \beta_c$, for some $\varepsilon' > 0$ and $\beta_c < \infty$; (a and b are finite positive constants). We use the ϕ-representation (2.11), (2.12), (2.1') of the hard core gas : Thus $\langle - \rangle_\Lambda(\beta,z)$ denotes the expectation in the measure, (signed for $z > 1$),

$$Z_\Lambda^{-1} \prod_{j\in\Lambda} (1+z\cos\phi(j))\, d\mu_{\beta V}(\phi), \tag{4.2}$$

and we impose free (i.e. insulating) b.c.. The more interesting case of Dirichlet b.c. is only slightly more difficult; see [15].

The lower bound in (4.1) has already been established in Section III, so we concentrate on the proof of the upper bound. We note that, by the $\phi \to -\phi$ symmetry of (4.2)

$$G_\Lambda(x) = Z_\Lambda^{-1} \int I_1(\phi_\Lambda)\, d\mu_{\beta V}(\phi), \tag{4.3}$$

where

$$\left.\begin{aligned} I_\alpha(\phi_\Lambda) &= \cos\phi(\alpha\,\delta\rho_{ox}) \prod_{j\in\Lambda} (1+z\cos\phi(j)), \\ \text{with}\quad \delta\rho_{ox}(j) &= \gamma(\delta_{jo}-\delta_{jx}), \quad 0<\gamma<1. \end{aligned}\right\} \tag{4.4}$$

Our proof of (4.1) is based on applying the following elementary identities to $I_\alpha(\phi_\Lambda)$:

$$\begin{aligned}(1+K_1\cos\alpha_1)(1+K_2\cos\alpha_2) &= \tfrac{1}{3}\sum_{\kappa=1,2}(1+3K_\kappa\cos\alpha_\kappa) \\ &\quad + \tfrac{1}{6}\sum_{\varepsilon=\pm1}(1+3K_1K_2\cos(\alpha_1+\varepsilon\alpha_2))\end{aligned} \tag{4.5}$$

$$\cos\alpha_0\,(1+K_1\cos\alpha_1) = \tfrac{1}{2}\sum_{\varepsilon=\pm1}(\cos\alpha_0+K_1\cos(\alpha_0+\varepsilon\alpha_1)) \tag{4.6}$$

and

$$\begin{aligned}&(\cos\alpha_0+K_1\cos(\alpha_0+\alpha_1))(1+K_2\cos\alpha_2) \\ &\quad = \tfrac{1}{3}(\cos\alpha_0+3K_1\cos(\alpha_0+\alpha_1)) \\ &\qquad + \tfrac{1}{6}\sum_{\varepsilon=\pm1}\{(\cos\alpha_0+3K_2\cos(\alpha_0+\varepsilon\alpha_2))+ \\ &\qquad\qquad (\cos\alpha_0+3K_1K_2\cos(\alpha_0+\alpha_1+\varepsilon\alpha_2))\}\end{aligned} \tag{4.7}$$

A function ρ of finite support contained in Λ with values in $\{+1,-1\}$ is called a charge density, and $Q(\rho) \equiv \sum_j \rho(j)$ is the total charge of ρ ; ρ restricted to a proper subset of supp ρ is called a constituent of ρ . A family of charge densities with disjoint supports is called an ensemble.

First (4.5) is applied to

$$I_\alpha(\phi_\Lambda) = \sum_m c_{\mathcal{E}_m^{(o)}} \cos\phi(\alpha\delta\rho_{0x}) \prod_{\rho\in\mathcal{E}_m^{(o)}} (1+K^{(o)}(\rho)\cos\phi(\rho)) \tag{4.8}$$

where $c_{\mathcal{E}_m^{(o)}} = \delta_{m1}$, each charge density $\rho \in \mathcal{E}_1^{(o)}$ has support on a single site, $j(\rho)$, where it takes the value 1, $\bigcup_{\rho\in\mathcal{E}_1^{(o)}} \{j(\rho)\} = \Lambda$, and $K^{(o)}(\rho) = z$.

The rules for applying (4.5) are as follows : Group all ρ's in $\mathcal{E}_1^{(o)}$ in pairs (ρ_1,ρ_2) supported on nearest neighbor sites in an otherwise arbitrary way. For a given pair (ρ_1,ρ_2) apply (4.5) to

$$(1+K^{(o)}(\rho_1)\cos\phi(\rho_1))(1+K^{(o)}(\rho_2)\cos\phi(\rho_2))$$

The result is then inserted on the r.s. of (4.8), for all possible (ρ_1,ρ_2) , and the resulting expression is expanded out. This yields

$$I_\alpha(\phi_\Lambda) = \sum_m c_{\mathcal{E}_m^{(1)}} \cos\phi(\alpha\delta\rho_{0x}) \prod_{\rho\in\mathcal{E}_m^{(1)}} (1+K^{(1)}(\rho)\cos\phi(\rho)) \tag{4.9}$$

with $c_{\mathcal{E}_m^{(1)}} > 0$, for all m , and each ρ on the r.s. of (4.9) is supported on one site, with a nearest neighbor site empty, or on a pair of nearest neighbors and takes values ± 1 . Notice that each application of (4.5) to a pair (ρ_1,ρ_2) produces a term, $1+3K^{(o)}(\rho_1)K^{(o)}(\rho_2)\cos\phi(\rho_1-\rho_2)$, with the property that the total charge $Q(\rho_1-\rho_2)$ vanishes. The density $\rho' \equiv \rho_1-\rho_2$ is interpreted as a neutral dipole. Another term that is produced is $1+3K^{(o)}(\rho_1)\cos\phi(\rho_1)$ which has the property that the charge ρ_2

has been eliminated. Thus the resulting ensembles, $E_m^{(1)}$, tend to contain neutral dipole densities and tend to be sparser than $E_1^{(o)}$. This is a feature common to all subsequent steps in an inductive expansion of $I_\alpha(\phi_\Lambda)$: During those steps charge densities ρ,ρ' are combined to larger densities $\rho\pm\rho'$, with a chance of 1/2 that the total charge is lowered, or one of the densities ρ,ρ' is eliminated which makes the ensemble sparser, at the prise of increasing the unrenormalized activities, $K(\rho)$, by factors of 3.

In the next step, the ρ's in each $E_m^{(1)}$ are paired among each other or with $\alpha\delta\rho_{ox}$, and identities (4.5), resp. (4.6) are applied to all such pairs, the resulting expression is expanded and yields a class $\{E_{m'}^{(2)}\}$ of ensembles derived from $\{E_m^{(1)}\}$ by combining densities ρ,ρ', with $\mathrm{dist}(\rho,\rho') \equiv \mathrm{dist}(\mathrm{supp}\rho, \mathrm{supp}\rho') = 1$, to a larger density $\rho\pm\rho'$, resp. cancelling either ρ or ρ', for all m'. After a finite number of steps, depending on an integer $k = 1,2,3,\ldots,$ in this inductive scheme one obtains

$$I_\alpha(\phi_\Lambda) = \sum_m c_{\mathcal{E}_m^k} \prod_{\rho\in\mathcal{E}_m^k \sim \rho_m^*} \left(1+K^k(\rho)\cos\phi(\rho)\right)\cdot \\ \cdot\left(\cos(\alpha\delta\rho_{ox}) + K^k(\rho_m^*)\cos\phi(\rho_m^*+\alpha\delta\rho_{ox})\right) \tag{4.10}$$

where $c_{E_m^k} > 0$, for all m, and each ensemble E_m^k is the union two sub-ensembles N_m^k and J_m^k, defined as follows :

Define the diameter, $d(\rho)$, of a charge density ρ in some ensemble E to be the smallest integer of the form 2^ℓ, $\ell = 1,2,3,\ldots,$ such that $\mathrm{supp}\,\rho$ can be covered by a single square in $\mathbb{Z}^2$ with sides of length $d(\rho)$. Furthermore, $d(\rho_m^*)$ is defined as the diameter of $\rho_m^*+\alpha\delta\rho_{ox}$.

The sub-ensembles N_m^k are now defined by the properties :

a) $N_m^k \supseteq N_m^{k-1}$, $N_m^o = \emptyset$;

b) each $\rho \in N_m^k$ is neutral, i.e. $Q(\rho) = 0$;

c) For all ρ,ρ' in N_m^k , $\rho \neq \rho'$,
$\mathrm{dist}(\rho,\rho') \geqq M \min(d(\rho),d(\rho'))^\alpha$, for some constants $M > 0$ and $\alpha \in (3/2,2)$, e.g. $\alpha = 5/3$, to be chosen appropriately, [15]. (Here $\mathrm{dist}(\rho,\rho') \equiv \mathrm{dist}(\mathrm{supp}\,\rho, \mathrm{supp}\,\rho')$) .

d) $\mathrm{dist}(\rho,\rho'') \geqq Md(\rho)^\alpha$, for all $\rho \in N_m^k$ and all $\rho'' \in J_m^k$. □

The expansion described between (4.9) and (4.10) terminates for all ρ's in N_m^k , for all m , because the ρ's in N_m^k are neutral, see b), and far separated from other charge densities, see c), d).

The sub-ensembles J_m^k are defined by

i) $J_m^k \cap N_m^k = \emptyset$, $J_m^k \cup N_m^k = E_m^k$;

ii) for arbitrary densities ρ and $\rho' \neq \rho$ in J_m^k ,

$$\mathrm{dist}(\rho,\rho') \geqq 2^k ,$$

for all m . Thus, k labels a distance scale. Identity (4.10) and properties a) - d) , resp. i) - ii) are obvious for $k = 0$, and we have already outlined how they are checked for $k = 1$. In order to do the induction step, $k \to k+1$, we split J_m^k into two disjoint subsets, ${}^{>}J_m^k$ and ${}^{<}J_m^k$, where ${}^{<}J_m^k$ has the property that, given any $\rho \in {}^{<}J_m^k$, there exists some $\rho' \in J_m^k$, with $\mathrm{dist}(\rho,\rho') < 2^{k+1}$. We set ${}^{>}J_m^k = J_m^k \sim {}^{<}J_m^k$. Densities in ${}^{>}J_m^k$ are separated from other densities in J_m^k by a distance $\geqq 2^{k+1}$ and will participate in the expansion only on scales $\geqq 2^{k+1}$. Next, a charge density $\rho \in {}^{<}J_m^k$ is paired with some $\rho' \in {}^{<}J_m^k$ for which $\mathrm{dist}(\rho,\rho') < 2^{k+1}$. Then identity (4.5), resp. identity (4.7) is applied to the factors labelled by (ρ,ρ') , and the resulting expression is expanded. This operation increases the activity of the resulting density by a factor of 3. Subsequently a new pair of den-

sities within distance $< 2^{k+1}$ from each other is formed, etc... After finitely many operations, (4.10) is recovered, with k increased to $k+1$. See [15] for details. By induction in k and a series of combinatorial arguments one obtains

Theorem 1 [15]

(1)
$$I_\alpha(\phi_\Lambda) = \sum_m c_{\mathcal{N}_m} \prod_{\rho \in \mathcal{N}_m \sim \rho_m^*} (1 + K(\rho)\cos\phi(\rho)) \cdot$$
$$\cdot \left(\cos\phi(\alpha\delta\rho_{0x}) + K(\rho_m^*)\cos\phi(\rho_m^* + \alpha\delta\rho_{0x})\right),$$

where $c_{N_m} > 0$, for all m; all ρ's in N_m, except possibly one density $\rho = \rho_c$ which is charged, are neutral and satisfy conditions b) - d) formulated above.

(2) For all $\rho \in N_m$, $\rho \neq \rho_c$,
$$K(\rho) \leq z^{|\rho|} \exp\left[c \sum_{n=0}^{n(\rho)} A_n(\rho)\right], \tag{4.11}$$

where $|\rho| = \sum_j |\rho(j)|$, $A_n(\rho)$ is the minimal number of $2^n \times 2^n$ squares necessary to cover the support of ρ, $n(\rho) \leq c' \ell n d(\rho)^\alpha$, and c, c' are finite constants independent of N_m.

(3) If some $\rho \in N_m$ contains a constituent ρ_1 such that $\text{dist}(\rho_1, \rho - \rho_1) \geq 2Md(\rho_1)^\alpha$ then $Q(\rho_1) \neq 0$. □

Remarks. Part (1) follows from (4.10) by induction in k, and (3) is a fairly simple consequence of conditions b) - d) satisfied by N_m. The hard part is (2) : Since, by condition a) above, each N_m is an inductive limit of a family $\{N^k_{n(m,k)}\}_{k=1,2,3,...}$, each neutral $\rho \in N_m$ belongs to some N^k_m, for a finite k. Thus, the term $1+K(\rho)\cos\phi(\rho)$ is produced after a finite number, N, of applications of identiy (4.5). That identity then yields

$K(\rho) \leqq z^{|\rho|} 3^N$, and a somewhat complicated estimate on N yields (4.11). (Same comment on term labelled by ρ_m^*) . The interpretation of the quantity $c \cdot \sum_{n=o}^{n(\rho)} A_n(\rho)$ is that of an entropy of ρ . See also (2.36). For details we refer to [15] .

Next, we note that

$$\left.\begin{aligned} Z_\Lambda &= \int I_{\alpha=0}(\phi_\Lambda)\, d\mu_{\beta V}(\phi), \\ Z_\Lambda G_\Lambda(x) &= \int I_{\alpha=1}(\phi_\Lambda)\, d\mu_{\beta V}(\phi). \end{aligned}\right\} \tag{4.12}$$

Since the algebraic structure of the expansion of $I_\alpha(\phi_\Lambda)$ is independent of α , the expansions of Z_Λ and $Z_\Lambda G_\Lambda$ involve the same N_m , ρ_m^* and c_{N_m} . For free b.c., the contribution of all terms containing a factor $K(\rho_c)\cos\phi(\rho_c)$, $Q(\rho_c) \neq 0$, to the r.s. of (4.12) vanishes; see (2.9). Thus

$$\int I_\alpha(\phi_\Lambda)\, d\mu_{\beta V}(\phi) = \sum_m c_{N_m'} \int \prod_{\rho \in N_m' \sim \rho_m^*} (1 + K(\rho)\cos\phi(\rho)) \cdot \tag{4.13}$$
$$\cdot \left(\cos\phi(\alpha\delta_{\rho ox}) + K(\rho_m^*)\cos\phi(\rho_m^* + \alpha\delta_{\rho ox})\right) d\mu_{\beta V}(\phi)$$

where all ρ's in N_m' and ρ_m^* are neutral, for all m .

Our goal is now to replace $\prod_{\rho \in N_m' \sim \rho_m^*} (1+K(\rho)\cos\phi(\rho))$ by a new product

$$\prod_{\rho \in N_m' \sim \rho_m^*} (1 + \zeta(\rho,\beta)\cos\phi(\bar\rho)), \tag{4.14}$$

where $\bar\rho$ is a renormalized charge density, and $\zeta(\rho,\beta)$ a renormalized activity, without changing the values of Z_Λ and $Z_\Lambda G_\Lambda$. An essential ingredient in the proof of (4.1) is that (4.14) defines a positive function of ϕ . This is manifestly true if

$$\zeta(\rho,\beta) < 1 \text{ , for all } \rho \in N_m' \text{ and all } m . \tag{4.15}$$

One of the main technical results of [15] is that $\bar\rho$ can be chosen

such that (4.15) holds for $z < e^{\varepsilon'\beta}$, for some $\varepsilon' > 0$, and β large enough.

The idea of the renormalization $\rho \to \bar{\rho}$ is as follows : First, we carry out the renormalization transformation described in Section II.2, (2.32) - (2.34). Next, let $\rho \in N'_m$ contain a charged constituent ρ_1 separated from $\rho-\rho_1$ by a distance $\geq 2M$. Then ρ_1 is replaced by a new charge density σ_1 concentrated near the surface of a domain Σ_1 containing $\operatorname{supp}\rho_1$ but not intersecting $\operatorname{supp}(\rho-\rho_1)$, in such a way that the electrostatic interactions of ρ_1 and σ_1 with $\rho-\rho_1$ and all other $\rho' \in N'_m$ are unchanged. This is achieved by a change of variables on the r.s. of (4.13), $\phi(j) \mapsto \phi(j)+ia_{\rho_1}(j)$, for some real function a_{ρ_1}. This is the method of complex translations introduced in [20]. As a result, $K(\rho)\cos\phi(\rho)$ is replaced by $\exp[-\beta(E(\rho_1)-E(\sigma_1))]K(\rho)\cos\phi(\sigma_1+\rho-\rho_1)$, where $E(\rho')$ is the electrostatic energy of ρ' ; see (2.4). In order to locate the charged constituents of some $\rho \in N'_m$, one uses part (3) of Theorem 1. Let $S_n(\rho)$ be a minimal collection of $2^n\times2^n$ squares needed to cover ρ, and let $S'_n(\rho)$ be defined as

$$\{s \in \mathcal{S}_n(\rho) : \operatorname{dist}(s,s') \geqq 2M\,2^{\alpha n},\ \text{for all } s' \neq s \text{ in } \mathcal{S}_n(\rho)\}.$$

By Theorem 1, (3), each $s \in S'_n(\rho)$ covers a charged constituent, ρ_s, of ρ. The renormalization procedure described above is now applied to ρ_s, for each $s \in S'_n(\rho)$, in such a way that $\operatorname{supp}\sigma_s$ is contained in the interior of a $2^{n+1}\times2^{n+1}$ square covering s. This permits one to apply the same renormalization transformation inductively on all length scales 2^n, $1 \leq n \leq n(\rho)$. One obtains

<u>Theorem 2</u> [15]

For $\alpha > 3/2$, M large enough and for all m,

$$\int I_\alpha(\phi_\Lambda)\,d\mu_{\beta V}(\phi) = \int F_m(\phi_\Lambda)\,(\cos\phi(\alpha\,\delta\rho_{ox}) +$$

$$K(\rho_m^*)\cos\phi(\rho_m^* + \alpha\,\delta\rho_{ox})\,d\mu_{\beta V}(\phi)$$

where $$F_m(\phi_\Lambda) = \prod_{\rho\in\mathcal{N}_m' \sim \rho_m^*} (1 + \xi(\rho,\beta)\cos\phi(\bar{\rho})), \tag{4.16}$$

and

$$\xi(\rho,\beta) \leqq K(\rho)e^{-\varepsilon\beta|\rho|}\exp\Big[-c''\beta\sum_{n=1}^{n(\rho)}\text{card}(\mathcal{S}_n'(\rho))\Big], \tag{4.17}$$

with c'' independent of N_m'. □

We now sketch, how Theorems 1 and 2 are used to prove (4.15) and the upper bound on $G(x)$ in (4.1). First, one proves a combinatorial lemma saying that, for $\alpha < 2$, there exists some $c''' > 0$ such that

$$|\rho| + \sum_{n=1}^{n(\rho)}\text{card}(\mathcal{S}_n'(\rho)) \geqq c'''\sum_{n=0}^{n(\rho)} A_n(\rho), \tag{4.18}$$

with c''' independent of N_m'; see [15].

Moreover, $$\sum_{n=0}^{n(\rho)} A_n(\rho) \geqq \text{const.}\,\ell nd(\rho). \tag{4.19}$$

It now follows from (4.17), (4.11), (4.18) and (4.19) that for $z < e^{\varepsilon'\beta}$, for some $\varepsilon' > 0$, there exist constants $c > 0$ and $d < \infty$ such that

$$\xi(\rho,\beta) \leqq \exp[-(c\beta - d)\,\ell n\, d(\rho)], \tag{4.20}$$

which proves (4.15) for $\beta > d/c$. The inductive renormalization transformation described above can of course be applied to $\cos(\delta\rho_{ox})$ and $\cos(\rho_m^* + \delta\rho_{ox})$, too, for all m. Together with (4.15), i.e. $F_m(\phi) \geqq 0$, this yields

$$Z_\Lambda G_\Lambda(x) = \sum_m c_{n'_m} \int F_m(\phi_\Lambda)\,(\xi' \cos\phi(\overline{\delta\rho_{ox}}) +$$

$$\xi'' \cos\phi(\overline{\rho_m^* + \delta\rho_{ox}}))\, d\mu_{\beta V}(\phi)$$

$$\leqq (\xi' + \xi'') \int F_m(\phi_\Lambda)\, d\mu_{\beta V}(\phi), \qquad (4.21)$$

where

$$0 < \xi', \xi'' \leqq \exp[-(c_\gamma \beta - d) \ln \ell],$$

and

$$\ell \equiv \min\left(d(\delta\rho_{ox}), d(\rho_m^* + \delta\rho_{ox})\right) \geqq |x|, \qquad (4.22)$$

with $0 < c_\gamma < c$, for $0 < \gamma < 1$, (see (4.4)). It follows immediately from (4.16) and (2.9) that

$$\int F_m(\phi_\Lambda)\, d\mu_{\beta V}(\phi) \leqq$$

$$\int F_m(\phi_\Lambda)(1 + K(\rho_m^*) \cos\phi(\rho_m^*))\, d\mu_{\beta V}(\phi) = Z_\Lambda$$

so that, with (4.21) and (4.22),

$$Z_\Lambda G_\Lambda(x) \leqq 2 \exp[-(c_\gamma \beta - d) \ln |x|]\, Z_\Lambda$$

which, for β large enough, yields (4.1), by taking $\Lambda \nearrow \mathbb{Z}^2$. Together with the material in Section III, i), in particular (3.3) and (3.7), this completes the proof of existence of a Kosterlitz-Thouless transition in the hard core Coulomb gas, as (β, z) is varied.

V. Other Models with Kosterlitz-Thouless Transitions

Here is a list of models for which transitions of the Kosterlitz-Thouless type (as some thermodynamic parameters are varied) are established in [15] .

1) Hard core-, standard- and Villain Coulomb gas in two dimensions.

2) The two-dimensional plane rotator model. It is shown in [15] that, for large enough β , $\langle \vec{s}_o \cdot \vec{s}_x \rangle(\beta) \geqq a(1+|x|)^{-c}$, in zero external field, for some finite c . An upper bound with power fall-off was previously proven in [20] . Some further results may be found in [21] .

3) The two-dimensional $\mathbb{Z}_n$ models, for n large enough : Existence of a massless phase for $T_- < T < T_+$, for some finite, positive T_-, T_+ .

4) The two-dimensional solid-on-solid model, for which it is shown e.g. that $\langle(\phi(0)-\phi(x))^2\rangle \geqq a\ell n|x|$, at sufficiently high temperatures.

5) A three-dimensional, non-compact lattice Higgs model (a Landau-Ginsburg lattice theory) for which the existence of a transition from a superconducting, massive to a massless QED phase is verified; (see also [22] and refs.).

6) The four-dimensional, pure $U(1)$-lattice gauge theory : Breakdown of confinement for large β . This was first shown in [23] , by a more complicated argument.

Acknowledgements. Many interesting discussions about Coulomb systems with D. Brydges, J. Lebowitz, E.H. Lieb, Y.M. Park, E. Seiler and others are gratefully acknowledged. We thank G. Velo and A.S. Wightman for inviting us to present the material in these notes at the 1980 Erice school.

References

1. W. Pauli and M. Fierz, Nuovo Cimento 15, 167, (1938); J.M. Jauch and F. Rohrlich, "Theory of Photons and Electrons", Cambridge, Mass. : Addison-Wesley 1955. See also W. Hunziker, in "Mathematical Problems in Theoretical Physics", K. Osterwalder (ed.), Lecture Notes in Physics 116, Berlin-Heidelberg-New-York : Springer-Verlag 1980; ref. 7 below.

2. E.H. Lieb, Rev. Mod. Phys. 48, 553, (1976).

3. J. Fröhlich and T. Spencer, J. Stat. Phys., in press.

4. J. Fröhlich, Commun. math. Phys. 47, 233, (1976).

5. D. Ruelle, "Statistical Mechanics", New York : Benjamin 1969.

6. J. Fröhlich and Y.M. Park, Commun. math. Phys. 59, 235, (1978).

7. J. Fröhlich and Y.M. Park, J. Stat. Phys., in press.

8. E.H. Lieb and J. Lebowitz, Adv. Math. 9, 316, 1972.

9. R.B. Griffiths, Phys. Rev. 176, 655, (1968).

10. D. Brydges, Commun. math. Phys. 58, 313, (1978).

11. D. Brydges and P. Federbush, Commun. math. Phys. 73, 197, (1980).

12. Y.M. Park, J. Math. Phys. 18, 12, (1977).

13. A.J.F. Siegert, Physica 26, 30, (1960).

14. J. Fröhlich, R. Israel, E.H. Lieb and B. Simon, J. Stat. Phys. 22, 297, (1980).

15. J. Fröhlich and T. Spencer, "The Kosterlitz-Thouless Transition in the Two-Dimensional Coulomb Gas", subm. to Phys. Rev. Letts.; detailed paper to appear.

16. J. Glimm, A. Jaffe and T. Spencer, Commun. math. Phys. 45, 203, (1975).

17. Ch. Gruber and Ph. A. Martin, in : "Mathematical Problems in Theoretical Physics", loc. cit. (ref.1), and references given there.

18. R.L. Dobrushin, Theory Prob. Appl. 17, 582, (1972).
H. van Beijeren, Commun. math. Phys. 40, 1, (1975).

19. V.L. Berezinskii, Soviet Phys. JETP 32, 493, (1971); J.M. Kosterlitz and D.J. Thouless, J. Physics C6, 1181, (1973);

J.V. José, L.P. Kadanoff, S. Kirkpatrick and D.R. Nelson, Phys. Rev. B16, 1217, (1977).

20. O. McBryan and T. Spencer, Commun. math. Phys. 53, 99, (1977).

21. J. Bricmont, J.-R. Fontaine, J. Lebowitz, E.H. Lieb and T. Spencer, Preprint, Rutgers University, 1980, Commun. math. Phys., to appear.

22. D. Brydges, J. Fröhlich and E. Seiler, Nucl. Phys. B152, 521, (1979).

23. A. Guth, Phys. Rev. D21, 2291, (1980).

See also J. Glimm and A. Jaffe, Commun. math. Phys. 56, 195, (1977).

DEBYE SCREENING IN CLASSICAL COULOMB SYSTEMS

David C. Brydges
Department of Mathematics
University of Virginia
Charlottesville, VA 22903

Paul Federbush
Department of Mathematics
University of Michigan
Ann Arbor, MI 48109

0.0 INTRODUCTION

The standard approach to Coulomb systems in equilibrium classical statistical mechanics (at low density or high temperature) was invented by Debye and Hückel. It may be found under their names in most text books on physical chemistry. Typically one considers a charge fixed in a background sea of other charges which are approximated by a continuous charge distribution density. The Debye Hückel theory shows that this background charge distribution arranges itself in a spherically symmetric way about the fixed charge so that by Newton's theorem the view from distance r is equivalent to a single charge at r = 0 whose magnitude equals the total charge of the fixed charge plus that of the "cloud" of radius r about it. Furthermore this total charge behaves as exp(-const x r) so that the charge is "shielded" and its "effective potential" is a Yukawa, exp(-cr)/r.

From this, it is not unreasonable to imagine that each charge in a Coulomb system arranges its neighbors in a clever way so that the system may be viewed as a system of interacting "charge clouds", the interaction being a Yukawa. Then one would predict that correlations should decay not as 1/r but as a Yukawa.

The serious problem with this argument lies in the initial approximation by a continuous charge distribution because of course discrete charges cannot form spherical clouds (except in one dimension) and this is essential for Newton's theorem. Nevertheless exponential decay of correlations is also predicted at least in classical mechanics by resummed high temperature expansions [1]. Furthermore in 1962 Lenard [2] and Lenard and Edwards [3] solved

the one dimensional classical continuum coulomb model exactly! Among other things they verified that the correlations decay exponentially. [The intriguing properties of one dimensional coulomb systems are discussed in Aizenman's lectures].

In this article we are going to give an exposition of the methods that we used in [4] and [5] to prove that for a large class of classical coulomb models in three dimensions at low density or high temperature the correlations do indeed decay exponentially. The strong restrictions on density, temperature come from our use of expansions which only converge if some system parameter is small enough. Fröhlich, in his lectures appearing in this volume, describes work with Park and with Spencer which for the most part avoids expansions but pays the price of being restricted to "charge symmetric" systems. With these methods (correlation inequalities, integration by parts, changes of variables and block spin transformations) they are able to obtain existence of infinite volume correlations, power law decays, bounds on correlation lengths and even existence of the Kosterlitz Thouless phase transition two dimensional coulomb gas. Many of these results are obtained for optimal or close to optimal ranges of system parameters in contrast to our expansion methods.

The success of our methods, the Fröhlich, Park, Spencer arguments and the Lenard and Edwards paper are all based on the "sine gordon" transformation which represents the partition function as an average over external fields of ideal gases of particles. In a sense it is a rigorous replacement of mean field arguments in which a partition function of an interacting system is approximated by an ideal gas in an external field chosen to optimise the approximation. Unfortunately for our intuition the average is over the imaginary continuation of an external electric potential.

The other ingredient in our methods are expansions invented by Glimm, Jaffe and Spencer in constructive field theory [6,7,8,9]. In [6] is a fine exposition of these expansions, but nevertheless we have taken the liberty of giving another exposition [Section 1.7-2.8] in these notes. In doing this, our main hope is to provide a sense of the unity behind all the various expansions used in statistical mechanics and field theory. The form of the cluster expansion we have adopted is not that in [6], but based on an earlier paper [9] by Spencer. We have found that this version adapts more readily to the environment of statistical mechanics.

In Section 5 we conclude this article with a discussion of quantum coulomb systems for which our methods have run into difficulties. This is a problem of principle because it is not even clear that perturbation theory predicts exponential screening.

In concluding our introduction, we wish to thank: Professor N. Kuiper for the hospitality of the Institute des Hautes Etudes

Scientifiques (where part of these notes were prepared), Professor K. Gawedski and J. Fröhlich, for useful conversations and suggestions, and Professor T. Spencer for suggesting a gaussian approximation in the form of equation (1.7.1") in the text.

0.1 STATEMENT OF RESULTS, THE HARD CORE SYSTEM

In [5] we have proven screening for a very large class of systems in classical statistical mechanics. This work allowed arbitrary charge species, non charge symmetric situations, and essentially arbitrary short range forces. These short range forces were required only to ensure the stability of the system (the pure 1/r potential is unstable in the classical situation), and to fall off quickly enough so as not to interfere with the shielding. Rather than to here detail these two very physical conditions, we present a special case indicative of the general result and of interest in its own right. We have deferred some definitions to Section 1.1.

We consider a system of charged $e = \pm 1$, spheres of radius R, so that the potential V is given by

$$V = \begin{cases} \frac{1}{2} \sum\limits_{i \neq j} \frac{e_i e_j}{4\pi |\vec{x}_i - \vec{x}_j|} , & |\vec{x}_i - \vec{x}_j| \geq 2R \quad \text{all} \quad i \neq j \\ \infty , & \text{otherwise} \end{cases} \tag{0.1.1}$$

The activities, z_+, z_- are equal, $= z$. The "Debye length" is $\ell_D \equiv (2z\beta)^{-1/2}$. By a limiting procedure to be explained, infinite volume expectation values of products of smeared density operators, $J(x)$, are defined. For simplicity we state our main clustering theorem for the two point function (rather than an arbitrary product). We introduce dimensionless parameters $\xi_1 = \beta/R$ and $\xi_2 = zR^3$. Δ_1, Δ_2 are cubes of side ℓ_D, and $h(\vec{x})$, $g(\vec{x})$ are functions of absolute value less than 1.

Theorem 0.1.1 Given γ, $0 < \gamma < 1$, there is c_γ and a function $f_\gamma(\xi)$, $0 \leq \xi < \infty$, $f_\gamma(\xi) > 0$, f_γ decreasing, such that for $\xi_2 < f_\gamma(\xi_1)$

$$\left| \int_{\Delta_1} d^3x_1 \int_{\Delta_2} d^3x_2 \; g(x_1) h(x_2) \left\langle J(x_1) J(x_2) \right\rangle \right| \leq c_\gamma e^{-\frac{\gamma}{\ell_D} \text{dist}(\Delta_1, \Delta_2)} \tag{0.1.2}$$

c_γ, f_γ do not depend on the choice of g, h, or Δ_i.

Thus at any fixed temperature if the density is low enough there is screening. To specify the limiting procedure we define V_L as

$$V_L = \frac{1}{2} \sum_{i,j} \frac{e_i e_j}{4\pi |\vec{x}_i - \vec{x}_j|} \left(1 - e^{-|\vec{x}_i - \vec{x}_j| / \lambda \ell_D}\right) \tag{0.1.3}$$

λ a small parameter. V_s is chosen so that

$$V = V_L + V_s \tag{0.1.4}$$

We note

$$V_L = \frac{1}{2} \int J(x_1) \left(\frac{1}{-\Delta_0} - \frac{1}{(-\Delta_0 + {}^1/_{\lambda^2 \ell_D^2})} \right)(x_1, x_2)\, J(x_2) \tag{0.1.5}$$

Δ_0 is the free Laplacian on R^3. For a large box Λ, we define V_L^Λ as (0.1.5) with the integrals restricted to Λ, and with Δ_0 replaced by the Laplacian with Dirichlet data on $\partial\Lambda$. With Λ' a box containing Λ, we obtain our infinite volume limit by performing the following three steps in order

a) Calculating expectation values for the grand canonical ensemble defined in Λ', for potential $V^{\Lambda'} = V_L^\Lambda + V_s$.

b) Letting Λ' become infinite.

c) Letting Λ become infinite.

We will say a few words now about more general systems. If we were dealing with a non charge symmetric situation we would impose a condition on the activities: that the system described by the same activities and V_s alone, should be neutral in the limit. In a system with charges and activities (e_i, z_i) we have imposed the condition

$$z_i \big/ \max_i z_i \geq c > 0 \tag{0.1.6}$$

Our estimates are not uniform as $c \to 0$, if one species is of extremely low density compared to the other species, and of fractional charge compared to the other species. (e_f is fractional with respect to $e_1, e_2, \ldots, e_r$ if e_f is not a multiple of g.c.d. $(e_1, \ldots, e_r)$.) We have proven that fractional charges are shielded --this may suggest the limiting condition (0.1.6) is merely a weakness of our procedure, and not a physical requirement. There is

obviously more work to be done--to generalize the procedure for taking the infinite volume limit (and thereby dealing with surface charge), to decide the necessity of condition (0.1.6), to prove useful asymptotic expansions for correlation functions, to study the quantum situation (see Section 5) -- there are many other questions.

1.0 DISCUSSION

Since it is our objective to bring out the ideas rather than prove the best possible theorems, we are not going to prove Theorem 0.1.1 but instead create a Coulomb system with artificial short range forces by a procedure that amounts to dropping the term V_s occurring in (0.1.4). This will act as a showcase for the types of theorems and methods of proof that generalize to more complicated situations. This is a technical and not conceptual simplification. The general procedure involves a more complicated sine gordon transformation which is described in [5].

For the benefit of readers who are not willing to accept our word on this point and are interested in looking at [5] we make the following comments: whereas the usual sine gordon transformation exhibits the Coulomb system as an average over external fields of an ideal gas, the transformation in [5] exhibits the Coulomb system with hard cores as an average over external fields of a system with the interaction V_s (which we call the "reference system"). Our strategy in [5] is to show that this reference system is sufficiently near to ideal for the methods we are about to describe for the $V_s = 0$ case (<=> ideal gas) to be still applicable with technical modifications. To do this, we use a Mayer expansion on the reference system. This can be done because V_s is short range. If the λ in V_s is sufficiently small, the Mayer expansion converges uniformly in the external field and the range of parameters R, β, z, delineated in Theorem 0.1.1.

1.1 THE MODEL

The usual Coulomb interaction energy between two particles at x,y is $(\pm)\frac{1}{4\pi|x-y|}$ (the Greens function, $\frac{1}{-\Delta}$, for the Laplacian Δ). We change this first by blunting its singularity at x = y via the replacement

$$\frac{1}{4\pi|x-y|} \to \frac{1-e^{-\frac{|x-y|}{\lambda \ell_D}}}{4\pi|x-y|} = \left[\left(\frac{1}{-\Delta} - \frac{1}{-\Delta+\lambda^{-2}\ell_D^{-2}}\right)(x,y)\right]$$

where $\lambda, \ell_D > 0$ will be discussed below. Secondly we suppose the particles are inside a closed "grounded" box Λ which means that Dirichlet boundary conditions are imposed on the Greens function at $\partial\Lambda$. Thus our interaction energy is

$$u(x,y) \equiv \left(\frac{1}{-\Delta_\Lambda} - \frac{1}{-\Delta_\Lambda+\lambda^{-2}\ell_D^{-2}}\right)(x,y) \tag{1.1.1}$$

where Δ_Λ is the Laplacian with Dirichlet boundary conditions, on $\partial\Lambda$.

The partition function is

$$\tilde{Z} = \sum_{n=0}^{\infty}\sum_{m=0}^{\infty} \frac{z^n}{n!}\frac{z^m}{m!}\int_\Lambda d^{n+m}x e^{-\beta V} \quad . \tag{1.1.2}$$

There are n + particles with coordinates $x_1,\ldots,x_n \in \mathbb{R}^3$ and charges $e_1,e_2,\ldots,e_n = +1$ and m - particles with coordinates $x_{n+1},\ldots,x_{n+m} \in \mathbb{R}^3$, charges $e_{n+1},\ldots,e_{n+m}$ all -1. Each coordinate is integrated over Λ.

z is the "chemical activity" for both species. $\beta = \frac{e^2}{kT}$ and

$$V \equiv \tfrac{1}{2} \sum_{\substack{i,j \\ 1\leq i,j \leq n+m}} e_i e_j u(x_i,x_j) \quad . \tag{1.1.3}$$

<u>The Debye length</u>

$$\ell_D \equiv \frac{1}{\sqrt{2z\beta}} \tag{1.1.4}$$

is called the Debye length and will turn out to be the length scale for exponential decay in the correlation functions.

1.2 OBSERVABLES

We will for pedagogic reasons restrict outselves to the observable J(y) which measures the charge density at y, namely

$$J(y) \equiv \sum_{i\geq 1}^{n+m} e_i \delta(x_i - y)$$

on the sector with n+, m- particles. Thus the expected charge density at y is

$$\langle J(y) \rangle_\Lambda \equiv \frac{1}{Z} \sum_{n,m} \frac{z^n}{n!} \frac{z^m}{m!} \int_\Lambda e^{-\beta V} J(y) d^{n+m}x \quad .$$

1.3 THERMODYNAMIC LIMIT

Theorem 1.3.1: for $\lambda > 0$, z, $\beta \geq 0$ the infinite volume correlation function

$$\lim_{\Lambda \to \mathbb{R}^3} \langle \prod_{i=1}^{k} J(y_i) \rangle_\Lambda \equiv \langle \prod_{i=1}^{k} J(y_i) \rangle$$

exists for y_i non coincident. k is arbitrary.

This theorem follows from the correlation inequalities of Fröhlich and Park. See Fröhlich's lectures.

1.4 SCREENING

An example of this phenomenon is the next theorem

Theorem 1.4.1: fix λ, $0 < \lambda \leq \frac{1}{2}$, if $z\beta^3$ is sufficiently small depending on λ, there exist c, γ so that

$$\langle J(y_1) J(y_2) \rangle_\Lambda \leq c z^2 e^{-\gamma |y_1 - y_2| / \ell_D}$$

uniformly in Λ. $0 < \gamma < 1$, $y_1 \neq y_2$.

The point to notice, obviously, is that the correlation decays exponentially. Note that $\langle J(y) \rangle_\Lambda = 0$ by the "charge symmetry", (interchange of +,- particles).

The restriction $\lambda \leq \frac{1}{2}$ simplifies certain estimates but is not necessary.

1.5 GAUSSIAN MEASURES AND THE SINE GORDON TRANSFORMATION

Suppose $A \equiv (A_{ij})$ is a positive definite n×n matrix. The following formulas are well known and easy to prove:

$$\frac{1}{N}\int e^{-\frac{1}{2}\sum_{i,j}\phi_i A_{ij}\phi_j}\, e^{i\sum_j \phi_j f_j}\, d^n\phi = e^{-\frac{1}{2}\sum_{i,j} f_i A^{-1}_{ij} f_j}$$

$$\frac{1}{N}\int e^{-\frac{1}{2}\sum_{i,j}\phi_i A_{ij}\phi_j}\, \phi_i\phi_j d^n\phi = A^{-1}_{ij} \ .$$

A^{-1} is the matrix inverse. It is called the "covariance". N is the normalization implied by setting f = 0. In this section we will be concerned with infinite dimensional analogues of these formulas. In particular we wish to work with a measure dμ (on a function space) which heuristically is

$$d\mu_0(\phi) \equiv \text{"}\frac{1}{\text{Norm}}\, e^{-\frac{1}{2}\int_\Lambda \phi u^{-1}\phi dx} \prod_{x\varepsilon\Lambda} d\phi(x)\text{"} \ . \tag{1.5.1}$$

The expression in quotations makes no sense as it stands but if it did we would expect in analogy to the finite dimensional case that

$$\int d\mu_0(\phi) e^{i\int\phi(x) f(x) dx} = e^{-\frac{1}{2}\iint fuf} \ . \tag{1.5.2}$$

We refer the mathematically minded reader to Minlos' theorem (for example in [10]) to find out that there exists a measure (which we call $d\mu_0$) on the distribution space $S'(\mathbb{R}^3)$ satisfying (1.5.2) if f is in the test function space $S(\mathbb{R}^3)$ and supp f $\subset \Lambda$. Also

$$\int d\mu_0 \phi(f)\phi(g) = \iint fug \tag{1.5.3}$$

as would be expected from the finite dimensional analogy. (f,g ε $S(\mathbb{R}^3)$, supp f,g $\subset \Lambda$). u is called the "covariance of $d\mu_0$".

We will use the heuristic formula (1.5.1) as a basis for formal calculations which will serve to motivate various rigorously correct formulas (such as identities known as Cameron Martin formulas which tell you how to change variables in such functional gaussian measures). Some mathematical details which we do not supply can be found in [10].

A final important point is that since u^{-1} is a fourth order elliptic partial differential operator on functions of three variables, it turns out (see for example [11]) that $S'(\mathbb{R}^3)$ is unnecessarily large and in fact $d\mu_0$ can be restricted (losing only a set contained in a set of measure zero) to $C_0(\Lambda)$, the set of continuous functions on Λ vanishing at $\partial\Lambda$. Heuristically speaking, the $\int\phi u^{-1}\phi$ in (1.5.1) forces some regularity on ϕ.

In Fröhlich's lectures, the sine gordon transformation is discussed. We refer the reader to his lectures for a more detailed discussion of the formulas we are about to give. Applied to our situation, the transformation says

$$\tilde{Z} = \int d\mu_0(\phi) e^{2z\int_\Lambda \cos\beta^{1/2}\phi(x)\,dx} \tag{1.5.4}$$

[The integrand is in fact (exercise for the reader) the partition function for an ideal gas with two species with potential energies $+i\phi(x)$, $-i\phi(x)$ respectively from their interaction with an imaginary external field ϕ. Thus $\tilde{Z}$ has been represented as a superposition of ideal gases.]

Define

$$Z \equiv \tilde{Z}e^{-2z\,\mathrm{vol}(\Lambda)} \equiv \int d\mu_0(\phi) e^{2z\int_\Lambda(\cos\beta^{1/2}\phi - 1)} \;. \tag{1.5.5}$$

Expectations can also be transformed. For example if $y_1,\ldots,y_n$ are distinct points in Λ:

$$\left\langle \prod_{i=1}^{n} J(y_i)\right\rangle_\Lambda = \frac{1}{Z}\int d\mu_0(\phi) e^{2z\int_\Lambda(\cos\beta^{1/2}\phi - 1)} \prod_{i=1}^{n} \hat{J}(\phi(y_i)) \tag{1.5.6}$$

$$(\hat{J}(\phi(y)) \equiv 2iz\,\sin\beta^{1/2}\phi(y)).$$

There is a generalization of (1.5.4) which is worth discussing. Define

$$\tilde{Z}_f \equiv \sum_{n,m} \frac{z^n z^m}{n!m!} \int_\Lambda e^{-\beta V} e^{i\beta^{1/2}\sum_{j=1}^{n+m} f(x_j)} dx_1,\ldots,dx_{n+m} \;.$$

This has the physical interpretation of a partition function for a Coulomb system in an external imaginary electric potential if. The sine gordon transformation for this object is

$$\tilde{Z}_f = \int d\mu_0(\phi) e^{2z\int_\Lambda \cos\beta^{1/2}(\phi+f)} \;. \tag{1.5.4'}$$

This is a valuable formula because it allows us to compute expectations of charge densities as in (1.5.6). For example, here is how

one would find $\langle J(y)\rangle_\Lambda$: notice that

$$\sum_{j=1}^{n+m} e_j\, f(x_j) = \int \sum_{j=1}^{n+m} e_j\, \delta(y'-x_j) f(y')dy'$$

$$= \int f(y')J(y')dy' \equiv J(f)\ .$$

Therefore by the definition (above (1.5.4')).

$$\left\langle \int f(y')J(y')dy' \right\rangle_\Lambda = \tilde{Z}^{-1} \frac{1}{i\beta^{1/2}} \frac{d}{d\alpha} \tilde{Z}_{\alpha f}\Big|_{\alpha=0}\ .$$

Compute the right hand side using (1.5.4') and obtain $\langle J(y)\rangle_\Lambda$ by taking $f = \delta(y'-y)$ in the answer. The result will be (1.5.6) with $n = 1$. To recover (1.5.6) for $n > 1$ consider

$$\tilde{Z}^{-1} \left(\frac{1}{i}\frac{\partial}{\partial\alpha_1} \cdots \frac{1}{i}\frac{\partial}{\partial\alpha_n}\right) \tilde{Z}_{\alpha_1 f_1+\ldots+\alpha_n f_n}\ .$$

Particularly useful for our purposes is the formula

$$\langle J(f)J(g)\rangle_\Lambda = \frac{-1}{\beta}\frac{\partial}{\partial\alpha_1}\frac{\partial}{\partial\alpha_2} \log \tilde{Z}_{\alpha_1 f+\alpha_2 g}\Big|_{\alpha_1=\alpha_2=0}$$

$$= \frac{-1}{\beta}\frac{\partial}{\partial\alpha_1}\frac{\partial}{\partial\alpha_2} \log Z_{\alpha_1 f+\alpha_2 g}\Big|_{\alpha_1=\alpha_2=0} \qquad (1.5.6')$$

The dropping of the tilde means a change in normalization as in (1.5.5). This formula is easily derived by evaluating the derivatives and using the fact that $\langle J(f)\rangle = \langle J(g)\rangle = 0$ because of $\phi \to -\phi$ symmetry. This formula will be our avenue to a proof of Theorem 1.4.1.

We will in the rest of Section 1 be analysing the sine gordon formulas (1.5.5), (1.5.4') for the partition function. The idea is to discuss, rigorously and heuristically, the types of approximations which will lead to a proof of screening, i.e., Theorem 1.4.1.

A natural first attempt at obtaining screening is to make the approximation

$$2z[\cos\beta^{1/2}(\phi+f)-1] \simeq -\tfrac{1}{2}2z\beta(\phi+f)^2$$

$$\equiv -\frac{1}{2\ell_D^2}(\phi+f)^2$$

in Z_f, obtaining

$$Z_f \simeq \left(\int d\mu_0(\phi) e^{-\frac{1}{2\ell_D^2}\int\phi^2} e^{-\frac{1}{\ell_D^2}\int\phi f}\right) e^{-\frac{1}{2\ell_D^2}\int f^2}$$

$(\int \equiv \int_\Lambda)$.

The measure

$$d\mu(\phi) \equiv N^{-1} d\mu_0(\phi) e^{-\frac{1}{2\ell_D^2}\int_\Lambda \phi^2} \tag{1.5.7}$$

(where N normalizes $d\mu$) is gaussian and by considering the heuristic formula (1.5.1) we see that its covariance ought to be

$$(u^{-1} + \frac{1}{\ell_D^2})^{-1} \equiv C.$$

This conclusion is correct, i.e., (c.f., (1.5.2))

$$\int d\mu(\phi) e^{i\int\phi f} = e^{-\frac{1}{2}\int fCf} \quad .$$

We use these relations to find that

$$\log Z_f \simeq \log N + \frac{1}{2\ell_D^4} \int fCf - \frac{1}{2\ell_D^2} \int f^2$$

Now we take $f = \alpha_1 f + \alpha_2 g$ and compute as in (1.5.6') that

$$\langle J(f)J(g)\rangle_\Lambda \simeq \frac{-1}{\beta} \frac{\partial}{\partial\alpha_1} \frac{\partial}{\partial\alpha_2} \log Z_{\alpha_1 f+\alpha_2 g}\Big|_{\underset{\sim}{\alpha} = 0}$$

$$= \frac{-1}{\beta\ell_D^4} \int fCg + \frac{1}{\beta\ell_D^2} \int fg \quad .$$

We take limits $f(y) \to \delta(y-y_1)$, $g(y) \to \delta(y-y_2)$ and find for $y_1 \neq y_2$ that

$$|\langle J(y_1)J(y_2)\rangle_\Lambda| \simeq \frac{1}{\beta\ell_D^4} |C(y_1,y_2)|$$

$$\leq z^2\beta \frac{\text{const.}}{\lambda\ell_D} e^{-|y_1-y_2|} \quad .$$

The bound will be discussed in the next section. The essential point is that by taking the quadratic part of the cosine we have modified $d\mu_0$ so that it becomes gaussian with a "screened" covariance C.

The trouble with this approach is that it assigns a distinguished role to the cosine period centered on $\phi = 0$. It is true that the Dirichlet b.c.'s distinguish this period at the boundary $\partial\Lambda$ of Λ, but this effect does not persist strongly enough as $\Lambda \nearrow \mathbb{R}^3$ to use this approximation for obtaining results on correlation functions, (although it could be used to obtain asymptotic forms for the pressure analogous to the Debye Hückel expressions for the Free Energy).

It is worth noting that the periodicity of the integrand in (1.5.4) reflects the fact that the system we are considering contains particles whose charges are integral multiples of a fixed number, i.e. discrete commensurable charges. It is reasonable to expect that discrete charges will experience more difficulty screening than continuous charge distributions because a continuous distribution, by arranging itself in a spherically symmetric shell around a given fixed charge, can make it invisible from the outside by Newton's theorem. Discrete charges cannot do this. The periodicity of the cosine is connected with this difficulty. A mathematically interesting question which we can shed no light on is "can a system with non commensurable charges screen?". In this case the cosine would be replaced by an almost periodic function and parts of our methods will fail.

1.6 THE KERNEL OF C

The properties of the kernel of C which we have just been using, specifically (1.6.1) given below, are derived in outline.

$$0 \leq C(x,y) \leq \frac{\text{const.}}{\lambda \ell_D} e^{-|x-y|/\ell_D} \tag{1.6.1}$$

The most elementary way to obtain this is to note that by (1.1.1)

$$u^{-1} = \lambda^2 \ell_D^2 (-\Delta_\Lambda)(-\Delta_\Lambda + \frac{1}{\lambda^2 \ell_D^2})$$

and so

$$C \equiv \frac{1}{u^{-1} + 1/\ell_D^2} = \frac{1}{\lambda^2 \ell_D^2 (-\Delta_\Lambda)(-\Delta_\Lambda + \frac{1}{\lambda^2 \ell_D^2}) + \frac{1}{\ell_D^2}}$$

$$= \frac{1}{\lambda^2 \ell_D^2} \left(\frac{1}{-\Delta_\Lambda + r_1} - \frac{1}{-\Delta_\Lambda + r_2} \right) \left(\frac{1}{r_2 - r_1} \right)$$

for some r_1, r_2 obtained by partial fractions. The reader can check for himself that if $\lambda \le \frac{1}{2}$ r_2, r_1 are real, $r_2 > r_1$ and $r_1 > \frac{1}{\ell_D^2}$.

The required properties of C are easily read off from this formula because the kernel of $\frac{1}{-\Delta+r}$ is

$$e^{-\sqrt{r}|x-y|}\Big/(4\pi|x-y|)$$

and the kernel of $\frac{1}{-\Delta_\Lambda+r}$ is constructed from this by the method of images.

1.7 THE BASIC APPROXIMATION

If $z\beta^3$ is small, we claim that, under the $d\mu_0$ integral in (1.5.5), we can make the approximation (a Taylor expansion)

$$2z[\cos\beta^{\frac{1}{2}}\phi(x)-1] \simeq \frac{1}{2\ell_D^2}(\phi(x)-h(x))^2 \tag{1.7.1'}$$

where h(x) is a locally constant function with values in $2\pi\beta^{-\frac{1}{2}}\mathbb{Z}$. ($\mathbb{Z}$ ≡ set of integers). The notion of "locally constant" is quantified by introducing a lattice, called the "$L\ell_D$-lattice", in $\mathbb{R}^3$ with side $L\ell_D$. L is to be chosen later to make our expansions converge. The closed cubes of side $L\ell_D$ centered on this lattice are called $L\ell_D$-cubes and we will suppose Λ is a union of these cubes.

Instead of (1.7.1') we write, more precisely,

$$e^{2z\int_\Lambda(\cos\beta^{\frac{1}{2}}\phi-1)} \equiv \Big(\sum_h e^{-\frac{1}{2\ell_D^2}\int_\Lambda(\phi-h)^2}\Big)e^{G(\phi)} \tag{1.7.1''}$$

which defines G as an error which we will argue is small. h is summed over all functions on $\mathbb{R}^3$ which are constant inside $L\ell_D$-cubes, take values in $2\pi\beta^{-\frac{1}{2}}\mathbb{Z}$, are constant outside $\Lambda \cup \partial\Lambda$ and vanish at ∞. (Since Λ is simply connected this amounts to h = 0 outside $\Lambda \cup \partial\Lambda$. We word it in this complicated way in anticipation of replacing Λ by more complicated regions.)

It is perhaps not obvious that $z\beta^3 \simeq 0$ is the relevant criterion for the validity of (1.7.1'). Here is a preliminary argument in favor of this claim. Suppose $z\beta^3 \simeq 0$ and in addition $z\beta$ is fixed equal to $\frac{1}{2}$ or equivalently $\ell_D = 1$. In this case it follows

that β is very small and z is very large in such a way that the $O(z\beta^{3/2})$ in

$$2z(\cos\beta^{\frac{1}{2}}\phi-1) \simeq -\tfrac{1}{2}(\phi-h)^2 + O(z\beta^{3/2})$$

is very small. This suggests that it might be possible to prove theorem 1.4.1 with the additional restriction that $\ell_D = 1$. However, ℓ_D is a length whereas $z\beta^3,\lambda$ are dimensionless, which means that one can scale ℓ_D to be anything one likes without changing $\lambda, z\beta^3$. Observe that if $\tilde{Z}$ is rewritten in terms of the variable $x' = a^{-1}x$ then $z \to za^3$, $\beta \to \beta/a$, $\lambda \to \lambda$, $\ell_D \to \ell_D/a$, $\Lambda \to \Lambda/a^3$. Λ does not enter in Theorem 1.4.1 so you can change ℓ_D to anything you want which means that the restriction $\ell_D = 1$ is irrelevant.

1.8 GAUSSIAN INTEGRALS CONTINUED

Suppose for the moment that in (1.7.1") we make the (uncontrolled) approximation that G = 0, obtaining

$$Z \simeq \sum_h \int d\mu_0 e^{-\frac{1}{2}\ell_D^{-2}\int(\phi-h)^2} \tag{1.8.1}$$

Heuristically the integral is up to an (infinite!) normalization

$$\int \Pi d\phi(x) e^{-\frac{1}{2}\int\phi u^{-1}\phi}\, e^{-\frac{1}{2}\ell_D^{-2}\int(\phi-h)^2} \tag{1.8.2}$$

In complete analogy to finite dimensional gaussian integrals such integrals can be evaluated by translation:

$$\phi(x) = \psi(x) + g(x): \quad (g \in \mathcal{D}(u^{-1})) \tag{1.8.3}$$

We rewrite the integral in terms of ψ and obtain

$$\left(\int \Pi d\psi(x) e^{-\frac{1}{2}\int\psi u^{-1}\psi}\, e^{-\frac{1}{2}\ell_D^{-2}\int\psi^2}\, e^{-F_2}\right) e^{-F_1} \tag{1.8.4}$$

where (summarizing some simple algebra)

$$F_2 \equiv \int_\Lambda \psi u^{-1} g + 1/\ell_D^2 \int_\Lambda \psi(g-h) = \int_\Lambda \psi C^{-1}(g-g_c) \tag{1.8.5}$$

$$C^{-1} \equiv u^{-1} + 1/\ell_D^2 \tag{1.8.6}$$

$$g_c \equiv 1/\ell_D^2\, Ch$$

$$F_1 = \tfrac{1}{2}\ell_D^{-2}\int_\Lambda (g-h)^2 + \tfrac{1}{2}\int_\Lambda gu^{-1}g \tag{1.8.7}$$

We can rewrite (1.8.4) as

$$(\int d\mu_0(\psi)e^{-\frac{1}{2}\ell_D^{-2}\int\psi^2} e^{-F_2})e^{-F_1} \tag{1.8.8}$$

and then as

$$(N\int d\mu(\psi)e^{-F_2})e^{-F_1} \tag{1.8.9}$$

by absorbing the $e^{-\frac{1}{2}\ell_D^{-2}\int\psi^2}$ into $d\mu_0$ to create the gaussian measure $d\mu$ with covariance C which we discussed above in our first naive attempt. This whole analysis can be repeated even if $G \neq 0$ to obtain

$$Z = \sum_h e^{-F_1} N\int d\mu(\psi)e^{-F_2} e^{G(\phi)} \tag{1.8.10}$$

Let us pause to find out what (1.8.10) achieves. If $G(\phi)$ were indeed so small that a reasonable first approximation would be to neglect it, then by taking $g = g_c$ we would also have $F_2 = 0$ so that

$$Z \simeq (\sum_h e^{-F_1(g_c,h)})N \tag{1.8.11}$$

which means that the $d\mu(\psi)$ integral has been performed up to an overall (and unimportant) constant N. This "h" system is an important piece of the total system which we shall now try to understand by getting a picture of what $g_c = g_c(h)$ looks like and subsequently $F_1(g_c,h)$. We call g_c the "classical translation".

1.9 THE CLASSICAL TRANSLATION

We will explain some features of the classical translation $g_c(h)$ in its relationship with h. Basically, it is a smoothed out version of h which decays exponentially fast to h in regions where h is constant. We illustrate this by considering the special case

$$h(x) = 2\pi\beta^{-\frac{1}{2}} \qquad x \in X$$
$$= 0 \qquad \text{otherwise.}$$

X is a box which is a union of $L\ell_D$-cubes $\subset \mathbb{R}^3$. For simplicity we take $\Lambda = \mathbb{R}^3$ and replace Δ_Λ by Δ in the definition of C, i.e. drop the Dirichlet b.c.'s. The diagram below is drawn as if $\mathbb{R}^3$ were $\mathbb{R}$.

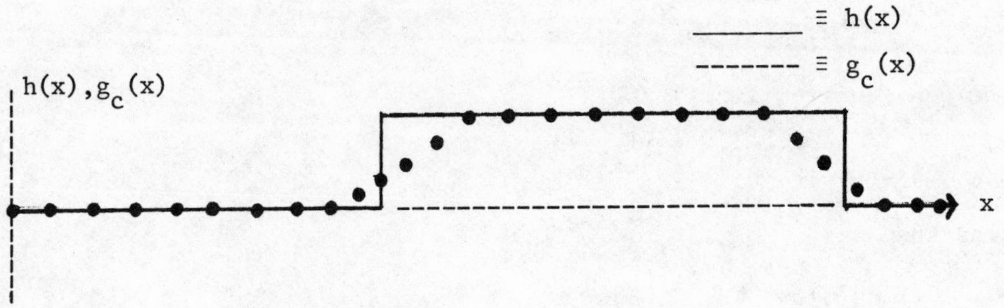

Comparison of $g_c(x)$ and $h(x)$

Or analytically, noting that the discontinuity set of $h(x)$, $\Sigma(h)$ is ∂X, we have

$$|g_c(x)-h(x)| \leq \text{const. } e^{-\text{dist}(x,\Sigma(h))/\ell_D} .$$

Proof: recall, that

$$g_c \equiv \frac{1}{\ell_D^2} \, Ch \equiv \frac{1}{\ell_D^2} \, \frac{1}{u^{-1}+\frac{1}{\ell_D^2}} \, h$$

u^{-1} is a fourth order differential operator which annihilates constants which implies that

$$\frac{1}{\ell_D^2} C1 = 1 \qquad \text{or} \qquad \int \frac{1}{\ell^2} C(x,y)dy = 1 \ \forall x$$

Let $\chi(x) = 1$ if $|x| < q$, 0 otherwise. Denote by $C\chi$ the operator with kernel $C(x,y)\chi(x-y)$. Let $x \in \mathbb{R}^3$ be at a distance greater than q from $\partial X = \Sigma(h)$. Then

$$g_c(x)-h(x) = \left(\frac{1}{\ell_D^2} Ch-h\right)(x)$$

$$=\left(\frac{1}{\ell_D^2} C\chi h-h\right)(x) + O(e^{-q/\ell_D})$$

(because $C(x,y) \leq \text{const. } \exp(-|x-y|/\ell_D)$)

$$= \left(\frac{1}{\ell_D^2} \int C\chi(x,y)dy\right)h(x)-h(x) + O(e^{-q/\ell_D})$$

(because $h(y) = \text{constant} = h(x)$ on supp. $\chi(x-y)$.)

$$= (1 + O(e^{-q/\ell_D}))h(x) - h(x) + O(e^{-q/\ell_D})$$
$$= O(e^{-q/\ell_D})$$

We have used the exponential decay of C and $\ell_D^{-2}\int C = 1$ as explained above. The proof is complete.

From this we also learn that the integrands of F_1 and F_2, when $g = g_c$, decay to zero exponentially fast away from $\Sigma(h)$.

We now describe an intuitive picture for the h system which ensuing estimates will justify.

A "typical" h configuration, h(x), is represented in the diagram. Inside $\partial\Lambda$ we have drawn "contours" which by definition are the connected components of the set $\Sigma(h)$ where h is discontinuous. If $z\beta^3 \simeq 0$ there will not be a high density of such contours in a typical h(x) because otherwise F_1 will be large which means that such configurations appear with a small weight in (1.8.11). The convergence of the expansions we describe in the next section will use the idea that if the h system is viewed as a "contour gas", then if $z\beta^3 \simeq 0$, it is very dilute: the contours are so far apart that their interaction (they are coupled by F_1 in (1.8.11)) is so small that contour configurations in regions A and B $\subset \Lambda$, separated by a large distance, are independently distributed up to errors exponentially small in dist(A,B). Indeed we will alter the classical translation slightly so that this exponential tail is replaced by an error that vanishes if dist(A,B) is larger than some fixed length.

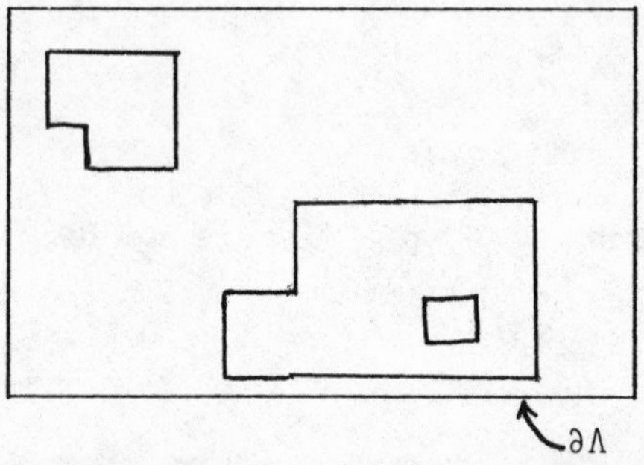

1.10 THE CHOICE OF g

$F_1 = F_1(g,h)$ is the "interaction" for our contour gas. If $g = g_c$, it falls off exponentially with distance, by which we mean that if $h = h_1 + h_2$ and $\Sigma(h) = \Sigma(h_1) \cup \Sigma(h_2)$ then

$$F_1(g_c(h),h) = F_1(g_c(h_1),h_1) + F_1(g_c(h_2),h_2) + O(e^{-\mathrm{dist}(\Sigma(h_1),\Sigma(h_2))/\ell_D}).$$

We alter $g_c(h)$ to a new translation $g_{L'}(h)$ picked so that the exponential tail vanishes if the distance between $\Sigma(h_1)$ and $\Sigma(h_2)$ is greater than $L'\ell_D$ where L' is a new parameter whose choice, which will be independent of λ, z, β, is deferred until Section 3. Thus we attempt to find a family of translations $g_{L'}(h)$ with the properties (focus on (3)).

(1) $g_{L'}(h) \varepsilon \mathcal{D}(u^{-1})$

(2) $g_{L'}(h)$ is linear in h

(3) If $h = h_1 + h_2$ with $\Sigma(h) = \Sigma(h_1) \cup \Sigma(h_2)$ and $\mathrm{dist}(\Sigma(h_1), \Sigma(h_2)) \geq L'\ell_D$ then

$$F_1(g_{L'}(h),h) = F_1(g_{L'}(h_1),h_1) + F_1(g_{L'}(h_2),h_2).$$

(4) $F_2(g_{L'}(h),h)$ is "small" if L' is large.

More precisely we require that for each $\varepsilon > 0$, $p \geq 0$ there exist L' so that

$$\int d\mu(\psi)\, e^{-pF_2(g_{L'}(h),h)} \leq Ce^{\varepsilon F_1(g_{L'}(h),h)}$$

uniformly in λ, z, β, L, h.

We require condition (4) in order that our hopefully "small" $G(\phi)$ not be drowned by a large F_2. We assert that it is possible to construct such translations $g_{L'}$. [We in fact achieve the additional property that $g_{L'}(x) = h(x)$ if x is further than $L\ell_D/4$ from $\Sigma(h)$. We shall not use this until Section 3.] A nice construction of such g's can be found in [5] section 7.

Anticipating our particular choice for L', $L' \gg 1$, we write

$$g \equiv g(h) \equiv g_{L'}(h)$$

and define the sphere of influence $\Sigma\hat{}(h)$ of the discontinuities of h:

$$\Sigma\hat{} \equiv \Sigma\hat{}(h) \equiv \bigcup_\alpha \Delta_\alpha \;:\; \mathrm{dist}(\Delta_\alpha, \Sigma(h)) < L'\ell_D.$$

Δ_α is an element of the set of "ℓ_D-cubes" $\{\Delta_\alpha\}$ which are closed cubes belonging to a lattice of side ℓ_D chosen so that Λ is a union of ℓ_D-cubes and each ℓ_D-cube is a union of the $L\ell_D$-cubes, $\{\Omega_\alpha\}$,

previously introduced. ($L \ll 1$ independent of z, β, λ).

Our contour gas can be made arbitrarily dilute by taking $z\beta^3$ small by virtue of the following lemma which shows that F_1 is large if h has many discontinuities.

Lemma 1.10.1: there exists $c = c(L)$ independent of λ, z, β so that

$$F_1(g,h) \geq c\ell_D \sum_{\alpha,\alpha'} (h_\alpha - h_{\alpha'})^2$$

where $h_\alpha \equiv$ the value of $h(x)$ in the interior of Ω_α and the sum is over all α,α' such that $\Omega_\alpha, \Omega_{\alpha'}$ are nearest neighbors.

For a proof of this lemma see the proof of Lemma 9.2 [5].

Since we will eventually fix L' independently of λ, z, β, and the density of the contour gas can be made arbitrarily small by taking $z\beta^3$ small, we can arrange that contour spheres of influence (connected components of $\Sigma^\wedge(h)$) hardly ever overlap so that the contour gas is essentially ideal.

We call connected components of $\Sigma^\wedge(h)$ "phase boundaries". They are obviously closely analogous to the Peierls contours of an Ising Model at low temperature.

Notation: From now on we will write

$$F_1 \equiv F_1(h),\ F_2 \equiv F_2(h) \equiv F_2(\psi,h)$$

because the translation g is assumed to be chosen according to the above criteria so that it is determined by h. F_1 and F_2 also depend on the parameter L' of course. We have not made this dependence explicit because the choice of L' will not depend on any of the system parameters z, β, λ.

1.11 LOCALITY

Definition 1.11: $R = R(\phi)$ is local (with respect to the ℓ_D-lattice $\{\Delta_\alpha\}$) if

$$R(\phi) = \prod_\alpha R(\phi, \Delta_\alpha)$$

where $R(\phi,\Delta_\alpha)$ is independent of $\phi(x)$ if $x \notin \Delta_\alpha$. For any such function we also define

$$R(X) \equiv R(X,\phi) \equiv \prod_{\alpha:\Delta_\alpha \subset X} R(\phi,\Delta_\alpha)$$

where X can be any union of ℓ_D-cubes.

For future reference we notice that e^{-F_2}, e^G are local functions. Define $F_2(X)$, $G(X)$ by

$$F_2(X) \equiv F_2(X,\psi,h) \equiv \int_X \psi u^{-1} g + \frac{1}{\ell_D^2} \int_X \psi(g-h) \tag{1.11.1}$$

$$e^{G_2(X)} \equiv e^{G_2(X,\phi)} \equiv \frac{e^{2z\int_X(\cos\beta^{\frac{1}{2}}\phi-1)}}{\sum_h \exp(-\frac{1}{2\ell_D^2}\int_X(\phi-h)^2)} \tag{1.11.2}$$

where X is any union of ℓ_D-cubes. h is (as indicated by the context) summed over all functions on $X \cup \partial X$, piecewise constant inside $L\ell_D$-cubes with values in $2\pi\beta^{-\frac{1}{2}}\mathbb{Z}$. Finally we will use $F_1(X,h)$ to denote the result of replacing Λ by X in (1.8.7).

2.0 CLUSTER EXPANSIONS

In this section we have tried to advertise the fact that high temperature expansions for lattice spin systems and cluster expansions used in constructive field theory can be developed from the same algebraic identities which come from an early paper by Spencer [9]. The Mayer expansion can also be developed in this way [12]. The basic object is to decompose a function depending on many variables into sums of products of functions depending on smaller subsets of variables called "clusters". One has great freedom in how to choose these "clusters" and therein lies the multiplicity of expansions. In the course of this we will develop the expansion for the Coulomb system.

We shall make the conventions that empty products are identically 1, empty sums, 0. In our analysis will appear many sums which are apparently infinite but the reader will be able to check for himself that in fact until Section 2.6, all such sums have only finitely many non zero terms. In this sense our manipulations are entirely algebraic until Section 2.6.

2.1 URSELL FUNCTIONS

The general principle behind all cluster expansions for systems with two body forces is to expand the Boltzman factor

$$\Psi(X) \equiv \exp\Big(\sum_{i<j,\, i,j \in X} v_{ij}\Big) \tag{2.1.1}$$

which occurs in the partition function according to

$$\begin{aligned}\Psi(X) &= \prod_{ij,\, i,j \in X} (e^{v_{ij}} - 1 + 1) \\ &= \sum_{G \text{ on } X} \prod_{ij \in G} (e^{v_{ij}} - 1)\end{aligned} \tag{2.1.2}$$

where ij denotes an ordered pair i,j with $i < j$, G on X means that G is summed over all graphs on X. A graph on X is a subset of $\{ij\colon i \in X,\ j \in X\}$. The pairs ij are referred to as "lines" or "bonds" in G.

We shall begin by assuming only that we have been presented with a set of numbers $\{v_{ij}\colon i = 1,2,\ldots,\ j = 1,2,\ldots\}$, $v_{ij} \in \mathbb{R}$. We supplement (2.1.1) by the conventions

$$\Psi(\phi) \equiv 1; \quad \Psi(\{i\}) \equiv 1, \qquad i = 1,2,\ldots$$

In a fairly standard way we use the set of equations

$$U(\{1\}) \equiv 1, \quad \Psi(\Lambda) = \sum_{\substack{X, X \subset \Lambda \\ 1 \in X}} U(X)\, \Psi(\Lambda \sim X) \tag{2.1.3}$$

to define the "connected parts", $U(X)$, of $\Psi(X)$ in terms of $\Psi \equiv (\Psi(X))$. (There is an equation for each choice of $\Lambda \subset \{1,2,\ldots\}$. The equations determine the $U(X)$ for all X recursively; $X, \Lambda \ni 1$).

<u>Example</u>. By taking $\Lambda = \{1,i\}$, the reader will see that $U(\{1,i\}) = e^{v_{1i}} - 1$. By taking $\Lambda = \{1,2,3\}$ he will obtain

$$U(\{1,2,3\}) = e^{v_{12}+v_{13}+v_{23}} - e^{v_{12}} - e^{v_{13}} - e^{v_{23}} + 2.$$

These equations can be solved for $U \equiv (U(X))$, the solution is, as we shall prove below,

$$U(X) = \sum_{G \in C(X)} \prod_{ij \in G} (e^{v_{ij}} - 1) \tag{2.1.4}$$

where $C(X)$ is the set of all graphs on X, <u>connected</u> on X. G is connected on X means that any two elements of X can be joined by a path made up of bonds in G.

<u>Proof of (2.1.4)</u>

We have by (2.1.2) that

$$\Psi(\Lambda) = \sum_{G \text{ on } \Lambda} I_G \,; \qquad I_G \equiv \prod_{ij \in G} (e^{v_{ij}} - 1)$$

We perform the sum over G in a special order. First hold $X \subset \Lambda$, $1 \in X$, fixed. (Without loss of generality we suppose 1 is the first element in Λ). We also fix a graph g on X, connected on X. Sum over all graphs G on Λ whose connected component which contains the vertex 1 is g. The result of this summation is $\Psi(\Lambda \sim X)$ by (2.1.1) with X replaced by $\Lambda \sim X$. Next sum over g and then X to obtain

$$\Psi(\Lambda) = \sum_{\substack{X \subset \Lambda \\ 1 \in X}} \Big(\sum_{g \in C(X)} I_g \Big) \Psi(\Lambda \sim X)$$

Comparison with (2.1.3) completes the proof.

2.2 EXPANDING Ψ INTO "CLUSTERS"

We can expand $\Psi(\Lambda)$ for any Λ in terms of these functions $U(X)$ by iterating (2.1.3). In more detail, expand the factor $\Psi(\Lambda \sim X)$ on the right hand side of (2.1.3) using (2.1.3) itself

but with Λ replaced by $\Lambda \sim X$. Repeat this process until there are no more factors of the form $\Psi(Y)$ with $Y \neq \emptyset$ to expand. At each stage one must choose a first element to play the role of the "1" in (2.1.3). This is most easily done by using the given order on $\Lambda \subset \{1,2,\ldots\}$. The result is

$$\Psi(\Lambda) = \sum_{n=0}^{\infty} \sum_{\substack{X_1,\ldots,X_n \subset \Lambda \\ X_1 \leq X_2 \leq \cdots \leq X_n,\ X_i \cap X_j = \emptyset\ \forall ij}} \prod_{i=1}^{n} U(X_i)$$

where $X \leq Y$ means that the first element in X is less than the first element in Y. This is more neatly written as

$$\Psi(\Lambda) = \sum_{n=0}^{\infty} \frac{1}{n!} \sum_{X_1,\ldots,X_n \subset \Lambda} \prod_{i=1}^{n} U(X_i) \tag{2.2.1}$$

("cluster decomposition")

where the sum is over all $X_i \subset \Lambda$, $i = 1,\ldots,n$, such that $X_i \cap X_j = \emptyset$ for all i,j.

2.3 THE TREE GRAPH FORMULA

We owe the ideas for this important identity to an early paper by Spencer on the Cluster expansion for $P(\phi)_2$ [9]. O. Penrose discovered related formulas [13].

By relabelling if necessary we suppose that $X = \{1,2,\ldots,n\}$. We introduce $n - 1$ parameters $\underline{s} \equiv s_1,\ldots,s_{n-1}$ and define

$$\begin{aligned} v_{ij}(\underline{s}) &\equiv s_i \cdots s_{j-1} v_{ij} \\ v_{ij}'(\underline{s}) &\equiv (s_i \cdots s_{j-1})' v_{ij} \equiv \frac{d}{ds_{j-1}} (s_i \cdots s_{j-1}) v_{ij} \end{aligned} \tag{2.3.1}$$

Recall that $i < j$. A "tree graph" T on X is a set of bonds $\{ij\}$ of the form $\{(\eta(j),j): j = 2,\ldots,n\}$ where η is a map $\eta: \{2,\ldots,n\} \to \{1,2,\ldots,n-1\}$ such that $\eta(j) < j$. We give this pedantic definition because this is not the standard way to define a tree graph in that, for example,

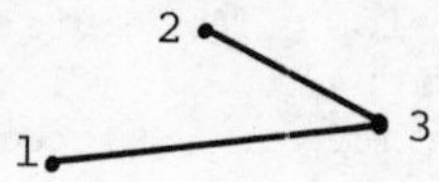

is not a tree graph according to this definition. Combinatorialists call our trees "label increasing trees". Given a function $f \equiv f(1,2,\ldots,n)$ of n "objects" (to be the subscripts on v_{ij}'s) let

$$Sf \equiv \frac{1}{n} \sum_{\pi} f(\pi(1), \pi(2), \ldots, \pi(n)) \tag{2.3.2}$$

where π is summed over all permutations of $\{1,2,\ldots,n\}$. (The normalization is weird and deliberate). Finally set

$$\Psi(X, \underline{s}) \equiv e^{\sum_{ij,\, i,j \in X} v_{ij}(\underline{s})} \tag{2.3.3}$$

<u>Lemma 2.3.1</u> (Tree graph Lemma):

$$U(X) = \sum_{T \text{ on } X} \int d\underline{s} \; S\left\{ \prod_{ij \in T} v'_{ij}(\underline{s}) \; \Psi(X, \underline{s}) \right\}$$

$\underline{s}$ is integrated over $[0,1]^{n-1}$. S permutes subscripts on v not $\underset{\sim}{s}$

Despite its complicated appearance, this is a most powerful identity. This will become apparent in the rest of this article, but there are two basic reasons. The first point is that it is a partial resummation of (2.1.4) in that many of the graphs of (2.1.4) are partly in $\Psi(X, \underline{s})$. If we have "stability bounds", i.e.,

$$\sum_{ij,\, i,j \in Y} v_{ij} \leq c|Y| \qquad \forall\, Y \subset \{1,2,\ldots\} \tag{2.3.4}$$

$|Y|$ is the number of elements in Y. Then these bounds are inherited by $v_{ij}(\underline{s})$, i.e. with the same constant

$$\sum_{ij,\, i \in Y,\, j \in Y} v_{ij}(\underline{s}) \leq c|Y| \tag{2.3.5}$$

so that

$$0 \leq \Psi(X, \underline{s}) \leq e^{c|X|} = e^{cn} \tag{2.3.6}$$

An explanation for (2.3.5) will be given in the proof of Lemma 2.3.1 in Appendix A. The next point is that

$$\sum_{T} \int d\underline{s} \prod_{ij \in T} (s_i \ldots s_{j-1})' \leq e^{n-1} \tag{2.3.7}$$

which will also be explained below. The estimate (2.3.7) shows that although the number of tree graphs is $(n-1)!$, the weighting by factors $s_1, \ldots,$ reduces this potential factorial divergence in n to something geometric in n which we will be able to control by arranging that the v_{ij}'s are small in a suitable sense. The

symmetrization, S, looks as if it will lead to another divergence, factorial in n, but it turns out not to be a problem. The reason is unfortunately buried in Section 3, but it has to do with the fact that for most permutations site $\pi(i)$ and site $\pi(j)$ are far separated which leads to small factors by virtue of conditions (not yet imposed) on v_{ij}'s.

The proof of the tree graph formula involves a considerable amount of notation which finds no other use in this discussion, so we postpone it to Appendix A.

We close this subsection by giving the proof of (2.3.7). We shall not be using this until Section 3 so it can be omitted on a first reading.

Proof of (2.3.7)

We shall actually show that

Lemma 2.3.2: If $A_1, \ldots, A_{n-1} \geq 0$,

$$\sum_T \int d\underline{s} \prod_{ij \in T} (s_i \ldots s_{j-1})' A_i \leq e^{A_1 + A_2 + \ldots + A_{n-1}}$$

where $\underline{s} = (s_1, \ldots, s_{n-1})$, T is summed over all tree graphs on $\{1,2,\ldots,n\}$.

If $A_1, \ldots, A_{n-1} = 1$ this reduces to (2.3.7).

Proof

The left hand side is smaller than

$$\sum_T \int d\underline{s}' \prod_{ij \in T} (s_i \ldots s_{j-1})' A_i \, e^{\sum_{i=1}^{n-1} s_i \ldots s_{n-1} A_i} \tag{2.3.8}$$

The proof is inductive. We begin by estimating the integral over s_{n-1}. We use T_{n-1} to denote a tree graph on $\{1,\ldots,n-1\}$ and observe that for any $F = F(T)$,

$$\sum_T F(T) = \sum_{T_{n-1}} \sum_{i=1}^{n-1} F(T_{n-1}, in)$$

because each tree graph T on $\{1,\ldots,n\}$ decomposes into a tree graph T_{n-1} on $\{1,\ldots,n-1\}$ and an extra bond "in". Also the

integrand in (2.3.8) factors into a part that involves s_{n-1} and a part that does not. In particular consider the expression (which is part of (2.3.8))

$$\sum_{i=1}^{n-1} \int ds_{n-1} (s_i \ldots s_{n-1})' A_i \, e^{(\sum_{i=1}^{n-1} s_i \ldots s_{n-1} A_i)}$$

$$= \int_0^1 ds_{n-1} \frac{d}{ds_{n-1}} (\sum_{i=1}^{n-1} s_i \ldots s_{n-1} A_i) \, e^{\sum_{i=1}^{n-1} (s_i \ldots s_{n-1}) A_i}$$

$$= e^{\sum_{i=1}^{n-2} s_i \ldots s_{n-2} A_i + A_{n-1}} - 1$$

$$\leq (e^{\sum_{i=1}^{n-2} s_i \ldots s_{n-2} A_i}) \, e^{A_{n-1}} .$$

Substituting this bound into (2.3.8) gives the upper bound

$$e^{A_{n-1}} \sum_{T_{n-1}} \int ds_1 \ldots ds_{n-2} \prod_{ij \in T_{n-1}} (s_i \ldots s_{j-1})' A_i \, e^{\sum_{i=1}^{n-2} s_i \ldots s_{n-2} A_i}$$

so that the same argument applies again with n replaced by $n - 1$.

QED.

2.4 AN APPLICATION TO FUNCTIONAL INTEGRALS

From Section 1 we have obtained functional integrals of the form

$$\int d\mu(\psi) \prod_{i=1}^{N} Q(Y_i); \quad (Q(Y) \equiv e^{G(Y)} \, e^{-F_2(Y)}) \qquad (2.4.1)$$

(e.g. see (1.8.10)). We are going to show how the analysis of cluster expansions just completed applies to such integrals. For this purpose we will be using the following assumptions:

(a) $d\mu$ is a Gaussian measure on the space of continuous functions on $\Lambda \subset \mathbb{R}^3$. (We continue to call its covariance C but it does not have to be the covariance of Section 1).

(b) There exist sets $Y_1,\dots,Y_N \subset \Lambda$ which have disjoint interiors and whose union is Λ. Some regularity condition (such as being a union of ℓ_D-cubes) must also be imposed. The functionals $Q(Y_i)$, $i = 1,2,\dots,N$, do not depend on $\Psi(x)$ for $x \notin Y_i$.

(c) We shall assume that $Q(Y_i)$ is a polynomial in Ψ, $(i = 1,\dots,N)$. A polynomial is a finite linear combination of monomials. Monomials have the general form

$$\int g(x_1,\dots,x_k)\,\Psi(x_1)\,\dots\,\Psi(x_k)\,dx_1\,\dots\,dx_k$$

with $g \in C^\infty(\Lambda^k)$, symmetric in $x_1,\dots,x_k$.

Assumption (c) clearly excludes the cases we are interested in. It will be clear that the formulas we derive will make sense for the integrands in Section 1, but we shall omit the approximation arguments that are needed to justify our applying the formulas we derive below to these integrands.

The artifice that allows us to apply our previous analysis to these integrals is the following formula for evaluating such integrals exactly, which we give for the one dimensional analogue first

$$\frac{1}{\text{Norm.}}\int e^{-\frac{1}{2}a\Psi^2}\,P(\Psi)d\Psi = e^{\frac{1}{2}a^{-1}\frac{d^2}{d\Psi^2}}\,P(\Psi)\Big|_{\Psi=0}$$

Ψ is a one dimensional variable, $a > 0$, P is a polynomial; the exponential of a derivative is defined by its formal power series which truncates to finitely many terms if applied to a polynomial and thus makes sense as an operator on the algebra of polynomials. The functional integral analogue is

<u>Theorem 2.4.1 (Wick's Theorem)</u>

$$\int d\mu(\Psi)\,P(\Psi) = e^{\frac{1}{2}O}\,P\Big|_{\Psi=0}$$

$$O \equiv \int_\Lambda\int_\Lambda {}^{\delta}/\delta\Psi(x)\;C(x,y)\;{}^{\delta}/\delta\Psi(y)\;dxdy$$

and the functional derivative ${}^{\delta}/\delta\Psi(x)$ is defined by giving the action on monomials, which is just as one would expect:

$${}^{\delta}/\delta\Psi(y)\int g(x_1,\dots,x_k)\,\Psi(x_1)\,\dots\,\Psi(x_k)\,d^kx$$

$$= k\int g(y,\,x_1,\dots,x_{k-1})\,\Psi(x_1)\,\dots\,\Psi(x_{k-1})\,d^{k-1}x$$

(recall g is symmetric). These formulas are a "Euclidean" version of Wick's theorem. We shall omit the proofs, but the one dimensional version is not difficult to prove.

We decompose the operator O according to

$$O = \sum_{i,j} O_{ij} \; ; \; O_{ij} \equiv \int_{Y_i} \int_{Y_j} {}^{\delta}/\delta\Psi \, C \, {}^{\delta}/\delta\Psi \tag{2.4.2}$$

As operators on the polynomial algebra, the O_{ij} commute so that

$$\int d\mu \, P = e^{\sum_{i<j} O_{ij}} e^{\frac{1}{2}\sum_i O_{ii}} P\Big|_{\Psi=0} \tag{2.4.3}$$

The first factor on the right hand side looks like the Boltzman factor $\Psi(\Lambda)$ (cf. (2.1.1)). A little thought shows that (2.1.3), the cluster decomposition (2.2.1), and the tree graph lemma, will remain valid as identities on the polynomial algebra if v_{ij} is replaced by O_{ij}, $i < j$. (Note that $\exp(\frac{1}{2}\Sigma O_{ii})P$ is a polynomial), so therefore by (2.2.1)

$$\int d\mu \prod_i Q(Y_i) = \sum_{n>0}^{\infty} \frac{1}{n!} \sum_{X_1,\ldots,X_n} \prod_{i=1}^{n} U(X_i) \cdot \tag{2.4.4}$$

$$\cdot \prod_i e^{\frac{1}{2} O_{ii}} \prod_i Q(Y_i)\Big|_{\Psi=0}$$

where although in the cluster decomposition (2.2.1) the sets X_i were subsets of $\{1,2,\ldots,n\}$, we are now identifying X_i with the subset of Λ obtained by taking the union of Y_j's labelled by the elements of X_i. Thus $X_1,\ldots,X_n$ are summed over all sets of n subsets of Λ which are each unions of Y_j's, $j = 1,\ldots, N$ and

$$\bigcup_{i=1}^{n} X_i = \Lambda, \qquad \dot{X}_i \cap \dot{X}_j = \emptyset \quad \text{all } i,j$$

$\dot{X} \equiv$ the interior of X. Each $U(X_i)$ is now of course an operator, but by the tree graph lemma, with v_{ij} replaced by O_{ij}, $U(X)$ only involves differentiations ${}^{\delta}/\delta\Psi(x)$ with $x \in X$. This means that we can reorganize as follows

$$\prod_{i=1}^{n} U(X_i) \prod_i e^{\frac{1}{2} O_{ii}} \prod_i Q(Y_i)\Big|_{\Psi=0}$$

$$= \prod_{i=1}^{n} [U(X_i)\, e^{\frac{1}{2}\sum_{k \in X} O_{kk}}\, Q(X_j)]_{\Psi=0} \tag{2.4.5}$$

($Q(X) \equiv \prod_{i \in X} Q(Y_i)$; $k \in X$ means $Y_k \subset X$)

because the $Q(Y_i)$'s only depend on $\psi(x)$ for $x \in Y_i$ by Assumption (b).

By appealing to the tree graph formula (with v_{ij} replaced by O_{ij}) we see that

$$U(X)\, e^{\frac{1}{2}\sum_{k \in X} O_{kk}}\, Q(X)\Big|_{\Psi=0} \tag{2.4.6}$$

($X = X_i$, some i) contains a factor

$$e^{\sum_{i<j} O_{ij} s_i \ldots s_{j-1}}\, e^{\frac{1}{2}\sum_i O_{ii}}\, Q(X)\Big|_{\Psi=0} \tag{2.4.7}$$

where $i,j \in X$. Since $Q(X)$ does not depend on $\Psi(x)$ for $x \notin X$ we can insert in (2.4.7) a further operator

$$\exp\Big(\frac{1}{2}\sum_{i,j \notin X} O_{ij}\Big)$$

without changing anything. In all we then have an exponent of the form

$$\sum_{\substack{i<j \\ i,j \in X}} O_{ij} s_i \ldots s_{j-1} + \frac{1}{2}\sum_{i \in X} O_{ii} + \frac{1}{2}\sum_{i,j \notin X} O_{ij}$$

$$\equiv \frac{1}{2}\int_\Lambda\int_\Lambda {}^{\delta}/\delta\Psi(x)\; C(x,y,\underline{s})\; {}^{\delta}/\delta\Psi(y) \tag{2.4.8}$$

(which defines $C(x,y,\underline{s})$). In Appendix A we show that $C(x,y,\underline{s})$ inherits the positivity properties of $C(x,y)$ so that by Minlos' theorem, there exists a Gaussian measure $d\mu_{\underline{s}}$ with covariance $C(x,y,\underline{s})$. Substitute into (2.4.6) using the tree graph formula, rewrite the resulting Boltzmann factor $\exp(\frac{1}{2}\iint \frac{\delta}{\delta\Psi} C(\underline{s}) \frac{\delta}{\delta\Psi})$ using Wick's Theorem 2.4.1 (in reverse) with $d\mu$ replaced by $d\mu_{\underline{s}}$, obtaining

$$\sum_{T} \int d\underline{s}\, S \{ \int d\mu_{\underline{s}} \prod_{ij \in T} (s_i \ldots s_{j-2}) O_{ij} \} Q(X) \equiv K(X) \qquad (2.4.9)$$

This, together with (2.4.4) rewritten as

$$\int d\mu \prod_{i=1}^{N} Q(Y_i) = \sum_{n=0}^{\infty} \frac{1}{n!} \sum_{X_1,\ldots,X_n} K(X_1) \cdots K(X_n) \qquad (2.4.10)$$

(where the sum is over all sets X_i, $i = 1,\ldots,n$, which are unions of sets Y_i, have disjoint interiors and exhaust Λ) is the cluster expansion for functional integrals which we promised.

We close this subsection by giving an explicit formula for $C(x,y,\underline{s})$. It should be quite clear from (2.4.8) how it comes about.

<u>The Kernel $C(x,y,\underline{s})$</u>

Given $Y_1,\ldots,Y_n$ and $\underline{s} = (s_1,\ldots,s_{n-1})$ then for $i = 1,\ldots,n+1$, $j = 1,\ldots,n+1$, $i \leq j$,

$$C(x,y,\underline{s}) = s_i \ldots s_{j-1}\, C(x,y) \text{ if either } \begin{cases} x \in Y_i, & y \in Y_j \\ & \text{or} \\ y \in Y_i, & x \in Y_j \end{cases} \qquad (2.4.11)$$

with the convention that $Y_{n+1} \equiv \Lambda \sim \bigcup_{i=1}^{n} Y_i$ and $s_n \equiv 0$.

2.5 MULTIPHASE EXPANSIONS

We have just been discussing an expansion suitable for the Gaussian integrals in

$$Z = \sum_{h} e^{-F_1(h)} N \int d\mu(\Psi)\, e^{-F_2(h)} e^{G(\phi)} \qquad (2.5.1)$$

which is where (c.f. (1.8.10)) Section 1 left off. We will now obtain an expansion for Z by combining what we now know about cluster expansions with a clever choice of the partition $Y_1,\ldots,Y_n$ on which the cluster expansion of Section (2.4) was based.

The choice of Y_i's

We are still free to choose the sets Y_i which partition Λ in the cluster expansion (2.4.10). In single phase expansions where there is no sum over h all the Y_i's can be taken to be cubes from a lattice whose side is of the same order as the exponential decay length of the $d\mu$ measure, ℓ_D-cubes in our case. The new idea in the multiphase expansion is to make the choice of partition $\{Y_1,\ldots,\} \equiv \overline{y}$ dependent on the h configuration.

Λ is the union of $\Sigma^\wedge$ (the set of points near discontinuities of h defined in Section 1.10) and $\Lambda \sim \dot{\Sigma}^\wedge$. $\Sigma^\wedge$ is partitioned into its connected components called "phase boundaries"; $\Lambda \sim \dot{\Sigma}^\wedge$ is partitioned into the ℓ_D-cubes it contains. The partition $\overline{y} \equiv \overline{y}(h)$ consists of the sets $Y_1,\ldots,Y_{N(h)}$ which are either phase boundaries or cubes in $\Lambda \sim \dot{\Sigma}^\wedge$.

Substituting our expansion (2.4.10) into (2.5.1) gives

$$Z = N \sum_{n=0}^{\infty} \frac{1}{n!} \sum_{h} \sum_{X_1,\ldots,X_n} e^{-F_1(h)} K(X_1,h)\ldots K(X_n,h) \qquad (2.5.2)$$

We write $K(X,h)$ for $K(X)$, defined in (2.4.9) because in this application the integrand $Q(X)$ will be h dependent. There is a further h dependence (not explicit) in the range of the sum over $(X_i)_{i=1,\ldots,n}$, beacause each X_i is a union of sets belonging to $\{Y_1,\ldots,Y_{N(h)}\} \equiv \overline{y}(h)$.

We invert the sum over (X_i) and h. For (X_i) fixed, the constraint on h that each X_i be a union of sets from $\overline{y}(h)$ says that h can be decomposed so that

$$h = h_1 + \ldots + h_n \quad \text{s.t} \quad \Sigma^\wedge(h_i) \subset X_i, \; i = 1,\ldots,n \qquad (2.5.3)$$

the decomposition is unique since as usual we require that $h_i(x) = 0$ if $x \notin \Lambda \cup \partial\Lambda$. Thus we can write (for (X_i) fixed)

$$\sum_h e^{-F_1(h)} K(X_1,h) \ldots K(X_n,h) = \sum_{h_1,\ldots,h_n} e^{-F_1(h_1)} \ldots e^{-F_n(h_n)} K(X_1,h)\ldots K(X_n,h) \qquad (2.5.4)$$

F_1 factors because of Property 3 in Section 1.10. Moreover for

each i, $K(X_i,h) = K(X_i,h_i)$. We leave the reader to check this for himself using (a) $G(\emptyset) \equiv G(\Psi + g) = G(\Psi + g - h)$ by periodicity. (b) Property (3) in Section 1.10 implies that if $\text{dist}(x,\Sigma(h)) \geq L'\ell_D$, $g(x) = h(x)$ which in turn implies that $g(h) - h\big|_{X_i} = g(h_i) - h_i\big|_{X_i}$. Thus we define

$$\tilde{K}(X) \equiv \sum_{h:\ \Sigma^{\wedge}(h) \subset X} e^{-F_1(h)} K(X,h) \tag{2.5.5}$$

and obtain

$$Z = N \sum_{n=0}^{\infty} \frac{1}{n!} \sum_{X_1,\ldots,X_n} \tilde{K}(X_1) \cdots \tilde{K}(X_n) \ . \tag{2.5.6}$$

In (2.5.6), the sum over $X_1,\ldots,X_n$ is constrained by $\cup X_i = \Lambda$. By REDEFINING $\tilde{K}(X)$ to be what was formerly $\tilde{K}(X) - 1$ whenever X is an ℓ_D-cube, we recover (2.5.6) without this constraint. [Since $X_1,\ldots,X_n$ are still required to have disjoint interiors, Z looks like the grand canonical partition function of a hard core gas].

Notice that we have just used periodicity which is a consequence of the charges in our system being commensurable. Our proof of screening gets into difficulties without this assumption.

2.6 AN EXPANSION FOR THE PRESSURE

We are really interested in obtaining an expansion for $\log Z$ because from this we will calculate the pressure. Also by taking suitable (functional) derivatives we obtain truncated correlation functions as was outlined in Section 1 for the correlation of two charge densities.

We define

$$Z(\xi) = N \sum_{n=0}^{\infty} \frac{\xi^n}{n!} \sum_{X_1} \cdots \sum_{X_n} \tilde{K}(X_1)\ldots\tilde{K}(X_n) \prod_{ij} (1 - \chi_{ij}) \tag{2.6.1}$$

We have dropped the constraints $\dot{X}_i \cap \dot{X}_j = \emptyset$ from the sum and reinserted it by means of the factors $1 - \chi_{ij}$:

$$\chi_{ij} \equiv \chi(X_i \cap X_j); \quad \chi(X) = \begin{cases} 0 & \text{if } \dot{X} = \emptyset \\ 1 & \text{otherwise} \end{cases} \tag{2.6.2}$$

If $\xi = 1$, $Z(\xi) = Z$. We shall obtain $\log Z$ by integrating, with respect to ξ, an expansion we are about to find for $Z^{-1}(\xi) \frac{d}{d\xi} Z(\xi)$.

We have

$$\frac{d}{d\xi} Z(\xi) \equiv Z'(\xi) = N \sum_{n=1}^{\infty} \frac{\xi^{n-1}}{(n-1)!} \sum_{X_1} \cdots \sum_{X_n}$$
$$\tilde{K}(X_1) \ldots \tilde{K}(X_n) \prod_{ij} (1 - \chi_{ij}) \tag{2.6.3}$$

We want to reorganize this in such a way that division by $Z(\xi)$ becomes trivial. To do this we follow [14] at least in spirit. The idea is that it would be trivial without the Boltzmann factor $\pi(1 - \chi_{ij})$ being present so we expand it into its connected part. We return to (2.1.3) and, letting $-\chi_{ij}$ play the role of $(e^{v_{ij}} - 1)$, write

$$\prod_{\substack{ij \\ 1 \leq i,j \leq n}} (1 - \chi_{ij}) = \sum_{\substack{1 \in C \\ C \subset \{1,2,\ldots,n\}}} U(C) \left(\prod_{\substack{ij \\ i,j \notin C}} (1 - \chi_{ij}) \right) \tag{2.6.4}$$

Then

$$Z'(\xi) = N \sum_{n=1}^{\infty} \frac{\xi^{n-1}}{(n-1)!} \sum_{C} \left[\sum_{(X_i)_{i \in C}} U(C) \prod_{i \in C} \tilde{K}(X_i) \right] \circ$$
$$\circ \left[\sum_{(X_i)_{i \notin C}} \prod_{i \notin C} \tilde{K}(X_i) \prod_{\substack{ij \\ i,j \notin C}} (1 - \chi_{ij}) \right] \tag{2.6.5}$$

The terms in square brackets depend on C only through its cardinality $|C| \equiv k$, therefore we can evaluate them on a "standard" C, $\{1,2,\ldots,k\}$, and rewrite the sum over C according to

$$\sum_{1 \in C \subset \{1,2,\ldots,n\}} (\cdot) = \sum_{k=1}^{n} \binom{n-1}{k-1} (\cdot)$$

We denote by $K(k)$ the first square bracket in (2.6.5) evaluated on $C = \{1,2,\ldots,k\}$, i.e.

$$\tilde{\tilde{K}}(k) \equiv \sum_{X_1} \cdots \sum_{X_k} U(\{1,\ldots,k\}) \, \tilde{K}(X_1) \ldots \tilde{K}(X_k) \tag{2.6.6}$$

Then we obtain

$$Z'(\xi) = N \sum_{n=1}^{\infty} \sum_{k=1}^{n} \frac{\xi^{k-1}\xi^{n-k}}{(n-1)!} \frac{(n-1)!}{(n-k)!(k-1)!} \tilde{\tilde{K}}(k) \cdot \tag{2.6.7}$$

$$\cdot \left[\sum_{X_{k+1}} \cdots \sum_{X_n} \tilde{K}(X_{k+1}) \cdots \tilde{K}(X_n) \prod_{\substack{ij \\ k+1 \leq i,j \leq n}} (1 - \chi_{ij}) \right]$$

We invert the k, n sums:

$$= N \sum_{k=1}^{\infty} \frac{\xi^{k-1}}{(k-1)!} \tilde{\tilde{K}}(k) \left[\sum_{n=k}^{\infty} \frac{\xi^{n-k}}{(n-k)!} \sum_{X_{k+1}} \cdots \sum_{X_n} \right. \tag{2.6.8}$$

$$\left. \tilde{K}(X_{k+1}) \cdots \tilde{K}(X_n) \prod_{\substack{ij \\ k+1 \leq i,j \leq n}} (1 - \chi_{ij}) \right]$$

The term in square brackets is, after some relabelling ($n - k \to m$, $X_{k+1}, \ldots X_n \to X_1 \ldots X_m$), simply $Z(\xi)/N$ by comparison with (2.6.1) so

$$\frac{Z'(\xi)}{Z(\xi)} = \sum_{k=1}^{\infty} \frac{\xi^{k-1}}{(k-1)!} \tilde{\tilde{K}}(k) \tag{2.6.9}$$

and integration with respect to ξ gives us

$$\log Z = \log N + \sum_{k=1}^{\infty} \frac{1}{k!} \tilde{\tilde{K}}(k). \tag{2.6.10}$$

Equation (2.6.5) is the first appearance of a series with infinitely many non vanishing terms. In Section 2.8 we shall state estimates on $K(k)$ which justify the manipulations leading from (2.6.5) to (2.6.10).

2.7 SUMMARY: THE CLUSTER EXPANSION FOR LOG Z

As we have tried to emphasize, the formulas developed in Section 2 can be assembled in different ways to fit different applications. Here we collect in one place the definitions and identities, as they apply to the Coulomb system, which we shall be needing in our discussion of convergence.

We shall give the expansion not for

$$Z = \int d\mu_0(\phi)\, e^{2z \int_\Lambda (\cos \beta^{1/2} \phi - 1)}$$

but for the more general object

$$Z_f = \int d\mu_0(\phi)\, e^{2z \int_\Lambda (\cos \beta^{1/2} [\phi + f] - 1)}$$

where $f = f(x)$ is a measurable function.

The expansion is:

$$\log Z_f = \int d\mu_0(\Psi) e^{-1/2\ell_D^{-2} \int_\Lambda \Psi^2} + \sum_{k=1}^{\infty} \frac{1}{k!} \tilde{\tilde{K}}_f(k) \tag{2.7.1}$$

where $d\mu_0$ has covariance u.

$$\tilde{\tilde{K}}_f(k) \equiv \sum_{X_1} \dots \sum_{X_k} U(X_1,\dots,X_k)\, \tilde{K}_f(X_1)\dots\tilde{K}_f(X_k) \tag{2.7.2}$$

$X_1,\dots,X_k$ are summed over all subsets of Λ which are unions of ℓ_D - cubes.

$$U(X_1,\dots,X_k) = \sum_{\substack{G \text{ connected on} \\ \{1,\dots,k\}}} \prod_{ij \in G} (-\chi(X_i \cap X_j)) \tag{2.7.3}$$

$\chi(X) = 1$ if $\dot{X} \neq \phi$, 0 otherwise.

$$\tilde{K}_f(X) \equiv \sum_h e^{-F_1(h)} K_f(X,h) \; ; \text{if } X \notin \{\ell_D\text{-cubes}\} \tag{2.7.4}$$

$$F_1(h) \equiv 1/2\ell_D^{-2} \int_\Lambda (g - h)^2 + 1/2 \int_\Lambda g u^{-1} g \tag{2.7.5}$$

h is summed over all functions $h(x)$ with values in $2\pi\beta^{-1/2}\mathbb{Z}$, constant on the interior of $L\ell_D$ - cubes, vanishing outside $\Lambda \cup \partial\Lambda$, with $\Sigma\hat{}(h) \subset X$. $\Sigma\hat{}(h)$ is the union of all ℓ_D-cubes whose distance from discontinuities of h is strictly less than $L'\ell_D$. $g = g_{L'}(h)$. See Section 1.10. If X is an ℓ_D-cube set $\tilde{K}_f(X) \equiv \int d\mu(\psi) \exp(G_f(X,\phi)) - 1$. c.f.(2.7.10).

$$K_f(X,h) \equiv \sum_{T \text{ on } \overline{y}(h)} \int d\underline{s} \; S\{\int d\mu_{\underline{s}}(\Psi) \quad (2.7.6)$$

$$(\prod_{ij \in T} (s_i \ldots s_{j-2})\, O_{ij})\}\, Q_f(X,h)$$

$$O_{ij} \equiv \int_{Y_i} \int_{Y_j} {}^{\delta}/\delta\Psi(x)\; C(x,y)\; {}^{\delta}/\delta\Psi(y)\; dx\, dy \quad (2.7.7)$$

$d\mu_s$ Gaussian, covariance $C(x,y,\underline{s}) \equiv C(\underline{s})$.

For each h,X, there is a partition $\overline{y}(h) \equiv \{Y_1,\ldots,Y_{n(h)}\}$ of X. The elements of $\overline{y}(h)$ are connected components ("phase boundaries") of $\Sigma\hat{}(h)$ and all ℓ_D-cubes contained in $\Lambda \sim \Sigma\hat{}(h)$.

$$Q_f(X,h) = e^{-F_2(X,h,\Psi)}\; e^{G_f(X,\, \Psi + g)} \quad (2.7.8)$$

$$F_2(X,h,\Psi) \equiv \int_X \Psi u^{-1} g \;+\; 1/\ell_D^2 \int_X \Psi(g - h) \quad (2.7.9)$$

$$e^{G_f(X,\phi)} \equiv \frac{e^{2z \int_X (\cos \beta^{1/2}\, [\phi + f] - 1)}}{\sum_h e^{-1/2\ell_D^{-2} \int_X (\phi - h)^2}} \quad (2.7.10)$$

In (2.7.10), h, $(h(x) \in 2\pi\beta^{-1/2}\mathbb{Z})$ is constant inside $L\ell_D$-cubes. $\underline{s} \equiv (s_1,\ldots,s_{n(h)-1})$ is integrated over $[0,1]^{n(h)-1}$ in (2.7.6). The symmetrization S in (2.7.6) is defined by

$$Sf(Y_1,\ldots,Y_{n(h)}) \equiv \frac{1}{n(h)} \sum_{\pi} f(Y_{\pi(1)},\ldots,Y_{\pi(n(h))}) \quad (2.7.11)$$

π is summed over all permutations of $\{1,\ldots,n(h)\}$. f is the argument of S in (2.7.6) regarded as a function of $Y_1,\ldots,Y_n$.

The covariance $C(\underline{s})$ was defined in (2.4.11). For our purposes all we shall need to know about it is that it is a convex combination of covariances of the form

$$\chi(x)\; C(x,y)\; \chi(y) \;+\; (1 - \chi(x))\; C(x,y)\,(1 - \chi(y)) \quad (2.7.12)$$

where χ is a characteristic function. An analogous statement is proved at the beginning of Appendix A. This, in a (coco-)nut shell is our expansion.

2.8 CONVERGENCE ESTIMATES. PROOF OF SCREENING; THEOREM 1.4.1

For pedagogic reasons we will give a proof of screening not in the form of Theorem 1.4.1 but in the following form

Theorem 1.4.1': fix λ, $0 < \lambda \leq 1/2$, if $z\beta^3$ is sufficiently small depending on λ, there are $c > 0$ and $\gamma, 0 < \gamma < 1$ such that

$$|<J(f_1)J(f_2)>_\lambda| \leq c[z\ell_D^3]^2 \cdot e^{-\gamma\frac{1}{\ell_D}\mathrm{dist}(\Delta_1,\Delta_2)}$$

f_1 and f_2 are characteristic functions of ℓ_D-cubes Δ_1 and Δ_2 with disjoint interiors. This estimate is uniform in Λ. γ may be picked arbitrarily close to 1 by picking $z\beta^3$ sufficiently small (depending on λ and γ). Recall

$$J(f) \equiv \int f(x)J(x) \tag{2.8.1}$$

We state an estimate on $\tilde{K}_f$ which implies convergence of the expansion 2.7.1 for $\log Z_f$. The proof of screening, assuming this estimate, follows by making a special choice for f and differentiating $\log Z_f$ with respect to parameters in f.

For any union of ℓ_D-cubes, X, we define a measure of the "spread" of X by

$$||X|| = \inf_{\hat{X}} |\hat{X}| \tag{2.8.2}$$

where the infimum is taken over all connected sets $\hat{X}$ which are unions of ℓ_D-cubes and contain X. $|\hat{X}|$ is the number of ℓ_D-cubes in $\hat{X}$.

Theorem 2.8.1: if L is fixed sufficiently small and L' is fixed sufficiently large then for each λ, $0 < \lambda \leq 1/2$, for each γ, $\gamma < 1$, there is a constant $c(\lambda,\gamma)$ such that if $z\beta^3 < c(\lambda,\gamma)$

$$\sum_{k=1}^{\infty} \frac{1}{k!} \sum_{X_1,\ldots,X_k:\, y_i \in \cup X_i} |U(X_1,\ldots,X_k)| \cdot \sup_f |\tilde{K}_f(X_1)\ldots\tilde{K}_f(X_k)| e^{\gamma(||X_1||+\ldots+||X_k||)} \tag{2.8.3}$$

converges uniformly in Λ. The supremum extends over functions f that are small depending on $z\beta^3$ in the sense

$$\sup_x \beta^{1/2}|f(x)| \le (z\beta^3)^{1/2} \tag{2.8.4}$$

y can be any point in Λ .

$\tilde{K}_f$ is analytic in f .

Proof of Screening, Theorem 1.4.1$'$

Take f of the form $\alpha_1 f_1 + \alpha_2 f_2$ and obtain an expansion for (c.f. (1.5.6'))

$$\frac{\partial}{\partial\alpha_1}\frac{\partial}{\partial\alpha_2} \log Z_f\Big|_{\alpha_1,\alpha_2=0} = -<J(f_1)J(f_2)>_\Lambda \beta \tag{2.8.5}$$

by differentiating under the sums in (2.7.1). Use analyticity in α_1,α_2 of $\tilde{K}_f$, i.e., Cauchy's formula for derivatives, and the bound of Theorem 2.8.1 to justify the differentiations under the sums. The result is, in absolute value, less than

$$\sum_{k=1}^{\infty} \frac{1}{k!} \sum_{X_1,\ldots,X_k} |U(X_1,\ldots,X_k)| \cdot \\ \cdot \Big| \frac{\partial}{\partial\alpha_1}\frac{\partial}{\partial\alpha_2} \tilde{K}_f(X_1)\ldots\tilde{K}_f(X_k)\Big|_{\underset{\sim}{\alpha}=0} \tag{2.8.6}$$

Take f_1,f_2 to have support within ℓ_D -cubes Δ_1,Δ_2 respectively. In this case the derivatives will annihilate terms which do not have the property that $\bigcup X_i$ contains Δ_1,Δ_2 . Assume that the L_∞ norms of f_1,f_2 are each unity and that the smallness restriction (2.8.4) on f has been satisfied by taking α_1,α_2 small, e.g., α_1,α_2 are constrained by $\alpha\beta^{1/2} \le (z\beta^3)^{1/2}$ ($\alpha = \alpha_1$ or α_2). We estimate the non zero derivatives using Cauchy's formula according to

$$\Big|\frac{\partial}{\partial\alpha_1}\frac{\partial}{\partial\alpha_2} F(\alpha_1,\alpha_2)\Big|_{\underset{\sim}{\alpha}=0} = \Big|\frac{1}{(2\pi i)^2} \oint\oint d\alpha_1 d\alpha_2 \cdot \\ \cdot \frac{F(\alpha_1,\alpha_2)}{\alpha_1^2\,\alpha_2^2}\Big| \le (z\beta^2)^{-1} \sup_{\alpha_1,\alpha_2} |F(\alpha_1,\alpha_2)| \tag{2.8.7}$$

where the contours were taken to be circles centered on the origin with radius $(z\beta^2)^{1/2}$ so that the smallness constraint on α_1,α_2 is satisfied. F denotes the product of $\tilde{K}$'s in (2.8.6). Putting this together we obtain, using the definition of ℓ_D

$$\left|\frac{\partial}{\partial\alpha_1}\frac{\partial}{\partial\alpha_2}\log Z_f\right|_{\underset{\sim}{\alpha}=0} \le 8\beta(z\ell_D^3)^2 \left\{ \sum_{k=1}^{\infty} \frac{1}{k!} \cdot \right.$$

$$\cdot \sum_{X_1,\ldots,X_k:\ \Delta_1,\Delta_2 \in \bigcup X_i} |U(X_1,\ldots,X_k)| \cdot \qquad (2.8.8)$$

$$\left. \cdot \sup_{\alpha_1,\alpha_2} |F(\alpha_1,\alpha_2)| e^{\gamma(\|X_1\|+\ldots+\|X_k\|)} \right\} .$$

$$\cdot\ e^{-\gamma d}$$

where d is the infimum of $\|X_1\|+\ldots+\|X_k\|$ over all X's such that $\Delta_1,\Delta_2 \in \bigcup X_i$. The extra two exponentals in (2.8.8) have been inserted by virtue of the fact that by definition of d their product is greater or equal to 1 .

The bound in Theorem (2.8.1) shows subject to the hypotheses of Theorem 1.4.1' that the quantity in curly brackets in (2.8.8) is bounded uniformly in Λ , α_1,α_2 so we have shown that

$$|\langle J(f_1)J(f_2)\rangle_\Lambda| \le c(z\ell_D)^3 e^{-\gamma d} \qquad \text{(see (1.5.6'))}$$

c depends on λ,γ . The proof of Theorem 1.4.1' is completed by showing that

$$d \ge \operatorname{dist}(\Delta_1,\Delta_2)/\ell_D \qquad (2.8.9)$$

For (2.8.9): the factor U in (2.8.6) forces $X_1,\ldots,X_k$ to overlap in such a way that

$$\|X_1\|+\ldots+\|X_k\| \ge \|X_1 \cup \ldots \cup X_k\| \qquad (2.8.10)$$

The constraint $\Delta_1,\Delta_2 \in \bigcup X_i$ implies that

$$\|X_1 \cup \ldots \cup X_k\| \ge \operatorname{dist}(\Delta_1,\Delta_2)/\ell_D$$

(2.8.9) follows because d is the infimum over the left hand side of (2.8.10). End of proof of Theorem 1.4.1'.

We conclude this Section 2 by discussing why one might expect Theorem 2.8.1 to be true. We do this by describing which small factors beat various potential divergences.

(1) Why doesn't the sum over $X_1, \ldots, X_k$ diverge because the sets $X_1, \ldots$ can be arbitrarily far apart? Because they can't be -- the factor U forces the sets $X_1, \ldots$ to overlap.

(2) The set X is not necessarily connected. It is a union of connected sets Y_i. Why doesn't the sum over a particular X diverge because its components $Y_1, Y_2, \ldots$ can be arbitrarily far apart? Because the covariances $C(x,y)$ inside the O_{ij}'s must have their arguments x,y in different Y_i's if any two Y_i's are far separated we pick up small factors because $C(x,y) = 0\ (\exp(-|x-y|/\ell_D)$.

(3) Why doesn't the sum over a particular Y_i diverge? Because if Y_i is large it must contain a large discontinuity set, which by Lemma 1.10.1 means that e^{-F_1} is very small.

(4) Why doesn't the sum over the number of X's or Y's diverge? Because if the number of such terms is large, there must be many factors O_{ij} and therefore many derivatives of $e^G e^{-F_2}$. If $z\beta^3$ is small, derivatives of $e^G e^{-F_2}$ are small. (see Section 1.7).

3.1 COMBINATORICS

The object of this section is to bound the numerous sums in our convergence estimate for the expansion, Theorem 2.8.1, by combinatoric arguments so that convergence is reduced to proving a bound on functional integrals. All the analysis will be in then proving this bound and that will be done in Section 4. The results of this section are summarized in Proposition 3.1.1 for the benefit of those readers who may wish to skip combinatorics. Whenever the symbol † appears, we are referring the reader to the footnote which appears after Appendix B, on page 438.

Some Conventions

Subsets of $\mathbb{R}^3$, in the absence of further qualifications, are to be understood to be unions of ℓ_D-cubes.

The set of ℓ_D-cubes is denoted $\{\Delta_\alpha\}$ and Δ_α, Δ_β,..., denote typical ℓ_D-cubes.

The term, "universal constant", means a constant independent of λ, z, β, $\underset{\sim}{s}$ and L, L'. In Section 4, L, L' will themselves be fixed independently of λ, z, β, $\underset{\sim}{s}$ and then constants are universal if independent of λ, z, β, $\underset{\sim}{s}$. Constants denoted by the same letter are not always the same unless occurring in the same equation.

Define a formal operation D_α, indexed by an ℓ_D-cube Δ_α, by its action on a functional $F = F(\psi)$, namely

$$D_{\alpha_1} \cdots D_{\alpha_N} F \equiv \ell_D^{-\frac{N}{2}} \int_{\Delta_{\alpha_1}} \cdots \int_{\Delta_{\alpha_N}} \left| \frac{\delta}{\delta\psi(x)} \cdots \frac{\delta}{\delta\psi(x_N)} F \right| \tag{3.1.1}$$

The objective of this section is to prove:

<u>Proposition 3.1.1</u> The bound in Theorem 2.8.1 is implied by (a) and (b)

(a) There exist constants $\bar{\gamma}$, p, c with $\gamma < 1$ so that

$$\int d\mu_{\underset{\sim}{s}}(\psi)\, e^{-F_2(h,\psi)} \left(\prod_{i=1}^{N} D_{\alpha_i}\right) e^{G_f(X,\phi)} \le (z\beta^3)^{\frac{N}{12}}\, e^{\bar{\gamma}\, F_1(h)} \prod_\alpha (n_\alpha!)^p\, e^{c|X|}$$

holds uniformly in z, β, $\underset{\sim}{s}$, X, h such that $\Sigma^\wedge(h) \subset X$, N, $\Delta_{\alpha_1}, \ldots, \Delta_{\alpha_N} \subset X$, f restricted according to (2.8.4). n_α is the number of times Δ_α appears in the list $\Delta_{\alpha_1}, \ldots, \Delta_{\alpha_N}$.

(b) $z\beta^3$ is sufficiently small depending on $\bar{\gamma}$, p, c, λ, L,L', γ.

The strategy of the proof is to construct a comparison series which converges and the structure of whose sums is the same as the structure of the sums in the estimate of Theorem 2.8.1. The extimate (a) then results by comparing the corresponding terms in these two series.

3.2 THE COMPARISON SERIES

It is a well known fact (a proof is outlined in [6]) that the number of connected sets which are unions of N ℓ_D-cubes and contain the origin is less than exp(cN) where c is universal and therefore if q is sufficiently large

$$\sum_{Y \ni 0} e^{-q|Y|} < c_1 e^{-q} \tag{3.2.1}$$

where Y is summed over all connected sets. In all our estimates the role of the origin, 0, could be played by an arbitrary point, y, in Λ. A simple generalization of this is: for any connected set Y_1.

$$\sum_{Y_2} e^{-q|Y_2|} w_\varepsilon(Y_1,Y_2) < c_1 e^{-q} \frac{c_2}{\varepsilon^3}$$

Y_2 is summed over connected sets not intersecting Y_1 and

$$w_\varepsilon(Y_1,Y_2) \equiv |Y_1|^{-1} \sum_{\alpha \in Y_i} \sum_{\beta \in Y_j} e^{-\varepsilon \mathrm{dist}(\alpha,\beta)/\ell_D} \tag{3.2.2}$$

$\alpha \in Y$ means $\Delta_\alpha \subset Y$. $\mathrm{dist}(\alpha,\beta) \equiv \mathrm{dist}(\Delta_\alpha,\Delta_\beta)$.

Proof: We estimate the sum over Y_2 containing Δ_β using (3.2.1), bound the sum over Δ_β by c_2/ε^2 using the exponential decay in (3.2.2), sum over Δ_α contained in Y_1 which gives a factor $|Y_1|$ to cancel the one in (3.2.2). End of Proof.

Now suppose we have n connected sets $Y_1,\ldots,Y_n$ with disjoint interiors and also a tree, T, on $\{1,\ldots,n\}$. Define

$$W_{\varepsilon,T}(Y_1,\ldots,Y_n) \equiv \prod_{ij\in T} w_\varepsilon(Y_i,Y_j) \qquad (3.2.3)$$

Then by iterating the estimate just proved, starting with extreme bonds in the graph T and working inwards we obtain (3.2.4) below. It is easiest to assume first that $Y_1 \ni 0$ and then relax it to $\cup Y_i \ni 0$ by an easy argument. For q sufficiently large

$$\sum_{Y_1,\ldots,Y_n : \cup Y_i \ni 0} W_{\varepsilon,T}(Y_1,\ldots,Y_n) \cdot e^{-q|\cup Y_i|} < (c_1 e^{-q})^n \left(\frac{c_2}{\varepsilon^3}\right)^{n-1} \frac{c_3}{\varepsilon} \qquad (3.2.4)$$

which is an estimate uniform in T. If n = 1, define $W_{\varepsilon,T}(Y_1) \equiv 1$ and set

$$W_\varepsilon(X) \equiv \sum_{n=1}^{\infty} \sum_{Y_1,\ldots,Y_n : \cup Y_i = X} W_{\varepsilon,T_n}(Y_1,\ldots,Y_n) \qquad (3.2.5)$$

$Y_1,\ldots,Y_n$ are each summed over all subsets of X such that Y_i and Y_j have disjoint interiors for all $i \neq j$. W_ε depends on the assignment $n \to T_n$ of a tree graph to each n, which is to be done randomly†. The result of summing (3.2.4) over n can be written in the form

$$\sum_{X, X\ni 0} W_\varepsilon(X)\, e^{-q|X|} < C\, e^{-q} \qquad (3.2.6)$$

for q sufficiently large depending on ε. This is uniform in the assignment of tree graphs.

Next, we argue that if q is sufficiently large

$$\sum_{k=1}^{\infty} \sum_{X_1,\ldots,X_k : \cup X_i \ni 0} \frac{1}{k!} |U(X_1,\ldots,X_k)| \cdot \prod_{i=1}^{k} W_\varepsilon(X_i) e^{-q|X_i|} < C\, e^{-q} \qquad (3.2.7)$$

where $X_1,\ldots,X_k$ are summed over all unions of ℓ_D-cubes without the condition that their interiors be disjoint. $(U(X) \equiv 1)$.

The proof will use the lemma 3.2.1 given below. A "standard tree graph", t, on {1,...,k} is a connected graph on all the vertices {1,...,k} with no closed loops.

Lemma 3.2.1: if $A_1,\ldots,A_{k-1} \geq 0$

$$\frac{1}{k!}\left|U(X_1,\ldots,X_k)\right| \leq e^{\Sigma A_i} \int d\omega(t) \prod_{ij\in t} \frac{1}{A_i} \chi_{ij}$$

where $\chi_{ij} \equiv \chi(X_i \cap X_j)$, $d\omega$ is a normalized measure on the set of standard tree graphs on {1,...,k}.

This lemma is proved in Appendix B using the tree graph lemma. The proof explicitly constructs $d\omega$ but all we shall need is that $d\omega$ is normalized.

Note that $U(X_1,\ldots,X_k)$ vanishes unless the X_i's intersect in such a manner that if we consider each X_i as connected then $\cup X_i$ is connected. This can be seen either from the definition of U in (2.7.3) or lemma 3.2.1.

Proof of (3.2.7)

We choose $A_i = |X_i|$ in Lemma 3.2.1 and substitute the bound of the lemma into the left hand side of (3.2.7). The $d\omega(t)$ integral is deferred. Let X_j be an extreme point of t. It intersects X_i (else the factor χ_{ij}, $ij \in t$, vanishes). We hold X_i fixed and estimate the sum over all X_j such that $X_j \cap X_i \neq \phi$. This is done by first estimating the sum over all X_j containing an ℓ_D-cube, Δ_α, using (3.2.6), then summing over all $\Delta_\alpha \subset X_i$ which gives a factor $|X_i|$ which is cancelled by the choice of A_i in Lemma 3.2.1. Next we repeat the procedure just described for another extreme point of the graph t' obtained from t by deleting the bond ij. We continue until all sums have been estimated, obtaining (3.2.7). End of Proof.

It should be obvious from the preceding arguments that by increasing q some more and choosing $\varepsilon > \gamma$ we can strengthen (3.2.7) to: for q sufficiently large depending on ε and γ,

$$\sum_{k=1}^{\infty} \frac{1}{k!} \sum_{X_1,\ldots,X_k : \cup X_i \ni 0} |U(X_1,\ldots,X_k)| .$$

$$\prod_{i=1}^{k} W_\varepsilon(X_i)\, e^{-q|X_i|}\, e^{\gamma \|X_i\|} < C\, e^{-q} \qquad (3.2.8)$$

(The norm $\| \ \|$ was defined in eq. (2.8.2)).

We have in (3.2.8) constructed the comparison series promised in Section 3.1.

3.3 COMPARISON

By comparing the terms in the summands of (3.2.8) with their analogues on the left hand side of the estimate in Theorem 2.8.1 we shall obtain a criterion (eq. (3.3.5)) that implies (2.8.3). For pedagogic reasons we will specialize to $\gamma = \frac{8}{10}$ in Theorem 2.8.1 and estimate (3.2.8). It is not difficult to trace an arbitrary $\gamma < 1$ through the estimates. We now fix q large enough that (3.2.8) holds for $\varepsilon = \frac{9}{10}$ $\gamma = \frac{8}{10}$. We will continue to use ε, γ, q, but they are now universal constants. We compare this estimate with the one in Theorem 2.8.1 and find the latter is implied by

$$|\tilde{K}_f(X)| < W_\varepsilon(X)\, e^{-q|X|} \qquad (3.3.1)$$

for q sufficiently large. We suppose $X \notin \{\ell_D\text{-cubes}\}$.*

Turning now to the definition of $\tilde{K}_f(X)$ given in (2.7.4) we see that the sum over h implies a sum over partitions $\overline{y}(h) = \{Y_1,\ldots,Y_n\}$ of X, i.e.,

$$\sum_h (\cdot) = \sum_{n=1}^{\infty} \frac{1}{n!} \sum_{Y_1,\ldots,Y_n : \cup Y_i = X} \ \sum_{h:\overline{y}(h)=\{Y_1,\ldots,Y_n\}} (\cdot) \qquad (3.3.2)$$

The $\frac{1}{n!}$ is there because the sum over $Y_1,\ldots,Y_n$ will reproduce the set $\{Y_1,\ldots,Y_n\}$ n! times. Thus if we compare (2.7.4) and the definition (3.2.5) of $W_\varepsilon(X)$, we see that (3.3.1) is implied by,

*See page 438.

$$\sum_{h:y(h)=\{Y_1,\ldots,Y_n\}}^{-} e^{-F_1(h)} K_f(X,h)$$
$$< n\, S\left[W_{\varepsilon,T}(Y_1,\ldots,Y_n)\right] e^{-q|X|} \qquad (3.3.3)$$

(We used the fact that $\sum_{Y_1,\ldots,Y_n} (\cdot) = \sum_{Y_1,\ldots,Y_n} \frac{1}{(n-1)!} S(\cdot)$ to insert the symmetrication S which was defined in (2.7.11).

In fact this estimate is not the one we need, but if the reader will recall in (3.3.5) the tree graph T_n could be assigned randomly for each n which means that we can insert in the right hand side of (3.3.3) $\int d\nu(T)$ where $d\nu$ is any normalized measure on the set of tree graphs of n vertices. For $d\nu$ we choose the measure defined by

$$\nu(\{T\}) = \Omega^{-1} \int d\underset{\sim}{s} \prod_{ij\in T} s_i \ldots s_{j-2} |Y_i| \qquad (3.3.4)$$

where Ω is the normalization.† We insert $\int d\nu(T)$ inside S in the right hand side of (3.3.3) and compare the definition of $K_f(X,h)$ given in (2.7.6) and the definition of $W_{\varepsilon,T}$ to conclude that (3.3.1) is implied by

$$\Big| \sum_{h:y(h)=\{Y_1,\ldots,Y_n\}}^{-} e^{-F_1(h)} \int d\mu_{\underset{\sim}{s}} \Big(\prod_{ij\in T} O_{ij} \Big).$$
$$\cdot\, Q_f(X,h) \Big| \le e^{-q|X|} \prod_{ij\in T} \sum_{\alpha\in Y_i} \sum_{\beta\in Y_j} e^{-\varepsilon \mathrm{dist}(\alpha,\beta)/\ell_D} \qquad (3.3.5)$$

The normalization Ω has been absorbed in $e^{-q|X|}$ which can be done by lemma 2.3.2. (q changes by 1).

Summary so far: there exists a universal constant q so that if (3.3.5) holds with $\varepsilon = \frac{9}{10}$, then (3.3.1) holds which in turn implies that the estimate in Theorem 2.8.1 holds with $\gamma = \frac{8}{10}$.

3.4 BEATING LEIBNIZ RULE

We have chosen $\varepsilon = \frac{9}{10} < 1$ so that the left hand side of (3.3.5) has better exponential decay (in the propagators contained in the operators O_{ij}. c.f. (1.6.1)) than the terms on the right. We need

the extra $(1-\varepsilon) = \frac{1}{10}$ exponential decay to dominate the large number of terms that will result from using Leibniz rule to evaluate the 0_{ij}'s. For this reason by trading in the extra exponential decay we will modify the right hand side of (3.3.5) to include a $\prod_{\alpha}(n_{\alpha}!)^{p}$. See (3.4.3) which is the first objective of this section. Lemma 3.4.1 is the means whereby we accomplish this.

The second accomplishment of this section is to show that functional derivatives of F_2 can be suitably dominated by F_1. The estimate (3.4.10) is the end product of this subsection.

Lemma 3.4.1: for any p and $\delta > 0$, there exists a constant $C_{p,\delta}$ so that

$$(m!)^{p} \le C_{p,\delta}^{m} \prod_{i=1}^{m} e^{\delta\, \mathrm{dist}(\Delta,\Delta_i)/\ell_D}$$

for all m. Δ is any given ℓ_D-cube and (Δ_i), $i = 1,\ldots,m$, is any sequence of disjoint ℓ_D-cubes in an order of distance from Δ.

Proof: easy consequence of Sterlings formula.

Each operator 0_{ij} in (3.3.5) starting with the last one to be applied can be expanded and estimated according to (3.4.1) below. There are $n-1$ operators 0_{ij} because a tree graph on n vertices has $n-1$ lines.

$$\begin{aligned}\left|0_{ij} F\right| &\equiv \left|\int_{Y_i}\int_{Y_j} \delta/_{\delta\psi}\, C\, \delta/_{\delta\psi}\, F\right| \\ &= \left|\sum_{\alpha\in Y_i}\sum_{\beta\in Y_j}\int_{\Delta_\alpha}\int_{\Delta_\beta} \delta/_{\delta\psi}\, C\, \delta/_{\delta\psi}\, F\right| \\ &\le \sum_{\alpha\in Y_i}\sum_{\beta\in Y_j} \frac{c}{\lambda\ell_D} e^{-\mathrm{dist}(\alpha,\beta)/\ell_D}\end{aligned}$$

$$\cdot \int_{\Delta_\alpha}\int_{\Delta_\beta}\left|\frac{\delta}{\delta\psi}\frac{\delta}{\delta\psi}F\right| \equiv \frac{c}{\lambda}\sum_{\alpha\in Y_i}\sum_{\beta\in Y_j} e^{-\mathrm{dist}(\alpha,\beta)/\ell_D} D_\alpha D_\beta F \tag{3.4.1}$$

$F = F(\psi)$ is a functional of ψ, in particular the result of previous applications of operators O_{ij} to Q_f. The bound made use of the exponential decay of $C = C(x,y)$ as detailed in (1.6.1). We have also used definition (3.1.1) to rewrite in terms of D_α, D_β.

Having used this estimate on all the O_{ij}'s we see that the left hand side of (3.3.5) has been dominated by a 2(n-1) fold sum over ℓ_D-cubes. We compare the terms in this sum with corresponding terms in the sum on the right hand side of (3.3.5) and find that (3.3.5) is implied by

$$\sum_{h:\bar{y}(h)=\{Y_1,\ldots,Y_n\}} e^{-F_1(h)} \int d\mu_{\underset{\sim}{s}} \prod_{ij\in T} D_{\alpha_i} D_{\beta_j} Q_f(X,h)$$

$$\leq e^{-q_\lambda|X|} \prod_{ij\in T} e^{(1-\varepsilon)\text{dist}(\alpha_i,\beta_j)/\ell_D} \tag{3.4.2}$$

q has been changed to q_λ because the constant $(\frac{c}{\lambda})^{(n-1)}$ coming from repeated application of (3.4.1) has been absorbed by this change. This estimate is required to hold uniformly in the positions of the $2(n-1) \equiv N$ cubes $\Delta_{\alpha_1}, \Delta_{\beta_1}, \ldots, \Delta_{\alpha_{n-1}}, \Delta_{\beta_{n-1}}$. Let n_α be the number of times Δ_α occurs in this list. (3.4.2) is itself implied by

$$\sum_{h:\bar{y}(h)=\{Y_1,\ldots,Y_n\}} e^{-F_1(h)} \int d\mu_{\underset{\sim}{s}} \prod_{ij\in T} D_{\alpha_i} D_{\beta_j} Q_f(X,h)$$

$$\leq e^{-q_{\lambda,p}|X|} \prod_\alpha (n_\alpha!)^P \tag{3.4.3}$$

We have used Lemma 3.4.1 with $\delta = 1 - \varepsilon$ to show that the product on the right hand side of (3.4.2) dominates the product on the right hand side of (3.4.3). The change q_λ to $q_{\lambda,p}$ absorbs the constant in Lemma 3.4.1. p can be arbitrary in (3.4.3) but $q_{\lambda,p}$ of course depends on p. If we prove (3.4.3) for any particular p, we prove the estimate in Theorem 2.8.1 by tracing back through our chain of reasoning.

Although it has been a considerable labour to arrive at this marvelous estimate, we shall now be perverse and use a slightly different one. Let $\Delta_{\alpha_1}, \ldots, \Delta_{\alpha_N}$ be N arbitrary ℓ_D-cubes. Let n_α be the number of times Δ_α occurs in this list. The estimate that replaces (3.4.3) is

$$\sum_{h:\bar{y}(h)=\{Y_1,\ldots,Y_n\}} e^{-F_1(h)} \int d\mu_{\underset{\sim}{s}}\, e^{-F_2(h)} \prod_{i=1}^{N} (D_{\alpha_i} + 3F_1^{1/2}(\Delta_{\alpha_i}, h))$$

$$e^{G_f(X)} \le e^{-q_{\lambda,p}|X|} \prod_{\alpha} (n_\alpha!)^p \tag{3.4.4}$$

$F_1\,(\Delta,h)$ was defined in Section 1.11 along with $G_f(X) \equiv G_f(X,\phi)$.

As before, the assertion is that if we can prove this estimate for any particular p (with $q_{\lambda,p}$ a given constant depending on p), then we have proven the estimate of Theorem 2.8.1. This estimate differs from the previous one in that an ackward factor e^{-F_2} has been moved past the derivatives.

To move the e^{-F_2} we must, beginning at (3.4.1) replace 0_{ij} by $e^{F_2}\, 0_{ij}\, e^{-F_2}$. Since F_2 is linear in ψ this is easily done by replacing each $^{\delta}/\delta\psi$ by $\frac{\delta}{\delta\psi} - \frac{\delta F_2}{\delta\psi}$. Instead of (3.4.1) we proceed by

$$e^{F_2}\, 0_{ij}\, e^{-F_2} = \int_{Y_i}\int_{Y_j} \left(\frac{\delta}{\delta\psi} - \frac{\delta F_2}{\delta\psi}\right) C \left(\frac{\delta}{\delta\psi} - \frac{\delta F_2}{\delta\psi}\right)$$

$$\equiv \int_{Y_i}\int_{Y_j} \left(\frac{\delta}{\delta\psi(x)} - (u^{-1}g)(x) - \frac{1}{\ell_D^2}\,(g-h)(x)\right) C\,(x,y,\underset{\sim}{s})$$

$$\left(\frac{\delta}{\delta\psi(y)} - (u^{-1}g)(y) - \frac{1}{\ell_D^2}\,(g-h)(y)\right) dx\, dy \tag{3.4.5}$$

We multiply this out. Terms which do not involve $u^{-1}g$ are expanded using ($Y = Y_i$, some i)

$$\int_Y (\cdot) = \sum_{\alpha\in Y} \int_{\Delta_\alpha} (\cdot) \tag{3.4.6}$$

as before in (3.4.1) and dominated by taking absolute values and using the exponential decay estimate (1.6.1) for the C's as was done before.

Before doing this to terms involving $u^{-1}g$ we integrate by parts: recall that $u^{-1} \equiv -\Delta + \lambda^2\, \ell_D^2\, \Delta^2$ so therefore

$$\int_Y C(x,y)(u^{-1}g)(y)\,dy = \int_Y (\vec{\nabla}_y C)\cdot\vec{\nabla} g\,dy + \\ + \lambda^2 \ell_D^2 \int_Y (\Delta_y C)(\Delta_y g)\,dy \tag{3.4.7}$$

There are no boundary terms by virtue of our construction of g and C. We expand each integral into a sum over ℓ_D-cubes e.g.

$$\int_Y (\vec{\nabla}_y C)\cdot(\vec{\nabla}_y g)\,dy \le \sum_{\alpha\in Y}\int_{\Delta_\alpha} |\nabla_y C|\,|\nabla_y g| \\ \le c\,\ell_D^{-1/2} \sum_{\alpha\in Y} e^{-\mathrm{dist}(\Delta_x,\Delta_\alpha)/\ell_D}\left(\int_{\Delta_\alpha} |\nabla g|^2\right)^{1/2} \tag{3.4.8}$$

Δ_x is the ℓ_D-cube containing x. We have used Schwarz's inequality and a bound that follows easily from the discussion in Section 1-6, namely: for all $\lambda \ge 0$

$$\left|\nabla_y C(x,y)\right| \le \frac{c}{\ell_D^2}\, e^{-\mathrm{dist}\,(x,y)/\ell_D} \tag{3.4.9}$$

provided $|x-y| \ge \ell_D$. Provided $L' \gg 1$ (a universal restriction) we can assume this because g is constant (=h) on a scale determined universally by $L'\ell_D$ near ∂Y_i so $\nabla g = 0$ near ∂Y_i. This argument has to be slightly modified if part of ∂Y_i coincides with $\partial\Lambda$. Now note that

$$\int_{\Delta_\alpha} |\nabla g|^2 \le \int_{\Delta_\alpha} g\,u^{-1} g \le 2F_1(\Delta_\alpha)$$

$$(\lambda\ell_D)^2 \int_{\Delta_\alpha} |\Delta g|^2 \le \int_{\Delta_\alpha} g\,u^{-1} g \le 2F_1(\Delta_\alpha)$$

$$\frac{1}{\ell_D^2}\int_{\Delta_\alpha} (g-h)^2 \le 2F_1(\Delta_\alpha)$$

With these estimates all the terms in $\frac{\delta F_2}{\delta\psi}$ can be dominated by $F_1^{1/2}(\Delta_\alpha)$ factors. The other terms in (3.4.5) require, instead of (3.4.9), estimates on higher derivatives, for example

$$|\nabla_x \nabla_y C(x,y,\underset{\sim}{s})| \le \frac{c}{\ell_D^3}\, e^{-\mathrm{dist}(x,y)/\ell_D}$$

which also holds under the same conditions as were imposed for (3.4.9). End of proof of (3.4.4).

We finish this subsection by expanding out the product in (3.4.4) into a sum of 2^N terms which is dominated by taking the largest term times $2^{|X|}$. In this largest term the factors of $F_1(\Delta_\alpha)$ are bounded using

$$F_1^{\frac{1}{2}}(\Delta_\alpha)^m \leq C\,\bar{\varepsilon}^{-m/2}\left(\frac{m}{2}\right)!\;e^{\bar{\varepsilon}\,F_1(\Delta_\alpha)}$$

which is valid for all $\bar{\varepsilon} > 0$. We find that (3.4.4) is implied by

$$\sum_{h:\bar{y}(h)=\{Y_1,\ldots,Y_n\}} e^{-(1-\bar{\varepsilon})F_1}\int d\mu_{\underset{\sim}{s}}\; e^{-F_2}\left(\prod_{i=1}^{N} D_{\alpha_i}\right).$$

$$\cdot e^{G_f} \leq e^{-q_{\lambda,p,\bar{\varepsilon}}|X|}\prod_\alpha (n_\alpha!)^p \tag{3.4.10}$$

The factor $\bar{\varepsilon}^{-m/2}$ and other constants are absorbed by changing $q_{\lambda,p}$ to $q_{\lambda,p,\bar{\varepsilon}}$. p and N are not the same as in (3.4.4). Thus if we prove this holds for some p and $\bar{\varepsilon}$ uniformly in all other variables then we prove the estimate of Theorem 2.8.1.

3.5 THE PROOF OF PROPOSITION 3.1.1

When the estimate in Proposition 3.1.1 is inserted into the combinatoric estimate (3.4.10) we find that the estimate of Theorem 2.8.1 is implied by

$$\left(\sum_{h:\bar{y}(h)=\{Y_1,\ldots,Y_n\}} e^{-(1-\bar{\gamma}-\bar{\varepsilon})F_1(h)}\right)(z\beta^3)^{\frac{N}{12}} < e^{-(q_{\lambda,p,\bar{\varepsilon}}+c)|X|} \tag{3.5.1}$$

($N = 2(n-1)$). We have only to argue that this holds if $z\beta^3$ is sufficiently small in order to complete the proof of Proposition 3.1.1.

Let m be the number of elements of $\bar{y}(h)\equiv\{Y_1,\ldots,Y_n\}$ which are not ℓ_D-cubes, i.e., are phase boundaries. Suppose the phase boundaries are $Y_1,\ldots,Y_m$. The finite range property (3) of Section 1.10 implies that

$$\sum_{h:\bar{y}(h)=\{Y_1,\ldots,Y_n\}} e^{-\mu\,F_1(h)} = \left(\prod_{i=1}^{m}\;\sum_{h:\Sigma^\wedge(h)=Y_i} e^{-\mu\,F_1(h)}\right)$$

$\mu = 1 - \overline{\gamma} - \overline{\varepsilon}$. Choose $\overline{\varepsilon} > 0$ so small that $\mu > 0$. If we take $z\beta^3$ small then (3.5.1) is true if for any set Y, (in particular $Y = Y_1,\ldots,Y_m$)

$$\sum_{h:\ \Sigma^\wedge(h) = Y} e^{-\mu F_1(h)} < e^{-(q_{\lambda,p,\overline{\varepsilon}} + c)|Y|} \tag{3.5.2}$$

Using Lemma 1.10.1 it is not particularly difficult (see Lemma 9.3 in [5]) to show that

$$\sum_{h:\ \Sigma^\wedge(h) = Y} e^{-\mu F_1(h)} < C_\mu^{|Y|L^{-3}} e^{-C'_\mu \; L^{-3}L'^3(z\beta^3)^{-1/2}|Y|}$$

which means that (3.5.2) holds if $z\beta^3$ is small enough. End of proof of Proposition 3.1.1.

4.1 THE PROOF OF THEOREM 2.8.1

By considering Proposition 3.1.1 in conjunction with the statement of Theorem 2.8.1 we find that the Proposition 4.1.1 given below is the missing link in our proof of Theorem 2.8.1.

Up to this point we have been fairly complete in our proofs. We abandon this policy for this last Proposition 4.1.1. The reason is that its proof is quite lengthy even though most of it is not particularly difficult. It is also quite well explained in [5].

Instead of a proof we give a "discussion" which has three main objectives. (1) to explain the choices of L and L' (2) to show how the various factors on the right of the bound in Proposition 4.1.1 arise (3) to guide the reader to the appropriate places in [5].

The conventions presented in Section 3.1 will still be in force.

PROPOSITION 4.1.1

For L sufficiently small, L' sufficiently large: for each λ, $0 < \lambda \leq 1/2$, there exists c_λ so that

$$\int d\mu_{\underset{\sim}{s}}\, e^{-F_2} \left(\prod_{i=1}^{N} D_{\alpha_i}\right) e^{G_f(X)}$$

$$\leq (\dot{z}\beta^3)^{\frac{N}{12}}\; e^{\gamma F_1}\; e^{c_\lambda |X|} \prod_\alpha (n_\alpha!)^P$$

p, γ are universal constants. $\gamma < 1$. N = 2(n-1) where n = card

$\bar{y}(h)$. The estimate is uniform in the choice of cubes $\Delta_{\alpha_1},\ldots,\Delta_{\alpha_N}$, N, X, $\underset{\sim}{s} \in [0,1]^{n-1}$, h such that $\Sigma^\wedge(h) \subset X$, z, β, f subject to (2.8.4).

4.2 DISCUSSION OF PROPOSITION 4.1.1

For pedagogic reasons we shall take f = 0. The smallness constraint (2.8.4) was chosen so that the estimates discussed below will continue to hold if f is non zero. (The values of universal constants may change).

Define a field A by

$$A(x) \equiv (L\ell_D)^{-3} \int_{\Omega} \phi(x)\, dx$$

where Ω is the $L\ell_D$ - cube which contains x. Thus A is the piecewise constant field obtained by averaging $\phi(x)$ over $L\ell_D$ - cubes. We also define a "fluctuation" field, δ, by

$$\phi(x) \equiv A(x) + \delta(x)$$

Recall that

$$e^{G(\phi)} \equiv \exp[2z \int(\cos\beta^{1/2}\phi-1)]/(\sum_h \exp[\frac{\cdot 1}{2\ell_D^2} \int (\phi - h)^2])$$

In this section $\int$ is to be read as $\int_X$. If we make the approximation $\phi \simeq A$, then this becomes $\prod_\alpha r(A_\alpha)$ where r(A) is the <u>function</u> (as opposed to functional):

$$r(A) \equiv \exp[2zL^3\ell_D^3 (\cos\beta^{1/2}A-1)]/.(\sum_{n\varepsilon Z} \exp\{-\frac{1}{2\ell_D^2} L^3\ell_D^3(A-n\tau)^2])$$

$$(\tau \equiv 2\pi\beta^{-1/2}) \qquad (4.2.1)$$

The damage of this approximation is in a factor we call e^{G_2}, i.e.,

$$e^{G(\phi)} \equiv (\prod_\alpha r(A_\alpha))\, e^{G_2(\phi)}$$

defines e^{G_2}.

We are, as per Proposition 4.1.1, interested in bounding

$$|\int d\mu_{\underset{\sim}{S}}(\psi)\; e^{-F_2} \left(\prod_{i=1}^{N} D_{\alpha_i}\right) \prod_{\alpha} r_{\alpha}(A_{\alpha}) e^{G_2(\phi)} \tag{4.2.2}$$

(Recall that $\phi = \psi + g$). We shall argue below that

$$|(\prod_i D_{\alpha_i})\;(\prod_{\alpha} r(A_{\alpha}))\; e^{G_2(\phi)}| \leq P \exp\left[\frac{\gamma}{2\ell_D^2}\int (A-h)^2\right]\cdot \exp\left[\frac{2}{\ell_D^2}\int \delta^2\right] \tag{4.2.3}$$

where γ is a universal constant, $\gamma < 1$. P is a local polynomial in ψ and $g - h$ whose coefficients behave no worse than $\prod_{\alpha}(n_{\alpha}!)^p$ for some p which is universal. Being local, it is a product of polynomials P_{α} which are independent of $\psi(x)$ and $(g - h)(x)$ if $x \notin \Delta_{\alpha}$. The degree of P_{α} is less than cn_{α}, c is universal. [In the absence of any derivatives the terms on the left hand side are bounded by the right hand side without the polynomial P.]

By Hölders inequality and (4.2.3), (4.2.2) is bounded by

$$\left(\int d\mu_{\underset{\sim}{S}}(\psi)|P|^{p_1}\right)^{p_1^{-1}} \left(\int d\mu_{\underset{\sim}{S}}(\psi)\; e^{-p_2 F_2}\right)^{p_2^{-1}} \cdot$$

$$\cdot\left(\int d\mu_{\underset{\sim}{S}}(\psi) \exp\left[\frac{p_3\gamma}{2\ell_D^2}\int (\psi + g - h)^2\right] \exp\left[\frac{2p_3}{\ell_D^2}\int \delta^2\right]\right)^{p_3^{-1}} \tag{4.1.4}$$

where $\Sigma p_i^{-1} = 1$. We shall choose p_1 to be even (and large). We have also used the inequality

$$\int (A - h)^2 \leq \int (\phi - h)^2 = \int (\psi + g - h)^2$$

which comes from the fact that A is obtained by applying a projection, self adjoint with respect to the inner product $u,v \to \int uv$, to ϕ .

We bound each of the factors in (4.1.4). This is the point at which the choices of L, L' are made. We shall do this by quoting three lemmas from [5]. Since $\gamma < 1$, we can and do choose p_3 so that $p_3\gamma < 1$. p_3 is now universal.

LEMMA 4.2.1

if L is sufficiently small and L' sufficiently large then for each λ, $0 < \lambda \leq 1/2$, there exists c_{λ} so that

$$(\int d\mu_{\underset{\sim}{S}} (\psi)\ e^{P_3 \gamma B})^{P_3^{-1}} \leq e^{c_\lambda |X|}\ e^{\gamma' F_1(h)}$$

where h is such that $\Sigma\hat{}(h) \subset X$ and

$$B \equiv \frac{1}{2\ell_D^2} \int_X (\psi + g - h)^2 + \frac{2}{\gamma \ell_D^2} \int_X \delta^2$$

γ' is a universal constant, $\gamma' < 1$. The restrictions on L, L' are also universal.

The choice of P_2 and P_1 is now made, consistently with $P_1^{-1} + P_2^{-1} + P_3^{-1} = 1$, but otherwise arbitrarily. [There is a choice which optimises our estimates but this estimation procedure is inherently so terrible that we shall not bother!]

LEMMA 4.2.2

Given $\varepsilon>0$, if L' is sufficiently large, for each λ, $0< \lambda \leq 1/2$, there exists c_λ so that

$$(\int d\mu_{\underset{\sim}{S}} (\psi)\ e^{-P_2 F_2(X,h)})^{P_2^{-1}} \leq e^{c_\lambda |X|}\ e^{\varepsilon F_1(h)}$$

provided $\Sigma\hat{}(h) \subset X$. Once ε is fixed, the choice of L' is universal.

We choose ε so that $\gamma' + \varepsilon < 1$. The polynomial P is bounded using

LEMMA 4.2.3

Suppose N not necessarily distinct arbitrary points $x_1,\ldots,x_N \in \Lambda$ are given. For each ℓ_D-cube Δ_α let n_α be the number of such points in Δ_α. Then for each λ, $0 < \lambda \leq 1/2$, there exists c_λ so that

$$\int d\mu_{\underset{\sim}{S}}(\psi) \prod_{i=1}^{N} \ell_D^{1/2} \psi(x_i) < \prod_\alpha c_\lambda^{n_\alpha} (n_\alpha !)$$

Proposition 4.1.1 now follows by collecting our estimates from (4.2.2) onwards. N, n_α are not the same in all places.

4.3 THE BOUND (4.2.3)

This is lemma 9.7 of [5] along with explicit computations with e^{G_2}. See equations (6.5) (6.6) in [5]. Take $\rho = z$ in those equations to recover our situation.

Derivatives of e^{G_2} yield factors like $\sin\beta^{1/2}\delta - \beta^{1/2}\delta$ or $\cos\beta^{1/2}\delta - 1$ which are bounded by $1/2\beta\delta^2$. We also use $|\cos|$, $|\sin| \leq 1$. With these types of bound trigonometrical quantities can be converted to polynomials. Lemma 9.7 of [5] tells us that

$$\left|\left(\ell_D^{-1/2}\frac{d}{dA}\right)^N r(A)\right| \leq (z\beta^3)^{N/12}(N!)^p \exp\left[\frac{\gamma}{2}\ell_D L^3(A-h)^2\right]$$

for some p, γ, $\gamma < 1$, both universal. Notice that this is a perfectly classical bound with nothing to do with functional integrals. This is the estimate that makes the approximation of Section 1.7 work. It is very important that $\gamma < 1$ otherwise

$$\int d\mu_{\underset{\sim}{s}}(\psi)\exp\left[\frac{\gamma}{2\ell_D^2}\int_X(\phi-h)^2\right] \tag{4.3.1}$$

would diverge as $\Lambda \to \infty$. (Recall $d\mu_{\underset{\sim}{s}}$ depends on Λ). This can be heuristically understood by noting that $d\mu_{\underset{\sim}{s}}$ is formally (take $\underset{\sim}{s} = (1, 1, \ldots)$, $h = 0$)

$$(\text{Norm.})^{-1}\prod_{\alpha} d\phi(x)\exp\left(-1/2\int\phi\, u^{-1}\phi - \frac{1}{2\ell_D^2}\int\phi^2\right) \tag{4.3.2}$$

and if $\gamma > 1$ the $\int\phi^2$ term will be overpowered inside X.

LEMMA 4.2.1

This is a special case of Lemma 9.5 in [5]. (The covariance C of Lemma 9.5 differs from the covariance C used here in that it contains a non local perturbation which is small enough to be controlled if λ is small. In our case this is not present so that the restriction on λ is weaker and all parts of the proof of Lemma 9.5 concerning this perturbation (which is called ν) drop out.

This lemma contains the most important restriction on L. It uses the idea that if L is very small, the minimum wavelength of the fluctuation δ is so small that its contribution to the functional integral is very small because of the $\int\phi u^{-1}\phi$ in (4.3.2). In this respect it justifies the idea of replacing ϕ by A mentioned at the beginning of section 4.2.

LEMMA 4.2.2

This is discussed in Section 9.4 in [5]. This contains the most important restriction on L', namely by taking L' large we get to choose a translation g so close to the "classical" translation g_c (for which the linear term F_2 in Lemma 4.2.2 vanishes) that F_2 is negligible.

LEMMA 4.2.3

This is a standard type of bound in respect to the appearance of $\Pi n_\alpha!$ which is a factor that counts the number of terms (weighted by exponential decay of factors C(x,y)) which Leibniz rule produces when Wicks Theorem (Theorem 2.4.1) is used to evaluate the multiple moment. The counting involved can be found in Lemma 2.6 of [15].

5. SOME REMARKS ON THE QUANTUM SITUATION

It is natural to try to extend the shielding results from the classical statistical to the quantum statistical mechanics setting. This section discusses some ideas and difficulties, in this direction. It is disjoint from the rest of the paper, and may be skipped till the reader feels ready to go beyond the present work.

We first indicate an obvious generalization of the sine gordon transformation to the quantum setting -- again dealing with two equally charged, $\pm$ 1, particles, with the same mass and activity. In the classical case, starting with the two body Boltzmann factor

$$e^{-\beta V(x,y)} \tag{5.1}$$

with V positive definite, we pass to the representation

$$\int d\mu_0(\phi) e^{2z\int_\Lambda (\cos \beta^{1/2}\phi - 1)} \tag{5.2}$$

for the partition function. In the quantum case (in the path space language) the two body Boltzmann factor is

$$e^{-\int_0^\beta dt V(x(t),y(t))} \tag{5.3}$$

We introduce a set of fields $\{\phi_\gamma(x)\}$, that may be viewed as time Fourier components of a field $\phi(x,t)$

$$\phi(x,t) = \sum_\gamma \phi_\gamma(x)\psi_\gamma(t) \tag{5.4}$$

$$\equiv \phi_0 \frac{1}{\sqrt{\beta}} + \sum_1^\infty \phi_{cn}\sqrt{2/\beta}\, \cos \frac{2\pi nt}{\beta} + \sum_1^\infty \phi_{sn}\sqrt{2/\beta}\, \sin \frac{2\pi nt}{\beta} \tag{5.5}$$

Each of the ϕ_γ is given the same covariance as the ϕ in the classical case. One then has the following representation for the partition function.

$$\prod_\gamma \left(\int d\mu_0(\phi_\gamma)\right) e^{\sum_i z_i \int d\mu_i \left(e^{ie_i "\int_0^\beta dt" \sum_\gamma \psi_\gamma(t)\phi_\gamma(x_p(t))} - 1\right)} \tag{5.6}$$

Here z_i and e_i are the activities and charges of species i (actually $z_i = z$, $e_i = \pm 1$). $d\mu_i$ is the single particle path space measure (signed for fermions), a measure on paths $t \to x_p(t)$. statistics causes a technicality: the paths may be on the intervals $[0,\beta]$, $[0,2\beta]$, $[0,3\beta]$, ...; the "$\int_0^\beta dt$" integral is understood

as over whichever interval the path specifies. In the classical limit $d\mu_i$ is concentrated on constant paths ($x_p(t)$ independent of t) and the classical representation (5.2) is obtained.

The physicist's traditional approach to shielding is by summing up certain infinite classes of diagrams. In the classical case the diagrams for the propagator (two point correlation function)

give contributions

$$\frac{1}{k^2} + \frac{1}{k^2}\left(-\frac{1}{\ell_D^2}\right)\frac{1}{k^2} + \frac{1}{k^2}\left(-\frac{1}{\ell_D^2}\right)\frac{1}{k^2}\left(-\frac{1}{\ell_D^2}\right)\frac{1}{k^2} + \cdots \tag{5.7}$$

that are summed up to $\dfrac{1}{k^2 + \frac{1}{\ell_D^2}}$. In the quantum case the corresponding diagrams are

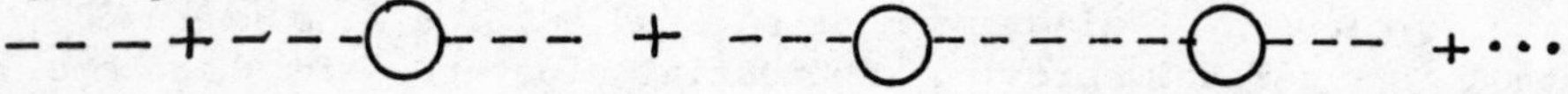

for the propagator of the field $\phi_{\gamma'}$. These give contributions

$$\frac{1}{k^2} + \frac{1}{k^2}\left(-P(k^2,\omega)\right)\frac{1}{k^2} + \frac{1}{k^2}\left(-P(k^2,\omega)\right)\frac{1}{k^2}\left(-P(k^2,\omega)\right)\frac{1}{k^2} + \cdots \tag{5.8}$$

that are summed to

$$\frac{1}{k^2 + P(k^2,\omega)} \tag{5.9}$$

We have written

$$\{\phi_{\gamma'}\} = \{\phi_0, \frac{\phi_{cn} + i\phi_{sn}}{\sqrt{2}}, \frac{\phi_{cn} - i\phi_{sn}}{\sqrt{2}}\} \tag{5.10}$$

and correspondingly

$$\{\omega\} = \{\omega(\gamma')\} = \left\{0, \frac{2\pi n}{\beta}, \frac{-2\pi n}{\beta}\right\} \tag{5.11}$$

The functions $P(k^2,\omega)$ are complicated, but can be shown to have the properties

$$P(0,0) = \frac{1}{\ell^2} \tag{5.12}$$

$$P(0, \frac{2\pi n}{\beta}) = 0 \quad n \neq 0 \tag{5.13}$$

From (5.12) we deduce "screening" for the ϕ_0 mode with screening length ℓ. From (5.13) we deduce the other modes do not screen.

We have envisioned two possible avenues of departure to study the quantum mechanical situation

1) Try to show screening (exponential clustering) for time averaged currents only, i.e. objects like

$$\int_0^\beta dt \int_0^\beta dt' \langle J(x,t)J(y,t')\rangle \tag{5.14}$$

Unfortunately we do not know if these objects are screened even by the perturbation theory arguments of the last paragraph. Complicated classes of diagrams must be studied. We also see no method of rigorously proving exponential clustering in this case -- should perturbation arguments indeed indicate a positive result. On the other hand probably most physicists would believe the study of such time averaged currents is the right place to seek screening.

2) Abandon the idea of exponential clustering, and settle for weak power law fall offs. The $\phi_{\gamma'}$ $(\neq \phi_0)$ may be expected to behave like the sine gordon field in a dipole system. Ideas are around to study the weak power clustering of such fields. A difficult, but theoretically possibly accessible, proof of the existence and weak power law clustering of the correlation functions in a matter system lies down this direction. But this is not likely to excite physicists.

We have presented (5.6) as a possibly useful point of departure. One may want to modify this formula, treating the short range forces differently. We point out that in [16] , we have handled the short range force problem in the study of matter systems. The techniques ofthat paper may hopefully be combined with the sine gordon transformation (some variation on (5.6)) to carry through the program of 2) above.

APPENDIX A

We give a proof of the tree graph formula, Lemma 2.3.1. That is, we show that the coefficients U(X), which are determined recursively by solving

$$U(X) = 1 \text{ if } X \text{ has only one element}$$

$$\Psi(\Lambda) = \sum_X U(X)\, \Psi(\Lambda \sim X) \tag{A.1}$$

(where X is summed over all subsets of Λ containing the first element in Λ), and the coefficients $\overline{U}(X)$ given by the tree graph formula

$$\overline{U}(X) \equiv \sum_{T \text{ on } X} \int d\underset{\sim}{s}\; S\, \{ \prod_{ij\varepsilon T} v'_{ij}(\underset{\sim}{s}) \,. \\ .\Psi(X, \underset{\sim}{s})\} \tag{A.2}$$

are the same. The notation is explained in Section 2.3.

We also discuss the stability properties (see (2.3.4), (2.3.5)) of the s dependent interactions:

$$\sum_{ij,i,j\varepsilon X} s_i \ldots s_{j-1}\, v_{ij} \equiv V(X; \underset{\sim}{s}) \tag{A.3}$$

introduced in Section 2.3. [This formula is written under the assumption that X has the form {m, m+1,...,n} with m < n. (m was always 1 in Section 2). If X does not have this form, its elements are to be relabeled m, m+1,...,n in such a way that their order is preserved].

We begin with the stability properties of V(X; s), proving that if for all X, there is a constant c so that

$$V(X) \equiv \sum_{ij,i,j\varepsilon X} v_{ij} \leq c|X| \tag{A.4}$$

(where $|X|$ is the cardinality of X) then with the same c

$$V(X; \underset{\sim}{s}) \leq c|X| \tag{A.5}$$

First we claim that $V(X; \underset{\sim}{s})$ is a convex combination of its "extreme points" where $\underset{\sim}{s} \equiv (s_1,\ldots,s_{n-m})$ has each $s_i = 0$ or 1 for $i = 1,\ldots,n-m$. This follows instantly by induction using

$$V(X; s_1,\ldots,s_i,\ldots,s_{n-m}) = \tag{A.6}$$

$$= s_i V(X; s_1,\ldots,1,\ldots,s_{n-m}) + (1-s_i)\, V(X; s_1,\ldots,0,\ldots S_{n-m})$$

which itself is easy to prove. From this it follows that it is enough to prove (A.5) when each s_i is either 1 or 0. Now we notice that if $s_i = 0$, all interactions between sites m,m+1,... m+i-1 (≡ the set Y say) and the remaining sites in X ~ Y are shut off in such a way that

$$V(X; s_1,\ldots,s_{n-m}) = V(Y;s_1,\ldots,s_{i-1}) + V(X \sim Y;s_{i+1},\ldots,s_{n-m}) \tag{A.7}$$

By repeatedly using this fact we see that in the case that all s_i are either one or zero $V(X; \underset{\sim}{s})$ has the form

$$V(X_1) + \ldots + V(X_p)$$

where $X_1,\ldots,X_p$ is a partition of X. Thus (A.4) implies (A.5) when $\underset{\sim}{s}$ has this special form and therefore in the general case.

Now we will prove the tree graph formula. Our procedure is to use the identity

$$e^{V(\Lambda;\, s_1,\ldots,s_i,1,1,\ldots,1)} = e^{V(\Lambda;s_1,\ldots,s_i,0,1,\ldots,1)} + \int_0^1 ds_{i+1} \frac{\partial}{\partial s_{i+1}} e^{V(\Lambda\,;\, s_1,\ldots,s_{i+1},1,\ldots,1)} \tag{A.8}$$

(which is just the fundamental theorem of calculus) repeatedly in order to generate a series for

$$\Psi\,(\Lambda) \equiv e^{V(\Lambda)} \equiv e^{V(\Lambda;\, 1,1,\ldots,1)}$$

We compare this series with (A.1) to prove the result. The reader might find the presentation in [12] of a closely related procedure helpful.

We will without loss of generality suppose that

$$\Lambda = \{1,2,\ldots,N\} \tag{A.9}$$

In order to describe how our series is obtained, define

$$E_o \equiv \Psi(\Lambda), \qquad E_{N+1} \equiv 0$$

$$E_k \equiv \int ds_1\ldots ds_{k-1}\; P_k[\, v'_{12}(v'_{23} + v'_{13})\ldots \ldots (\sum_{j=1}^{k-1} v'_{jk})\; e^{V(\Lambda;s_1,\ldots,s_{k-1},1,\ldots,1)}] \tag{A.10}$$

for k=1,...,N. $v'_{ij} \equiv v'_{ij}(\underset{\sim}{s})$ was defined in Section 2.3. P_k is a formal operator. It sums over all quantities obtained by transposing site 2 with site $\lambda(2)$, site 3 with site $\lambda(3)$,..., site k with site $\lambda(k)$ where $\lambda(2),\ldots,\lambda(k)$ is an arbitrary choice of k-1 distinct sites of Λ. For uniformity of notation we write $\lambda(1) \equiv 1$. For example

$$P_k[e^{V(\Lambda;s_1,\ldots,s_{n-1})}] \equiv P_k[\exp(\sum_{ij,i,j=1,\ldots,N} s_i\ldots s_{j-1} v_{ij}]$$

$$\equiv \sum_{\lambda} \exp(\sum_{ij,i,j=1,\ldots,N} s_i\ldots s_{j-1}\, v_{\lambda(i)\lambda(j)})$$

Note that P only affects labels on v**'s.**

We will argue using (A-8) that these quantities satisfy, for k=0,...,N, a relation of the form

$$E_k = \sum_X W(X)\ \Psi(\Lambda \sim X) + E_{k+1} \qquad (A.11)$$

where X is summed over k element subsets of Λ containing 1. By iterating (A·11) we find that

$$\Psi(\Lambda) \equiv E_0 = \sum_X W(X)\ \Psi(\Lambda \sim X) \qquad (A-12)$$

where X is now summed over all subsets of Λ which contain 1. Comparison of (A.12) and (A.1) proves that $U(X) = \overline{W}(X)$, so therefore it is enough to prove (A.11) and $W(X) = \overline{U}(X)$.

To prove (A.11) we define

$$Y_k \equiv \{1,\ldots,k\};\ \sim Y_k \equiv \Lambda \sim Y_k \qquad (A-13)$$

and first concentrate on proving that

$$E_k = P_k[\int ds_1\ldots ds_{k-1}\, v'_{12}\ldots(\sum_{j=1}^{k-1} v'_{jk}) \cdot e^{V(Y_k;s_1,\ldots,s_{k-1})} \cdot$$

$$\cdot\ \Psi(\sim Y_k)] + E_{k+1} \qquad (A.14)$$

which will turn out to be just another way of writing (A.11).

<u>Proof of (A.14):</u>

In the definition of E_k, substitute using (A.8) with i = k-1. We rewrite the term corresponding to the first term in (A.8) using a form of (A.7), namely

$$e^{V(\Lambda;s_1,\ldots,s_{k-1},0,1,\ldots,1)} = e^{V(Y_k;s_1,\ldots,s_{k-1})}\cdot e^{V(\sim Y_k)} \tag{A.15}$$

Since the second factor in (A.15) is $\Psi(\sim Y_k)$, the first term in (A.14) is explained, as arising from the first term in (A.8). To see that the second term in (A.8) gives E_{k+1}, we substitute using

$$\frac{\partial}{\partial s_{k+1}} V(\Lambda;s_1,\ldots,s_{k+1},1,\ldots,1) = \sum_{i=1}^{k}\sum_{j=k+1}^{N} v'_{ij}(\underset{\sim}{s}) \tag{A.16}$$

The sum over j in (A.16) is subsumed by the P_{k+1} operation in E_{k+1}. End of proof of (A.14).

<u>Proof of (A.11)</u>

We substitute the easy identity

$$v'_{12}\cdots\left(\sum_{j=1}^{k-1} v'_{jk}\right) = \sum_{T \text{ on } Y_k}\ \prod_{ij\varepsilon T} v'_{ij} \tag{A.17}$$

into the first term in (A.14). Next we rewrite the operator P_k using

$$P_k[F(1,2,\ldots,k)] = \sum_X \sum_\pi F(\pi(x_1),\ldots,\pi(x_k)) \tag{A.18}$$

where X is summed over all k element subsets of Λ containing 1. $x_1=1$, $x_1,x_2,\ldots x_k$ are the elements of X, written in increasing order. F is the argument of P_k in (A.14). π is summed over all permutations of k objects such that $\pi(1)$ $(= \pi(x_1)) = 1$. (A.18) is obtained by considering the map λ: $1,\ldots,k \to \lambda(1),\ldots,\lambda(k)$ introduced in the definition of P_k. (A.18) states that the sum over such λ can be performed by first fixing a set $X = \{x_1,\ldots,x_k\}$ with $x_1 = 1$, $x_2 < x_3 < \ldots < x_k$, and summing over all λ such that $\{\lambda(1),\ldots,\lambda(k)\} = X$, i.e. summing over all permutations, π, of $x_1,\ldots,x_k$ such that $\pi(x_1) = x_1$, and then summing over X.

We find using (A.17) and (A.18) that the first term in (A.14) is

$$\sum_{X} \sum_{T \text{ on } X} \int d\underset{\sim}{s} \sum_{\pi} \pi[\prod_{ij\varepsilon T} v'_{ij} \Psi(X, \underset{\sim}{s})]. \quad \Psi(\Lambda \sim X) \qquad (A.19)$$

where $\pi[.]$ indicates that the sites $x_1,\ldots,x_k$ occuring in the expression in square brackets are to be permuted according to π. Thus we have proved (A.11) with

$$W(X) = \sum_{T \text{ on } X} \int d\underset{\sim}{s} \sum_{\pi} \pi[\prod_{ij\varepsilon T} v'_{ij} \Psi(X, \underset{\sim}{s})]$$

To finish off the proof of the tree graph lemma we need to see that

$$W(X) = \overline{U}(X)$$

By recalling the definition of the symmetrization S in the tree graph formula, we see that it is sufficient to argue that the right hand side is unchanged if we arbitrarily relabel the sites in X so that a new site is labelled "1". If this is so, we can omit the restriction $\pi(1) = 1$ in the sum over π if we divide through by $\frac{1}{k}$ to normalize out the extra permutations. The resulting $\frac{1}{k}\sum_{\pi} \pi[.]$ is, by definition, $S\{.\}$. We have already shown that $W(X) = U(X)$ and the formula (2.1.4) shows that $U(X)$ and therefore $W(X)$ is not dependent on which site is labeled "1". This concludes the proof of the tree graph formula.

APPENDIX B

We will describe a variant of the tree graph formula of Appendix A and use it to prove Lemma 3.2.1.

Let $\{v_{ij}\}$ be a collection of negative numbers which will be allowed to assume the value $-\infty$. $i,j \in \Lambda = \{1,2,\ldots,N\}$. X is a subset of Λ. Formulas are written for X of the form $\{m,m+1,\ldots,n\}$ with $m < n$. If X is not of this form, relabelling is necessary.

Define (for $s_i, s_{i+1},\ldots,s_{j-1} \in [0,1]$)

$$v_{ij}(\underset{\sim}{s}) \equiv \log[s_i \ldots s_{j-1} (e^{v_{ij}} - 1) + 1]$$

$$v'_{ij}(\underset{\sim}{s}) \equiv \frac{\partial}{\partial s_{j-1}} v_{ij}(\underset{\sim}{s})$$

Note that $v_{ij}(\underset{\sim}{s}) = 0$ if $s_i \ldots s_{j-1} = 0$, $= v_{ij}$ if $s_i \ldots s_{j-1} = 1$.

With these properties, the reader can verify that the structure of the proof of the tree graph formula permits it to be repeated with $v_{ij}(\underset{\sim}{s})$ redefined in this manner. As before

$$\Psi(X, \underset{\sim}{s}) \equiv e^{V(X;\, \underset{\sim}{s})}$$

$$V(X; \underset{\sim}{s}) \equiv \sum_{ij,\, i,j \in X} v_{ij}(\underset{\sim}{s})$$

We find that in analogy to our former tree graph formula

$$U(X) = \sum_{T \text{ on } X} \int d\underset{\sim}{s}\; S\{\prod_{ij \in T} v'_{ij}(\underset{\sim}{s})\;\; \Psi(X, s)\}$$

which by substituting in definitions, reads

$$U(X) = \sum_{T \text{ on } X} \int d\underset{\sim}{s}\; S\{\prod_{ij \in T} s_i, \ldots s_{j-2} \cdot$$

$$\cdot\, (e^{v_{ij}} - 1) \prod_{ij \notin T} [s_i \ldots s_{j-1} (e^{v_{ij}} - 1) + 1]\,\} \qquad \text{(B.1)}$$

(Empty products are 1 by convention). This is our variant of the tree graph formula. It is only useful when v_{ij} is negative. We demonstrate why, by proving an estimate. Take absolute values inside the sums and integrals using

$$0 \le s_i \ldots s_{j-1} (e^{v_{ij}} - 1) + 1 \le 1$$

We obtain

$$|U(X)| \leq \sum_{T \text{ on } X} S\{\int d\underset{\sim}{s} \prod_{ij\varepsilon T} A_i(s_i \ldots s_{j-2}) \cdot$$

$$\cdot \prod_{ij\varepsilon T} |e^{v_{ij}} - 1| A_i^{-1}\} \tag{B.2}$$

where $\{A_i\}$ is a set of arbitrary positive numbers that have been inserted in such a way as to cancel out. The symmetrization S permutes the subscripts on v's and A's but not on $\underset{\sim}{s}$. Define a <u>normalised</u> measure $dv(T)$ on tree graphs by

$$\int dv(T)\ F(T) = N^{-1} \sum_{T \text{ on } X} \int d\underset{\sim}{s} \cdot \prod_{ij\varepsilon T} A_i s_i \ldots s_{j-2} F(T)$$

and note by Lemma 2.3.2 that the normalisation satisfies

$$N \leq e^{A_1 + \ldots + A_{k-1}}$$

where $k = \text{card }(X)$. Thus we deduce

$$|U(X)| \leq e^{A_1 + \ldots + A_{k-1}} S\{ \int dv(T) \prod_{ij\varepsilon T} A_i^{-1} |e^{v_{ij}} - 1|\} \tag{B.3}$$

We recall that a "standard tree graph" (as opposed to "tree graph") is a connected graph on all the (labelled) vertices in X with no closed loops. t will denote a standard tree graph. A permutation of the labels of vertices in general destroys the label increasing property we have put in the definition of "tree graph" and changes a tree graph into a standard tree graph. Thus by recalling the definition (2.3.2) of S we see that

$$S\{\int dv(T) \frac{1}{(k-1)!} \prod_{ij\varepsilon T} A_i^{-1} \quad |e^{v_{ij}} -1|\}$$

can be rewritten as

$$\int d\omega(t) \prod_{ij\varepsilon t} A_i^{-1} |e^{v_{ij}} - 1|$$

where $d\omega$ is a measure on standard tree graphs. The $\frac{1}{(k-1)!}$ is inserted so that $d\omega$ is <u>normalised</u>. Then (B.3) beomes

$$\frac{1}{k!} |U(X)| \leq \frac{e}{k}^{A_1 + \ldots + A_{k-1}} \int d\omega(t) \prod_{ij\varepsilon t} A_i^{-1} |e^{v_{ij}} - 1| \tag{B.4}$$

This is the estimate we promised. It is valid if $v_{ij} \in [0, -\infty]$. By taking

$$e^{v_{ij}} = 1 - \chi_{ij}$$

we prove Lemma 3.2.1.

---

*If X is an ℓ_D-cube (3.3.1) holds if $z\beta^3$ is sufficiently small-one of the hypotheses of Theorem 2.8.1. A hasty argument would be that $\exp(G_f) \to 1$ as $z\beta^3 \to 0$ so $\tilde{K}(X) \equiv \int d\mu[\exp(G_f) - 1] \to 0$ by dominated convergence uniformly in X by translation invariance. This is wrong because the boundary conditions at $\partial\Lambda$ in the covariance of $d\mu$ break translation invariance. Nevertheless by comparing $d\mu$ with Λ and $\Lambda \to \mathbb{R}^3$ using a "change of covariance formula", this argument can be made to work. See (A 4.1.8) in [5].

†We have omitted to remark that estimate (3.2.4) continues to hold if the choice of tree graph T is dependent on $|Y_i|$, i=1,...,n, and is uniform over all such choices. This can be seen by splitting each sum over Y_i into a sum over Y_i constrained by $|Y_i| = n_i$ followed by a sum over n_i, which we defer in estimates (3.2.1) through (3.2.4). This allows us in (3.2.5) to make a random choice of T_n for each n, $|Y_1|,\ldots, |Y_n|$, so that our use in (3.3.4) of a probability measure that depends on n and $|Y_i|$, i=1,...,n is valid.

References

[1] Friedman, H.L.: Ionic Solution Theory. Interscience 1962. Publishers, J. Wiley and Sons. New York.

[2] Lenard, A.: J. Math. Phys. 2, 682 (1961). See also Prager, S.: In: Advances in Chemical Physics, Vol. IV (edited by I. Prigogine). Interscience Publishers, Inc., New York 1961.

[3] Edwards, S.F., Lenard, A.: J. Math. Phys. 3, 778 (1962)

[4] Brydges, D.C.: Comm. Math. Phys. 58, 313 (1978).

[5] Brydges, D.C., Federbush, P.: Comm. Math. Phys. 73,197 (1980).

[6] Glimm, J., Jaffe, A., Spencer, T.: The particle structure of the weakly coupled $P(\phi)_2$ model and other applications of high temperature expansions, Part II. The cluster expansion. In: Constructive quantum field theory (eds G. Velo, A. Wightman). Lecture Notes in Physics, Vol. 25. Berlin, Heidelberg, New York: Springer 1973.

[7] ______________________: Ann. Math. 100, 585 (1974).

[8] ______________________: Ann. Phys. 101, 610 and 631 (1975).

[9] Spencer, T.: unpublished.

[10] Simon, B.: Functional integration and quantum physics. Academic Press, New York, San Francisco, London 1979.

[11] Garsia, A.M.: Continuity properties for multi-dimensional gaussian processes. In: Proc. 6th Berkeley Symposium on Math. Statistics and Probability. 2, 369 (1976).

[12] Brydges, D., Federbush, P.,: J. Math. Phys. 19 2064 (1978).

[13] Penrose, O.: Convergence of fugacity expansions for classical systems. In: Statistical mechanics, foundations and applications, Proc. of the I.U.P.A.P. meeting, Copenhagen 1966, (ed. Thor A. Bak), Benjamin, New York, 1967.

[14] Kunz, H., Souillard, B.: unpublished.

[15] Dimock, J., Glimm, J.: Adv. Math. 12, 58 (1974).

[16] Brydges, D., Federbush, P.: Comm. Math. Phys. 49, 233 (1976) and 53, 19 (1977).

INTERNAL STRUCTURE OF COULOMB SYSTEM IN ONE DIMENSION

Michael Aizenman*

Department of Physics
Princeton University
Princeton, N.J. 08544

1. Introduction (or "Should we consider a one-dimensional system?")

It is generally recognized that one dimensional systems have rather special features. Frequently, a one dimensional model is exactly soluble ; alas, its properties differ significantly from those of the corresponding models in spaces of more realistic dimensions. In this respect, the one dimensional Coulomb systems are no exception. Nevertheless, I consider them interesting and worth our attention for a number of pedagogical reasons.

i) It is of an intrinsic interest that, even in one dimension, Coulomb systems exhibit symmetry breaking. This is manifested in the formation of the Wigner lattice in the jellium, and in the occurrence of "θ-states" in the two component system. The two phenomena can be treated in a unified way.

ii) The phenomenology of the θ-states is closely analogous (and, in fact, related) to that of the θ-vacua of gauge field theories in higher dimensions. These have been widely discussed in relation to the problem of confinement of fractional charges, which is also exhibited in one dimension.

*Supported in part by U.S. National Science Foundation grant PHY-7825390 A01.

iii) The wealth of results presented by J. Fröhlich, and D. Bridges, forms a most convincing case for the extreme utility of the "Sine-Gordon" transformation. However another impression with which I shall leave this school is a fascination by the wide range of approaches and techniques which may be applied to systems with Coulomb interactions. Thinking about some aspects of what we have heard, I cannot suppress the feeling that there are yet other beautiful arguments lying just below the turf, waiting to be brought up. I modestly hope that, for some of you, such feelings would be reinforced by a discussion of some very simple systems.

No attempt is being made to represent in this lecture the vast literature on the subject. (For an interesting account of early works, see [1]). The material discussed here is mostly contained in two papers, [2] with Ph. Martin and [3] with J. Fröhlich. I am indebted to both authors for a very instructive and enjoyable collaboration. Since the two papers are quite accessible, the notes prepared for the proceedings contain mainly qualitative discussions.

2. The Two-Component Coulomb Gas

The one dimensional Coulomb potential energy of a system of charges σ_i, located at positions q_i, is

$$V(\sigma,q) = -\tfrac{1}{2}\sum_{i,j} \sigma_i\sigma_j |q_i - q_j| \tag{2.1}$$

We shall be mainly concerned with a two-component gas, of particles with charges $\sigma_i = \pm e$.

The study of the bulk properties of such a system typically starts with the analysis of the free energy, $P(\beta,z)$ derived from the partition function

$$P(\beta,z) = \lim_{L\to\infty} \frac{1}{2L} \ln \Xi_L(\beta,z), \tag{2.2}$$

and of the related correlation functions. For a neutral gas of charges $\sigma_i = \pm e$, with the partition function

$$\Xi_L(\beta,z) = \sum_{n=0}^{\infty} \frac{z^n}{n!} \sum_{\sigma_i=\pm e} \int_{-L}^{L}\cdots\int dq_i\ldots dq_n\, e^{-\beta V(\sigma,q)}\, \delta_{\sum_1^n \sigma_i, 0} \tag{2.3}$$

(β-inverse temperature, z-fugacity), the free energy was found by Lenard [4] and by Prager [5]. The correlation functions for this

system were expressed in terms of an explicit "transfer matrix" in Edwards and Lenard [6]. Their derivation utilizes what is nowadays called the "Sine-Gordon" transformation.

One starting point for latter developments was the following puzzle. Consider a completely specified configuration of a very large system of charges. One finds the internal forces by calculating the electric field, which is

$$E(x) = \sum_i \sigma_i \operatorname{sgn}(x - q_i) \tag{2.4}$$

This expression is highly non-local. In particular, it becomes meaningless in the usual approximation of a large system by an infinite one. Furthermore, in this limit, with any configuration of charges there will be associated various electric fields, differing by a constant. This raises the following question.

Question 1: What are the forces in a specified infinite configuration of charges?

Or, equivalently: can the electric field in a finite, but large, system be computed quasi-locally? On a more formal level, the question is whether the standard infinite-system formalism would require additional degrees of freedom, e.g. describing the "charges at infinity", to provide a complete description of the dynamics in the thermodynamic limit?

The infinite-system formalism which we have in mind is, for a classical system, a description in a phase space, Ω [7]. The phase space consists of countable configurations, $\omega = \{(\sigma_i, q_i, p_i)\}$, of points in one particle phase space, mod. permutations of the indices. States of the system are described by probability measures, μ, on Ω. Equilibrium states are limits of grand canonical ensembles, possibly with constraints like the neutrality condition.

For a system with the Hamiltonian

$$H(\omega) = \sum P_i^2/2 + \tfrac{1}{2}\sum_{i,j} V(\sigma_i, x_i; \sigma_j, x_j), \tag{2.5}$$

where V is a short range interaction, the equilibrium states are characterized by the Dobrushin-Lanford-Ruelle condition. If states that for any specified configuration, ω_{Λ^c}, in the compliment of a domain Λ, the conditional distribution of the configuration in

Λ, ω_Λ, is

$$\mu(d\omega_\Lambda|\omega_{\Lambda^c}) = \frac{e^{-\beta H_\Lambda(\omega_\Lambda|\omega_{\Lambda^c})}}{\text{Norm.}}\, d\omega_\Lambda \,. \tag{2.6}$$

H_Λ is defined as the sum of only those terms in (2.5) which involve ω_Λ, and $d\omega_\Lambda$ is the corresponding a-priori ("free") measure. A similar condition holds for the states in the configuration space, obtained by integrating out the momenta $\{p_i\}$.

In formulating the DLR condition for Coulomb systems we encounter the problem of the divergence of the electric potential, which is even more severe than that of the electric field. Thus we are led to a second problem.

Question 2: What is the structure of the equilibrium conditional distributions $\mu(d\omega_\Lambda|\omega_{\Lambda^c})$, in the thermodynamic limit?

The explicit correlation functions do not provide evident answers to the above questions.

3. The Electric Field Ensemble

To answer the above questions it is useful to consider the distribution of the electric field in equilibrium ensembles. Once this is understood, it provides a rather direct pathway for the derivation of various features of the Coulomb system.

The equilibrium distribution of $E(\cdot)$ for a finite system, in $[-L,L]$, is determined by the following three factors.

1) The "free measure", $\nu_z(dE)$, which is associated with the a-priori distribution of charges. It describes a piecewise constant function with independent jumps, of the magnitude $\pm 2e$, which occur with equal densities, z.

2) The Gibbs factor, $\exp(-\beta H)$. Its description is facilitated by the electrostatic identity

$$V(\{\sigma_i,q_i\}) = \tfrac{1}{4}\int_{-L}^{L} dx\,|E(x)|^2 - \tfrac{1}{2}(\Sigma\sigma_i)^2 L \,, \tag{3.1}$$

for a system of charges in $[-L,L]$.

3) The neutrality condition, $\Sigma\sigma_i = 0$. In a system with no external potential this has two implications for E. One, is the boundary condition

$$E(-L) = E(L) = 0 . \tag{3.2}$$

The other, is the following constraint which is satisfied throughout the system.

$$E(x) \in 2e\,\mathbb{Z}\,, \quad \forall\, x \in [-L,L]. \tag{3.3}$$

Thus, the equilibrium measure is formally described by

$$\mu_o(dE) = \frac{1}{\text{Norm.}}\, \nu_z(dE)\, e^{-\frac{\beta}{4}\int dx|E(x)|^2}\, \delta[E(\cdot) \in 2e\,\mathbb{Z}] \tag{3.4}$$

where $\delta[\ \ ]$ is supported on the configurations on which the condition in [ ] is satisfied.

The corresponding distribution of $E(\cdot)$ in the thermodynamic limit was first studied by Lenard [8], who showed how it can be used to rederive the free energy and the correlation functions. He proved the following result.

<u>Proposition 1</u>: The distribution of the electric field, in the neutral ensemble, converges to a limit as $L \to \infty$. The limiting state is Markovian, translation invariant, and has the exponential clustering property.

The proof is simplified by the fact that $\nu_z(dE)\delta[E(\cdot) \in 2e\,\mathbb{Z}]$ may be viewed as the distribution of the paths of a random jump process, with values in $2e\,\mathbb{Z}$, for which x is the "time parameter". This process is associated with the semigroup $e^{-x(-z\Delta)}$, on $\ell^2(2e\,\mathbb{Z})$, whose generator is a multiple of the operator defined by

$$\Delta f(u) = f(u+1) + f(u-1) - 2f(u) . \tag{3.4}$$

By an application of the Feynman-Kac formula, one can see that the measure $\mu_o(dE)$ describes the paths of another Markov process whose generator is

$$\mathcal{L} = -z\,\Delta + \frac{\beta}{4}\,\tilde{E}^2 + \text{const.}\,, \tag{3.5}$$

where $\tilde{E}$ is the multiplication operator:

$$\tilde{E}\, f(u) = u\, f(u) \; . \tag{3.6}$$

In particular, the free energy is

$$P(z,\beta) = \text{inf.spec.}[\mathcal{L};\ \text{on}\ \ell^2(2e\,\mathbb{Z})] = \\ = \text{inf.spec.}[\mathcal{L};\ \text{on}\ L^2(\mathbb{R})] \tag{3.7}$$

(with a slight abuse of notation).

We shall skip further details and the exact formulae. These can be found in [8,2], and are not that relevant for the following discussion.

Since there is an obvious correspondence (see fig. 1):

$$\text{electric field } (E(\cdot)) \to \text{charge configuration } (\{\sigma_i, q_i\}) \tag{3.8}$$

a state of $E(\cdot)$ determines a state of the system of charges. To answer Question 1 one should find a way of inverting this correspondence. The basic observation made in [2] was that, despite the manifest noninvertibility, the association (3.8) is <u>essentially 1-1</u>; in the sense that it may be inverted by a quasi-local "prescription" which gives the right answer with probability one. Physically, the reason is that the various electric field configurations which differ by a constant have different bulk energies. It is the one with the minimal energy density (out of all the $E(\cdot)$ fields differing by $2e\,n$, $n \in \mathbb{Z}$) which would occur in conjunction with a given charge configuration. The explicit reconstruction of $E(\cdot)$, from the positions of its singularities, is done as follows.

Let $y > x$. Assuming there are no particles at x or y:

$$E(y) = E(x) + 2 \sum_{q_i \in [x,y]} \sigma_i \tag{3.9}$$

By the ergodic theorem, which is applicable by proposition 1, for μ_o-almost every (a.e.) $E(\cdot)$:

$$\lim_{r\to\infty} \frac{1}{r} \int_x^{x+r} dy\, E(y) = \langle E(0)\rangle_o \; , \tag{3.10}$$

$\langle - \rangle_o$ denotes here the expectation value in the state μ_o. Substituting (3.9) in (3.10) we obtain the desired relation

$$\begin{aligned} E(x) &= \lim_{r\to\infty} -2 \sum_{q_i \in [x,x+r]} \left(1 - \frac{q_i - x}{r}\right) \sigma_i + \langle E(0)\rangle_o \\ &= \lim_{r\to\infty} 2 \sum_{q_i \in [x-r,x]} \left(1 + \frac{q_i - x}{r}\right) \sigma_i + \langle E(0)\rangle_o , \end{aligned} \tag{3.11}$$

which holds for "almost every" (a.e.) $\{\sigma_i, q_i\}$. By the charge conjugation symmetry of $\langle - \rangle_o$,

$$\langle E(0)\rangle_o = 0 . \tag{3.12}$$

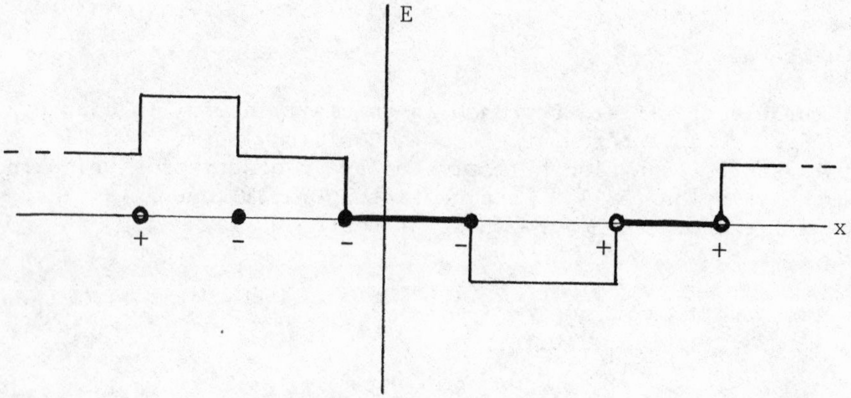

Fig. 1: The electric field in relation to a charge configuration

One main feature of the relation (3.11) is not restricted to one-dimension. Question 1 may be raised (although less dramatically) with respect to Coulomb systems in any dimension, since the electric field of a single charge is never integrable at ∞. For such a system, let $\underline{E}_i(\cdot)$ denote the electric field produced by the charge at q_i. Then, as a consequence of Newton's theorem,

$$\frac{1}{|B_r|}\int_{B_r} d\underline{y}\ \underline{E}_i(\underline{y}) = \underline{E}_i(0)\ \min(1,|\frac{\underline{x}}{r}|^d) \qquad (3.13)$$

for any ball $B = \{\underline{x} \in \mathbb{R}^d \mid |\underline{x}| \leq r\}$. Thus a possible generalization of (3.11) could be based on the following relation

$$\underline{E}(0) = \sum_{\underline{q}_i \in B_r} \underline{E}_i(0)(1 - |\frac{\underline{q}_i}{r}|^d) + \frac{1}{|B_z|}\int_{B_r} d\underline{y}\ \underline{E}(\underline{y}) \qquad (3.14)$$

(For $d = 1$, (3.14) corresponds to the average of the two expressions in (3.11)).

A feature of (3.11) which seems restricted to one dimension is that it permits the reconstruction of $E(x)$ from the distribution of charges on either side of x. We shall discuss the implication of this observation in section 7i).

4. θ-states

The electric-field approach is specially useful in considering the effect of an external field on a Coulomb system. The discussion of this subject can also be motivated by an observation which follows from the relation (3.11). Exponentiating (3.11) one finds that with probability one, in the free neutral system,

$$\lim_{r\to\infty} \exp[i\ 2\pi \sum_{q_i \in [0,r]} (\frac{q_i}{r})(\frac{\sigma_i}{e})] = 1 \qquad (4.1)$$

This is puzzling since for any fixed r the above quantity can take any unitary value. One is led to ask whether the system has other equilibrium states, characterized by a phase. As we shall see, an external field produces such states.

When a finite neutral system is placed in a constant external field, D, the total Hamiltonian of the system is still described by the right side of (3.1), where $E(\cdot)$ denotes the total electric field. The distribution of $E(\cdot)$ in a finite ensemble is as described above, except for the change of the boundary condition (3.2) to

$$E(-L) = E(L) = D\ , \qquad (4.2)$$

and the corresponding change of (3.3) to

$$E(x) \in D + 2e\ \mathbb{Z} \tag{4.3}$$

As for the case $D = 0$, large fluctuations of $E(\cdot)$ are strongly suppressed by the Gibbs factor, and it should be no surprise that the distribution of $E(\cdot)$ converges, as $L \to \infty$, to a translation-invariant limit. The proof of this assertion does not differ from the case $D = 0$. The system maintains $\exp[-|x|(\mathcal{L} + \text{const.})]$ as its transfer matrix; however, because of (4.3), $\mathcal{L}$ should be viewed as acting on $\ell^2(D + 2e\ \mathbb{Z})$. Correspondingly, the free energy in a constant external field is:

$$P(z,\beta;D) = \text{inf.spec.}[\mathcal{L};\ \text{on}\ \ell^2(D + 2e\ \mathbb{Z})] \tag{4.4}$$

In the Fourier-transform representation the operator $(-z\Delta + \frac{\beta}{4}\tilde{E}^2)\restriction \ell^2(D + 2e\ \mathbb{Z})$ is

$$e^2\beta(i\frac{d}{dx} + \frac{D}{2e})^2 + 2z(\cos x + 1)\ , \tag{4.5}$$

in $L^2([-\pi,\pi])$, defining $i\frac{d}{dx}$ with the periodic boundary conditions. Under the unitary transformation $U = e^{-ixD/(2e)}$ this operator is transformed to

$$-e^2\beta\,\Delta^{(\theta)} + 2z(\cos x + 1)\ , \tag{4.6}$$

where the Laplacian $\Delta^{(\theta)}$ is defined with the boundary conditions $f(-\pi) = f(\pi)e^{i\pi D/(2e)}$.

The above considerations lead to the following result ([2,3]).

Proposition 2: For each value of D, $E(\cdot)$ has a limiting probability distribution, obeying (4.3), which depends only on $\theta = D_{\text{mod.}2e}$. Furthermore, these θ-states are Markovian, translation invariant, and clustering in x. The free energy in the external field, $P(\beta,z;\theta)$, is given by the lowest-band eigenvalue of $-e^2\beta\Delta + 2z(\cos x + 1)$ (on $L^2(\mathbb{R})$), at the Bloch momentum $k = \theta/(2e)$.

Thus, the dependence of the state of the system on the external field D is periodic, with the period 2e. The periodicity is explained by the partial screening of D by opposite charges accumulating at the two sides of the boundary. The screening, however, is not com-

plete, since the effect of the fractional part of the field cannot be canceled by integral charges.

The persistence of the effect of D on $E(\cdot)$ is manifested by (4.3). Still, (4.3) in itself does not imply that the distributions of the charges are distinct for the various θ-states. (E.g., (4.3) is also satisfied in a family of states which differ only by an overall shift in E?). At this point the observation made in [2] becomes instrumental. Repeating for the θ-states the argument which led to (3.13) we see that the total $E(\cdot)$ can be reconstructed from the configuration of the charges. In fact, (3.13) holds also for the θ-states, with $\langle E(\cdot)\rangle_\theta$ replacing $\langle E(\cdot)\rangle_0$. Since in a θ-state $e^{2i\pi E(0)/(2e)} = e^{2i\pi\theta/(2e)}$, we obtain the following characterization of the typical configurations.

Lemma 1: In each θ-state, with probability one:

$$\lim_{r\to\infty} \exp[i\, 2\pi \sum_{q_i \in [0,r]} (\frac{q_i}{r})(\frac{\sigma_i}{e})] = \exp[i2\pi g(\theta)], \qquad (4.7)$$

for $g(\theta) = [\theta - \langle E(0)\rangle_\theta]/(2e)$.

The function $\langle E\rangle_\theta$ is explicitly expressible in terms of the ground state of $\mathcal{L}$ on $\ell^2(\theta + 2e\,\mathbb{Z})$. Its analysis shows that $g(\theta)$ is strictly monotone in $\theta \in [0,2e)$, taking values in $[0,1)$. (4.7), and the invertibility of $g(\theta)$, imply ([2]):

Proposition 3: The distributions of the charges in the various θ-states are mutually singular.

It may amuse some to observe that the θ=e-state is also produced as the limit of free ensembles, D = 0, with the total charge $\Sigma\sigma_i$=e, or any other odd multiple of e. The determining factor is the range of values of $E(x)$ is the ensemble.

Regarding the behavior in external field the system may be viewed as a dielectric (an observation made also in [9]). At low values of $\frac{z}{\beta e^2}$ the charge configurations have the appearance of a gas of dipoles, which interact only when overlapping. The effect of D is to polarize the dipoles, disassociating a finite number of them. The average electric field $\langle E\rangle_\theta$ is non-zero except at θ = 0,e. The system has no phase transition, however in the "plasma limit", $\frac{z}{\beta e^2} >> 1$, the above description is less effective due to the large overlap of the dipoles.

We shall return to the θ-states in section 8. The phase non-uniqueness, which they represent, is replaced by the translation symmetry breaking in the system which will be discussed next.

5. Translation-Symmetry Breaking in Jellium

Jellium is a Coulomb system where only one of the species (of charge -e) is mobile, simulating the situation in solids. The distribution of the other charges is replaced by a uniformly charged background. A special feature of the system is translation-symmetry breaking, interpreted as the formation of the "Wigner lattice", in which the charges oscillate around periodically arranged sites. The phenomenon was proven by Kunz [10] for classical systems, and by Brascamp and Lieb [11] for quantum systems at low enough temperatures. The method discussed here was applied to the system in [2], where the translation-symmetry breaking was given a structural proof.

The interaction energy of a configuration of charges -e at the positions q_i, with a uniform background charge of density ρ in $[-L,L]$, is

$$V(\{q_i\}) = -e^2 \sum_{i<j} |q_i - q_j| + e\rho \sum_i (q_i^2 + L^2) + \text{const.}(L) \tag{5.1}$$

(computed from (2.1)).

For strict neutrality one should impose the condition

$$\Sigma\sigma_i = 2L\rho \ , \tag{5.2}$$

which requires $2L\rho/e \in \mathbb{Z}$. It would follow from our analysis that even with this constraint the limiting state, as $L \to \infty$, is not pure. To obtain an extremal state (with respect to the decomposition to translation invariant states) L should be further restricted, e.g. by:

$$L\,\rho/e \in \mathbb{Z} \ . \tag{5.3}$$

Throughout the system the electric field has constant slope $dE/dx = 2\rho$, for $x \notin \{q_i\}$, and discrete jumps, by $-2e$, at $x = q_i$ (see fig. 2). (5.2) and (5.3) imply the boundary condition

$$E(-L) = E(L) = 0 \tag{5.4}$$

and the persistent (as $L \to \infty$) restriction

$$E(x)/(2e) \in x/x_o + \mathbb{Z} , \tag{5.5}$$

with $x_o = e/\rho$.

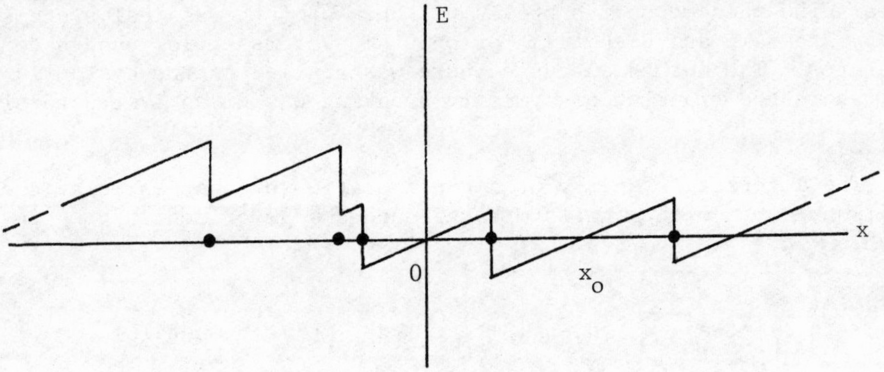

Fig. 2: The electric field in jellium

The manifest non-invariance of (5.5) was used in [2] in conjunction with the methods outlined in section 3. These lead to the following result.

Proposition 4: i) The neutral-ensemble distribution of $E(\cdot)$ converges to a limit, as $L \to \infty$ satisfying (5.3), in which (5.5) is still valid. The limiting state is invariant under shifts by x_o, and has exponential clustering.

ii) In the above limit $E(\cdot)$ can be recovered, with probability one, from the positions of charges by:

$$E(x) = \lim_{r \to \overset{+}{(-)}\infty} \overset{+}{(-)} 2e \sum_{q_i \in [x,x+r]} \left(1 - \frac{q_i - x}{r}\right) - e\, r/x_o . \tag{5.6}$$

Exponentiating (5.6), and using (5.5), we obtain a non-invariant value for the following limit, which exists for "almost every" configuration.

$$\lim_{n\to\infty} \exp[2\pi i \sum_{q_i \in [x, x+2nx_o]} [(q_i - x)/(2nx_o)]] = \exp(-2\pi i x/x_o) \tag{5.7}$$

Corollary: The above limiting state is not invariant under continuous translations.

To summarize: electric-field considerations directly exhibit the translation symmetry breaking, as soon as one shows that $E(\cdot)$ attains a limiting state. The necessary compactness is a consequence of the suppression of fluctuations by the Gibbs factor. This approach may also be applied to quantum systems, where the configuration of point charges on a line is replaced by a configuration of Brownian paths which, for the Maxwell-Boltzmann statistics, are periodic in time,with the period β. Bose-Einstein statistics are obtained by relaxing the periodicity to that of the configuration as a whole. Fermi-Dirac systems are obtained (in one-dimension,!) by excluding particle collisions. One should, however, work out the necessary compactness estimates.

6. Other Special Features

The Coulomb interaction produces some special features, which will be discussed here for the two component system. We find the electric-field formalism very suitable for their analysis.

i) Unboundedness of the potential

The compactness of the distribution of $E(\cdot)$ may be more appreciated if one notices its lack for the electric potential $\phi(\cdot)$. If $\phi(\cdot)$ exists, in the limit, then

$$\phi(0) - \phi(x) = \int_o^x du\, E(u) \tag{6.1}$$

It is easy to show (e.g. using the Markov property and conditioning on the first site where $E = 0$) that

$$\langle E(u)\, E(0)\rangle_o > 0 \, . \tag{6.2}$$

Therefore

$$\frac{1}{x} \langle |\phi(0) - \phi(x)|^2 \rangle_o \xrightarrow[x\to\infty]{} \int_{\infty}^{\infty} du \;\langle E(u)\, E(0)\rangle_o > 0 \, . \tag{6.3}$$

With a little more effort one may show that the whole distribution of

$\left| \int_0^x du\ E(x) \right|$ is diverging to ∞, as $x \to \infty$. Thus there is no field

$\phi(\cdot)$, satisfying (6.1) and having a translation invariant distribution.

ii) Suppression of charge fluctuations

The total charge in the interval $[0,x]$ is

$$Q_{[0,x]} = \sum_{q_i \in [0,x]} \sigma_i = [E(x) - E(0)]/2 \; . \tag{6.4}$$

This implies that as $x \to \infty$ $Q_{[0,x]}$ attains a well defined distribution. In particular, we have:

Lemma 2:

$$<|Q_{[0,x]}|^2>_\theta \xrightarrow[(x\to\infty)]{} [<E(0)^2>_\theta - E(0)>_\theta^2]/2 < \infty \tag{6.5}$$

(6.5) shows that the bulk charge in the region $[0,x]$ is only of the order of the region's boundary. This might be a general feature of Coulomb systems. In [12,13] it was shown that such "abnormal" charge fluctuations are implied by fast enough clustering of the correlation functions. Some results on charge fluctuations in higher dimensions were reported in Fröhlich's lecture. Notice that (6.4) is just the one-dimensional Gauss theorem, which should be useful in any dimension.

iii) Decomposition to neutral clusters

Typical configurations of charges may be partitioned to neutral clusters. For concreteness, we define such a partition using the boundary points of the set $\{x \in \mathbb{R} | \ E(x) > 0\}$. Let $|\mathcal{C}_n|$, $n \geq 1$, denote the length of the n-th cluster lying entirely to the right of the origin.

Lemma 3: Denoting by $<\{-\}>_\theta$ the probability in a θ-state, we have:

$$<\{|\mathcal{C}_n| > x\}>_\theta \leq e^{-\lambda x} \tag{6.6}$$

for some $\lambda > 0$ (independent of θ).

The assertion can be proved using the exponential clustering property of $E(\cdot)$, and the fact that $|\zeta_n| > x$ implies that $E(\cdot)$ does not change its sign over an interval of length x.

The bound (6.6) is another reflection of the suppression of charge fluctuations. For comparison, we note that for the free state (i.e. the two component ideal-gas)

$$\text{Prob.}(|\zeta_n| > x) \geq c/\sqrt{zx} \ . \tag{6.7}$$

In particular, $\langle|\zeta_n|\rangle_{\text{free}} = \infty$. The proof is left to the reader, with the hint that $|\zeta_n|$ has the distribution of the recurrence time for a random walk (corresponding to $E(\cdot)$).

The decomposition of charge configurations to neutral clusters was studied and applied by Lenard. Recently Fröhlich and Spencer [14] developed an efficient decomposition to neutral clusters for Coulomb systems in higher dimensions.

iv) Shielding

The ratio of the correlation functions $\rho_2(\underset{(-)}{+} e,x;\sigma,o)/\rho_1(\sigma,0)$ describes the density of charges at x, in an ensemble in which a charge $\sigma = \pm e$ is observed at the origin. The fixed charge polarizes the medium, attracting the total charge:

$$Q_{\text{induced}} = \int_{-\infty}^{\infty} dx[\rho_2(e,x;\sigma,0) - \rho_2(-e,x;\sigma,0)]/\rho_1(\sigma,x) \tag{6.8}$$

The observed charge is shielded if

$$Q_{\text{induced}} = -\sigma \ . \tag{6.9}$$

Lemma 4: Charges $\pm e$ (and their multiples) are shielded in each of the θ-states.

To prove it, one may use Gauss theorem, (6.4), by which

$$\sigma + Q_{\text{induced}} = \lim_{x\to\infty} \langle E(x) - E(-x)|\sigma,0\rangle_\theta /2 \ , \tag{6.10}$$

where $\langle\cdot|\sigma,0\rangle_\theta$ corresponds to conditioning on the existence of a charge σ at $q = 0$. Due to the mixing property of $\langle-\rangle_\theta$, the

expression in the limit converges to $\langle E\rangle_\theta - \langle E\rangle_\theta = 0$.

Similar considerations show that, in general, non-integral charges, in units of e, are not shielded (in one dimension). This fact is related to the "confinement", discussed next.

v) Confinement of fractional charges

Let us consider (following the discussion in [3]) the equilibrium distribution of a pair of charges $\pm\alpha$ inserted into a neutral Coulomb gas, which is in one of the θ-states.

If the charge $+\alpha$ is fixed at the origin, the density at x of the distribution of the charge $-\alpha$ is proportional to the following function:

$$\rho_{\alpha;\theta}(x) = \lim_{L\to\infty} \frac{\Xi_L(\beta,z,0;(\alpha,0),(-\alpha,x))}{\Xi_L(\beta,z,\theta)} \quad . \tag{6.11}$$

$\Xi_L(\beta,z,\theta;\{(\sigma_i,q_i)\})$ is the partition function of the gas in the the presence of external charges σ_i at the positions q_i, and an external field θ.

There are now two qualitatively different possibilities.

i) If $\int_{-\infty}^{\infty} dx\,\rho_{\alpha;\theta}(x) = \infty$, then the charge $-\alpha$ would drift away and would not be observed in any finite region.

ii) If $\int_{-\infty}^{\infty} dx\,\rho_{\alpha;\theta}(x) < \infty$, then the charge $-\alpha$ has a probability distribution on the line, whose density is $\rho_{\alpha;\theta}(x)/\int_{-\infty}^{\infty} dy\,\rho_{\alpha;\theta}(y)$.

In the second case, the charges exhibit confinement. Such a situation would not occur in systems with short range interactions.

The asymptotic behavior of $\rho_{\alpha;\theta}$ is described by the following result.

Lemma 5: For any $\theta \in [0,2e)$, $\alpha \in \mathbb{R}$

$$\lim_{x\to \underset{(-)}{+}\infty} P_{\alpha;\theta}(x)\, \exp\{|x|[P(\beta,z;\theta \underset{(-)}{+} 2\alpha) - P(\beta,z;\theta)]\} = C_{\underset{(-)}{+}} \tag{6.12}$$

for some, explicitly known, constants $C_+, C_- \in (0,1]$.

To explain it, we observe that the total electric field in the system satisfies

$$E(y) = \begin{cases} \theta + 2e\,\mathbb{Z} & y \notin [0,x] \\ \theta + 2\alpha\,\mathrm{sgn}\,x + 2e\,\mathbb{Z} & y \in [0,x] \end{cases} \tag{6.13}$$

The effect of the external charges is, therefore, to shift the distribution of $E(\cdot)$ in $(0,x)$ (see fig. 3).

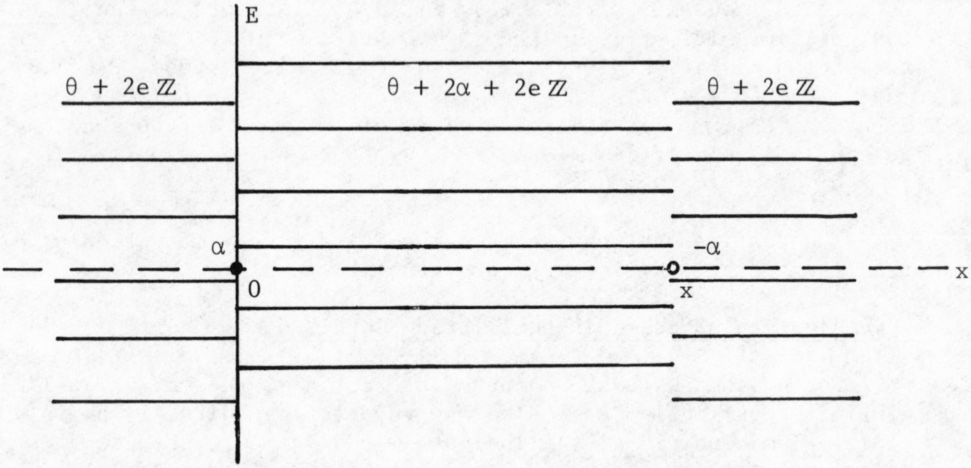

Fig. 3 The range of values of $E(\cdot)$ in a system with a pair of fractional charges.

The formalism discussed in section 3 leads now to the following expression for the partition function

$$\Xi_L(\beta,z,\theta;(\alpha,0),(-\alpha,x)) = \int \nu(dE)\, e^{-\frac{\beta}{4}\int_{-L}^{L} dy\,|E(y)|^2}\, \delta[E(\cdot) \text{ obeys } (6.13)] \tag{6.14}$$

(6.12) follows from (6.14) after taking care (e.g. using the transfer matrix) of the finite contribution due to transition at 0 and at x.

In fact, reflection positivity [15] (which follows from the reflection invariance of $\langle\{dE\}\rangle_\theta$ and the Markov property) implies

that

$$P_{\alpha;\theta}(x) \le \lim_{n\to\infty} P_{\alpha;\theta}(nx)^{\frac{1}{n}} = \tag{6.15}$$

(by 6.12)

$$= \exp\{-|x|[P(\beta,z;\theta + 2\alpha\ \mathrm{sign}(x)) - P(\beta,z;\theta)]\} .$$

$P(\beta,z;\theta)$ is related by proposition 2 to a well studied, periodic, function of θ [16]. It has its minima at $\theta \in 2e\mathbb{Z}$, is strictly increasing in θ on $[0,e]$ and is symmetric about $\theta = e$. These properties, together with (6.15), imply confinement of fractional charges, in units of e, in the $\theta = 0$ state. In the $\theta = e$ state, the second charge is always expelled to infinity, since $P(\beta,z,\cdot)$ is maximal at $\theta = e$. In other states we would find both behaviors, depending on x. For suitable values of θ and α the second charge is confined only if restricted to $\{x > 0\}$, and is expelled in $\{x > 0\}$ (or vice versa).

The confinement results from the inefficiency of screening of a fractional charge by integral charges (in $d = 1$).

7. Adjustments Needed in Equilibrium Conditions

For systems with short range interactions, thermodynamic limits of finite equilibrium ensembles may be described directly by various equilibrium conditions. As indicated in section 2, some of these criteria cannot even be formulated for Coulomb systems. In this section we discuss the required modifications (following [2]).

i) DLR condition

In the DLR condition (see Section 2) a reference is made to the potential induced in a finite volume by the charges which lie outside. However, as explained in Section 6i), this quantity is not well defined. This problem brought us to Question 2, which we shall now answer.

Let us consider the conditional distribution of charges in a region $[a,b]$, for a given configuration of charges in $\mathbb{R}\setminus[a,b]$.

Lemma 1 permits the reconstruction of $E(b)$ and $E(a)$ from the position of the charges in (b,∞) and, correspondingly in $(-\infty,a)$. Since the total charge in $[a,b]$ is $Q_{[a,b]} = [E(b+) = E(a-)]/2$, it follows that the conditional distribution of $Q_{[a,b]}$ is restricted to a single value (!).

Furthermore, as a consequence of the Markov property of $E(\cdot)$,

the external charge configuration has no other effect on the distribution in [a,b], except through the boundary values of $E(\cdot)$. Considering the equilibrium distribution of $E(\cdot)$ in [a,b] subject to the constraints at the boundary, one obtains the following result.

Proposition 5: For any θ-state, the conditional distribution of charges in [a,b] is an equilibrium ensemble in the external field $D = [E(b+) + E(a-)]/2$, subject to the charge constraint $Q_{[a,b]} = [E(b+) - E(a-)]/2$. $E(b+)$ and $E(a-)$ are explicit functions of the configuration in $\mathbb{R}\setminus[a,b]$.

While the DLR condition describes the conditional distributions by grand-canonical ensembles, for Coulomb systems the distributions are microcanonical with respect to the charge. (Still, the states are "canonical" in the terminology of [17].)

An implication of the above analysis is that adding a single charge to a "typical" configuration produces one which is "a-typical". Similar singularity was investigated by Dyson and Mehta [18] for the distribution of eigenvalues of random matrices, which is given by a Gibbs state with the logarithmic potential. They suggested an application of such an effect to the study of nuclear interactions.

ii) KMS condition

Equilibrium states defined over the phase-space can be characterized by the (classical) Kubo-Martin-Schwinger condition. The adaptation of this condition to classical systems was made in [19], and further studied in [20,21]. It states that for any pair of local C^2 functions $f,g: \Omega \to \mathbb{R}$:

$$\langle\{f,g\}\rangle = \beta\langle f\{H,g\}\rangle \quad . \tag{7.1}$$

{ , } denotes here the Poisson bracket:

$$\{f,g\} = \sum_i \frac{\partial}{\partial p_i} f \frac{\partial}{\partial q_i} g - \frac{\partial}{\partial q_i} f \frac{\partial}{\partial p_i} g \ . \tag{7.2}$$

For systems with finite range interactions $\{H,\cdot\}$ makes sense even in the thermodynamic limit, where H can no longer be defined as a function.

Since $-\frac{\partial}{\partial q_i} H$ is the force on the i-th particle, the formulation of the KMS condition for Coulomb systems is not trivial. However, by now we have a method of evaluating the forces in a given configuration (lemma 1). Indeed, as follows from the structure of the conditional distributions discussed above, the corresponding KMS condition is satisfied in all the θ-states ([2]).

iii) BBGKY hierarchy

The correlation functions of an equilibrium state are characterized by the stationary Bogoliubov-Born-Green-Kirkwood-Yvon hierarchy. The Lausanne group suggested to study Coulomb systems in external field via this hierarchy of equations, by adding an effective-field term, [12]. The resulting equations, with the momentum integrated out, are:

$$\frac{\partial}{\partial q_i} \rho_n (\{\sigma_k, q_k\}_{k=1,\ldots,n}) =$$

$$= \beta[E_{eff.} \sigma_i + \sum_{j \neq i} F_{i,j}] \rho_n (\{\sigma_k, q_k\}_{k=1,\ldots,n}) +$$

$$+ \beta \lim_{r\to\infty} \int_{r}^{r} dq_{n+1} \sum_{\sigma_{n+1}=\pm e} F_{i,n+1} \rho_{n+1} (\{\sigma_k, q_k\}_{k=1,\ldots,n+1}) \tag{7.3}$$

where F is the force

$$F_{i,j} = \sigma_i \sigma_j \operatorname{sgn}(q_i - q_j) .$$

With the summation first, the integral in (7.3) is well defined; and it is interesting to see what value should be taken for E_{eff}. The answer is ([2]) that the correlation functions in a θ-state satisfy (7.3) with

$$E_{eff} = \langle E \rangle_\theta . \tag{7.4}$$

This can be understood by considering the forces on the particles in $[-r,r]$. In addition to the mutual interactions, there are the effects of the external field and the other charges. As explained by proposition 4, the magnitude of the total field due to sources not in $[-r,r]$ is $[E(-r) + E(r)]/2$, whose average value is $\langle E \rangle_\theta$.

8. Relation to "θ-vacua"-Phenomenology of Gauge Fields in Higher Dimensions

As we saw in the lectures of Fröhlich and Brydges, the Sine-Gordon representation offers a powerful tool for the study of systems with pair interactions of positive type (specially when there is charge symmetry). It also provides us with a link between the statistical-mechanical system of Coulomb gas and a field. It turns out that the θ-states are analogous, and related, to θ-vacua of an Abelian

gauge field theory in two (space-time) dimensions. (A point clarified by J. Fröhlich). The relation reflects the fact that the Abelian gauge field theory is a covariant theory of electromagnetizm, and it reduces to the instantaneous Coulomb interaction in the non-relativistic limit. The confinement of the fractional charge which we have discussed in Section 6 is related, physically and mathematically, to the confinement of fractionally charged Wilson loop in the gauge theory. These are but the simplest examples of the phenomenology of θ-vacua and confinement, which has been widely discussed for non-Abelian fields in higher dimensions, in the quest for the understanding of quark confinement. Our discussion follows that of [3].

i) The Sine-Gordon representation

Let $\rho(x) = \sum \sigma_i \delta(x-q_i)$ denote the charge density. The Gibbs factor in (2.3), with the imposed neutrality condition, can be expressed by the following functional integral

$$e^{-\beta V(\sigma,q)} \delta_{\sum_1^n \sigma_i, 0} \equiv \exp[-\frac{\beta}{2} \iint dx\, dy\, \rho(x)\, |x-y|\, \rho(y)]\, \delta_{\int dx \rho(x), 0}$$

$$= \langle\langle \exp[i \int dx\, \rho(x)\, \phi(x)] \rangle\rangle \quad , \qquad (8.1)$$

Here $\langle\langle - \rangle\rangle$ represents expectation value for the Guassian random field, ϕ, with the covariance $(\beta(-\Delta)^{-1} + \varepsilon^2)^{-1}$, in the $\varepsilon \downarrow 0$ limit. Plainly speaking, $\langle\langle - \rangle\rangle$ is the average over Brownian paths, $\phi(x)$ with x as "time" parameter, whose initial point, $\phi(0)$, is uniformly averaged over $\mathbb{R}$.

Inserting (8.1) in (2.3) one obtains the following expression for the partition function.

$$\Xi_L(\beta,z) = \langle\langle \exp 2z \int_{-L}^{L} dx \cos[e\, \phi(x)] \} \rangle\rangle \qquad (8.2)$$

The correlation functions have a related expression,

$$\rho_n(\{\sigma_j, q_j\}_{i=1,\dots,n}) = \langle \sum_1^n e^{i\sigma_j \phi(q_j)} \rangle_{(\phi)} \quad , \qquad (8.3)$$

where

$$\langle - \rangle^{(\phi)} = \lim_{L\to\infty} \overline{\underline{\ -\ }} (\beta,z)^{-1} \langle\langle - \exp\{2\,z \int_{-L}^{L} dx \cos[e\,\phi(x)]\}\rangle\rangle \quad . \tag{8.4}$$

This is the Sine-Gordon representation. One may view it as describing the system via the canonically conjugate field (ϕ) to the charge field (ρ). The state $\langle - \rangle^{(\phi)}$ is formally described by the measure

$$\text{"}\prod_x d\phi(x) \exp[-\mathcal{A}(\phi)]/\text{Norm."}$$

with the action

$$\mathcal{A}(\phi) = \int dx\{\frac{\beta^{-1}}{2} |\nabla\phi(x)|^2 - 2\,z \cos[e\,\phi(x)]\} \tag{8.5}$$

The variables $\exp[i \int dx\; g(x)\; \phi(x)]$ act as shift operators, $\rho \to \rho + g$, which create charge of density $g(x)$ in the "vacuum" $\langle - \rangle^{(\phi)}$.

The representation (8.3) was used by Edwards and Lenard [6] to solve the two-component system (soluble by the Markov property of $\langle - \rangle^{(\phi)}$). The positivity of the measure (8.5) is a consequence of the charge symmetry. It has been used by Fröhlich and Park [22] to derive bounds on the correlation functions and prove their existence in the thermodynamic limit in any dimension.

Since $\cos(e\phi)$ is bounded, ϕ in the state $\langle - \rangle^{(\phi)}$ is locally as regular as the Brownian motion (in particular continuous). The state is invariant under shifts of ϕ by $2\pi/e$, with a bias towards the values $\phi = 2\pi/e\,\mathbb{Z}$. The action-minimizing transitions are described by the solutions of the variational equation

$$-\beta^{-1}\, \Delta\,\phi(x) + 2\,z\,e\,\sin(e\,\phi(x)) = 0 \tag{8.6}$$

with the boundary conditions

$$\phi(-\infty) = 0,\ \phi(+\infty) = \begin{cases} +2\pi/e & \text{"instanton"} \\ -2\pi/e & \text{"anti-instanton"} \end{cases}$$

The transition length is $\approx (z\beta e^2)^{-\frac{1}{2}}$, to be compared with the distance $\approx (\beta e^2)^{-1}$ over which the Brownian fluctuations cause a drift of ϕ by $2\pi/e$. This implies that in the plasma limit, $\frac{z}{\beta e^2} \gg 1$, the typical configurations of ϕ resemble a gas (superposition) of "instantons" and "anti-instantons". In the region $\frac{z}{\beta e^2} \ll 1$, the

fluctuations of ϕ obscure this picture (compare it with the remarks at the end of Section 4).

ii) θ-vacua

The θ-states of Section 4 are produced by an application of external field $D = \theta$. The same effect is obtained by placing a pair of charges $\theta/2, -\theta/2$ at the positions $-L, L$; $L \to \infty$. Thus the correlation functions of the θ-states are still given by (8.3), with $\langle - \rangle^{(\phi)}$ replaced by the states:

$$\langle - \rangle_\theta^{(\phi)} = \lim_{L\to\infty} \frac{\langle - \, e^{i\,\theta/2\,[\phi(-L) - \phi(L)]} \rangle_\theta^{(\phi)}}{\langle e^{i\,\theta/2\,[\phi(-L) - \phi(L)]} \rangle_\theta^{(\phi)}} \tag{8.7}$$

Up to a uniformly small term, $[\phi(L) - \phi(-L)]\, e/(2\pi)$ is the total instanton charge in $[-L,L]$, i.e. the number of instantons minus the number of anti-instantons. Thus the states $\langle - \rangle_\theta^{(\phi)}$ may be viewed as describing the gas of instantons at a complex fugacity.

Formally, the states $\langle - \rangle_\theta^{(\phi)}$ are obtained by adding to the action in (8.5) the term

$$\delta \mathcal{A} = i\,\theta/2 \int dx \frac{d}{dx} \phi(x) \; . \tag{8.8}$$

This procedure is strictly analogous to the addition of the term

$$\delta \mathcal{A} = -i\,\theta/2 \int d^2x \; \mathrm{curl}\, \underline{A}\,(x) \tag{8.9}$$

to the action of a two dimensional (Euclidean) gauge field theory, with a vector potential $\underline{A}$ and some matter fields. It was suggested in [23] that the two dimensional spinor QED should have a family of vacua, parametrized by an angle, with a nonvanishing electric field. These θ-vacua are formally obtained by the above construction. The idea was extended in [24] (on a formal level) to the two space-time dimensional Higgs model, whose action has stationary vortex solutions. (In this case, the added term counts the total vortex charge.) In its lattice version, the model was rigorously analyzed in [25,26].

iii) Confinement in θ-vacua

The interaction of matter with the electromagnetic field is via the coupling of the total current $J^\mu(x)$, described by the term

$$\delta \mathcal{A} = i \int dx \; J^\mu(x)\, A_\mu(x) \tag{8.10}$$

in the total (Euclidean) action. In order to analyze the confinement of external charges, as described in Section 6iii), one evaluates

$$\langle e^{i \int dx\, J^\mu(x)\, A_\mu(x)} \rangle_\theta^{(A,\ldots)}$$

for a fixed external current. The case of two charges $\pm\alpha$, separated to a distance L for a time T, is described by a current $\hat{J}$ which spans a loop in the space time, forming the boundary of a $L \times T$ rectangle-Λ. $\hat{J}$ is singular with $\int d^2x \ \hat{J}^\mu(x) A_\mu(x) = \alpha \oint_{\partial\Lambda} dx_\mu A^\mu(x)$. One may use this <u>Wilson-loop</u> to define an effective potential ,

$$e^{-T\, U_{eff}(L)} = \langle e^{-i\,\alpha \oint_{\partial\Lambda} dx^\mu\, A_\mu(x)} \rangle_\theta^{(A,\dots)}$$

$$= \langle e^{-i\,\alpha \int_\Lambda d^2x \ \mathrm{curl}\, A(x)} \rangle_\theta^{(A,\dots)} \ , \qquad (8.11)$$

where the last equality follows by integration by parts.

The case discussed in section 6iii) can be viewed as a particular example of this procedure (in one dimension). The function defined by (6.11) can also be expressed as:

$$\rho_{\alpha;\theta}(L) \equiv e^{-U_{eff}(L)} = \langle e^{-i\,\alpha\,[\phi(L) - \phi(-L)]} \rangle_\theta^{(\phi)}$$

$$" = \langle e^{-i\,\alpha \int_{-L}^{L} dx \frac{d}{dx}\, \phi(x)} \rangle_\theta^{(\phi)} " \qquad (8.12)$$

In [25,26] chessboard estimates were used to prove that in the lattice Higgs model:

$$(0\le) < e^{i\,\alpha \oint_{\partial\Lambda} dx^\mu A_\mu(x)} >_\theta \ \le e^{-[\varepsilon(\theta+2\alpha) - \varepsilon(\theta)]\, L \cdot T} \qquad (8.13)$$

where $\varepsilon(\theta)$ is the energy density in the θ-vacuum ($\theta \in [0,2e)$). Exactly the same argument proves (6.15), which was derived above by a similar method, applied however to the $E(\cdot)$ field.

Furthermore, using Ginibre-type inequalities it was shown in [25] that

$$\langle e^{i\,\alpha \oint_{\partial\Lambda} dx^\mu A_\mu(x)} \rangle_\theta^{\mathrm{Higgs}_2} \le \langle e^{i\,\alpha[\phi(L) - \phi(-L)]} \rangle_\theta^{(\phi)}, \qquad (8.14)$$

which establishes a mathematical relation of the Coulomb gas to this model. The physical origins of this relation were mentioned in the introduction to this section.

The simplest criterion for a strong confinement is

$$U_{eff}(L) \geq - c\ L$$

(for $c > 0$). We see that the one-dimensional Coulomb gas may be used to exhibit such confinement (of fractional charge) in one space, and two space-time, dimensions.

The interest in θ-vacua is far from being restricted to low dimensions. In higher dimensions "instantons" occur in non-Abelian gauge theories, raising the possibility of θ-vacua and the question of their effect on quark confinement [27].

References

[1] E. H. Lieb and D. C. Mattis: Mathematical Physics in One Dimension, Academic Press, N.Y. (1966).

[2] M. Aizenman and Ph.A. Martin: "Structure of Gibbs States of One Dimensional Coulomb Systems". Commun. Math. Phys. 78, 99 (1980).

[3] M. Aizenman and J. Fröhlich: "States of One-Dimensional Coulomb Systems as Simple Examples of θ-vacua and Confinement". To appear in J. Stat. Phys.

[4] A. Lenard: J. Math. Phys. 2, 682 (1961).

[5] S. Prager: in Advances in Chemical Physics IV, Wiley (Interscience), N.Y. (1962); p. 201.

[6] S. Edwards and A. Lenard: J. Math. Phys. 3, 778 (1962).

[7] O. E. Lanford: in Dynamical Systems, Theory and Applications, Lecture Notes in Physics 38, Springer-Verlag (1975).

[8] A. Lenard: J. Math. Phys. 4, 533 (1963).

[9] J. Zittartz: Zeit. Physik B31, 63 and 79 (1978).

[10] H. Kunz: Ann. Phys. (N.Y.) 85, 303 (1974).

[11] H. J. Brascamp and E. H. Lieb: in Functional Integration and its Applications, A. M. Arthurs, ed. Clarendon Press, Oxford (1979).

[12] Ch. Gruber, Ch. Lugrin and Ph.A. Martin: J. Stat Phys. 22, 193 (1980).

[13] Ph.A. Martin and T. Yalcin: J. Stat. Phys. 22, 435 (1980).

[14] J. Fröhlich and T. Spencer: In preparation.

[15] J. Fröhlich and B. Simon: Ann. Math. 105, 493 (1977) and J. Fröhlich, R. Israel, E. H. Lieb and B. Simon: Commun. Math. Phys. 62, 1 (1978).

[16] M. Reed and B. Simon: Methods of Modern Mathematical Physics, IV, Academic Press, N.Y. (1978), Section XIII.16.

[17] M. Aizenman, S. Goldstein and J. Lebowitz: Commun. Math. Phys. 62, 279 (1978).

[18] M. L. Mehta, F. J. Dyson: J. Math. Phys. 4, 713 (1963).

[19] G. Gallavotti, E. Verboven: Nuovo Cimento 28, 274 (1975).

[20] M. Aizenman, S. Goldstein, Ch. Gruber, J. Lebowitz and Ph.A. Martin: Commun. Math. Phys. 53, 209 (1977).

[21] Ch. Gruber, Ch. Lugrin and Ph.A. Martin: Helv. Phys. Acta 51, 829 (1979).

[22] J. Fröhlich, Y. M. Park: Commun. Math. Phys. 59, 235 (1978).

[23] S. Coleman, R. Jackiw and L. Susskind: Ann. Phys. (N.Y.) 93, 267 (1975).

[24] C. Callan, R. Dashen and D. Gross: Phys. Lett. 63B, 334 (1976).

[25] D. Brydges, J. Fröhlich and E. Seiler: "On the Construction of Quantized Gauge Fields, I." To appear in Ann. Phys. (N.Y.).

[26] R. Israel and C. Nappi: Commun. Math. Phys. 68, 29 (1979).

[27] C. Callan, R. Dashen, D. Gross: in Quantum Chromodynamics (La Jolla Inst., 1978) W. Frazer and F. Henyey, ed., AIP Conf. Proc. 55. Amer. Inst. Phys., N.Y. (1979).

S. Coleman: in Proc. International School of Subnuclear Physics (Ettore Majorana 1977), A. Zichichi, ed.

M. Lüscher: Phys. Lett. 78B, 465 (1978).

FREE ENERGY AND CORRELATION FUNCTIONS OF COULOMB SYSTEMS *

Joel L. Lebowitz**

Institut des Hautes Etudes Scientifiques

Bures-sur-Yvette, France

I. INTRODUCTION

These notes, like my lectures in Erice consist of several parts which are loosely related by the fact that they all deal with some aspects of statistical mechanics of Coulomb systems. In some sense of course all of statistical mechanics, which is the microscopic theory of macroscopic matter, deals with Coulomb systems. The properties of the materials we see and touch are almost entirely determined by the nature of the Coulomb force as it manifests itself in the collective behavior of interacting electrons and nuclei. In most applications of statistical mechanics however this fact is not explicit at all. One starts with an "effective" short range microscopic Hamiltonian appropriate to the problem at hand. For example, to discuss the superfluidity of He^4 we describe the fluid as a collection of neutral atoms represented by point masses interacting via Lennard-Jones pair potentials. This description seems very adequate. Statistical mechanics of Coulomb systems therefore usually refers to those investigations in which the Coulomb potential is explicitly considered as a part of the starting microscopic Hamiltonian. (Appropriate quantum statistics are always assumed).

There are two reasons for considering explicitly systems with such Hamiltonians. The first is primarily a theoretical one - we would like to understand in a more precise way how the Coulomb forces give rise to the effective interactions. The second reason is more practical. There are many systems, e.g. plasmas, molten salts, ionic crystals, etc. where bare Coulomb interactions are part of the appropriate effective Hamiltonian. We shall review here, very briefly, the status of some selected topics in both of these categories.

Since the aim of the two types of studies mentioned above are different, precise vs. pragmatic understanding of macroscopic behavior, their methods are also different. Workers in the first vineyard state their results as mathematical theorems while those in the second make and exploit various physically reasonable approximations. Of course the good work in either category, such as that reported by Aizenman, Brydges, Fröhlich, Lieb, Seiler, Thirring and other speakers on this subject here, is relevant to both areas. They are the interpid brave explorers leading the attack against ignorance and sloth.

II. THERMODYNAMIC LIMIT FOR COULOMB SYSTEMS (LIEB-LEBOWITZ)

The rigorous derivation of effective Hamiltonians from first principles is at the present time mostly beyond the reach of our mathematical abilities. I simply have no ideas of how to show that the Lennard-Jones description of He^4 is a good approximation, in certain ranges of temperature and density, to an overall neutral system of α-particles and electrons interacting via the Coulomb potential and satisfying the proper statistics, Bose for the α-nuclei, Fermi for the electrons with spins (see however Remarks later). We have therefore to be satisfied at present with much more modest goals : Prove that the most basic fact of macroscopic thermodynamics, extensivity and stability of the free energy of neutral systems, follows from the prescriptions of statistical mechanics for computing this free energy, $-\beta a(\beta,\underline{\rho})$

$$a(\beta,\underline{\rho}) = \lim_{|\Lambda|\nearrow \mathbb{R}^3} |\Lambda|^{-1} \ell n[\mathrm{tr}\, \exp[-\beta H(\underline{N};\Lambda)] \tag{2.1}$$

Here $H(\underline{N};\Lambda)$ is the Coulomb Hamiltonian of k species of charged particles, with charges e_α and numbers N_α, $\alpha = 1,\dots,k$ contained in a box Λ, $\Lambda \subset \mathbb{R}^3$ with volume $|\Lambda|$. The thermodynamic limit $\Lambda \nearrow \mathbb{R}^3$ is taken along a "reasonable" sequence of boxes Λ_j and particle numbers $\underline{N}_j$ such that

$$\underline{N}_j\cdot\underline{E} \equiv \sum_{j=1}^{k} N_j^\alpha e_\alpha = 0 \;,\quad N_j^\alpha \geq 0 \;,\ \text{(neutrality)} \tag{2.2}$$

and

$$N_j^\alpha/|\Lambda_j| = \rho_j^\alpha \underset{j\to\infty}{\to} \rho^\alpha \tag{2.3}$$

The existence and thermodynamic stability of the limit function $a(\beta,\underline{\rho})$ was proven by Elliott Lieb and myself [1,2] under the assumption that at least one type of charge, the positive or negative ones, obey Fermi statistics - as indeed electrons do. This condition

was proven by Dyson and Lenard [3] to be both sufficient and necessary for the extensivity of the ground state energy of the system, H-stability,

$$\min_{\underset{\sim}{N}} H(\underline{N};\Lambda) \geq -bN \ , \ b < \infty \tag{2.4}$$

$N = \sum_{\alpha=1}^{k} N^{\alpha}$. A simple elegant derivation of (2.4) with a greatly improved constant b was later given by Lieb and Thirring; see lectures by these authors in this volume.

The actual Hamiltonian for which (2.1) was proven in [1] is

$$H(\underline{N}) = \sum_{i=1}^{N} p_i^2/2m_i + \frac{1}{2}\sum_{i\neq j} e_i e_j/|x_i - x_j| + U(x_1,\ldots,x_N) \tag{2.5}$$

where $e_i = e_\alpha$ and $m_i = m_\alpha$ when particle i is of species α. Dirichlet boundary conditions are used for the wave functions on $\partial\Lambda$ and the negative charges are Fermions. U is any "standard" short range ℓ-body interaction, $\ell = 2,3,\ldots$ finite, which satisfies classical H-stability [4]

$$\min_{\{x_i\}} U(x_1,\ldots,x_N) \geq -bN \tag{2.6}$$

U is essentially such that if $e_\alpha = 0 \ \forall\alpha$, then the thermodynamic limit in (2.1) would exist without any restrictions on the statistics (superstability is not required when $e_\alpha \neq 0$) - in fact the system could be treated by classical statistical mechanics with the trace in (2.1) replaced by the appropriate symmetrized integrals. In fact if U contains a hard core, i.e. $U(x_1,\ldots,x_N) = \infty$ whenever $|x_i - x_j| < d$, $d > 0$, then classical and hence also quantum mechanical H-stability for $H(N)$ in (2.5) was proven by Onsager [5] in 1939. The results in [1] then hold also for such classical systems (a suitable representation in many cases).

Onsager's proof is based on the fact that in the presence of hard cores the charge on each particle can be considered (as far as the interactions are concerned) to be smeared out on the surface of a sphere of radius d. This has a self energy $\varepsilon(d)$. It is then a basic fact of electrostatics that

$$\begin{aligned}\frac{1}{2}\sum_{i\neq j} e_i e_j/|x_i - x_j| &= \frac{1}{2}\int_{\mathbb{R}^3} E^2(x)d^3x - \Sigma \text{ self energy} \\ &\geq -N\varepsilon(d) \quad , \quad |x_i - x_j| \geq d \ .\end{aligned} \tag{2.7}$$

Here $\underline{E}(x)$ is the electric field at the point x and the integral is obviously non-negative.

To the best of my knowledge, Onsager was the first one to consider the problem of H-stability. This expresses the "saturation" of the interactions between a particle and the "rest of the universe". This is an absolute necessity for the existence and extensivity of the free energy in macroscopic matter. Onsager's proof is not conceptually satisfactory because it requires the existence of a hard core. Even if such a hard core (or something equivalent to it) were to be true in nature as a consequence of other non-Coulombic (strong, weak) interactions, the lower bound $\varepsilon(d)$ in (2.7) would be, for any reasonable d, orders of magnitude larger than the actually observed energies : see article by Thirring. Thus Fermi statistics of electrons are a central ingredient in the stability of matter.

Once we have H-stability there still remains the problem of how to deal with the long range nature of the Coulomb potential : if all the charges were of the same sign then clearly the energy would be non-negative but the system would "explode" - all particles going to the surface of Λ . The reason this does not happen for neutral systems is of course screening . Just because the Coulomb forces are long range they are also very strong (their integral diverges at infinity) and cause the system to be locally neutral. The "effective" interaction between different regions of a macroscopic system is therefore greatly attenuated. This idea was turned (with some effort) into a formal proof of (2.1) for neutral systems. Since I have nothing to add on this point to what is already described in references [1,2] I shall not discuss this further here. Instead, I shall turn, after a few remarks, to the situation in which the system is not strictly neutral. Elliott Lieb has discovered a flaw in one point of our reasonings in [1] regarding some charged systems which leaves a (small) gap in our proof there. The question raised also concerns general mixtures in which the interactions are short range. I hope that this will soon be resolved.

Remarks : (i) We note that the results described above are for the three dimensional Coulomb system specified by the non-relativistic Hamiltonian (2.5). For results in two dimensions we refer the reader to the article by Fröhlich and Spencer in this volume [6]. For the one dimensional Coulomb system see the article by Aizenman (see also remark iii).

It is not clear at present how to deal with inclusion of relativistic effects. H-stability is the basic problem here - magnetic dipolar forces behave as $|x|^{-3}$ and their inclusion makes even the hydrogen atom unstable against collapse. Until this is resolved - which might involve getting a good theory of all the forces in nature - there seems little incentive to consider the thermodynamic

limit problem for relativistic Hamiltonians. This is also the only answer I know to a question raised by E. Wigner in one seminar - if the Coulomb ground state energy per particle for a neutral collection of $2N$ bosons, with charges $\pm e$, is unbounded below, $E_o(N) \sim -N^{7/5}$ according to Dyson [3], then why is there not spontaneous generation of positive and negative π-mesons.

I believe, on the other hand, that the problem of deriving effective (e.g. Lennard-Jones) interactions between neutral atoms from the Coulomb Hamiltonian (2.5) while very difficult, may not be entirely hopeless. I regard the work described by Fröhlich and Spencer on the two dimensional charged lattice gas as having (despite its being two-dimensional and entirely classical) the right flavor for this problem. They show, in a precise way, that in a certain range of temperature and density the weight of the Gibbs measure is concentrated entirely on "neutral molecules of finite extent". We don't expect this to happen in three dimensions where the Coulomb force is weaker and there should always be a finite density of loose charges (electrons and ions) around - we do expect that these play a negligable role at low temperatures and moderate densities. (At very very low densities the system will always be almost completely ionised.)

(ii) The Hamiltonian (2.5) corresponds to the box Λ having insulating boundaries. The existence of the thermodynamic limit for a Coulomb system in three dimensions with super-conducting boundary conditions was proven by Penrose and Smith [7]. It is not known whether the free energy is the same in both cases.

(iii) A model often used by physicists to describe certain kinds of plasmas or solids is the so-called jellium system. In this model one of the charges, say electrons, are assumed to form a uniform background of constant negative charge density in which the ions move. We shall not go into the justification of this model, which neglects fluctuations in the density of one of the species, or into any details of its properties referring the reader to the excellent recent review by Baus and Hansen [8]. The Hamiltonian for this model is obtained from (2.5) by adding there the potential produced by the uniform background of charge density ρ_B

$$H'(\underline{N},\Lambda) = H(\underline{N},\Lambda) + \sum_{i=1}^{N} e_i \rho_B v_\Lambda(x_i) + \frac{1}{2}\rho_B^2 \int_\Lambda v(x) d^3x \tag{2.8}$$

with

$$v_\Lambda(x) = \int_\Lambda d^3y/|x-y| \quad .$$

Charge neutrality is now given by the relation

$$\sum_{\alpha=1}^{k} N^{\alpha} e_{\alpha} + \rho_B |\Lambda| = 0 \tag{2.9}$$

Lieb and Narnhoffer [9] proved the existence of the thermodynamic limit for the free energy of this system in three dimensions. In two dimensions the electrostatic potential is, of course, $\ell n|x|$ rather than $|x|^{-1}$ and the proof needs changing [9b].

In one dimension the Jellium model with electrostatic potential $|x|$ and one moving component, the so-called OCP (One component plasma) can be solved exactly classically as can the two component (no background) charged system : see article by Aizenman. The Jellium system is known to form a crystal, periodic state with spacing $-e_1/\rho_B$ (k = 1 here) [10] at all temperatures $\beta > 0$. This is true in both the classical and quantum description and is the only system with translation invariant forces (in the thermodynamic limit or periodic b.c.) for which a crystalline Gibbs state is known to exist. There is strong theoretical and numerical (computer simulation) evidence that in three (and two) dimensions the OCP forms a crystal at low temperatures (Wigner crystal) which melts as the temperature is raised [8] .

This system is in some sense both simple and interesting - since its interactions are exact rather than approximate or modeled - and therefore presents a worthwhile challenge to the theorist. It would be particularly interesting to know whether its behavior in three dimensions is "physical" despite of the suppression of fluctuations in one component. As already noted the behaviour in one-dimension is artificial. In two dimensions the system is exactly solvable at one temperature, $\beta e_1 = 2$, where the truncated correlation functions are found to be Gaussian [11] a much faster than typical decay.

It may also be worth mentioning here another problem in this area which is a challenge to theorists. There is some evidence [12] that when there is more than one species of positively charged ions in a uniform negative background, say protons and α-particles, then the system will segregate into a "hydrogen" and "helium" phase at low temperatures. This can be understood to be a consequence of the systems attempt to stay locally neutral which, since the background is assumed at a uniform density, requires a large spacing between the more highly charged species. If this is true <u>and</u> persists also in real plasmas it may be of great relevance in astro-physics where it might cause the highly charged heavy ions, e.g. iron, in a star to clump together - rather than be uniformly distributed. This might even help explain the apparent deficiency in the number of observed solar neutrinos [12]. (The calculated rate assumes a homogeneous distribution of the heavy ions.)

Finally I want to mention here the question of the existence of the thermodynamic limit of "real" jellium, that is of a regular Coulomb system, confined to a two dimensional layer, i.e. the particles move in $\mathbb{R}^2$ (and ρ_B is a surface charge density) but the interaction potential is $|x|^{-1}$, $x \in \mathbb{R}^2$. Such systems are used [13] as models for layers of electrons on the surface of fluids like helium. The methods used in [1] make essential use of the harmonic nature of the Coulomb potential and do not therefore apply here. The requirement on the regularity of domain shapes Λ_j in [1] rules out the consideration of this case as a limit.

(iv) As mentioned earlier classical H-stability, Eq. (2.6), can be proven for more or less all potentials commonly used to describe the interactions between neutral molecules. There is a little known simple proof of this fact in a very complicated paper by Morrey [14] for pair potentials $v(r)$ satisfying the inequalities

$$\begin{aligned} v(\underline{r}) &\geq C_1 \, r^{-(\nu+\varepsilon)}, \quad r < r_o \\ v(\underline{r}) &\geq -C_2 \, r^{-(\nu+\varepsilon)}, \quad r > r_o \end{aligned} \tag{2.10}$$

where $\underline{r} \in \mathbb{R}^\nu$, $r = |\underline{r}|$, and $0 < C_i$, $\varepsilon < \infty$. It seems worthwhile presenting his argument here.

Consider a configuration $\{x_1,\ldots,x_N\}$ with $x_i \in \mathbb{R}^\nu$. Let λ be the smallest distance between any pair, $\lambda = \mathrm{Min}|x_i - x_j|$, $i \neq j$. Call the pair for which this minimum is achieved x_1 and x_2. Write then

$$\begin{aligned} U(x_1,\ldots,x_N) = \frac{1}{2} \sum_{i\neq j} v(x_i - x_j) &= [v(x_1-x_2) + \sum_{j=3}^{N} v(x_1-x_j)] \\ &\quad + U(x_2,\ldots,x_N) \end{aligned} \tag{2.11}$$

Since the spacing between x_i and x_j is at least λ we have by (2.10) that

$$V_1 = v(x_1-x_2) + \sum_{j=3}^{N} v(x_1-x_j) \geq A_1/\lambda^{\nu+\varepsilon} - A_2/\lambda^{\nu} = G(\lambda)$$

where $0 < A_i < \infty$. Let now $-b = \mathrm{Min}_{\lambda}\, G(\lambda) > -\infty$ and repeat the procedure on the interaction between the N-1 particles $\{x_2,\ldots,x_N\}$ to obtain (2.6). N.B. The theorem does not give a lower bound for the interaction energy of <u>any</u> particle with the rest of the system. This is clearly impossible if $v(r) < 0$ for some $\underline{r}$ and there is no hard core, e.g. for the Lennard-Jones potential.

The paper by Morrey sets out to prove that in a certain limit the Liouville equation leads to the Euler equations of hydrodynamics. It is not clear whether the paper ever achieves this goal (I doubt it) but it does prove on the way the convergence of the Mayer expansion for the pressure at small fugacities for potentials satisfying (2.10). This is a remarkable achievement since the author had apparently never heard of the Mayer expansion before, so he derives it as an aside (in fifty pages). His proof ante-dates by many years the proofs of Groeneveld, Ruelle, and Penrose [4] .

III. SYSTEMS WITH NET CHARGE

it is intuitively clear that the condition of strict charge neutrality, $\underline{N}_j \cdot \underline{E} = 0$, is unnecessarily restrictive. We expect that a "small" amount of uncompensated charge will have no effect on the free energy density in the thermodynamic limit while a "large" amount of uncompensated charge will lead to a divergent free energy density in that limit. The dividing line between "small" and "large" occurs when the excess charge Q_j , in a domain Λ_j , increases in proportion to the "surface area" of Λ_j as $j \to \infty$. In this case we expect the thermodynamic limit of the free energy density to exist but that its value depends also on the limiting shape of the domains Λ_j .

These expectations come from macroscopic electrostatic theory [15] which shows that the lowest energy configuration for any net charge Q confined to a domain Λ is obtained when Q is concentrated at the boundary of Λ . This configuration of the charge is described in electrostatics by a two dimensional charge density $\sigma(\underset{\sim}{x})$, $\underset{\sim}{x} \in S_\Lambda$, where S_Λ is the surface of Λ . This surface charge density will be such as to make the electrostatic potential constant in the interior of Λ , i.e., there will be no electric field in Λ . The electrostatic energy of this surface layer is equal to $\frac{1}{2} Q^2/C(\Lambda)$ where $C(\Lambda)$ is the capacitance of Λ .

For a given domain shape, $C(\Lambda)$ is proportional to $[V(\Lambda)]^{1/3}$ $(V(\Lambda) = |\Lambda|)$ and the electrostatic energy per unit volume will thus be proportional to $[Q/V^{2/3}]^2$, the square of the "average surface charge density". Hence for sequence of domains $\{\Lambda_j\}$ with volumes $\{V_j\}$ and capacitances $\{C_j\}$ each containing a net charge Q_j such that as $j \to \infty$, $V_j \to \infty$, $C_j/V_j^{1/3} \to c$, the minimum electrostatic energy per unit volume $\underset{\sim}{e}_j$ will also approach a limit $\frac{1}{2} \sigma^2/c$.

We therefore considered in [1] a sequence of domains Λ_j (ellipsoidal for technical reasons) with particle numbers $\underline{N}_j + \underline{n}_j$ such that $\underline{N}_j \cdot \underline{E} = 0$, $\underline{n}_j \cdot \underline{E} = Q_j$. It was then claimed that as $j \to \infty$ the free energy density

$$a_j(\beta,\underline{\rho}_j,\underline{n}_j) \to a(\beta,\underline{\rho}) - \frac{1}{2}\sigma^2/c \ , \quad \underline{\rho} = \lim \underline{N}_j/V_j \ . \qquad (3.1)$$

where $a(\beta,\underline{\rho})$ is given in (2.1) and c is shape dependent. In the proof of (3.1), for arbitrary $\underline{n}_j$, we used the continuity of the neutral free energy density $a(\beta,\underline{\rho})$ in each ρ_α . The argument used for this was the standard one - concavity of $a(\beta,\underline{\rho})$.

The difficulty discovered by Lieb in the proof of (3.1) is that, as is well known, concavity does not guarantee continuity at the boundary of the domain of the function, i.e. at $\rho^\alpha = 0$ for some α . This is taken care of in the usual considerations for one component systems [4] by showing explictly that $a(\rho) \sim \rho\log\rho$ near $\rho \sim 0$. A similar form is true for multi-component systems, including Coulomb ones [1,6]in the neighborhood of $\underline{\rho} = 0$. We have however been unable to come up with a proof of continuity at a general point of the boundary of the positive cone $\rho^\alpha \geq 0$. This requires that we add some conditions on the sequences $\underline{n}_j$ for which (3.1) can be shown to hold rigorously (at the present time). Namely we need to assume that there exists $m_j^\alpha \geq 0$ such that a) $(\underline{n}_j+\underline{m}_j)\cdot\underline{E} = 0$ and b) $\rho^\alpha > 0$ for all those α , $\alpha = 1\cdots k$ for for which $n_j^\alpha+m_j^\alpha > 0$. In other words we require that the excess charge and what it requires to neutralize it be available inside the system at a non-vanishing density. If a) and b) are not satisfied then (3.1) would still be a lower bound but in the upper bound $a(\beta,\rho)$ would have to be replaced by $a(\beta,\underline{\rho}+)$ the limiting value of $a_j(\beta,\underline{\rho}_j,\underline{n}_j+\underline{m}_j)$ (It might be necessary as in [1] to introduce a "new" charged species, $\alpha = 0$, to satisfy the strict neutrality condition $(\underline{n}_j+\underline{m}_j) \cdot \underline{E} = 0$).

As I already mentioned before the difficulty here has really nothing to do with the specifics of the Coulomb interaction. Consider, for example, a classical system of two kinds of "neutral atoms" interacting with Lennard-Jones type potentials, i.e. they satisfy (2.10). It is then a standard (by now so familiar as to be trivial) argument [4] to show that for a sequence of 'regular domains' $\Lambda_j \nearrow \mathbb{R}^3$ and densities $\underline{\rho}_j \to \underline{\rho}$

$$\lim_{j\to\infty} a_j(\beta,\rho_j^{(1)},\rho_j^{(2)}) \to a(\beta,\rho^{(1)},\rho^{(2)}) \qquad (3.2)$$

exists with $a(\rho^{(1)},\rho^{(2)})$ defined and concave in the positive quadrant of the $\underline{\rho}$ plane ($\beta > 0$ fixed). It follows from this that $a(\rho^{(1)},\rho^{(2)})$ is continuous in the open quadrant $\mathbb{R}^2_+$. The method of proof also yields,

$$a(\underline{\rho}) \sim -\Sigma\rho^{(i)}\ell n\rho^{(i)} \quad \text{as} \quad \underline{\rho} \to 0 \tag{3.3}$$

and the chemical potentials $\mu_i = -\partial a/\partial\rho^{(i)}$ exist and are monotone for almost all $\underline{\rho} \in R^2_+$. Independent arguments further prove that $a(\underline{\rho}) + \Sigma\rho^{(i)}\ell n\rho^{(i)}$ is analytic in some domain $D \subset \mathbb{C}^2$ around $\underline{\rho} = 0$ (convergence of the density expansion [16] .) What is not known however is whether, outside the region $D_+ = D \cap \mathbb{R}^+_2$,

$$\lim_{\rho^{(2)}\searrow 0} a(\rho^{(1)},\rho^{(2)}) \overset{?}{=} a(\rho^{(1)}) \tag{3.4}$$

the free energy density of the one component system. (The same problem arises when one considers the grand canonical pressure as a function of the fugacities z_1 and z_2). Indeed I do not even know how to prove when $N^{(2)}_j$ is fixed, say one, that $a_j(\beta,\rho^{(1)}_j, 1/V_j)$ approaches the same limit as $a_j(\beta,\rho^{(1)}_j)$. Help please !

The difficulty here lies entirely with obtaining an upper bound to a_j (lower bound on the free energy). This requires proving (the "obvious" fact) that configurations in which a finite fraction of particles of species one (A particles) which sit in the "minimum" of the potential of the species two (B) particle make a vanishing contribution to a_j as $j \to \infty$. There is of course no such problem if the particles have hard cores in which case (for interactions decaying faster than $r^{-(\nu+\varepsilon)}$) the interaction energy of any specified particle with the rest of the system is uniformly bounded, e.g. lattice systems. A similar situation obtains when the interaction potential is non-negative - but this still leaves open the general case.

To be a bit more specific let the Hamiltonian of the system, in Λ , be written as

$$H(N_A,N_B,\Lambda) = H(N_A,\Lambda) + H(N_B,\Lambda) + W(N_A,N_B,\Lambda) \tag{3.5}$$

The standard techniques [1,4] will then give, for the partition function Z ,

$$Z(N_A,N_B,\Lambda) > Z(N_A,\Lambda_A)Z(N_B,\Lambda_B)\exp[-\beta\langle W\rangle'] \tag{3.6}$$

where Λ_A and Λ_B are non overlapping regions contained in Λ

and $\langle W\rangle'$ is the expectation value of W in the ensemble specified by having all A-particles in Λ_A and all B-particles in Λ_B. Using the decay of the interaction given in (2.10) the "distance" between Λ_A and Λ_B can be increased, as $\Lambda \nearrow \mathbb{R}^\nu$, in such a way [1,4] that in the limit $j \to \infty$.

$$a(\beta,\rho_A,\rho_B) \geq \rho_A/\rho_A' a_A(\beta,\rho_A') + \rho_B/\rho_B' a_B(\beta,\rho_B') + c\rho_A'\rho_B' \qquad (3.7)$$

where ρ_A' and ρ_B' are the (increased) limiting densities in Λ_A and Λ_B and $c < \infty$ is some constant. Letting $\rho_B' \to 0$ gives $\rho_B' \to 0$, $\rho_A' \to \rho_A$ and the right side of (3.7) then approaches $a(\beta,\rho_A)$ by the continuity of $a_A(\beta,\rho)$, $\rho > 0$, and the bound on $a_B(\beta,\rho)$ as $\rho \to 0$. (When $N_B/V(\Lambda) \to 0$ as $\Lambda \nearrow \mathbb{R}^\nu$ the right side is just $a_A(\beta,\rho_A)$). The same kind of argument, suitably modified for Coulomb systems [1,2], yields (3.1) as a lower bound.

The difficulty arises in getting an upper bound on a. A general method frequently used, for doing this is to define a new Hamiltonian $H_o = H-G$ where H is given in (3.5). For $G = W$, H_o is just the energy of the A and B particles in Λ without any mutual interaction. It is however possible to make H_o be the Hamiltonian of quite a different system, e.g. by adding some particles to the system (as was done in [1]). What is important is that Z_o be somehow "controllable" in the sense that we can split G into two parts, $G = G_1 + G_2$ and then use the Peierls-Bogolyubow inequality

$$Z_o = Z(N_A,N_B,\Lambda) < \exp\beta G> \geq Z_1 \exp[<G_2>_1] \qquad (3.8)$$

to yield a useful bound on Z. Here $< >$ is the expectation taken with the Gibbs measure specified by H in Λ while Z_1 and $< >_1$ are obtained from $H_1 = H+G_1$. To be useful we will want $\ell nZ_1/V(\Lambda) \to a(\beta,\rho_A,\rho_B)$ and the lower bound obtained from (3.8) to coincide with $a_A(\beta,\rho_A)$ when $\rho_B \to 0$.

In the case where $G = W = G_2$ we find

$$a(\beta,\rho_A;\rho_B) \leq a(\beta,\rho_A) + a(\beta,\rho_B) - \beta\,\overline{\langle w\rangle}$$

where $w(x)$ is the pair potential between A and B particles and

$$\begin{aligned} \langle\overline{w}\rangle &= \lim_{j\to\infty} \int [\frac{1}{V_j} \int_{\Lambda_j} \rho_{AB}(x_1, x_1+x;j)d^\nu x_1]\, w(x)d^\nu x \\ &= \lim_{j\to\infty} \int \overline{\rho}_{AB}(x;j)w(x)d^\nu x \end{aligned} \tag{3.9}$$

$\rho_{AB}(x_1,x_2;j)$ is the density of A,B pairs in Λ_j (for particle numbers $N_{A,j}$ and $N_{B,j}$) and the limit may have to be taken along subsequences. Writing $w(x) = w_+(x) - w_-(x)$, $w_+ \geq 0$, $w_- \geq 0$, we find, using (2.10) that to control the r.s. of (3.9) it is certainly sufficient to have a bound on the radial distribution function

$$-\langle\overline{w}\rangle \leq \rho_A\rho_B K \lim_{j\to\infty} [\sup_x \overline{\rho}_{AB}(x;j)/\rho_A(j)\rho_B(j)] \tag{3.10}$$

with $K = \int w_-(x)d^\nu x < \infty$ and the sup is over the support of w_-. It should certainly be possible to do this in general.

IV. CORRELATION FUNCTIONS AND SUM RULES

So far we have dealt only with the thermodynamic limit of the free energy - which is necessary and sufficient for the description of the macroscopic behavior of bulk matter. Statistical mechanics goes however beyond thermodynamics in that it also gives a prescription for computing results of microscopic experiments on bulk matter - experiments which directly probe the atomic structure of matter as they show themselves in the deviations from "average" behavior, e.g. fluctuations in the local density measured by x-rays and neutrons.

It is part of the basic dogma of statistical mechanics [4], whose validity has been confirmed experimentally beyond any reasonable doubt (despite the lack of completely convincing theoretical arguments), that the results of such microscopic experiments on macroscopic systems can be obtained in thermal equilibrium, from "appropriate" Gibbs ensembles. These are, for finite systems of N-particles, the canonical (micro, macro grand,...) ensembles. Results of observations are obtained as ensemble averages using the appropriate density matrix or, for classical systems, the measure on the phase space.

The passage to the infinite volume limit, required to obtain unambiguous, surface independent, results for bulk systems is much more delicate, and therefore mathematically more difficult, for these distributions than it is for the free energy. It is particu larly so for quantum systems [17] and I shall not discuss these

further here. For classical systems with rapidly decaying interactions between the particles the mathematical theory is based on the existence of infinite volume Gibbs states [4]. These are the generalization, via the Dobrushin, Lanford and Ruelle equations, of the grand canonical ensemble for a system with a given temperature and fugacity in a finite box Λ . They are (modulo some technical restrictions) the infinite volume thermodynamic limit of all the different ensembles, e.g. microcanonical, canonical, etc..., used to describe finite systems in equilibrium. This requires at the very least that the potential be integrable at infinity, e.g. if the particles interact via a pair potential then it should decay as fast as $r^{-(\nu+\varepsilon)}$, e.g. the Lennard-Jones potential which decays as r^{-6} .

A problem occurs however when we are dealing with systems in which the very long range Coulomb force is explicitly present (there is also some problems for dipoles). In this case, there is no conceivable way in which distant parts of a system would be sufficiently decoupled for arbitrary configurations. The usual treatment of infinite volumes Gibbs states, e.g. the DLR equations, will therefore not work here. We expect nevertheless as indicated earlier, that for typical configurations distant parts of an overall neutral system will sufficiently decouple, due to screening, to make the passage to the thermodynamic limit well defined. (For non-neutral systems the limit way present some extra problems). More precisely we expect the correlation functions to have well defined infinite volume limits. This holds indeed at sufficiently high temperatures and low densities where, as described by Brydges here [18] the correlation functions cluster exponentially and for "charge symmetric systems" where it follows from the Fröhlich-Park correlation inequalities [19]. These results however still leave open the question of how to characterize directly the infinite volume Gibbs states of general Coulomb systems in the absence of well defined DLR equations ?

In a recent series of papers [20] Gruber, Martin and their associates have explored the consequences of assuming that the equilibrium correlation functions of Coulomb systems satisfy the stationary BBGKY, (Bogolyubov, Born, Green, Kirkwood, Yvon) hierarchy. These equations are obtained from the time dependent BBGKY hierarchy which describes the time evolution of the n-particle spatial and momentum distribution functions of a classical system $f_r(x_1p_1,\dots x_np_njt)$. These have the form [21]

$$\frac{\partial f_n}{\partial t} = H_n f_n + C_{n,n+1} f_{n+1} , \quad n = 1,2,\dots \qquad (4.1)$$

where H and C are linear operators. Assuming that f_n

is a product of a Maxwellian momentum part and a purely spatial part and setting $\frac{\partial f_n}{\partial t} = 0$ leads to an integro-differential equation for the f_n. These stationary eqs. are identically satisfied by the finite volume canonical and grand canonical distribution functions. It is also known that for short range potentials these eqs. are essentially equivalent to the DLR eqs. for the Gibbs states [22]. It is thus expected that the infinite volume limit of the correlations for charged systems will also satisfy them; in fact, this can be rigorously shown to hold for Coulomb systems in one dimension [20] and can presumably be established in all cases where the existence of the infinite volume limit has been proven [23] . It is therefore reasonable to expect that this is always true and explore the consequences which are non-trivial. I now describe briefly this work including extensions in which I participated [24].

Description of the system

We consider as before a system consisting of k species of charged particles which move either in the whole ν-dimensional space $\mathbb{R}^\nu$, or in a restricted domain $\mathcal{D}$ defined by appropriate walls. The only condition we impose on $\mathcal{D}$ is that it extends to infinity in at least one direction. Typically $\mathcal{D}$ can be the half-space $\{x \in \mathbb{R}^\nu;\ x^1 \geq 0\}$ of the electrode problem.

The particles interact by means of a two body force of the form

$$F^s_{\alpha_1\alpha_2}(x_1-x_2) + e_{\alpha_1} e_{\alpha_2} F(x_1-x_2) \quad \text{where} \quad F^s_{\alpha_1\alpha_2}(x)$$

is short range and $F(x)$ is the Coulomb force,

$$F(x) \sim \frac{x}{|x|^\nu} \quad , \ |x| \to \infty \ .$$

The thermodynamic equilibrium state of the system at a temperature T is assumed to be described by means of correlation functions $\rho_{\alpha_1}(x_1)$, $\rho_{\alpha_1\alpha_2}(x_1,x_2)\ldots$, which have their usual meaning, $\rho_{\alpha_1}(x_1)$ being the density of species α_1 at x_1 , etc... We shall write these as $\rho(q_1)$, $\rho(q_1,q_2)$, $\rho(Q)$, using the abbreviated notation $q_i = (\alpha_i,x_i)$, $Q = (q_{i_1},\ldots,q_{i_n})$. These functions are assumed to satisfy the stationary BBGKY equation

$$kT\nabla_1\rho(q_1,Q) = [e_{\alpha_1}E(x_1) + \sum_{j=2} F(q_1-q_j)]\rho(q_1,Q) \tag{4.2}$$

$$+ \int_{\mathcal{D}} dqF(q_1-q)[\rho(q_1,q,Q) - \rho(q_1,Q)\rho(q)]$$

Here $E(x)$ is the electric field due to all the charges, i.e. all the system's and all the external charges, $F(q_1-q_2)=e_{\alpha_1}e_{\alpha_2}F(x_1-x_2)$.

In order to understand (4.2) and the structure of $E(x)$, it is useful to consider first the equilibrium distribution of finite systems in bounded regions $\{\Lambda\}$ and then take the thermodynamic limit $\Lambda \to \mathcal{D}$. It is easily found that the correlation functions of the finite systems satisfy eq. (4.2) with the electric field $E^{\Lambda}(x)$ given by the sum of an "external" and an "internal" field

$$
\begin{aligned}
E^{\Lambda}(x) &= D^{\Lambda}(x) + \int_{\Lambda} dy F(x-y) C^{\Lambda}(y) , \\
C^{\Lambda}(x) &= \sum_{\alpha} e_{\alpha}\rho_{\alpha}^{\Lambda}(x)
\end{aligned}
\tag{4.2}
$$

being the charge density at x . Let now $\Lambda \to \mathcal{D}$ and assume that the state of the finite system converges to a state of the infinite system defined by (4.2). We then have

$$
\begin{aligned}
E(x) &= D(x) + \lim_{\Lambda\to\mathcal{D}} \int dy F(x-y) C^{\Lambda}(y) \\
D(x) &= \lim_{\Lambda\to\mathcal{D}} D^{\Lambda}(x)
\end{aligned}
\tag{4.3}
$$

Writing

$$
\begin{aligned}
C(x) &= \lim_{\Lambda\to\mathcal{D}} C^{\Lambda}(x) = \sum_{\alpha} z_{\alpha}\rho_{\alpha}(x) , \\
E(x) &= G(x) + \lim_{\Lambda\to\mathcal{D}} \int_{\Lambda} F(x-y) C(y) ,
\end{aligned}
\tag{4.4}
$$

defines the effective external field $G(x)$:

$$
G(x) = D(x) + \lim_{\Lambda\to\mathcal{D}} \int_{\Lambda} dy F(x-y) (C^{\Lambda}(y) - C(y)) \tag{4.5}
$$

Clearly the second term in (4.5) represents the field due to the system's charges located at infinity when Λ becomes $\mathcal{D}$. Therefore, one must remember that $G(x)$ has its origin both in external and in the system's charges which reside at "infinity" and are thus invisible in $C(x)$ for any x (see Appendix A in [24] for illustrative example).

It turns out that (4.2) together with certain clustering assumptions on the truncated correlation functions $\rho_T(q_1,\dots q_n)$ imply the abzence of non-translation invariant Gibbs states [25] as well as, certain sum rules. I only describe the latter here.

Electrostatic sum rules

Define the excess charge density at x in presence of charges located in $Q = (q_1,\dots q_n)$ as :

$$C(x|Q) = \sum_\alpha e_\alpha \left(\frac{\rho(\alpha x,Q)}{\rho(Q)} + \sum_j^n \delta_{\alpha\alpha_j} \delta(x-x_j) - \rho(\alpha,x)\right) \quad (4.6)$$

and consider the following relations.

<u>0-sum rule : neutrality</u>

$$\int C(x|Q)\,dx = 0 \quad . \quad (4.7)$$

More explicitly,

$$\left(\sum_j^n e_{\alpha_j}\right)\rho(Q) + \int dq\,e(\rho(qQ) - \rho(q)\rho(Q)) = 0 \quad .$$

<u>1-sum rule : absence of dipolar moment</u>

$$\int x C(x|Q)\,dx = 0 \quad , \quad (4.8)$$

or

$$\left(\sum_j^n e_{\alpha_j} x_j\right)\rho(Q) + \int dq\,e x(\rho(qQ) - \rho(q)\rho(Q)) = 0 \quad .$$

<u>2-sum rule : isotropy</u>

$$\int x^r x^s C(x|Q)\,dx = \delta_{rs} \frac{1}{\nu} \int |x|^2 C(x|Q)\,dx \quad (4.9)$$

<u>Proposition</u>

Let $\rho_T^n(\alpha_1 x_1, \alpha_2 x_2 \dots \alpha_n, x_n) = O\left(\frac{1}{|x_1|^{\nu+\ell+\varepsilon}}\right) \quad \varepsilon > 0 \quad (4.10)$

unif.with respect to x_2 when $n \geq 3$. If (4.11) holds with $\ell = 0$, the 0-sum rule follows. If (4.11) holds with $\ell = 1,2$, the 1-2 sum rules follow respectively, except in one dimensional Coulomb systems.

<u>Sketch of proof of $\ell = 0$ sum rule for $Q = q_2$</u>

Combination of first and second BBGKY equation and definition of truncated functions gives the identity

$$e_{\alpha_1} \rho(q_1)\rho(q_2) \int F(x_1-x)C(x|q_2)\,dx = \quad (4.11)$$

$$kT\nabla_1 \rho_T(q_1q_2) - (e_{\alpha_1} E(x_1) + F(q_1q_2))\rho_T(q_1q_2) - \int dq F(q_1 q)\rho_T(q_1q_2q)$$

Clustering hypothesis with $\ell = 0$ implies that the left side of (4.11) has the form

$$\int F(x_1 - x) C(x|q_2)dx = \frac{\hat{x}_1}{|x_1|^{\nu-1}} \int C(x|q_2)dx + o\left(\frac{1}{|x_1|^{\nu-1}}\right) \quad (4.12)$$

and that all terms on the right hand side of (4.11) are

$$o\left(\frac{1}{|x_1|^{\nu-1}}\right)$$

It follows therefore that

$$\int C(x|q_2)dx = 0 \ .$$

The $\ell = 1,2$ sum rules are obtained in a similar way—they are trivial for invariant states when only one particle is kept fixed. In the latter case the right side of (4.9) takes the form $\delta_{rs} I_\gamma/(\nu\rho_\alpha)$, where

$$I_\alpha = \int dx|x|^2[\sum_\alpha e_\alpha \rho^T_{\alpha,\gamma}(|x|,0)] \quad (4.13)$$

is the second moment of the spherically symmetric charge density cloud surrounding an ion of type α located at the origin. This quantity was investigated by Stillinger and Lovett [26] who showed, on the basis of certain physically very reasonable assumptions, that for Coulomb systems

$$\sum_{\alpha=1}^{k} I_\alpha = -2/(\beta\omega_\nu) \quad (4.14)$$

is a universal constant, independent of the short range interactions,

$$\omega_1 = 2, \ \omega_2 = 2\pi, \ \omega_3 = 4\pi, \text{ see also [27].}$$

Eq. (4.14) is known as the Stillinger-Lovett second moment condition (the first being the corresponding zero moment condition) and can be shown to hold in the high temperature clustering region of Brydges and Federbush but may fail at low temperatures in two dimensions [23]. An interesting open question is whether (4.14) holds at low temperatures (or at critical points) in three dimensions. This may be related to the question of screening of non-integer charges [6, 24].

V. APPROXIMATIONS

In Erice I spent some time describing certain commonly used schemes for calculating the pair correlation functions, and from them the thermodynamic functions, of a charged fluid, i.e. a molten salt. Among these the Mean Spherical Approximation [28] can be solved easily in a closed form. It yields substantial improvements (when compared with computer and real experiments) over the Debye-Huckel theory at not too low concentrations. A more accurate description is given by the Hyper Netted Chain Approximation which can however only be solved numerically. Since that time there have appeared two review articles covering this material, references [8] and [29]. I refer the interested reader to these excellent expositions. (I should also mention here some interesting recent work on the critical region of the two dimensional Coulomb system by Hoye and Olaussen [30].

* Research supported in part by NSF Grant PHY 78-15920

** Permanent address: Dept. of Math. and Phys., Rutgers University, New Brunswick, N. J.

Acknowledgements: I would like to thank J. Fröhlich, Ch. Gruber, E. H. Lieb and Ph. Martin for many discussions and N. H. Kuiper for gracious hospitality at IHES.

References

[1] J. L. Lebowitz and E. H. Lieb, Phys. Rev. Lett. 22, 631 (1969); E. H. Lieb and J. L. Lebowitz, Adv. Math., 9, 316 (1972).

[2] The main results (with proofs) of [1] are given in E. H. Lieb and J. L. Lebowitz, Statistical Mechanics and Mathematical Problems, LNP 20, Springer (1973), A. Lenard, editor. An excellent review of all aspects of Coulomb systems is E. H. Lieb, Rev. Mod. Phys. 48, 553 (1976).

[3] F. J. Dyson, J. Math. Phys. 8, 1538 (1967); F. J. Dyson and A. Lenard, J. Math. Phys. 8, 423 (1967); A. Lenard and F. J. Dyson, J. Math. Phys. 9, 698 (1968).

[4] D. Ruelle, Statistical Mechanics: Rigorous Results, Benjamin (1969); O. Lanford, in LNP 20, Springer (1973).

[5] L. Onsager, J. Phys. Chem. 43, 189 (1939).

[6] J. Fröhlich and T. Spencer, Phys. Rev. Lett. (to appear); article in this volume.

[7] O. Penrose and E. R. Smith, Comm. Math. Phys. 26, 53 (1972).

[8] M. Baus and J. P. Hansen, Phys. Rep. 59, 1 (1980).

[9] E. H. Lieb and H. Narnhofer, J. Stat. Phys. 12, 291 (1975); 14, 465 (1976). R. R. Sari and D. Merlini, J. Stat. Phys. 14, 91 (1976).

[10] H. Kunz, Ann. Phys. 85, 303 (1974); H. J. Brascamp, E. Lieb in Functional Integration and its Applications, Clarendon Press (1975), A. M. Arthurs, editor; M. Aizenman, Ph. A. Martin, Comm. Math. Phys., to appear.

[11] B. Jancovici, Phys. Rev. Lett. (1981).

[12] D. J. Stevenson, Phys. Rev. B 12, 3999 (1975); E. L. Pollock, B. J. Alder, Nature 275, 41 (1978); J. P. Hansen, J. de Phys. 41, C2-43 (1980); B. J. Alder, E. L. Pollock and J. P. Hansen, Proc. Natl. Acad. Sci. 77, 6272 (1980).

[13] C. C. Grimes, G. Adams, Phys. Rev. Lett. 42, 795 (1979); P. S. Crandall, R. Williams, Phys. Lett. 34A, 404 (1971).

[14] C. B. Morrey, Comm. Pure Appl. Math. 8, 279 (1939).

[15] O. D. Kellog, Foundations of Potential Theory, Dover (1953).

[16] J. L. Lebowitz and O. Penrose, J. Math. Phys. 5, 841 (1964).

[17] c.f. H. Araki, Colloq. Int. CNRS 248, 61 (1976).

[18] D. Brydges and P. Federbush, Comm. Math. Phys. 73, 197 (1980).

[19] Y. M. Park, J. Math. Phys. 18, 12 (1977); J. Fröhlich and Y. M. Park, Comm. Math. Phys. 59, 235 (1978); J. Stat. Phys. 23, 701 (1980).

[20] Ch. Gruber, Ph. A. Martin, Ch. Lugrin, Helv. Phys. Acta 51, 829 (1979); J. Stat. Phys. 22, 193 (1980).

[21] c.f. R. Balescu, Equilibrium and Non-equilibrium Statistical Mechanics, Wiley (1975); O. E. Lanford, in Dynamical Systems, LNP 38, Springer (1975).

[22] G. Gallavotti, E. Verboven, Nuevo Cim. 28, 274 (1975); M. Aizenman, S. Goldstein, Ch. Gruber, J. L. Lebowitz, Ph. A. Martin, Comm. Math. Phys. 53, 209 (1977).

[23] J. Fröhlich, private communication.

[24] Ch. Gruber, J. L. Lebowitz, Ph. A. Martin, J. Chem. Phys., to appear.

[25] Ch. Gruber, Ph. A. Martin, Phys. Rev. Lett. 45, 853 (1980).

[26] F. Stillinger, R. Lovett, J. Chem. Phys. 49, 1991 (1968).

[27] D. J. Mitchell, D. A. McQuarrie, A. Szabo, J. Groeneveld, J. Stat. Phys. 17, 15 (1977).

[28] E. Waisman and J. L. Lebowitz, J. Chem. Phys. 56, 3086 and 3093 (1972).

[29] Bjørn Hafskjold and George Stell, <u>The Equilibrium Statistical Mechanics of Simple Ionic Liquids</u>, to appear in Studies in Statistical Mechanics, North Holland (1981); J. L. Lebowitz and E. Montroll, eds.

[30] J. S. Hoye and K. Olaussen, Physica <u>104A</u>, 435, 447 (1981).

SEMINARS

S. Graffi Resonances in the Stark Effect in Atomic Systems For further information see the paper of the same title by S. Graffi and V. Grecchi, to appear in Communcations in Mathematical Physics.

P. Perry Absence of Singular Continuous Spectrum in N-Body Systems For further information see the paper of the same title by I. Segal, P. Perry, and B. Simon to appear in Bull. of Amer. Math. Soc.

T. Hoffmann-Osterhof Asymptotic Properties of Atomic Bound States For further information see the paper of the same title by R. Ahlrichs, J. Morgan, M. Hoffmann-Osterhof, to appear in Physical Review A.

H. Silverstone L-dependent, Large r Behavior of Atomic and Molecular Bound States For further information see Asymptotic Behavior of Atomic Hartree-Fock Orbitals by G.S. Handler, D.W. Smith, and H.J. Silverstone, J. Chem. Phys. 73, 3936 (1980), and Long Range Behavior of Electronic Wave Functions; Generalized Carlson-Keller Expansion by H.J. Silverstone (preprint).

INDEX

Absorbed states, 128
Activities, 373
Activity, 337, 338, 347
Admissible density matrix, 257
Analyticity of eigenvalues in
Stark effect, 165-168
Asymptotic completeness, 38,
48, 57, 90
Asymptotic condition, 43ff,
56ff
Asymptotic evolution, 37ff
Asymptotic sequence, 113
Asymptotically free motion,
37ff, 43ff, 48ff
Asymptotics, 223, 239, 289
Atomic radius, 240, 273
Atomic surface, 266

Banach-Alaogulu theorem, 218
Barrier penetration, 132
Baxter's theorem, 243
BBGKY hierarchy, 460
Benguria's theorem, 249
Binding, 232, 241ff, 291ff
in magnetic fields, 172, 175-
181
in one dimension, 172
Boltzmann factor, 391, 398
Bonds, 391
Borel summability of eigenvalues
in Stark effect, 165,
168
Bound state energy, 63ff
Bound states, 13ff, 64, 125
Boundary, 222
conditions, 330, 344, 359

Center of mass removal in
Stark effect, 135, 146
Charge clouds, 371, 402
Charge densities, 360, 379
Charge fluctuations, 454
Charge symmetry, 372, 377
Chemical activity, 376
Chemical potential, 221, 237,
239
asymptotics, 239
Clamped nuclei, 189ff
Classical approximations to
molecular energies, 202ff
Classical translation, 385-387
Cluster decomposition, 391-393
Cluster expansion, 372, 391-410
Collapse of a star, 312
Compact operator, 8ff
Compactness criterion, 10ff
Comparison series, 412-415
Completeness, 44, 48ff, 57
Complex translation, 365
Compressibility, 247
Confinement of fractional charge,
456, 464
Connected graph, 391
Connected parts, 392
Convexification, 270
Convexity, 217
Core, atomic, 265
Correlation functions, 342,
443, 460, 461
Correlation inequalities, 336,
344, 377
Coulomb force, 55
Coulomb interaction energy, 376

Coulomb potential, 215
  periodic, 227
Coulomb systems, 328, 329
  in external potential, 449, 460
  jellium, 451
  two-component gas, 442
Covariance, 378, 382, 383
Critical density, 219
Critical mass, 312

Debye-Hückel, 371
Debye length, 376
Decay at infinity, 7, 11ff, 37ff
Density, 214, 283
  function, 101
  matrix, admissible single particle, 359
  operators, 373
Dilation, 246-248
  analyticity of Stark Hamiltonians, 133, 135-156
  generator, 20ff, 30ff, 38ff, 53
  operator, 125
Dipolar phase, 350
Dipole-dipole interaction, 255
Dipole layers, 356
Dipole potential, 331
Dirac correction, 216
Distance scale, 362
DLR condition, 444, 459
Domain of the energy functional, 216, 277, 301
Double layer, 228

Effective potential, 343
Effective system of charges and masses, 146
Eigenvalues of the Hamiltonian, 63ff
Electric field, 132, 172
  in equilibrium, 445, 449, 452
Electronic contribution, 245, 246
Electronic number, 216
Electrostatic energy, 338, 365

Energy functional, 214, 215, 216, 268, 277
Ensemble, 360, 362
Entropy, 349, 364
Equally accelerated clusters, 146, 153
Equilibrium, distribution, 341
Equilibrium states, 335
Exclusion principle, 310
Exponential decay in Stark effect, 134, 171
Exponential falloff, 266
External field, 372, 379

Fatou's lemma, 219, 280
Finite range property, 17
Firsov's principle, 230
Fractional charge, 374
  correlation, 341, 353, 354
Free boundary problem, 222
Free energy, 443, 449, 461
Free time evolution, 24ff
  modified, 56ff
Function of positive type, 311
Functional derivatives, 397
Functional integrals, 396

Gauge fields, 463, 465
Gauge theory, 368
Gauss theorem application, 454
Gaussian measures, 378-382, 384, 385
Geometric characterization, 13ff
Geometric method, 1ff
Graph, 391
Ground state, approximate, 262, 265

H-stability, 470
H_2^+, the hydrogen molecular ion, 188
Hardcore, 324
Heavy atom, 265
Higgs model, 464
High temperature expansion, 371

Ichinose's lemma, 153
Ideal gas, 372, 375

Infinite atom, 264, 295
Infinite volume limit, 374
Inner core, 265
Instantons, 462
Ionization potential, 239, 240
Ions in magnetic fields, 176-181

j-model, 270, 273
Jellium, 471

Kinetic energy, 216, 247
KMS condition, 459
Kosterlitz-Thouless transition, 329, 336, 358, 367

L^p space, 217
Label increasing trees, 393
Landau orbits in magnetic field, 173, 175, 178
Landau-Ginzburg model, 329
Lifetime of hydrogenic state in electric field, 133, 171
Limiting absorption principle, 74
Lippman-Schwinger equation, 73
Liquid crystal, 329
Liquid phase, 351
Local adsorption, 11
Local compactness, 9ff, 13ff, 53
Local singularity, 6, 10ff, 40, 49
Locality, 389-390
Long-range interaction, 253-255

Magnetic fields, 172-181
Many-body potentials, 30
Mayer expansion, 375, 391
Meromorphic continuation of matrix elements, 133, 155, 171
Minimization, 217
Minimizing density, 218
Minlos theorem, 378
Multiphase expansions, 400
Mutual subordination, 11ff

Neutron star, 313
Newton's theorem, application of, 447
No-binding theorem, 232, 242, 314
Nuclear charge, 215
Nuclear coordinates, 215
Nuclear potential, 215
Nuclear repulsion, 215

Observables, 377
Operator inequalities, 309
Ordered states, 350
Outer shell, 265
Over-screening, 229

Parabolic coordinates, 131
Partition function, 376
Partition $\bar{y}(h)$, 401, 406
Periodicity, 382, 402
Perturbation theory, 372, 429
Phase boundaries, 389, 406
Phase diagram, 345
Phase-space decomposition, 18ff, 27ff, 59
Plane rotator, 368
Plasma phase, 329, 348
Polynomials (monomials), 397
Positive type, 344
Potential
  amenable, 274
  fluctuations of, 453
  scattering, 1ff, 92
  theory, 232
Potentials
  absorbing, 114ff
  long range, 37ff, 55ff
  long range with bumps, 113ff, 118ff
  many-body, 244
  patching, 114
  periodic, 113
  short range, 7ff, 37ff, 45, 49ff, 57
  short range, with bumps, 123
  singular, 114, 124
Pressure, 226, 246, 247, 402

Q states, 449
Q vacua, 463
Quantum mechanics, 428-430
Quantum theory, 256ff

Rayleigh-Schrödinger perturbation series in Stark effect, 165
References, 439
Regularity, 379
Renormalization transformation, 37, 38, 343, 347, 348, 350, 365, 366
Renormalized activity, 364
Repulsive electrostatic energy, 215
Resonances in Stark effect, 133, 134, 143
Resonances in Stark ladder, 171
Roughening transition, 330, 350, 351, 357

Saturation, 309, 313
Scaling, 225, 226
Scattering state, 13, 41, 43, 126
Scattering, two-cluster, 65
Schrödinger equation, 107
Schrödinger operator, 1ff
Scott correction, 263
Screening, 228, 335, 341, 349, 350, 351, 380-382, 407-410, 449, 456, 464
Semigroup in Stark effect, 139, 142, 153, 162
Separated atom limit, 194ff
Shielding, 371, 429, 455
Short-range forces, 373, 375
Sine-Gordon, 372, 375, 379, 380
  quantum case, 428
  representation, 462
  transformation, 337, 358
Single phase expansions, 401
Singularities, 224
Sobolev inequality, 278
Solid on solid model, 368
Solids, 226
Sommerfeld radiation condition, 72, 81
Spectral function, 107-109
Spectral measure, 108, 110
Spectral projection, 101
Spectral subspace, 9
Spectral theorem, 101
Spectral trace, 74
Spectrum, 99
  absolutely continuous, 9, 17, 47, 50, 63, 114ff
  continuous, 9, 13ff, 39ff
  essential, 12
  of Stark Hamiltonians, 133, 139, 147, 154
  of Zeeman Hamiltonians, 174, 175, 177
  point, 9, 13ff, 63ff
  singular continuous, 9, 49ff, 63, 101, 118ff
  spin state number (q), 216, 256
Stability, 330, 394, 431, 432
  of eigenvalues in Stark effect, 156-164, 171
Stark effect, 132-172
Strong resolvent convergence, 9
Strong singularity, 224, 255
Subadditive, 235
Subharmonic, 234
Subspace, 101
  absolutely continuous, 102
  continuous, 101, 106
  discrete, 101
  interpretation, 104, 105, 122-124
  singular continuous, 119
Sum rules, electrostatic, 481
Superadditive, 242, 251
Superharmonic, 221
Superconductivity, 336
Surface, atomic, 265
Surface, charge, 375
Surface tension, 357
Symmetric decreasing function, 224
Symmetry breaking, 442, 453

Teller's lemma, 235
Teller's theorem, 232, 242
Test functions, 378
Thermodynamic functions, 334
Thermodynamic limit, 345, 377
  free energy, 468ff
  correlation functions, 478ff
Thomas-Fermi
  differential equation, 222

Thomas-Fermi (continued)
  energy, 217
  equation, 221
  equation, generalized, 233
  potential, 220, 221, 233
  theory, 310, 317
Thomas-Fermi-Dirac equation, 233, 271
Thomas-Fermi-Dirac-von Weizsäcker equation, 302
Thomas-Fermi-von Weizsäcker equation, 281
Three-body problem, 94
Time averages, 106
  of compact operators, 15
Time-dependent method, 1ff
Total shielding, 273
Trace class, 103-105
Transfer matrix, 115, 347
Tree graphs, 393
Tree graph formula, 393-396, 432-438
Truncated correlations, 402

Ursell functions, 391

Variational principle, 230, 233, 237
Virial theorem, 225
von Weizsäcker correction, 216

Wave operators, 43, 48ff, 125
  modified, 56ff
Weak convergence, 105
Weinberg-van Winter equation, 147, 155
Wicks theorem, 397
Wilson loop, 464

Z^2 correction in TFW theory, 294
$Z \to \infty$ limit, 256ff